The Collected Works of Eugene Paul Wigner

Part B · Volume VI

Eugene Paul Wigner

The Collected Works of Eugene Paul Wigner

Part A

The Scientific Papers

Editor: Arthur Wightman

Annotated by

Nandor Balazs   Herman Feshbach   Brian Judd   Walter Kohn
George Mackey   Jagdish Mehra   Abner Shimony   Alvin Weinberg
Arthur Wightman

Part B

Historical, Philosophical, and Socio-Political Papers

Editor: Jagdish Mehra

Annotated by
Conrad Chester   Gérard Emch   Jagdish Mehra

The Collected Works of Eugene Paul Wigner

Part A

The Scientific Papers

Volume I
Part I: Eugene Paul Wigner – A Biographical Sketch
Part II: Applied Group Theory 1926–1935
Part III: The Mathematical Papers

Volume II
Nuclear Physics

Volume III
Part I: Particles and Fields
Part II: Foundations of Quantum Mechanics

Volume IV
Part I: Physical Chemistry
Part II: Solid State Physics

Volume V
Nuclear Energy:
Part I: Eugene Wigner and Nuclear Energy
Part II: Memoir of the Uranium Project
Part III: Articles, Reports, and Memoranda on Nuclear Energy
Part IV: The Wigner Patents

Part B

Historical, Philosophical, and Socio-Political Papers

Volume VI
Philosophical Reflections and Syntheses

Volume VII
Historical and Biographical Reflections and Syntheses

Volume VIII
Socio-Political Reflections and Civil Defense

The Collected Works of

Eugene Paul Wigner

Part B

Historical, Philosophical, and Socio-Political Papers

Volume VI

Philosophical Reflections
and Syntheses

Annotated by Gérard G. Emch
Edited by Jagdish Mehra

Springer-Verlag
Berlin Heidelberg NewYork
London Paris Tokyo
Hong Kong Barcelona
Budapest

Jagdish Mehra

Department of Physics
Grimsley Hall, The Citadel
Charleston, SC 29409-0270, USA

Arthur S.Wightman

Department of Physics, Princeton University
Joseph Henry Laboratories, Jadwin Hall
Princeton, NJ 08544, USA

Gérard G. Emch

Department of Mathematics
University of Florida
Gainesville, FL 32611, USA

ISBN-13:987-3-540-63372-3   e-ISBN-13:978-3-642-78374-6
DOI: 10.1007/978-3-642-78374-6

Library of Congress Cataloging-in-Publication Data
(Revised for volume 6)
Wigner, Eugene Paul, 1902–
The collected works of Eugene Paul Wigner.
Includes bibliographical references.
Contents: pt. A. The scientific papers – v. 5. Nuclear energy. pt. B Historical, philosophical, and socio-political papers –
v. 6. Philosophical reflections and syntheses. 1. Mathematical physics. I. Weinberg, Alvin Martin, 1915-.
QC19.3.W54   1992   539.7   92-38376
ISBN-13:987-3-540-63372-3

Typesetting of the annotation and the reset contributions: Springer TEX in-house system
SPIN 10101573   55/3140-543210 - Printed on acid-free paper

Editors' Preface

The papers have been divided, necessarily somewhat arbitrarily, into two parts

Part A: Scientific Papers
Part B: Historical, Philosophical, and Socio-Political Papers

Within each part, the papers have been divided by subject, and within each subject printed chronologically. With some exceptions, every scientific paper is reprinted in its original form. One class of exceptions consists of papers that are simply translations into Hungarian from German or English; they are omitted, but listed in the bibliographies. Scientific papers originally in Hungarian have been translated into English. Some of the papers of Volume V/ Part III, Articles, Reports, and Memoranda on Nuclear Energy, have been reset and the figures redrawn. The originals were declassified reports, some in nearly illegible shape. Some reports and patents in Volume V/ Part III and Part IV are listed by title only. In contrast to the scientific papers where the coverage is essentially complete, in Part B, a selection has been made. We believe it is representative of Wigner's far ranging concerns. The five books in which Wigner was involved as author, co-author, or lecturer are not reprinted in the Collected Works, but are noted in the annotations and bibliographies.

Jagdish Mehra
Arthur S. Wightman

Contents

Philosophical Reflections and Syntheses
Annotation by Gérard G. Emch 1

PART I

Epistemology of Quantum Mechanics

Discussion: Comments on Professor Putnam's Comments
(with H. Margenau) 31
Two Kinds of Reality 33
Epistemology of Quantum Mechanics 48
Epistemological Perspective on Quantum Theory 55
Reality and Quantum Mechanics / Realität und Quantenmechanik 72
Interpretation of Quantum Mechanics 78
The Limitations of Determinism 133
The Nonrelativistic Nature of the Present Quantum Mechanical
Measurement Theory 139

PART II

Quantum-Mechanical Measuring Process

The Measurement of Quantum-Mechanical Operators / Die Messung
quantenmechanischer Operatoren 147
Theory of Quantum-Mechanical Measurement / Theorie der quanten-
mechanischen Messung 155
The Problem of Measurement 163
Some Comments Concerning Measurements in Quantum Mechanics
(with J.M. Jauch and M.M. Yanase) 181
On the Change of the Skew Information in the Process of
Quantum-Mechanical Measurements (with A. Frenkel and M. Yanase) 189
The Subject of Our Discussions 199
The Philosophical Problem 218
Questions of Physical Theory 221

On Bub's Misunderstanding of Bell's Locality Argument
    (with S. Freedman) 223
Review of the Quantum-Mechanical Measurement Problem 225

PART III
Consciousness

Remarks on the Mind-Body Question 247
The Place of Consciousness in Modern Physics 261
New Dimensions of Consciousness 268
The Existence of Consciousness 274

PART IV
Symmetries

Invariance in Physical Theory 283
On the Law of Conservation of Heavy Particles 294
Symmetry and Conservation Laws 297
The Role of Invariance Principles in Natural Philosophy 311
Events, Laws of Nature, and Invariance Principles
    (Nobel address) 321
Events, Laws of Nature, and Invariance Principles 334
Violations of Symmetry in Physics 343
Symmetry Principles in Old and New Physics 359
Symmetry in Nature 382

PART V
Relativity

Relativistic Invariance and Quantum Phenomena 415
Relativistic Equations in Quantum Mechanics 446

PART VI
Nuclear Physics

On the Development of the Compound Nucleus Model 459
Summary of the Conference (Properties of Nuclear States, Montreal 1969) 476
Summary of the Conference (Polarization Phenomena, Madison 1971) .. 487
Introductory Talk (Statistical Properties of Nuclei) 494
Concluding Remarks (Symmetry Properties of Nuclei,
    Solvay Conference 1970) 508

PART VII

Broader Philosophical Essays

The Limits of Science 523
The Unreasonable Effectiveness of Mathematics
   in the Natural Sciences 534
The Growth of Science – Its Promise and Its Dangers 550
Physics and the Explanation of Life 564
On Some of Physics' Problems 578
Physics and Its Relation to Human Knowledge 584
The Problems, Future and Limits of Science 594
The Extension of the Area of Science 603
The Glorious Days of Physics 610
Some Problems of Our Natural Sciences 616

Bibliography 627

Philosophical Reflections and Syntheses

Annotation by Gérard G. Emch

Introduction

Other volumes in this series cover the contributions of Eugene Wigner, the Scientist and the Statesman; the present volume is devoted to the contributions of the Natural Philosopher that place him in a lineage starting in the 17th century, gaining legitimacy in the Enlightenment, and leading through the general methodology of the Positivist School. The tenets of the latter could still be maintained during his early formative years: "... The first physics book I read said: 'Atoms and molecules may exist, but this is irrelevant from the point of view of physics'; and this was entirely correct – at that time physics dealt only with macroscopic phenomena and Brownian motion was sort of a miracle" [1].

Boltzmann's classical statistical mechanics, in the third quarter of the 19th century, and most drastically the advent of quantum theory in the first quarter of this century certainly contributed to bringing the microscopic world into the scope of the physicists. In a sense, quantum theory is a case where the solution came before the problem. Indeed, while the emerging paradigm (the Bohr atom, the photon in the photoelectric effect and in Einstein's fluctuation formula for electromagnetic radiation, the phonon in the computation of the temperature dependence of the specific heat in solids) already presented challenges of interpretation, the crisis came to a head with the advent of the quantum mechanics of Heisenberg and Schrödinger. Eugene Wigner, born in 1902, came of age at that time. Under the impetus of the Göttingen School around Born and Hilbert, quantum theory achieved an internally consistent mathematical status in the hands of von Neumann [2]; together with the work of London and Bauer [3], von Neumann's book became Eugene Wigner's standard reference for the "orthodox theory".

Having acquired a definite mathematical form, quantum mechanics could now be submitted to an epistemological critique. Eugene Wigner's general commentaries on the ensuing controversy are grouped in the first eight papers in this volume (Part I); these papers are primarily addressed to philosophers of science. A more technical, but still generally accessible, version of these criticisms can be found in the papers of Eugene Wigner grouped in Part II of the present volume. The peculiar role of the observer in the quantum measuring process also led him to question the nature of human consciousness; see Part III.

One of Eugene Wigner's major contributions to the practice of quantum theory was his pioneering work on symmetries and invariance principles. Among these leading achievements are his recognition of the physical meaning of unitary group representations, and his explicit computation of these representations for specific groups, such as: for atomic theory, the group of rotations in three dimensions; and for the relativistic theory of elementary particles, the inhomogeneous Lorentz group. While the more technical papers on these subjects are reproduced in other volumes, the papers of Eugene Wigner selected for the present volume explore and explain the relations between invariance principles and conservation laws; see Part IV.

Two papers on the interplay of relativity and quantum mechanics properly are an extension of the above category; they are written in terms that are so generally understandable that it was decided to include them as Part V of the present volume.

Between the purview of atomic spectroscopy and the concerns of elementary particle theory is the theory of the nuclear structure, to which Eugene Wigner also contributed. Five general review articles are reproduced in Part VI of this volume; note that the short time that spans the writing of four of them belies Eugene Wigner's life-long interest in nuclear physics.

Finally, we placed at the end of this volume (Part VII) ten papers which are witness to Eugene Wigner's broadest concerns with the role of the natural sciences in the 20th-century intellectual landscape.

In this commentary we will endeavor to introduce the reader to each of these seven classes of articles. The task of this commentator has been made both easy and hard by the fact that all of these papers are already overviews, written by the hand of a master of style as well as of content. In all reasonable senses, there is no substitute for reading the originals reprinted here.

I. Epistemology of Quantum Mechanics

The papers selected in this part present Eugene Wigner's assessment of the general significance of the quantum revolution and his reflections on the possible interpretations proposed for the formalism of quantum mechanics.

Practitioners from diverse disciplines brought their own specialized training into the search for interpretations; these cultural differences often resulted in a discourse impeded by basic semantic misunderstandings. Eugene Wigner was called in to help remove the shackles locked onto the debate by scholars who were either less sensitive to these diverse backgrounds or less intimate with the practice and limitations of this specific body of physical theory. While he kindly obliged, his entrance on the stage was also accompanied by a directness that was to focus the dialogue. The paper [4] opening this part is a case in point; it certainly brings to the fore the view that mutual explanations are necessary, and it illustrates the style characteristic of many of Eugene Wigner's interventions in scientific meetings.

In [5] Eugene Wigner draws the philosophical outline of the debate, and alludes to the role of consciousness in the natural sciences, an issue covered more specifically in the papers of Part III. The reader will note, in particular, the references he gives here as either seminal, authoritative or challenging: Freud, Poincaré, Hadamard; Heisenberg, Schrödinger, von Neumann, London and Bauer, Einstein; Bridgman, Margenau.

The successful formalism, from which Eugene Wigner produced his critical appraisal of the principles of quantum mechanics, is summarized in his papers [6,7,8]. It emerges from the book of von Neumann [2] along the following lines.

To each quantum system is associated a complex Hilbert space $\mathcal{H}$; namely, a vector space with complex numbers as scalar coefficients, that is equipped with an inner product, i.e., a rule attributing to every pair of vectors, ψ_1, ψ_2 in $\mathcal{H}$, a complex number, denoted by (ψ_1, ψ_2), in such a manner that:

(1.1)
$$\begin{aligned}
&\text{(a)} \quad (\psi_1, \psi_2)^* = (\psi_2, \psi_1) \\
&\text{(b)} \quad (\psi, c_1\psi_1 + c_2\psi_2) = c_1(\psi, \psi_1) + c_2(\psi, \psi_2) \\
&\text{(c)} \quad (\psi, \psi) \geq 0 \quad \text{with} \quad (\psi, \psi) = 0 \text{ iff } \psi = 0 .
\end{aligned}$$

We should add, for mathematical rigor's sake, that $\mathcal{H}$ is complete with respect to the norm (1.1.c).

Physically, each nonzero element ψ of $\mathcal{H}$ describes a 'state' of the system, in a way which is made more precise by (1.4) below. These vectors ψ play in the quantum theory a role analogous to that played, in the classical theory, by the points ξ of the phase-space Ω.

Typically, for a single particle moving on a straight line R, $\mathcal{H} = \mathcal{L}^2(R, dx)$ is the space of square-integrable complex-valued functions on the configuration space R, with inner product

(1.2)
$$(\psi_1, \psi_2) = \int dx\, \psi_1(x)^* \psi_2(x) .$$

The 'observables' of the system are represented as linear, self-adjoint operators, i.e., maps from $\mathcal{H}$ into itself, that satisfy:

(1.3)
$$\begin{aligned}
&\text{(a)} \quad A(c_1\psi_1 + c_2\psi_2) = c_1 A\psi_1 + c_2 A\psi_2 \\
&\text{(b)} \quad (\psi_1, A\psi_2) = (A\psi_1, \psi_2) .
\end{aligned}$$

Here again, we should be mathematically more precise, and distinguish between bounded observables which are defined exactly as above, and unbounded observables for which domain restrictions should be specified; we shall not bother with these mathematical niceties here, although the observable (1.5) defined below is precisely one of those unbounded observables.

Physically, the 'expectation value' of the observable A, when the system is in the state ψ, is given by the real number:

(1.4)
$$\langle A \rangle_\psi = \frac{(\psi, A\psi)}{(\psi, \psi)} .$$

The reader should compare this with the familiar situation encountered in classical theory, where the observables are represented by smooth, real-valued functions f on the phase-space Ω of the system; and where the value of f, when the system is in the state ξ, is the real number $f(\xi)$.

Notice that (1.4) implies immediately that the correspondence between states of the quantum system and vectors in its Hilbert space is not one-to-one. Indeed, two vectors ψ_1 and $\psi_2 = c\psi_1$ (where c is an arbitrary, nonzero complex number) give the same number in (1.4). It is therefore customary to assume, without loss of generality, that the state-vectors ψ are normalized, i.e., satisfy $(\psi, \psi) = 1$; this evidently still leaves an ambiguity in the specification of the vector ψ, namely a "phase", i.e. a complex number of modulus 1. This ambiguity implies that symmetry groups (see Part IV below) are manifest in quantum mechanics not only through their ordinary representations, but more generally through their projective representations. Very early in the development of quantum mechanics, Eugene Wigner not only recognized the importance of this fact which allows us, for instance, to integrate naturally in the theory the non-classical spin $1/2$ of the electron; but he also pioneered the classification of the projective representations of groups of particular interest in physics: see already [9] and [10].

Going back to the general presentation of the theory, we also note that in the typical case considered above, where $\mathcal{H} = \mathcal{L}^2(R, dx)$, the observable 'position' is represented by the operator Q defined by:

$$(1.5) \qquad (Q\psi)(x) = x\,\psi(x)$$

so that we obtain from (1.4):

$$(1.6) \qquad \langle Q^n \rangle_\psi = \int dx\, |\psi(x)|^2 x^n \ .$$

Hence $|\psi(x)|^2 = \psi(x)^* \psi(x)$ is to be interpreted as the probability density that the particle is at the position x. This remark is the basis of the *statistical interpretation* Born [11] gave to the 'wave-function' ψ appearing in the Schrödinger formulation of quantum mechanics.

Let us give still another example to reinforce this statistical interpretation of (1.4). Given the state vector ψ_1, with $(\psi_1, \psi_1) = 1$, let P_1 be the self-adjoint operator defined on $\mathcal{H}$ by:

$$(1.7) \qquad P_1\psi = (\psi_1, \psi)\,\psi_1$$

which is interpreted as the special observable, or proposition, corresponding to the question: "Is the system in the state ψ_1?" From (1.4), we compute the expectation that the system, prepared in a state ψ_0 (with $(\psi_0, \psi_0) = 1$), be found in the state ψ_1 (with $(\psi_1, \psi_1) = 1$); and we obtain:

$$(1.8) \qquad \langle P_1 \rangle_{\psi_0} = |(\psi_0, \psi_1)|^2 \ .$$

This number, the value of which is between 0 and 1, is referred to as the 'transition probability' between the state vectors ψ_0 and ψ_1; it is equal to 1 exactly when $\psi_1 = c\psi_0$, i.e., when ψ_1 and ψ_0 represent the same state of the system; and it is 0 whenever ψ_1 and ψ_0 are orthogonal; for all other state-vectors ψ_0, $\langle P_1 \rangle_{\psi_0}$ is strictly larger than 0 and strictly smaller than 1.

When one attempts to think in classical terms about what has just been said, it soon appears that the description of quantum mechanics given so far raises serious epistemological questions, which Eugene Wigner reviewed in the papers considered here.

The most pressing of these questions is whether any 'reality' can be assigned to ψ beyond Born's statistical interpretation, i.e., beyond the view that ψ is merely a summary of our knowledge about (the state of) the system considered; see e.g. [12, 13]. The address [14] places this problem in perspective: "This interpretation states that the wave-function does not describe reality but is merely a tool to determine the statistical relations between successive observations. I must admit, however, that I am not satisfied with this interpretation." The argument then proceeds to show, with the help of what has become known as the archetypical 'Wigner's friend', that a strict adherence to this interpretation leads to "dreadful" solipsisms. This is brought into sharp focus in the analysis of the measurement process, a detailed discussion of which is given in the papers reprinted in the next part. Before we study these papers, the reader may still want to look at [15] and gain further insight into Eugene Wigner's persistent struggle with his positivist background: he seems indeed to be on the verge of abandoning some form of causality to preserve the possibility of using at least a modified form [14,1] of the standard quantum Liouville equation of motion.

We now list a few of the other epistemological problems raised in the papers under review in this part.

A. To understand the vector space structure of the Hilbert space $\mathcal{H}$ [7]; in the physical literature, this structure is referred to as the 'superposition principle', a term borrowed from the study of classical linear differential equations. In the context of quantum mechanics, this principle asserts that to every pair of states ψ_1, ψ_2 and every pair of complex numbers c_1, c_2, one can associate another physically realizable state, namely:

$$(1.9) \qquad \psi = c_1\psi_1 + c_2\psi_2 .$$

In particular, this means (in the simple case where ψ_1 and ψ_2 are normalized and mutually orthogonal, and where $|c_1|^2 + |c_2|^2 = 1$), that the state ψ, described by (1.9), carries not only the information contained in the transition probabilities $|c_1|^2$ and $|c_2|^2$, but also the information which is contained in the correlations $c_1^* c_2$ and $c_1 c_2^*$. We will come back to this in Part II when we discuss the 'collapse of the wave vector'.

B. To understand the role of the complex numbers on which the Hilbert space $\mathcal{H}$ is constructed. Specifically, to examine whether something is lost, or could be gained, when one tries to construct an internally consistent quantum theory involving only Hilbert spaces built on real numbers, or allowing for Hilbert spaces built on the quaternions [16] (recall that the quaternions are the third of the three finite-dimensional division algebras on the real number field, of which the real and the complex are the two abelian representatives). The first consideration of such possible generalizations of quantum mechanics appeared in [17]; the idea was then pursued much later by the Geneva School under the direction of Stueckelberg [18] and Jauch [19, 20, 21].

C. To sharpen the concept of observable in order to give an operational meaning to the mathematical object appearing under that name in the formalism purporting to describe the physical theory. Eugene Wigner addresses three facets of this problem.

The first facet [7, 8] is what has become known as the Dirac problem, namely to establish an unambiguous correspondence between the 'observables' of the quantum theory and those of the classical theory. For instance, let P and Q be the self-adjoint operators that represent respectively the momentum and the position of a quantum particle, and thus satisfy the Heisenberg 'canonical commutation rule':

$$(1.10) \qquad\qquad [P, Q] = -i\hbar I$$

(where $[A, B]$ generally denotes the commutator $AB - BA$ of the quantum observables A and B); and let p and q be the classical observables that correspond to P and Q, and thus satisfy:

$$(1.11) \qquad\qquad \{p, q\} = 1$$

(where $\{f, g\}$ generally denotes the Poisson bracket $\partial_p f \, \partial_q g - \partial_p g \, \partial_q f$ of the classical observables f and g). What is, then, the self-adjoint operator A corresponding to a given functions f of p and q ? For instance, already for the fourth degree monomial $p^2 q^2$, there seems to be *a priori* no reason to choose one over the others of the three symmetric operators $PQ^2P = QP^2Q$, $\frac{1}{2}(P^2Q^2 + Q^2P^2)$ and $\frac{1}{2}(PQPQ + QPQP)$; the ambiguities become increasingly worse for higher polynomials. This problem was first studied by Born, Heisenberg and Jordan [22]; yet, the severe limitations inherent to any such attempt were established much later by Chernoff [23]. Largely for purely pragmatic reasons of convenience, physicists adopted over the years various 'ordering' rules; seizing on the problem, mathematicians have subsequently devised some justification for these choices; one of these approaches is the polarization method encountered in the geometric quantization programme [24, 25, 26].

The second facet of problem (C) is to give an empirical, i.e., actually implementable, measurable meaning to the quantities that are called, perhaps too lightly 'observables'. Eugene Wigner [27] noticed that: "Only quantities that commute with all additive conserved quantities are precisely measurable" [8, 27]. This point is most evident for what one would like to call a 'position' observable [1]; see also (E) below, and the commentary on problem (B) of the next part.

The third facet of problem (C) is to ask whether every self-adjoint operator corresponds to any observable quantity at all. The answer to this question turned out to be negative when Wick, Wightman and Wigner [28] discovered the 'superselection rules' associated to the existence of 'essential observables', i.e., observables that commute with all other observables on the system considered [27, 12]. Note that this analysis also puts a limitation to the heretofore unrestricted validity of the superposition principle discussed in (A) above; equivalently, more than a phase ambiguity remains in the specification of the state-vector ψ.

D. To determine whether the statistical aspect of quantum mechanics could be of a nature akin to the statistical processes introduced in the program initiated by D. Bernoulli and culminating, still in the classical realm, with the Boltzmann approach to thermodynamics. Specifically, could it be that quantum mechanics is an incomplete theory that admits a finer description in terms of classical, but 'hidden', variables [7, 8] and [29, 1]? While it seemed that von Neumann had disposed of the idea, this line of thought was given a new life by Bohm [30]. However, Bell [31] gave an empirical criterion, the inequalities that now bear his name, to test all such theories which satisfy some 'locality' or 'causality' requirement; the laboratory evidence subsequently adduced [32, 33, 34] ruled out these 'hidden variables' theories on the basis of Bell's inequalities; for a review of these issues, see [35].

E. To understand how quantum theory and relativity can be reconciled [36]. Even at the level of special relativity, one of the problems is to give a covariant description of the measuring process that takes into account the fact that measurements are not instantaneous [14]. Moreover, difficulties already appear at this level in defining some of the most basic observables, e.g., the position of a space-time point [37]; see also the paper [16] reprinted in this volume, and [38, 39, 40, 41]. As for general relativity, one of the questions is then to define empirically the metric tensor in a manner that does not idealize away the difficulties one encounters, as a matter of principle, in the measurement of distances and curvatures on a very small scale [42, 1]; see in particular [43].

The main outstanding problem that played a central motivating role in Eugene Wigner's reflections on the epistemology of quantum mechanics, namely the measuring process, is discussed in the next part.

II. Quantum Mechanical Measuring Process

Before entering the description of the measuring process itself, we briefly recall two additional tenets of the orthodox formalism for quantum mechanics reviewed in Part I.

The first is that an 'isolated' system evolves according to a one-parameter group $\{U_t \mid t \in R\}$ of unitary operators. Specifically, the time-evolution of the system, from a state ψ_0 at time t_0 to a state ψ_1 at time t_1, is given by

(2.1) $\psi_0 \quad \rightarrow \quad \psi_1 = U_{t_1 - t_0} \psi_0$,

the infinitesimal expression of which is the differential equation

(2.2) $-i\hbar\, \partial_t \psi_t = H\psi_t$

where the generator H of the evolution is the self-adjoint operator corresponding to the Hamiltonian of the system, i.e., the observable 'energy' (typically the
sum of its kinetic energy $K = p^2/2m$ and its potential energy V). For instance,
the quantum description of the motion of a particle in a potential $V(x)$ is given
by the Schrödinger equation, which we write here for a one-dimensional system:

(2.3) $-i\hbar\, \partial_t \, \psi(x;t) \;=\; \left[-\dfrac{\hbar^2}{2m} \partial_x^2 + V(x) \right] \psi(x;t)$.

The second of the tenets of the quantum mechanical formalism which we
want to recall here is that if two quantum systems $S^{(1)}$ and $S^{(2)}$ are described
in the Hilbert spaces $\mathcal{H}^{(1)}$ and $\mathcal{H}^{(2)}$ respectively, then the combined system
$\{S^{(1)}, S^{(2)}\}$ is described in the Hilbert space

(2.4) $\mathcal{H} = \mathcal{H}^{(1)} \otimes \mathcal{H}^{(2)}$.

In particular, if $S^{(1)}$ [resp. $S^{(2)}$] is in the state $\psi^{(1)}$ [resp. $\psi^{(2)}$], and $A^{(1)}$ [resp.
$A^{(2)}$] is an observable for $S^{(1)}$ [resp. $S^{(2)}$], then the combined system is in the
state

(2.5) $\psi = \psi^{(1)} \otimes \psi^{(2)}$

while the observable $A^{(1)}$ [resp. $A^{(2)}$], considered as an observable on the combined system is described by the self-adjoint operator $A^{(1)} \otimes I$ [resp. $I \otimes A^{(2)}$].
Two comments have to be made here.

Firstly, this formalism is internally consistent in the sense that the expectation values do match, i.e.,

(2.6)
$$\langle A^{(1)} \otimes I \rangle_\psi = \langle A^{(1)} \rangle_{\psi^{(1)}}$$
$$\langle I \otimes A^{(2)} \rangle_\psi = \langle A^{(2)} \rangle_{\psi^{(2)}}$$.

More generally, we have also:

(2.7) $\langle A^{(1)} \otimes A^{(2)} \rangle_\psi = \langle A^{(1)} \rangle_{\psi^{(1)}} \langle A^{(2)} \rangle_{\psi^{(2)}}$.

Secondly, however, for an arbitrary state-vector $\psi \in \mathcal{H}$ on the combined
system, i.e.,

(2.8) $\psi = \displaystyle\sum_{\mu,\nu} c_{\mu\nu} \psi_\mu^{(1)} \otimes \psi_\nu^{(2)}$

where $\{\psi_\mu^{(1)} \mid \mu = 1, 2, \ldots\}$ [resp. $\{\psi_\nu^{(2)} \mid \nu = 1, 2, \ldots\}$] is an orthonormal basis
in $\mathcal{H}^{(1)}$ [resp. $\mathcal{H}^{(2)}$], it is *not* true in general that there exists $\psi^{(1)} \in \mathcal{H}^{(1)}$ and

$\psi^{(2)} \in \mathcal{H}^{(2)}$ such that (2.7) is valid. Consider for instance the case where both $\mathcal{H}^{(1)}$ and $\mathcal{H}^{(2)}$ are two-dimensional; choose

$$(2.9) \qquad \psi_1^{(1)} = \begin{pmatrix} 1 \\ 0 \end{pmatrix} = \psi_1^{(2)} \; ; \quad \psi_2^{(1)} = \begin{pmatrix} 0 \\ 1 \end{pmatrix} = \psi_2^{(2)}$$

and consider the state vector:

$$(2.10) \qquad \psi = \frac{1}{\sqrt{2}} \left[\psi_1^{(1)} \otimes \psi_2^{(2)} - \psi_2^{(1)} \otimes \psi_1^{(2)} \right] \;.$$

We obtain then for any observable $A^{(2)}$ on $\mathcal{S}^{(2)}$ (the same argument can be made for an observable $A^{(1)}$ on $\mathcal{S}^{(1)}$):

$$(2.11) \qquad \langle I \otimes A^{(2)} \rangle_\psi = \frac{1}{2} \sum_{\nu=1}^{2} (\psi_\nu^{(2)}, A^{(2)} \psi_\nu^{(2)}) \;.$$

The right-hand side of this expression can be rewritten as

$$(2.12) \qquad \mathrm{Tr}\big(\rho^{(2)} A^{(2)}\big) \quad \text{with} \quad \rho^{(2)} = \frac{1}{2} \begin{pmatrix} 1 & 0 \\ 0 & 1 \end{pmatrix}$$

where $\rho^{(2)}$ is *not* a one-dimensional projector of the form (1.7). The quantum phenomenon just illustrated is in stark contrast with the situation encountered in classical mechanics, where every point ξ of the combined phase space $\Omega^{(1)} \times \Omega^{(2)}$ can always be visualized as a pair $(\xi^{(1)}, \xi^{(2)})$ with $\xi^{(1)} \in \Omega^{(1)}$ and $\xi^{(2)} \in \Omega^{(2)}$. This departure between the classical and quantum theories plays an important role in the discussion of the quantum measuring process, see e.g., [13].

Let us now turn to what Eugene Wigner refers to as the 'orthodox description' of the quantum measurement process as it emerges from the book of von Neumann [2].

We have a quantum system $\mathcal{S}$, described with the help of a Hilbert space $\mathcal{H}^{(s)}$; its vector-states are denoted by $\psi^{(s)}$, and its observables by $A^{(s)}$. Suppose that we are interested in 'measuring' a particular observable $A_0^{(s)}$ on $\mathcal{S}$. For the sake of simplicity, we shall assume that $A_0^{(s)}$ has non-degenerate, discrete spectrum, i.e., that there exists an orthonormal basis $\{\psi_\nu^{(s)} \mid \nu = 1, 2, \ldots\}$ of $\mathcal{H}^{(s)}$, such that

$$(2.13) \qquad A_0^{(s)} \psi_\nu^{(s)} = a_\nu \, \psi_\nu^{(s)} \quad \text{with} \quad a_\mu \neq a_\nu \quad \text{whenever} \quad \mu \neq \nu \;.$$

According to the orthodox theory, to describe an apparatus $\mathcal{A}$, that 'measures' the observable $A_0^{(s)}$ on $\mathcal{S}$, is to exhibit two things: a quantum system $\mathcal{A}$ and an interaction between $\mathcal{A}$ and $\mathcal{S}$ that satisfy the prescriptions listed below.

$\mathcal{A}$ is a quantum system described with the help of a Hilbert space $\mathcal{H}^{(a)}$; its vector states are denoted by $\psi^{(a)}$, and its observables by $A^{(a)}$. We then mimic the existence of pointers in $\mathcal{A}$ for the eigenvalues a_ν of $A_0^{(s)}$ by the following requirements.

To every a_ν corresponds a vector-state $\psi_\nu^{(a)} \in \mathcal{H}^{(a)}$ and an observable $A_\nu^{(a)}$ on $\mathcal{A}$, such that:

(2.14)
$$(\psi_\nu^{(a)}, \psi_\nu^{(a)}) = 1 \quad \text{for all} \quad \nu = 1, 2, \ldots$$
$$(\psi_\mu^{(a)}, \psi_\nu^{(a)}) = 0 \quad \text{whenever} \quad \mu \neq \nu$$

and

(2.15)
$$A_\nu^{(a)} \psi_\nu^{(a)} = \psi_\nu^{(a)} \quad \text{for all} \quad \nu = 1, 2, \ldots$$
$$A_\nu^{(a)} \psi_\mu^{(a)} = 0 \quad \text{whenever} \quad \mu \neq \nu .$$

The first step in setting up the measurement is to prepare the apparatus in a state $\psi_0^{(a)}$; this preparation must evidently be independent of the initial, unknown state $\psi_0^{(s)}$ of $\mathcal{S}$; however $\psi_0^{(a)}$ will depend in general on the specific observable $A_0^{(s)}$ which we want to measure.

The second, and most delicate, part of the set-up is to provide an interaction between $\mathcal{A}$ and $\mathcal{S}$ in such a manner that the initial state of the combined system $\{\mathcal{A}, \mathcal{S}\}$ evolves according to the equation:

(2.16)
$$\psi_0 = \psi_0^{(a)} \otimes \psi_0^{(s)} \quad \rightarrow \quad \psi_1 = \sum_\nu c_\nu \psi_\nu^{(a)} \otimes \psi_\nu^{(s)}$$

for *every* initial state

(2.17)
$$\psi_0^{(s)} = \sum_\nu c_\nu \psi_\nu^{(s)}$$

of the system $\mathcal{S}$ under consideration.

The final step in the measurement process is to separate the apparatus from the system considered.

We can now make more explicit the sense in which (2.16) transfers to $\mathcal{A}$ the information on $A_0^{(s)}$ contained in the state (2.17). To this effect, we compute from the above equations:

(2.18)
$$\langle A_\nu^{(a)} \otimes I \rangle_{\psi_1} = |c_\nu|^2 .$$

This result tells us that, in the final state ψ_1, the pointer $A_\nu^{(a)}$ of the apparatus 'lights up' with probability $|c_\nu|^2$, thus giving us the value of the coefficients to insert in the computation of the expectation value of the observable $A_0^{(s)}$ when the system $\mathcal{S}$ is in the initial state $\psi_0^{(s)}$, namely

(2.19)
$$\langle A_0^{(s)} \rangle_{\psi_0^{(s)}} = \sum_\nu |c_\nu|^2 a_\nu .$$

Note also that, during the measurement, this information has not been lost on the system $\mathcal{S}$ since we have, before the measurement has taken place:

$$(2.20) \qquad \langle I \otimes A_0^{(s)} \rangle_{\psi_0} = \sum_\nu |c_\nu|^2 \, a_\nu = \langle A_0^{(s)} \rangle_{\psi_0^{(s)}} \; ;$$

and after the measurement:

$$(2.21) \qquad \langle I \otimes A_0^{(s)} \rangle_{\psi_1} = \sum_\nu |c_\nu|^2 \, a_\nu = \langle A_0^{(s)} \rangle_{\psi_1^{(s)}} \; .$$

However, the same computation gives us, for an arbitrary observable $A^{(s)}$ on $\mathcal{S}$:

$$(2.22) \qquad \langle I \otimes A^{(s)} \rangle_{\psi_0} = \sum_{\mu,\nu} c_\mu^* c_\nu \, (\psi_\mu^{(s)}, \, A^{(s)} \psi_\nu^{(s)}) = \langle A^{(s)} \rangle_{\psi_0^{(s)}}$$

and

$$(2.23) \qquad \langle I \otimes A^{(s)} \rangle_{\psi_1} = \sum_\nu |c_\nu|^2 \, (\psi_\nu^{(s)}, \, A^{(s)} \psi_\nu^{(s)}) = \langle A^{(s)} \rangle_{\psi_1^{(s)}} \; .$$

Hence the measuring process has erased the initial correlations present in the initial state $\psi_0^{(s)}$, although (2.20–21) shows that the states ψ_0 and ψ_1, as seen through the observable $A_0^{(s)}$ only, are equivalent [44].

An elementary example of this phenomenon is provided by the Stern-Gerlach experiment measuring a chosen component of the spin of an electron, say

$$(2.24) \qquad A_0^{(s)} = \frac{1}{2} \begin{pmatrix} 1 & 0 \\ 0 & -1 \end{pmatrix} \; .$$

As far as the spin variables are concerned, $\mathcal{H}^{(s)}$ is two-dimensional with a basis of eigenvectors given by:

$$(2.25) \qquad \psi_1^{(s)} = \begin{pmatrix} 1 \\ 0 \end{pmatrix} \quad \text{and} \quad \psi_2^{(s)} = \begin{pmatrix} 0 \\ 1 \end{pmatrix} \; .$$

In this basis a general state-vector takes the form:

$$(2.26) \qquad \psi_0^{(s)} = c_1 \psi_1^{(s)} + c_2 \psi_2^{(s)} \quad \text{with} \quad c_1 c_1^* + c_2 c_2^* = 1 \; ;$$

and we can write (2.22) in the form:

$$(2.27) \qquad \langle I \otimes A^{(s)} \rangle_{\psi_0} = Tr \, \rho_0^{(s)} A^{(s)}$$

with

$$(2.28) \qquad \rho_0^{(s)} = \begin{pmatrix} c_1 c_1^* & c_1 c_2^* \\ c_2 c_1^* & c_2 c_2^* \end{pmatrix}$$

which is the one-dimensional projector on $\psi_0^{(s)}$ (see 1.7). We have however from (2.23):

$$(2.29) \qquad \langle I \otimes A^{(s)} \rangle_{\psi_1} = Tr \, \rho_1^{(s)} A^{(s)}$$

with

(2.30)
$$\rho_1^{(s)} = \begin{pmatrix} c_1 c_1^* & 0 \\ 0 & c_2 c_2^* \end{pmatrix} .$$

The difference between (2.28) and (2.30) is precisely that the correlations $c_1 c_2^*$ and $c_2 c_1^*$ between $\psi_1^{(s)}$ and $\psi_2^{(s)}$, present in (2.26–28), have disappeared in (2.29–30). The equivalence between $\rho_0^{(s)}$ and $\rho_1^{(s)}$, when viewed from $A_0^{(s)}$ only, is expressed by

(2.31)
$$\mathrm{Tr}\,\rho_0^{(s)} A_0^{(s)} = \mathrm{Tr}\,\rho_1^{(s)} A_0^{(s)} .$$

The states $\rho_0^{(s)}$ and $\rho_1^{(s)}$ are however different in general (i.e., except when either c_1 or c_2 vanishes): $\rho_0^{(s)}$ is pure (i.e., is a vector state, the coherent superposition of the states $\psi_1^{(s)}$ and $\psi_2^{(s)}$), whereas $\rho_1^{(s)}$ is a mixture (i.e., a non-correlated convex combination of the two states $\psi_1^{(s)}$ and $\psi_2^{(s)}$). In this context, the Stern-Gerlach experiment, some of its variants, and its epistemological implications, are discussed in [13, 1].

The measuring process has thus introduced a further probabilistic element in the description of the system, or more precisely a further departure from the classical idea of reality. Indeed, whereas there exists a complete set of commuting observables $B_0^{(s)}$ for which the pure state $\psi_0^{(s)}$ (of the system *before* the measurement) is dispersion-free, i.e., for which

(2.32)
$$\left\langle \left(B_0^{(s)} - \langle B_0^{(s)} \rangle_{\psi_0^{(s)}} \right)^2 \right\rangle_{\psi_0^{(s)}} = 0 ,$$

this is not the case anymore for the mixture $\rho_1^{(s)}$ (describing the state of the system *after* the measurement); see e.g., [12, 45, 46, 15]. This is at the root of the problems collectively known in the theory of the quantum measuring process as the 'collapse of the state-vector'; see (E and F) below.

In Eugene Wigner's view this peculiarity of quantum theory, although of paramount importance for the contrasted epistemologies of quantum and classical theories, is nevertheless only one of the several (related) problems raised by the quantum theory of the measurement process. We list a few of these problems below.

A. The most immediate of these problems was already recognized by von Neumann [2]: the acquisition of knowledge on the state of the system requires an infinite regression in the sense that a second apparatus $\mathcal{A}'$ must be called upon to measure the state of the first apparatus $\mathcal{A}$, and so on; see e.g., [46]. One possible escape is to require that $\mathcal{A}$ be macroscopic; Eugene Wigner attributes this proposal to Fock [47], and his own early involvement with such a requirement is witnessed explicitly in [2]. This proposal has the advantage of bringing the description of $\mathcal{A}$ closer to the more familiar categories prevalent in the realm of classical physics, in compliance with the views advocated by Bohr, see e.g. [48].

B. Eugene Wigner moreover adduced an original argument according to which this proposal is not only convenient, but also necessary [27]: "no observable which does not commute with the additive conserved quantitites (such as linear and angular momenta, or electric charge) can be measured precisely; and in order to increase the accuracy of the measurement one has to use a very large measuring apparatus" [13]; this argument was subsequently refined, see [49, 50], and revisited repeatedly by Wigner himself, see e.g. [6, 46, 51, 8].

C. The macroscopic nature of the measuring apparatus in turn raises several other questions, the first of which being that it becomes futile to pretend that we know precisely the initial state of the apparatus: the state vector $\psi_0^{(a)}$ must be replaced by a mixture [13, 6, 51, 52]; see also (E) below.

D. An even more serious problem, connected with the required large size of the measuring apparatus, was raised by Zeh [53] who pointed out that a macroscopic quantum system must not be considered to be isolated. This discovery is discussed in [1]; see also [7, 14].

E. In [12, 13, 46] Eugene Wigner reviews a question raised already by von Neumann, namely whether the quantum measurement process implies or not that there be two types of time-evolution in quantum theory: the deterministic evolution described by (2.1) and the probabilistic evolution issued from (2.16), for which, in contrast with (2.1), one expects the information content of the state of the system to be affected; [12] and [45] reexamine critically in this context a concept of 'skew information'.

F. Attempts have been made to resolve the dichotomy between (2.1) and (2.16) by using, for the analysis of the quantum measurement process, ideas (in particular asymptotic states) developed originally for the purposes of scattering theory; this approach, coupled with the concept of equivalence classes of macroscopic states, has particular epistemological appeal [46], and its self-consistency has been established within the framework of quantum statistical mechanics [54, 55, 56]. These successes however do not answer, in a manner that would fully satisfy a (classical) physicist, the even more vexing question of how to interpret the probabilistic statement (2.16) in terms of *individual* events, in particular events such as those which one traditionally pictures as taking place at the times when the results of the measurement are registered.

III. Consciousness

The papers in this part were admittedly among the hardest ones this commentator had to review for this volume, as they impinge on areas of concern that are foreign to present-day physics; they are addressed to audiences that most practitioners of 'normal' [57] physics would locate outside the paradigms of their science, although Eugene Wigner reminds us that "regions of enquiry, which were long considered to be outside the province of science, were drawn into this province ... The best known example is the interior of the atom, which was considered to be a metaphysical subject ... When the province of physical theory was extended to encompass microscopic phenomena, through the creation of quantum mechanics, the concept of consciousness came to the fore again: it was not possible to formulate the laws of quantum mechanics in a fully consistent way without reference to consciousness" [58].

The documentary evidence indicates how Eugene Wigner's interest in the nature and role of consciousness was triggered by his efforts to cope with the theory of the quantum measurement process. To understand his position, consider indeed the contrast between classical and quantum physics that must have hit the Founders of the new theory. On the one hand, the shadows on the walls were not affected by whether the platonic philosopher observed them or not; and the classical physicist thought that he could, at least in principle, rig his measuring devices in such a manner as to make arbitrarily small any influence a measurement might have on the system under investigation, thus making possible an effective empirical separation between the observer and the observed. On the other hand, in quantum mechanics "even though the dividing line between the observer ... and the observed ... can be shifted ..., it cannot be eliminated" [58]. Specifically, Eugene Wigner focuses on the inability of the orthodox theory to resolve the dichotomy between the two types of evolution encountered in quantum theory: a system, starting in a *pure state,* and evolving according to a deterministic equation (e.g. the Schrödinger equation), will be in a *pure state* at all subsequent times, while the same system, prepared identically in a *pure state,* but now evolving through the description given of the measuring process will end up in a state that is a *mixture.* Quantum theorists could thus not escape viewing as most unsettling a situation for which the paradoxical Schrödinger cat was developed as an archetype [59]. Eugene Wigner pushed the argument even further by the interposition of a "friend" who opens the cat's box to see whether the poor thing is dead or alive, and only *subsequently* reports his finding, thus raising the question of precisely when the measurement is completed; [58] gives a vivid description of this now proverbial 'Wigner's friend'.

[60] appears more as a compilation of arguments Eugene Wigner presented elsewhere, although it also comprises a few editorial comments. Yet, it does actually provide insights into what Eugene Wigner saw as the role of consciousness. It also contains an incidental remark on the interest there would be in trying to gain experimental information about the mental process by the "observation of infants where we may be able to sense the progress of the awakening of consciousness", an avenue opened by Jean Piaget who researched the formation, in the child's mind, of specific concepts (language, causality, construction of physical quantities, motion, geometry, ...) in a series of monographs dating back to 1923 and extending over more than fifty years; two general references are given in [61].

[62] extends further out of the province of physics, its leading metaphor being "to incorporate the description of the observer into physics, just as Faraday and Maxwell incorporated light into that description." The metaphor implies

that what were once thought to be negligible circumstances may later become an indispensable part of the general representation. Eugene Wigner gives an example of this: the liberties physicists traditionally take, when talking about 'isolated' systems, are disallowed by Zeh's argument [53]; see also [1].

While Eugene Wigner readily admits, and in fact emphasizes, that "life and consciousness cannot be described in terms of present-day physics", he argues indefatigably that a theory of consciousness should be attempted to provide a description of the observation process and of the "mysterious collapse of the wave packet" [63].

IV. Symmetries

The papers collected in this part were written in the course of about a quarter of a century (1949–1972); all can be read without technical preparations, and the purpose of this commentary is therefore reduced to pointing out some of the highlights. In summary, these papers travel along two lines of inquiry. From the first line, broad views are obtained over the conceptual development of invariance principles, of group theory, and of their representations; clearly, the vantage point to which we are invited in these papers is provided by a mathematical physicist who contributed original ideas to the subject, and who also reflected on their antecedents and broader meanings. The second line gives focussed vignettes of more immediately contemporary concerns.

One of the general themes, running through most of the papers of this part (see in particular [64, 65, 66, 67, 68]), is that much of the methodology in the natural sciences is predicated on a separation between the accidental (the 'initial conditions') and the expression of regularities (the 'laws of motion'). Eugene Wigner traces back the implementation of this basic distinction to the work of Newton, which he contrasts with the aims of Kepler who, although "we owe [him] the three precise laws of planetary motion, tried to explain also the size of the planetary orbits and their periods." Incidentally, Leibniz was reacting also to Kepler's aspirations when he proposed to distinguish between 'contingent truths' and 'necessary truths', i.e., between 'truths of fact' and 'truths of reason'; the point of this remark is not to suggest another sterile quarrel of priorities, but rather to emphasize with Eugene Wigner that the recognition of this type of distinction is intimately associated with the birth of modern science. Moreover, he brings this into the focus of contemporary scientific language by introducing here *invariance principles* as "the laws which the laws of nature must obey" [66].

Furthermore, in [64, 65, 66], Eugene Wigner points out another important methodological distinction, namely whether invariance principles are derived from the laws of motion (e.g., Poincaré's derivation of the geometric invariance of the Maxwell equations, which following his own usage, we refer to as the group of the Lorentz transformations), or whether the laws of nature are derived from invariance principles (the outstanding example here is Einstein's invention of general relativity, while his discovery of special relativity involves both approaches). Other illustrations of the distinction between these two modes of enquiry can be found in Eugene Wigner's own contributions. On the one hand, he showed us how to exploit rotational symmetries in spectroscopy (the 'Gruppenpest', of which [9] is one of the foremost expressions); he also used the established principle of invariance under the inhomogeneous Lorentz group to give us his classification of elementary particles as irreducible projective representations of this group, characterized by mass and spin [10]. On the other hand, upon reflecting on the conservation laws one associates to dynamical invariance principles [64] (for elaborations, see [69] and [70]), he postulated an analogy between the conservation law for electric charges, associated to the gauge invariance of electromagnetic interactions, and the conservation laws for baryons and leptons, to be associated to the group of invariance for strong and weak interactions; note the credit given to the early contribution of Stueckelberg [71] concerning the baryon conservation law.

Eugene Wigner's original contributions to the understanding of invariance groups and their physical consequences place him in a privileged position when it comes to present the history of the unfolding of these concepts: for the notion of invariance, see [64, 70, 72]; and for the theory of groups and their projective representations [73]. The latter paper is the text of his 1968 AMS Gibbs Lecture, and it is thus primarily addressed to mathematicians. Beyond general statements, such as the Galilean acknowledgment that "mathematics is the language of physics"; and remarks of interest to the student of intellectual history, such as the fact that Hessel's determination of the 32 point groups in R^3 and Galois' precise formulation of the concept of groups are ideas that came to light within two years of each other (1830–1832); Eugene Wigner directs attention to specific cases of interactions between mathematicians and physicists (his life-long association with von Neumann is also worth recalling here, and so are his lasting collaborations with Bargmann and with Wightman), and he documents the role physics played in delineating some purely mathematical problems; it is with the particular authority of one of the prime movers that he mentions, for instance, the theories of projective representations of groups and of unitary representations of non-compact Lie groups. The reader will also find in this, and other papers reprinted here [70, 65, 67] some of Eugene Wigner's views on the invariance groups that emerge in the theory of elementary particles, in particular on broken (or approximate) symmetries and their hierarchy. Among these, [67] vividly conveys the "astonishment" of the physics community at the discovery of parity violation (predicted theoretically in 1956, and verified experimentally within the next few months), and discusses the subsequent revisions this discovery stimulated in our views on the CPT symmetries: the charge conjugation C (the symmetry between matter and antimatter), the space inversion P (the symmetry between left and right orientations, or parity), and the time reflection T (the symmetry resulting from the reversal of directions of motion). In January 1957, Eugene Wigner had already given an early assessment of these momentous events in his retirement address as President of the American Physical Society, reprinted here as [42] (cf. the third part of that paper and its Appendix II).

V. Relativity

Of the two revolutionary ideas which, during the first quarter of this century, took hold of the way we think about the physical world, each has firmly established itself in its own immediate purview: quantum theory deals exquisitely with the microscopic phenomena in the physicist's laboratory, and general relativity accounts remarkably well for the large-scale structures of the cosmologist's universe. The satisfaction one may derive from this upbeat assessment hides however an important epistemological question, namely whether it is possible to devise a unified conceptual scheme from which both quantum mechanics and general relativity would emerge as limiting cases. The two papers reprinted in this part address issues that anticipate this question.

The extent of what has been firmly established so far seems to be that no conceptual incompatibility has yet surfaced between the basic tenets of quantum theory (i.e., quantum mechanics and quantum field theory) and those of *special* relativity. To restate this assertion more positively, let it be recalled here that Eugene Wigner himself contributed two essential pillars to the existing structure; the first is his mathematical classification of the irreducible projective representations of the inhomogeneous Lorentz group [10], which he identified in physical terms with the elementary quantum particles; the second of his contributions to this theory is the relation between these representations and the quantum relativistic wave equations [74]. Both are reviewed in [16], and so is the status of the identification of a position operator in this context [37]. A related aspect of the issues to be faced when quantum mechanics and (special) relativity are to be harmonized, namely the localizability of collisions between particles, is discussed in [42]. The first half of that paper also contains a review of the concept of intrinsic helicity for elementary particles of mass zero; and a discussion of the CPT symmetries (cf. the closing paragraph in Part IV of this commentary). In summary: "It is at least possible to formulate the requirements of special relativistic invariance for quantum theories and to ascertain whether these requirements are met. The fact that the answer is more likely *no* than *yes* is perhaps irritating. It does not alter the fact that the consistency of the two theories can at least be formulated ... [it is] more nearly ... a puzzle than ... a problem" [42].

The developments that occurred mostly in the twenty-five years after these lines were written have made this prophetic statement sound somewhat optimistic. Indeed, still within the realm of special relativistic quantum physics, the theory of interacting fields remains frustrated by considerable technical difficulties. Two axiomatic systems have been proposed that seem to encompass the essence of what we should call quantum fields in special relativity; the Wightman axioms [75] delineate sharply the analytic aspects of the theory, while its algebraic aspects are neatly formulated in the Haag–Kastler axioms [76]. Beyond the differences in methodological emphasis, these two axiomatic systems are mutually compatible. However, the implementation of these axioms in the construction of actual models of interacting quantum fields in $(3 + 1)$ space-time dimensions has fiercely resisted the best efforts [77].

The putative extension of these syntheses to *general* relativity presents the most vexing conceptual questions, one of which Eugene Wigner discusses in the second half of [42], namely the measurement (i.e., the pragmatic meaning or foundation of the concepts) of space-time distances and curvatures, and the essential limitations imposed thereupon by quantum theory. Nevertheless, the ultimate harmony of quantum theory and general relativity is still a lofty goal that has not receded beyond the hopes of the daring (for an update, see e.g., [78]).

VI. Nuclear Physics

The history of quantum mechanics and of relativity has been the subject of more studies and commentaries than one may perhaps even wish to count. Although these sources do not seem yet to have been exhausted, a more balanced view of the actual modes of thought that dominate in the physics community would surely require that some attention be devoted to the analysis of the inner dynamics of a domain such as nuclear physics. The papers collected in this part provide elements for such a reflection. Each of them presents specific case studies; although the more technical aspects of these studies may be of primary interest mostly to the specialist, some general features emerge that should open usable information to the generalist.

We should first note that these papers fall naturally into two sets. A public lecture [79], the text of which was published in 1955 by the widely circulated American Journal of Physics, is directed to a broad audience; while it focuses on one particular model, the compound nucleus approximation, it also documents, with the story of the development of this particular model, the phases through which many physical theories unfold. The other papers [80, 81, 82, 83] in this series appeared some fifteen years after [79] was published; they are overviews of four separate scientific gatherings held within the span of less than two years; they are thus more specialized; nevertheless, when read together, they provide a window on what the frontiers in nuclear physics looked like at the very end of the 1960's and in the early 1970's.

As for the general features that these five papers share, the following remarks are intended only to serve as beacons: they are not reviews of reviews.

First, and foremost, is the demanding interest and command Eugene Wigner, a theorist's theorist, maintained in the latest experimental techniques and results. The lofty considerations on mathematically consistent expression and on purely theoretical speculations, preeminent elsewhere in this collection, pale here before the concern that physics is, in the minds of most of its practitioners, a *laboratory* science. Second is the ensuing pragmatic necessity to delineate carefully in a theory the elements that are actually *observable*. Third is the use of the word 'model' in a sense that is rather different from the use made in mathematical physics circles, and thus in most of the reflections on the papers in Parts I, II, IV, and V of this collection; there, a model is meant to be a fully controlled mathematical construct that is designed to *verify* the consistency of a programme, body of doctrine, or system of axioms; here, a model describes a tentative approximation scheme, and much of the concern is to compare the results of different computational methods to *explore* the complex situations encountered in what physicists commonly refer to as 'the real world'. Fourth, this *complexity* of the situations that one wishes to account for in nuclear physics is truly daunting; indeed so daunting that even the design of specific approximation schemes sometimes seems to be a thankless task; statistical methods are then introduced already in the definition of the theory, see for instance [82] where the search is for the properties common to 'most' symmetric matrices, with the hope to hit on the few restrictive assumptions that will allow one to isolate those essential features of the nuclear levels distributions that are independent of the choice of a specific Hamiltonian.

The inclusion of these five papers in the present collection is thus motivated, at least in part, by a view to stimulate philosophers of science to come to terms with physics as it is practiced by so many physicists; we see also that these papers exemplify the sort of reviews Eugene Wigner (see Part VII) urges scientists to write in order to mitigate what he calls the "balkanization of science" [84].

VII. Broader Philosophical Essays

In contrast with the mainstream of 'normal' [57] physical sciences, one encounters much more often in philosophy, and to some extent also in mathematics, papers that are most interesting by the questions they choose to raise and the premises from which they are raised; all is for the best then if, in addition, the reader receives a sketch or a programme for the solution of some of these problems.

The contributions reprinted in this part address two types of concerns, which can be schematically identified as cognitive and social problems, although this separation is somewhat artificial since these two components do have implications on one another; consequently, representatives of both classes will be found together in several of these papers.

The reader who is primarily interested in getting an overview of Eugene Wigner's ideas, rather than in observing their development over the span of more than 35 years, may want to read first the last three papers of this part: [85] and [86] indeed paint the landscape in broad strokes, while [87] is argued for a more technically inclined audience. An even more focused overview also obtains from [88]; after a few introductory remarks on the dominion of science over our daily lives, material as well as cultural, this paper concentrates on the cognitive development of science, in particular physics, with an emphasis on the expansion of its territory, and on the acceleration of its growth; some of the themes of [89] on the unifying role of mathematics in the conceptualization of natural phenomena are taken up again in this paper.

[84] warns against 'whiggish' oversimplifications such as "the history of science reads like a success story". While the immediately perceptible aims of science can easily be listed (consistency, completeness, accessibility, conciseness, and the establishment of a hierarchy of concepts), it is important to delineate also the limits of science, lest one get trapped in "illusions". For instance, fundamental concepts often fall short of accounting for the most basic facts of experience. Even the usual physics curriculum recognizes that quantum mechanics is "more fundamental" than solid state physics; nevertheless, as Eugene Wigner notes, quantum mechanics has remarkably little to say on the very existence of crystals. More generally, he emphasizes that the role of the proper approximations would be worth closer attention in philosophical reflections on the nature of science. The failure to recognize this, together with the phenomenal growth of science in this century ("The American Physics Society has about 30,000 members now [1977] – it had about 3,000 in the thirties, it had 96 in 1900" [90]), led to a "balkanization of science" [84] against which Eugene Wigner protests repeatedly. "The problem of communication is not only external to physics; it threatens also to develop into an internal one" [91]. "It was possible, at that time [the nineteen-twenties and -thirties], to know physics; today it is difficult to know nuclear physics ... During my work on nuclear chain reactions, I already became scared by the increasing specialization of knowledge" [90]. To this problem, Eugene Wigner then advocates a remedy, as part of his "personalized version" of "Alvin Weinberg's injunctions" [92]: "to write our articles for a less specialized readership, to devote more time and energy to the composition of reviews and to reading of more of the reviews covering the results of sister sciences" [91]; short of this, we will soon fail to see that "science is an edifice, not a pile of bricks" [84].

One typical example of the disruption the contemporary neglect of a broad scientific culture has brought about is the misunderstandings that exist between mathematicians and natural scientists who often limit their view of the formers' relevance to that of keymakers, who could not care whether the door opens on a wall, their interest being restricted to the inner workings of the locks. Eugene Wigner wants to restore more than a sense of wonder, namely an understanding of the "unreasonable effectiveness of mathematics in the natural sciences" where "mathematical concepts turn up in entirely unexpected connections" [89]. For instance, commenting on the normal distribution in the introduction of this paper, Eugene Wigner imagines the surprize of the uninitiated: "surely the population has nothing to do with the circumference of the circle". To put this alienation in historical perspective, let it be allowed here to repeat Lord Kelvin's comment about the formula:

$$\int_{-\infty}^{+\infty} dx \; e^{-x^2} = \sqrt{\pi}$$

"A mathematician is one to whom *that* is as obvious as twice two makes four is to you. Liouville was a mathematician." As Lagrange or Euler before him, Liouville could perhaps grasp with a better sense of immediacy than we can today, how it is that the laws of nature are so simple or, as Eugene Wigner puts it, that we "got something out of the equations that we did not put in" [89]. It is nevertheless very much true that if one reflects on the ingredients that went into the inception of quantum theory, one cannot help but marvel at the result of the computation of the Lamb shift to an "accuracy not dreamed of at the time of the origin of the theory" [89]. As a corollary, and this is another limit of science, the overconfidence we acquire in the predictability of science becomes an "article of faith" which, however, can turn into a "nightmare" for the theoretician whose vigilance is allowed to lapse. We indeed conveniently forget that we *choose* the domain of applicability of our theories: the roulette wheel does not violate the laws of mechanics, but as Hadamard pointed out already at the end of the nineteenth century [93], the serene predictability of mechanics suddenly evaporates when one considers (non linear) systems with sensitive dependence on initial conditions; Hadamard's remark has blossomed today into the active field of research, starting with the pioneering work of the Russian school [94, 95], and popularly known as the study of 'chaos'. That it was mostly ignored, at the time Eugene Wigner wrote his article, confirms rather than denies his basic thesis: contrary to the misconception according to which mathematical techniques are relevant to the natural science merely as tools to evaluate the results of the laws of nature, he sees mathematics actually entering the picture at the much earlier stage when these laws are formulated. This thesis thus echoes Einstein's views: "nature is the realization of the simplest mathematical ideas ... the creative principle resides in mathematics" [96]; and it is curiously reminiscent of a position already taken by Leibniz, according to whom the laws of nature are those which must hold in a world having the greatest possible metaphysical perfection, calling to mind again the doctrine of preestablished universal harmony; however, Eugene Wigner's sustained interests in experimental observations and theoretical approximations (see for instance Part VI in this collection) would immediately exonerate him from the reproach of 'panlogism' sometimes addressed to Leibniz: the mode of work of the theoretical physicist differs from that of the mathematician in ways that were also discussed sensitively by von Neumann [97] and Bochner [98].

Another theme that recurs in Eugene Wigner's writings is his discomfort with a gap, that he finds increasingly unacceptable, between the physical sciences of the "inanimate" on the one hand, and on the other the sciences of "life" [99] and of the "mind". We already considered the source of this preoccupation in Parts II and III of this commentary: "present day physics, in particular the philosophical ideas forced on us by quantum mechanics, postulates the study of mental phenomena just as strongly as the fact that we obtain our information on the motion of objects, in particular of planets, by means of light rays postulated the study of the light phenomena" [90]. Eugene Wigner may even seem to push the argument further. Or does he? We already saw that, while classical physics assumed that measurements can be made that perturb only insignificantly the state of the system under observation, quantum mechanics has shown this idealization to be generally not justified. Now, Eugene Wigner uses this for the following metaphor: "experiments in the mental domain cannot easily be repeated – the initial conditions cannot be precisely reproduced. When an experiment is made on me, I probably remember it and will not be the same person when a repetition is attempted" [90]. When he calls for such a new synthesis, he nevertheless emphasizes that he advocates venturing on unsecured ground: "The drift toward new areas of science will not be easy because it requires less effort to add to a building, the foundations of which are already firmly laid, than to start new foundations on ground which appears to be soft" [100].

This is still another reflection on the balkanization of science, and it brings us to two specific societal problems of science that have retained Eugene Wigner's attention. The first is the professionalization of science, and in particular, the rise of "big science"; the second is the social responsibility of the scientist. With the former comes the questions of "whether this or that piece of research is worth doing" [84] (read perhaps: worth supporting); of the administration of collective research ([100] contains, inter alia, a summary of a specific proposal, designed by Eugene Wigner while he was a member of the US Government's Committee on Science and Information); and of the preservation of individual creativity [84] and of "little science"[100], echoing another of Einstein's concerns: "One can organize the application of a discovery already made, but one cannot organize the discovery itself"[101]. It is no accident that, in these same papers, Eugene Wigner also addresses such areas of the scientist's responsibility as the work on nuclear weapons and recombinant DNA. Here again, it may be worthwhile to add a reminder on historical premises which Eugene Wigner might have been too modest to mention there. He was, together with Leo Szilard, one of the initiators in July 1939 of the confidential approach that led Einstein to write his famous letter of August 2 to President Roosevelt on the consequences of the discovery of chain reactions made by Otto Hahn and his co-workers of the Kaiser Wilhelm Institute in Berlin, the work of Frederic Joliot in France, and the research conducted in the United States by Enrico Fermi and his collaborators.

References

1. E.P. Wigner: Review of the Quantum-Mechanical Measurement Problem. In: Quantum Optics, Experimental Gravity and Measurement Theory, P. Meystre and M.G. Scully (eds.), Plenum Press, New York 1983, pp. 43–63; in: Science, Computers and the Information Onslaught, D.M. Kerr et al. (eds.), Academic Press, New York 1984, pp. 63–82; Vol. VI
2. J. von Neumann: Mathematische Grundlagen der Quantenmechanik, Springer, Berlin 1932; Mathematical Foundations of Quantum Mechanics (R.T. Beyer, transl.), Princeton University Press, Princeton, NJ 1955
3. F.W. London and E. Bauer: La Théorie de l'Observation en Mécanique Quantique, Hermann, Paris 1939; The Theory of Observation in Quantum Mechanics (A. Shimony, J.A. Wheeler, W.H. Zurek, J. McGrath, and S. McLean McGrath,

transls.), in: Quantum Theory of Measurement, J.A. Wheeler and W.H. Zurek (eds.). Princeton University Press, Princeton, NJ 1983, pp. 217–259

4. E.P. Wigner: Discussion: Comments on Prof. Putnam's Comments. Phil. Sci. **29** (1962) 292–293; Vol. VI

5. E.P. Wigner: Two Kinds of Reality. The Monist **48** (1964) 248–264; in: Symmetries and Reflections, Indiana Univ. Press, Bloomington, IN 1967, pp. 185–199; Vol. VI

6. E.P. Wigner: Epistemology of Quantum Mechanics. In: Contemporary Physics, Vol. II, Atomic Energy Agency, Vienna 1969, pp. 431–437; Vol. VI

7. E.P. Wigner: Epistemological Perspective on Quantum Theory. In: Contemporary Research in the Foundations and Philosophy of Quantum Theory, C.A. Hooker (ed.), Reidel Publ., Dordrecht 1973, pp. 369–385; Vol. VI

8. E.P. Wigner: Interpretation of Quantum Mechanics. In: Quantum Theory and Measurement, J.A. Wheeler and W.H. Zurek (eds.), Princeton University Press, Princeton, NJ 1983, pp. 260–314; Vol. VI

9. E.P. Wigner: Gruppentheorie und ihre Anwendungen auf die Quantenmechanik der Atomspektren, Vieweg, Braunschweig 1931; Group Theory and its Applications to the Quantum Mechanics of Atomic Spectra (J.J. Griffin transl.), Academic Press, New York 1959

10. E.P. Wigner: On Unitary Representations of the Inhomogeneous Lorentz Group. Ann. Math. **40** (1939) 149–204; Vol. I, Part III

11. M. Born: Zur Quantenmechanik der Stoßvorgänge. Z. Physik **37** (1926) 863–867; ibid. **38** (1926) 803–867; Göttingen Nachr. (1926) 146–160. Statistical Interpretation of Quantum Mechanics (Nobel Lecture, 1954), in: Nobel Lectures, Physics, 1942–1962, Nobel Foundation, Elsevier Publ., Amsterdam 1964, pp. 256–267

12. E.P. Wigner: Theorie der quantenmechanischen Messung. In: Physikertagung, Wien 1961, Physik Verlag, Mosbach/Baden 1962, pp. 1–8; Vol. VI

13. E.P. Wigner: The Problem of Measurement. Am. J. Phys. **31** (1963) 6–15; in: Symmetries and Reflections, Indiana Univ. Press, Bloomington, IN 1967, pp. 153–170; in: Theory of Measurement in Quantum Mechanics, Yanase, Namicki, and Machida (eds.), Phys. Soc. Japan, 1978, pp. 123–132; in: Quantum Theory and Measurement, J.A. Wheeler and W.H. Zurek (eds.), Princeton University Press, Princeton, NJ 1983, pp. 324–341; Vol. VI

14. E.P. Wigner: Realität und Quantenmechanik. Address in Lindau and Tutzing (1982) 7–17; Vol. VI

15. E.P. Wigner: The Limitations of Determinism. In: Absolute Values and the Creation of the New World (Proc. 11th ICUS, 1982), International Cultural Foundation Press, New York 1983, pp. 1365–1370; Vol. VI

16. E.P. Wigner: Relativistic Equations in Quantum Mechanics. In: The Physicist's Conception of Nature, J. Mehra (ed.), Reidel Publ., Dordrecht 1973, pp. 320–330; Vol. VI

17. P. Jordan, J. von Neumann, and E.P. Wigner: On an Algebraic Generalization of the Quantum Mechanical Formalism. Ann. Math. **35** (1934) 29–64; Vol. I, Part III

18. E.C.G. Stueckelberg: Quantum Theory in Real Hilbert Space. Helv. Phys. Acta **33** (1960) 727–752; (with M. Guenin) ibid. **34** (1961) 621–628; (with M. Guenin, C. Piron, and H. Ruegg) ibid. **34** (1961) 675–698; (with M. Guenin) ibid. **35** (1962) 673–695

19. J.M. Jauch, D. Finkelstein, S. Schiminovitch, and D. Speiser: Foundations of Quaternionic Quantum Mechanics, J. Math. Phys. **3** (1962) 207–220; Quaternionic Representations of Compact Groups, ibid. **4** (1963) 136–140; Principles of General Q-covariance, ibid. **4** (1963) 788–796

20. G.G. Emch: Mécanique Quantique Quaternionienne et Relativité Restreinte I. Helv. Phys. Acta **36** (1963) 739–769; II. Helv. Phys. Acta **36** (1963) 770–788; Representations of the Lorentz Group in Quaternionic Quantum Mechanics, in: Lectures in Theoretical Physics, Boulder 1964, Vol. VIIa, W.E. Britten and A.O. Barut (eds.), University of Colorado Press, Boulder, CO 1965, pp. 1–36

21. J.P. Eckmann and Ph. Zabey: Impossibility of Quantum Mechanics in a Hilbert Space over a Finite Field. Helv. Phys. Acta **42** (1969) 420–424

22. M. Born and P. Jordan: Zur Quantenmechanik. Z. Physik **34** (1925) 858–888; (with W. Heisenberg) ibid. **35** (1925) 557–615

23. P. Chernoff: Difficulties of Canonical Quantization. Lecture notes, Berkeley, CA, 1969; see also R. Abraham and J.E. Marsden, Thm 5.4.9, in: Foundations of Mechanics. Benjamin, Reading, MA, 1978

24. J.M. Souriau: Structure des Systèmes Dynamiques. Dunod, Paris 1970

25. A. Kirillov: Eléments de la Théorie des Représentations. Mir, Moscow 1974

26. G.G. Emch: Geometric Quantization: Regular Representations and Modular Algebras, in: Proc. XVIII Colloquium on Group Theoretical Methods, Moscow, 1990, M.A. Markov and V.I. Man'ko (eds), Springer, Heidelberg 1991, pp. 356–364; Geometric Quantization, Modular Reduction Theory and Coherent States, (with S.T. Ali), J. Math. Phys. **27** (1986) 2936–2948

27. E.P. Wigner: Die Messung quantenmechanischen Operatoren. Z. Physik **133** (1952) 101–108; Vol. VI

28. G.C. Wick, A.S. Wightman, and E.P. Wigner: Intrinsic Parity of Elementary Particles. Phys. Rev. **88** (1952) 101–105; Vol. III, Part II

29. S. Freedman and E.P. Wigner: On Bub's Misunderstanding of Bell's Locality Argument. Found. Phys. **3** (1973) 457–458; Vol. VI

30. D. Bohm: A Suggested Interpretation of the Quantum Theory in terms of Hidden Variables. Phys. Rev. **85** (1952) 166–179, 180–193

31. J.S. Bell: On the Einstein-Podolsky-Rosen Paradox, Physics **1** (1965) 195–200; On the Problem of Hidden Variables in Quantum Theory, Rev. Mod. Phys. **38** (1966) 447–452

32. J.F. Clauser, M.A. Horne, A. Shimony, and R. Holt: Proposed Experiment to Test Local Hidden-Variable Theories. Phys. Rev. Lett. **23** (1969) 880–884

33. S.J. Freedman and J.F. Clauser: Experimental Tests of Local Hidden-Variable Theories. Phys. Rev. Lett. **28** (1972) 938–941

34. A. Aspect: Trois Tests Expérimentaux des Inégalités de Bell par Measure de Corrélation de Polarisation de Photons. Thèse, Orsay 1983

35. D. Mermin: Quantum Mysteries Revisited. Am. J. Phys. **58** (1990) 731–734

36. E.P. Wigner: The Non-Relativistic Nature of the Present Quantum-Mechanical Measurement Theory. Ann. NY Acad. Sci. **480** (1986) 1–5; Vol. VI

37. T.D. Newton and E.P. Wigner: Localized States for Elementary Systems. Rev. Mod. Phys. **21** (1949) 400–406; Vol. I, Part III

38. A.S. Wightman: On the Localizability of Quantum Mechanical Systems. Rev. Mod. Phys. **34** (1962) 845–872

39. G.N. Fleming: Non-local Properties of Stable Particles. Phys. Rev. **B 139** (1965) 963–968

40. G.C. Hegerfeldt: Remark on Causality and Particle Localization. Phys. Rev. **D 10** (1974) 3320–3321

41. S.T. Ali and G.G. Emch: Fuzzy Observables in Quantum Mechanics. J. Math. Phys. **15** (1974) 176–182; and references quoted therein

42. E.P. Wigner: Relativistic Invariance and Quantum Phenomena, Rev. Mod. Phys. **29** (1957) 255–268; in: Symmetries and Reflections, Indiana Univ. Press, Bloomington, IN 1967, pp. 51–81; Vol. VI

43. H. Salecker and E.P. Wigner: Quantum Limitations of Measurement of Space-time Distances. Phys. Rev. **109** (1958) 571–577; Vol. III, Part I

44. J.M. Jauch, E.P. Wigner, and M.M. Yanase: Some Comments Concerning Measurement in Quantum Mechanics. Il Nuovo Cimento **48 B** (1967) 144–151; Vol. VI

45. A. Frenkel, E.P. Wigner, and M. Yanase: On the Change of the Skew Information in the Process of Quantum Mechanical Measurement. Mimeographed notes (ca. 1970); Vol. VI

46. E.P. Wigner: The Subject of our Discussions. In: Foundations of Quantum Mechanics (Intern'l School of Physics "Enrico Fermi" 1970), B. d'Espagnat (ed.), Academic Press, New York 1971, pp. 1–19; Vol. VI

47. V. Fock: Classical and Quantum Physics: ICTP Trieste **SMR 5/8** (1968)

48. N. Bohr: Discussion with Einstein on Epistemological Problems in Atomic Physics. In: Albert Einstein, Philosopher-Scientist, P.A. Schilpp (ed.), Library of Living Philosophers, Evanston, IL 1949, pp. 200–241

49. H. Araki and M. Yanase: Measurement of Quantum Mechanical Operators. Phys. Rev. **120** (1960) 622–626

50. M. Yanase: Optimal Measuring Apparatus. Phys. Rev. **123** (1961) 666–668

51. E.P. Wigner: The Philosophical Problem. In: Foundations of Quantum Mechanics, (Intern'l School of Physics "Enrico Fermi" 1970) B. d'Espagnat (ed.), Academic Press, New York 1971, pp. 122–124; Vol. VI

52. E.P. Wigner: Questions of Physical Theory. In: Foundations of Quantum Mechanics, (Intern'l School of Physics "Enrico Fermi" 1970) B. d'Espagnat (ed.), Academic Press, New York 1971, pp. 124–125; Vol. VI

53. H.D. Zeh: On the Interpretation of Measurement in Quantum Theory. Found. Physics **1** (1970) 69–76

54. K. Hepp: Quantum Theory of Measurement and Macroscopic Observables. Helv. Phys. Acta **45** (1972) 237–248

55. B. Whitten-Wolfe and G.G. Emch: A Mechanical Quantum Measuring Process. Helv. Phys. Acta **49** (1976) 45–55

56. J.A. Wheeler and W.H. Zurek: Quantum Theory of Measurement, (in particular Part V and Appendix II). Princeton University Press, Princeton, NJ 1983

57. T.S. Kuhn: The Structure of Scientific Revolutions (2nd ed.). Chicago University Press, Chicago, IL, 1970

58. E.P. Wigner: Remarks on the Mind-Body Question. In: The Scientist Speculates, I.J. Goog (ed.), Heinemann, London 1961, pp. 284–302, Basic Books, New York, 1962; in: Symmetries and Reflections, Indiana Univ. Press, Bloomington, IN 1967, pp. 171–184; in: Quantum Theory and Measurement, J.A. Wheeler and W.H. Zurek (eds.), Princeton University Press, Princeton, NJ 1983, pp. 168–181; Vol. VI

59. E. Schroedinger: Die Gegenwärtige Situation in der Quantenmechanik. Naturwissenschaften **23** (1935) 807–812, 823–828, 844–849; The Present Situation of Quantum Mechanics, A Translation of Schrödinger's Cat Paradox Paper (J.D. Trimmer, transl.), in: Quantum Theory of Measurement, J.A. Wheeler and W.H. Zurek (eds.), Princeton University Press, Princeton, NJ 1983, pp. 152–167

60. E.P. Wigner: The Place of Consciousness in Modern Physics. In: Conciousness and Reality, C.A. Muses and A.M. Young (eds.), Outerbridge and Lazard, New York 1972, pp. 132–141; Avon Books, 1974; Vol. VI

61. J. Piaget: Introduction à l'Epistémologie Génétique: I. La Pensée Mathématique; II. La Pensée Physique; III. La Pensée Biologique, Psychologique et Sociale, P.U.F., Paris 1950; L'Equilibration des Structures Cognitives, P.U.F., Paris 1975

62. E.P. Wigner: New Dimensions of Consciousness. Mimeographed notes (ca. 1978); Vol. VI

63. E.P. Wigner: The Existence of Conciousness. In: The Reevaluation of Existing Values and the Search for Absolute Values (Proc. 7th ICUS, 1978), International Cultural Foundation Press, New York 1979, pp. 135–143; in: Modernization, R.L. Rubinstein (ed.), Paragon House, New York 1982, pp. 279–285; Vol. VI

64. E.P. Wigner: Invariance in Physical Theory. Proc. Amer. Phil. Soc. **93** (1949) 521–526; in: Symmetries and Reflections, Indiana Univ. Press, Bloomington, IN 1967, pp. 3–13; Vol. VI

65. E.P. Wigner: Events, Laws of Nature, and Invariance Principles. Science **145** (1964) 995–998; in: Les Prix Nobel en 1963, Nobel Foundation, Stockholm, 1964; in: Symmetries and Reflections, Indiana Univ. Press, Bloomington, IN 1967, pp. 38–50; Vol. VI

66. E.P. Wigner: Events, Laws of Nature, and Invariance Principles. Mimeographed notes (ca. 1980); Vol. VI

67. E.P. Wigner: Violations of Symmetry in Physics. Scientific American **213** (1965) 28–36; Vol. VI

68. E.P. Wigner: Symmetry in Nature. In: Proc. R.A. Welsh Foundation on Chemical Research XVI. Theoretical Chemistry, Houston, Texas, 1972, W.O. Milligan (ed.), 1973, pp. 231–260; Vol. VI

69. E.P. Wigner: On the Law of Conservation of Heavy Particles. Proc. Natl. Acad. Sci. USA **38** (1952) 449–451; Vol. VI

70. E.P. Wigner: Symmetry and Conservation Laws. Proc. Natl. Acad. Sci. USA **51** (1964) 956–965; in: Symmetries and Reflections, Indiana Univ. Press, Bloomington, IN 1967, pp. 14–27; Vol. VI

71. E.C.G. Stueckelberg: Die Wechselwirkungskräfte in der Elektrodynamik und in der Feldtheorie der Kernkräfte I. Helv. Phys. Acta **11** (1938) 225–244; II und III, ibid. 299–328

72. E.P. Wigner: The Role of Invariance Principles in Natural Philosophy. In: Dispersion Relations and their Connection with Causality (Intern'l School of Physics "Enrico Fermi" 1963), E.P. Wigner (ed.), Academic Press, New York 1964, pp. ix–xvi; in: Symmetries and Reflections, Indiana Univ. Press, Bloomington, IN 1967, pp. 28–37; Vol. VI

73. E.P. Wigner: Symmetry Principles in Old and New Physics. Bull. Amer. Math. Soc. **74** (1968) 793–815; Vol. VI

74. V. Bargmann and E.P. Wigner: Group Theoretical Discussion of Relativistic Wave Equations. Proc. Natl. Acad. Sci. USA **34** (1948) 211–223; Vol. III, Part I

75. A.S. Wightman: Quantum Field Theory in terms of Vacuum Expectation Values. Phys. Rev. **101** (1956) 860–866

76. R. Haag and D. Kastler: An Algebraic Approach to Quantum Field Theory. J. Math. Phys. **5** (1964) 848–861

77. J. Glimm and A. Jaffe: Quantum Physics – a Functional Integral Point of View. Springer, New York 1981

78. A. Ashtekar: Recent Developments in Quantum Gravity. In: Fourteenth Texas Symposium on Relativistic Astrophysics, E.J. Fenyves (ed.), Ann. NY Acad. Sci. **571** (1989) 16–26

79. E.P. Wigner: On the Development of the Compound Nucleus Model. Am. J. Phys. **23** (1955) 371–380; in: Symmetries and Reflections, Indiana Univ. Press, Bloomington, IN 1967, pp. 93–109; Vol. VI

80. E.P. Wigner: Summary of the Conference. In: Proc. Intern'l Conference on Properties of Nuclear States. Les Presses de l'Université de Montréal, Montreal 1969, pp. 633–647; Vol. VI

81. E.P. Wigner: Summary of the Conference. In: Polarization Phenomena in Nuclear Reactions, H.H. Barschall and W. Haeberli (eds.). University of Wisconsin Press, Madison, WI 1971, pp. 389–395; Vol. VI

82. E.P. Wigner: Introductory Talk. In: Statistical Properties of Nuclei, J.B. Garg (ed.). Plenum Publ., New York 1972, pp. 7–23; Vol. VI

83. E.P. Wigner: Concluding Remarks. In: Symmetry Properties of Nuclei (Proc. of the 1970 Solvay Conference). Gordon and Breach, New York 1974, pp. 351–362; Vol. VI

84. E.P. Wigner: The Limits of Science. Proc. Amer. Phil. Soc. **94** (1950) 422–427; in: Symmetries and Reflections, Indiana Univ. Press, Bloomington, IN 1967, pp. 211–221; Vol. VI

85. E.P. Wigner: The Extension of the Area of Science. In: The Role of Consciousness in the Physical World, AAAS Symposium No. 57, R.G. Jahn (ed.). Westview Press, Boulder, CO 1981, pp. 7–16; Vol. VI

86. E.P. Wigner: The Glorious Days of Physics. In: The Unity of Fundamental Interactions (19th Course of the Intern'l School of Subnuclear Physics, 1981) A. Zichichi (ed.), Plenum Press, New York 1983, pp. 765–774; in: Quantum Optics, Experimental Gravity and Measurement Theory, P. Meystre and M.G. Scully (eds.), Plenum Press, New York 1983, pp. 1–7; Vol. VI

87. E.P. Wigner: Some Problems of our Natural Sciences. Intern'l Jour. Theor. Phys. **25** (1986) 467–476; Vol. VI

88. E.P. Wigner: Physics and its Relation to Human Knowledge. Hellenike Anthropistike Hetaireia, Athens 1977; Vol. VI

89. E.P. Wigner: The Unreasonable Effectiveness of Mathematics in the Natural Sciences. Comm. Pure and Applied Math. **13** (1960) 1–14; in: Symmetries and Reflections, Indiana Univ. Press, Bloomington, IN 1967, pp. 222–237; Vol. VI

90. E.P. Wigner: The Problems, Future and Limits of Science. In: The Search for Absolute Values in a Changing World (Proc. 6th ICUS, 1977). International Cultural Foundation Press, New York 1978, pp. 869–877; Vol. VI

91. E.P. Wigner: On some of Physics' Problems. Main Currents in Modern Thought **28** (1972) 75–78; in: Vistas in Physical Reality, E. Laszlo and E.B. Sellon (eds.), Plenum Publ., New York 1976, pp. 3–9; Vol. VI

92. A.M. Weinberg: Impact of Large–Scale Science on the United States. Science **134** (1961) 161–164

93. J. Hadamard: Les Surfaces à Courbures Opposées et leurs Lignes Géodésiques. J. Math. Pures Appl. **4** (1898) 27–74

94. D.V. Anosov: Geodesic Flows on Closed Riemannian Manifolds with Negative Curvature. Proc. Steklov Inst. Math. **90** (1967) 1–235; AMS (1969)

95. Ya.G. Sinai: Introduction to Ergodic Theory (V. Scheffer, transl.). Princeton University Press, Princeton, NJ 1976

96. A. Einstein: On the Methods of Theoretical Physics (Herbert Spencer Lecture, Oxford, 1933). In: Ideas and Opinions (S. Bargmann, transl. and ed.), Bonanza Books, New York 1954, pp. 220–276

97. J. von Neumann: The Mathematician. In: The Works of the Mind, R.B. Heywood (ed.), University of Chicago Press, Chicago 1947, pp. 180–196

98. S. Bochner: The Role of Mathematics in the Rise of Science. Princeton University Press, Princeton, NJ 1966

99. E.P. Wigner: Physics and the Explanation of Life. Found. Phys. **1** (1970) 33–45; in: Philosophical Foundations of Science, Proc. of an AAAS 1969 Program, R.J. Seeger and R.S. Cohen (eds.), Boston Studies in the Philosophy of Science XI, Reidel Publ., Dordrecht 1974, pp. 119–131; Vol. VI

100. E.P. Wigner: The Growth of Science – Its Promise and Its Dangers. In: Symmetries and Reflections, Indiana Univ. Press, Bloomington, IN 1967, pp. 267–280; Vol. VI
101. A. Einstein: Atomic War and Peace (Atlantic Monthly, Nov. 1945); in: Ideas and Opinions (S. Bargmann, transl. and ed.), Bonanza Books, New York 1954, pp. 118–131

PART I

Epistemology of Quantum Mechanics

Discussion:
Comments on Professor Putnam's Comments

H. Margenau and E. P. Wigner

Phil. Sci. *29*, 292–293 (1962)

Received March, 1962

Professor Putnam's comments [1] on David Sharp's paper [2], claiming to have discovered grave inconsistencies in the conceptual structure of quantum mechanics, evidently were not meant to be accepted at face value. The author surely realized that a contradiction as obvious as he claims to have found it, if real, would have been noticed by scores of physicists. Rather, we understand Professor Putnam's comments as a challenge to restate the theory of measurement in terms which do not use the mathematical formalism of quantum theory and to point to errors in his own argument.

According to von Neumann [3] and to London and Bauer [4], who gave the most compact and the most explicit formulations of the conceptual structure of quantum mechanics, every measurement is an interaction between an object and an observer. The state of the object, if one has the maximum possible knowledge about it, can be described by a state vector (also called wave function, or, as by Professor Putnam, a state function). The object cannot be the whole universe because the observer, at least, is distinct from it. The object obeys the laws of motion (the so-called Schrödinger equation) as long as it is "closed," that is, separated from the rest of the world. One can, and does, assume that this is the case during time intervals *between* measurements. It is evidently not true when a measurement takes place because the measurement is an interaction between the object and the observer. Hence, Professor Putnam's assumption 2 is not supportable in the form given in the text. On the other hand, if it is modified according to his footnote 1, pointing out that the object need not be the whole universe (as stated above, it *cannot* be the whole universe), one wonders how he arrived at conclusion 3, which states that "the whole universe never undergoes measurement" but which should read, if the footnote is taken into account, "the system which consists of an object *and* a measuring apparatus never undergoes measurement." This conclusion is, in our opinion, in error. In fact, if one wants to ascertain the result of the measurement, one has to observe the measuring apparatus, i.e., carry out a measurement on it.

The chain of transmission of information from the object to the consciousness of the observer may consist of several steps and one may pursue these steps to a greater or smaller extent. One cannot follow the transmission of information to the very end, i.e., into the consciousness of the observer, because present-day physics is not applicable to the consciousness. This point, which may be unpleasant from the point of view of certain philosophies, has been clearly recognized by both von Neumann and by London and Bauer.

DISCUSSION: COMMENTS ON PROFESSOR PUTNAM'S COMMENTS   293

As they express it, one must introduce a cut between object and observer and assume that the observer has a "direct knowledge" of what is on his side of the cut [4]. It is true that the dividing line between object and observer can be moved considerably [3]. Thus, a photographic plate, on which an electron may leave a track, can be treated as part of the object, or as part of the system of the observer with which the electron undergoes a measurement-interaction. In the former case, if the photographic plate is considered to be part of the system, the interaction may involve, principally, the observer and the photographic plate, rather than the electron directly. However, the dividing line, even though it can be shifted between observer and object, cannot be abolished. This is in conformity with the interpretation of the state vector as a description of the best possible knowledge available to the observer.

There are other instances of confusion in Professor Putnam's comments. How it follows from his assumption 4 that an electron subject to external forces can never be in a δ-function state (eigenstate of position) at two consecutive times, even if it obeys the Schrödinger equation, is obscure, and the conclusion is erroneous.

We must also reject the suggestion that quantum mechanics treats the universe as consisting of two qualitatively different kinds of things, "classical" objects and micro-objects. Indeed, when Bohr's correspondence principle and other features of quantum mechanics are properly interpreted, they imply that classical objects are included as proper limiting concerns of a probabilistic theory which, in this limit, reduces to classical physics.

Overall consistency of all parts of quantum mechanics, especially when that theory is forced to make reference to "the entire universe," has never been proved nor claimed. And it is not likely that any expert in the modern developments of logic will demand it.

We do not wish these remarks to be construed as affirming that all problems in the interpretation of quantum mechanics have been solved. As in any living part of science, reinterpretations and refinements are to be expected and are indeed hoped for in our understanding of that discipline. However, we could not convince ourselves of the validity of the particular strictures contained in reference [1].

REFERENCES

[1] PUTNAM, HILARY, Phil. of Science, 28, 234 (1961).

[2] SHARP, D. H., ibid., 28, 225 (1961).

[3] NEUMANN, J. VON, *Mathematical Foundations of Quantum Mechanics* (Princeton University Press, 1955); German original, Julius Springer, Berlin, 1932.

[4] LONDON, F. and BAUER, E., *La Theorie de l'Observation en Mecanique Quantique* (Hermann and Co., Paris, 1939).

Two Kinds of Reality

E. P. Wigner

Symmetries and Reflections. University Press, Bloomington, Indiana, 1967, pp. 185–199

The present discussion arose from the desire to explain, to an audience of non-physicists,[1] the epistemology to which one is forced if one pursues the quantum mechanical theory of observation to its ultimate consequences. However, the conclusions will not be derived from the aforementioned theory but obtained on the basis of a rather general analysis of what we mean by real. Quantum theory will form the background but not the basis for the analysis. The concept of the real to be arrived at shows considerable similarity to that of the idealist. As the title indicates, it is formulated as a dualism. It is quite possible that it will soon be rejected not only by the community of the philosophers but also by that of the scientists. If this should be the case, the attempt to derive an epistemology from physics will prove to have been premature. Naturally, the author hopes that this will not be the case because, quite apart from the quantum theoretical background, the concepts to be presented appear natural also as an outgrowth of common sense considerations. They have been arrived at by many (including Schrödinger) who did not accept the epistemology of quantum mechanics.

Disclaimer of Authority

The problems of the present inquiry have been grappled with by the keenest minds, and for much longer periods than the present writer has

Reprinted by permission from *The Monist*, Vol. 48, No. 2 (April, 1964).
[1] Conference at Marquette University, Summer 1961.

devoted to them. His only qualification for speaking about them is that he believes to represent the view at which most physicists would arrive if they were sufficiently pressed for their opinions on the subject. He realizes the profundity of his ignorance of the thinking of some of the greatest philosophers and is under no illusion that the views to be presented will be very novel. His hope is that they will appear sensible.

The discussion will be divided into two parts. The first part will describe two kinds of realities of very different characters. It is my conviction that the distinction to be made is valid and represents a large measure of consensus among physicists and perhaps even natural scientists. The second part of the discussion will be concerned with the relation of these two "realities" and will touch some of the thorniest problems which have puzzled each of us and which even the greatest philosophers have failed to solve completely. The second part of the discussion will not contain an attempt at the solution of these questions. It will be confined to the statement of some of the problems, and to questions as to what the prospect of their solution is and what such a solution would mean.

Even though it is not strictly relevant, it may be useful to give the reason for the increased interest of the contemporary physicist in problems of epistemology and ontology. The reason is, in a nutshell, that physicists have found it impossible to give a satisfactory description of atomic phenomena without reference to the consciousness. This had little to do with the oft rehashed problem of wave and particle duality and refers, rather, to the process called the "reduction of the wave packet." This takes place whenever the result of an observation enters the consciousness of the observer—or, to be even more painfully precise, my own consciousness, since I am the only observer, all other people being only subjects of my observations. Alternatively, one could say that quantum mechanics provides only probability connections between the results of my observations as I perceive them. Whichever formulation one adopts, the consciousness evidently plays an indispensable role.[2]

[2] This is not the proper place to give a detailed proof for this assertion since such a proof would have to be based on the mathematical formulation of quantum mechanics and more particularly on the superposition principle. It should suffice, therefore, to mention that the fact was pointed out with full clarity first by von Neumann (see Chapter VI of his *Mathematical Foundations of Quantum Mechanics*, [Princeton, N.J.: Princeton University Press, 1955], or the German original [Berlin: J. Springer, 1932]). It would not be difficult to mount a battery of authorities affirming the assertion of the text, the clearest being W. Heisenberg's statement: "The laws of

Two Kinds of Reality   *187*

In outline, the situation is as follows. The interaction between the measuring apparatus and the system on which the measurement should take place (the *object* of the measurement) results in a state in which there is a strong statistical correlation between the state of the apparatus and the state of the object. In general, neither apparatus nor object is in a state which has a classical description. However, the state of the united system, apparatus plus object, is after the interaction such that only one state of the object is compatible with any given state of the apparatus. Hence, the state of the object can be ascertained by determining the state of the apparatus after the interaction has taken place between them. It follows that the measurement of the state of the object has been reduced to the measurement of the state of the apparatus. However, since the state of the apparatus has no classical description, the measurement of the state of the apparatus is, from the conceptual point of view, not different from the measurement on the original object. In a similar way, the problem can be transferred from one link of a chain to the next, and so on. However, the measurement is not completed until its result enters our consciousness. This last step occurs when a correlation is established between the state of the last measuring apparatus and something which directly affects our consciousness.[3] This last step is, at the present state of our knowledge, shrouded in mystery and no explanation has been given for it so far in terms of quantum mechanics, or in terms of any other theory.

It would be, in my opinion, not only premature but even foolhardy to draw far-reaching ontological conclusions from our present way of expressing the laws of inanimate nature—in terms of measurements as described above—just as foolhardy, though less absurd, than was the attempt to consider the materialistic philosophy to be established on

nature which we formulate mathematically in quantum theory deal no longer with the particles themselves but with our knowledge of the elementary particles" [*Daedalus*, 87, 99 (1958)]. In this writer's opinion, the most readable exposition of the epistemological implications of quantum mechanics is F. London and E. Bauer's *La Théorie de l'observation en mécanique quantique* (Paris: Hermann and Co., 1939). See also the writer's article on "The Problem of Measurement," *Am. J. Phys.*, 31, 6-15 (1963), particularly the section "What Is the State Vector?" (reprinted in this volume), and a forthcoming article by P. A. Moldauer.

[3] London and Bauer (*loc. cit.*, reference 2, page 40) say: "Il [l'observateur] dispose d'une faculté caractéristique et bien familière, que nous pouvons appeler la 'faculté d'introspection': il peut se rendre compte de manière immédiate de son propre état." They could add that, from the point of view of quantum mechanics, the faculty in question is completely unexplained.

the basis of an earlier set of physical laws. We know far, far too little of the properties and the working of the consciousness to propose a philosophy on a scientific basis. In particular, the "reduction of the wave packet" enters quantum mechanical theory as a *deus ex machina*, without any relation to the other laws of this theory. Nevertheless, the fact that the laws of inanimate nature, at least at one stage of the development, could not be formulated without reference to the consciousness remains significant and provides a proper background for the rest of our discussion.

Two Kinds of Reality

It seems idle to think about the meaning of the existence of a material object which one has in one's hand, or can grasp at any minute. However, as a physicist, one is often confronted with more subtle questions of reality, and my point of departure will be such a question. Does a magnetic field in the vacuum exist? For many years, this was a burning question; it is now a forgotten one. Nevertheless, most of us physicists would answer the question in the affirmative. Then, if we analyze the meaning of the statement that a magnetic field exists in the vacuum, we find that the statement means for us that it is convenient to think of such a field, that it enters our calculations, that we can explain to others our calculations, and the conclusions resulting from these calculations, more easily if we may refer to the magnetic field, as given by Maxwell's equations, everywhere, even in interstellar space. The reality of the magnetic field in vacuum consists of the usefulness of the magnetic field concept everywhere; reality is in this case synonymous with the usefulness of the concept, both for our own thinking, and for communicating with others.[4]

One has to go but one step further to realize that the existence of a book which I am holding in my hand is of the same nature. The existence manifests itself in my inability to bring my fingers together, in my knowledge that I would hear a noise if I failed to press my fingers toward each other, in the possibility that I might open and read it. The book is a convenient expression for describing some of the sensations which I have and which codetermines further sensations which I could

[4] See P. W. Bridgman's article in *The Nature of Physical Knowledge* (Bloomington: Indiana University Press, 1960), page 20.

Two Kinds of Reality   189

have myself or cause in others by acting in certain ways—for instance, by throwing the book at them. The only difference between the existence of the book and of the magnetic field in interstellar space is that the usefulness of the concept of the book is much more direct, both for guiding my own actions, and for communicating with other people. It appears that there exists only one concept the reality of which is not only a convenience but absolute: the content of my consciousness, including my sensations.[5]

It also follows from the preceding discussion that there are two kinds of reality or existence: the existence of my consciousness and the reality or existence of everything else. This latter reality is not absolute but only relative. In a gathering of physicists, the existence of the aforementioned magnetic field is an almost absolute one. If I were cast upon an island abounding with poisonous snakes and had to defend my life against them, the reality of the magnetic field in vacuum would fade, at least temporarily, into insignificance. What I am saying is that, excepting immediate sensations and, more generally, the content of my consciousness, everything is a construct, in the sense in which, for instance, Margenau uses this term,[6] but some constructs are closer, some farther, from the direct sensations.

The First Kind of Reality

It is profoundly baffling that the existence of the first kind of reality could ever be forgotten. Yet one finds even now serious articles which completely disregard it or even relegate its existence into the realm of wish-dreams. The only explanation that I can conceive for this is that mankind was, for a long time, engaged in an intense struggle for survival and everyone had to concentrate his attention on the external, inimical

---

[5] Dr. S. A. Basri made this same point at the conference which is reported on in the book cited in reference 4 (page 131). It appears from the discussion that he was not understood. However, the most eloquent statement of the prime nature of the consciousness with which this writer is familiar and which is of recent date is on page 2 of E. Schrödinger's *Mind and Matter* (Cambridge: Cambridge University Press, 1958): "Would it (the world) otherwise (without consciousness) have remained a play before empty benches, not existing for anybody, thus quite properly not existing?"

[6] H. Margenau, *The Nature of Physical Reality* (New York: McGraw-Hill Book Co., 1950), particularly Chapters 4 and 5. The first sentence of *Mind and Matter* (reference 5) reads: "The world is a construct of our sensations, perceptions, memories." See also page 44.

forces. We learned somehow that our consciousness is extinguished unless we undertake certain actions and these actions, and the preparation for them, claimed all our attention. *Primum vivere, deinde philosophare* is the old adage. The philosopher is also a man and he became a victim of the universal preoccupation with survival.

It is, at first, also surprising that biologists are more prone to succumb to the error of disregarding the obvious than are physicists. The explanation for this may be similar to the one advanced for the more general phenomenon: as a result of the less advanced stage of their discipline, they are so concerned with establishing *some* regularities in their own field that the temptation is great to turn their minds away from the more difficult and profound problems which need, for their solution, techniques not yet available. Yet, it is not difficult to provoke an admission of the reality of the "I" from even a convinced materialist if he is willing to answer a few questions, and I suspect that, if carefully analyzed and followed to its conclusions, his philosophy becomes the most solipsistic of all.

The fact that the first kind of reality is absolute, and the circumstance that we discuss the realities of the second kind much more, may lead to the impression that the first kind of reality is something very simple. We all know that this is not the case. On the contrary, the content of the consciousness is something very complicated and it is my impression that not even the psychologists can give a truly adequate picture of it. There always seems to be some single sensation or thought at the center of my attention, but there are other sensations which cast shadows on this center, as if they were just outside my field of vision. Then, there is my whole store of knowledge and recollections which can enter the center of my attention at any time; there are subconscious processes which can suddenly jump into the center of my attention, such as that I should have reconfirmed my reservation. There is then the truly subconscious, discovered in different contexts by Freud, by the great writers, and by Poincaré.[7] Hence, the nature of the first kind of reality is already quite complex and the inadequacy of our appreciation of its

[7] As to the writers' and poets' instinctive realization of many of Freud's recognitions, the reader may be interested in Freud's letters to A. Schnitzler, *Letters of Sigmund Freud* (New York: Basic Books, 1960), particularly letter 123. Poincaré's realization of the role of the subconscious is the basis of J. Hadamard's *The Psychology of Invention in the Mathematical Field* (Princeton, N.J.: Princeton University Press, 1949).

properties may be one of the most potent barriers against establishing the nature of universal realities at the present time. The writer would like to underline this paragraph three times.

A second point worth keeping in mind is the complexity of our perceptions, the fact that when they enter our upper consciousness they are already sophisticated translations of our primitive sensations. Thus, if we pass a STOP sign, what enters into our consciousness is not an octagonal table, with four small figures on it, the first snake-like, the second consisting of two perpendicular lines, etc., but simply a STOP sign. More strikingly, most people do not know what the face of their watch looks like, even though they "take in" several times a day what the watch tells them. The complexity of the perceptive process was commented upon also by Schrödinger. He observed that the routine operations of the mind are relegated to the subconscious and only the learning process becomes conscious. Whether or not this last observation is accurate, it is clear that the content of the consciousness is difficult to specify precisely, that it depends on the part of the consciousness to which one refers, and that the boundaries of the first kind of reality may not be sharply defined. It is even possible that only a limiting case of the first kinds of realities should be called "absolute."

The Second Kind of Reality

If we deny the absolute reality of objects such as a book or, rather, attribute to them a different type of reality from that of sensations, are we in any way in conflict with the fact that we continue to act as if these objects were real? I do not believe so, if we admit that the usefulness of the concept of objects is so great that it would be virtually suicidal to refuse using it, in one form or another. As far as reality is concerned, there is a sharp division between the reality of my consciousness—which is absolute—and the reality of objects, which ranges over a wide spectrum. In order to doubt the usefulness and hence the validity of thinking in terms of the existence of this book, it would be necessary to consider extravagant improbabilities as, for instance, that I am now asleep. If I were, the book would be only part of my dream, and it could fade away as objects often do in a dream. In order to doubt the existence of—that is, the usefulness of assuming—a magnetic field where there is no matter, no such improbable circumstances need be assumed. One may not be interested in the way in which light from the stars reaches the Earth, or assume that light does not produce magnetic fields along its path. There are other phenomena, connected with Northern Lights and cosmic radiation, which one would have to forget about temporarily. Even less is lost if one denies the existence of a wave function describing the external world, and there are, of course, concepts of much smaller significance. This shows that there is a continuous spectrum of the reality of existence from absolute necessity for life to insignificance. A corollary to this statement is that the existence and validity of spiritual values is hardly different from that of objects or concepts as we have considered them. On the other hand, there are many useful concepts, such as mathematical ones, which one would not call "real."

Not only material objects and mental constructs have a reality of the second type: the sensations of other people also fall into this category. We all have had unpleasant experiences when we, I am sure through a lapse, forgot about this reality. It is, therefore, on the very cogent end of the spectrum. Even those who profess to an extremely materialistic point of view act, as a rule, as if the sensations of others were just as real as any material object. They are, in this regard, just as inconsistent as the absolute idealist, or the positivist, whose acts betray that he does believe in the reality of the material world. From the point of view adopted here, the sensations of others, and the material world, have the same *type* of reality and, I might add, also about the same *degree* of reality.

The recognition that physical objects and spiritual values have a very similar kind of reality has contributed in some measure to my mental peace. Apart from this point, however, there is a good deal of uneasiness in my mind—uneasiness that my point of view is so clearly correct that it is also uninteresting. At any rate, it is the only known point of view which is consistent with quantum mechanics. I will admit, on the other hand, that I do not always think or speak in terms of the picture presented.

Before going on to the discussion of the relation of the two realities to each other—this is the thorny problem mentioned in the introduction—I would like to make two remarks: one on the nature of scientific explanations, the other on the relation of a possible universal reality to the two kinds of realities of the present section. The first remark will be a very brief one.

The Nature of Scientific Explanations

There is an anecdote in the Preface to the second volume of Boltzmann's *Kinetic Theory of Gases* which illustrates the emptiness of the simple, naive concept of explanation better than I could in a number of learned paragraphs.[8] The fact is that what we call scientific explanation of a phenomenon is an exploration of circumstances, properties, and conditions of the phenomenon, a coordination thereof into a larger group of similar phenomena, and the ensuing discovery of a more encompassing point of view. This more encompassing point of view, the "theory," should permit one to describe not only the original phenomenon but also the phenomena related to it, give an account of the relations between these phenomena and their properties and circumstances. The "explanation" tells us why the phenomenon occurs only in terms of new postulates. It should give, on the other hand, a clear and accurate description not only of the phenomenon itself, but also of the circumstances surrounding it, its relation to other phenomena—some of which may not have been known before the explanation of the original phenomenon was discovered. This is, then, what we should expect eventually of the explanation of the relation of the two types of realities.

The Universal Reality

Once it has been admitted that the only absolute reality is a personal one, there is always some embarrassment in trying to develop a concept of any other type of reality. This need not be so. *The universal or impersonal reality as a concept is a reality of the second type* which, as all other concepts with a second-type reality, may be very useful for my own thinking and for communicating my ideas to others. It may have

[8] The story is that he (Boltzmann) was, in his youth, dissatisfied with the looseness of the logic in the books on physics. Finally, he heard about a physics book with an impeccable logic. He rushed to the library to get it but found to his dismay, first, that the book was out on loan (a rather frequent occurrence in Vienna libraries), second, that it was all in English. At that time, Boltzmann spoke no English. He went home, in low spirits, but his brother—much wiser than himself, according to Boltzmann—comforted him. If the book is really all that splendid, he said, it is surely worth waiting for a few weeks (Boltzmann's brother must have been an optimist). Further, if it is truly logical, the language can't really matter because the author will surely explain every term carefully before he uses it. Boltzmann implies that this event cured him from trying to be too deductive in his physics.

some features of a group photograph which includes oneself and which we like to keep even though it distorts us and is not as good as the original which, after all, we cannot possibly lose. The consideration of the universal or impersonal reality as a concept with a second-type reality may strike one as unnatural but it is hard to see from what other point of view the impersonal reality could be considered.

Even if considered from the point of view of usefulness, the validity of the concept of universal reality may not stand the test. It may evaporate just as the concept of every object's localization evaporated with the advent of quantum mechanics and had to be replaced by more subtle concepts. If this should *not* turn out to be the case, one might like to speak about a universal reality. It is clear, however, that it will not be possible to use this concept meaningfully without being able to give an account of the phenomena of the mind, which is much deeper than our present notions admit. This is a consequence of the fact that, clearly, from a non-personal point of view, other people's sensations are just as real as my own. In all our present scientific thinking, either sensations play no role at all—this is the extreme materialistic point of view which is clearly absurd *and*, as mentioned before, is also in conflict with the tenets of quantum mechanics—or my own sensations play an entirely different role from those of others. It follows that before we can usefully speak of universal reality, a much closer integration of our understanding of physical and mental phenomena will be necessary than we can even dream of at present. This writer sees no cogent reason to doubt the possibility of such an integration—with regard to this point he probably differs from many of his colleagues—but he does see that it has not yet taken place. For this reason it appears that the concept of universal reality is not a useful concept at present.

One can and does speak, of course, of physical realities, emotional and mental realities, and even of political realities. The qualifying adjectives show, however, that these are not universal realities and the concepts defined by them, though they are useful in their own domain, are not sufficiently deep to be interesting in the present context.

The Relation Between the Two Kinds of Reality

If my consciousness is the only absolute reality, one would expect it to be independent of the constructs which are the realities of the second type. This is not so. It is true that I can go into a dark and silent room and think—perhaps think better than in my office—but this would all cease if I stayed there without air or food and water for any length of time. The ideal of the Buddha, detached from all material support and worries, is possible only as a non-existent Buddha. Similarly, one would expect that my consciousness, the only absolute reality, should be permanent. It should have existed always and remain in existence forever. Again, this is clearly not so. On the contrary, there are realities of the second kind of which we think as permanent—electric charges, heavy particles. Surely the permanence of these objects after my death is meaningless; but as long as I live, it is useful to think of them as permanent.

One must admit that the absolute realities do not have the properties which one would expect. This is not a logical contradiction: scientific inquiry has shown many other relations to be very different from what the naive mind believed. It only means that our expectations in regard to the first kind of realities are not fulfilled, that it is useful for me to act and think as if my sensations had not existed always and my consciousness would dissolve into nothing some day. Then, there will be no absolute reality—and indeed there will be nothing. Of the two terminations of all reality, that in the future is much more useful to keep in mind, because its possible arrival affects my sensations much more than the one in the past and because I can influence its onset to some degree. In conformity with this, we think much more about our death than about our birth.

Even if the phenomena of birth and death do not invalidate the rather tautological description of the two types of realities, they do not fit into a satisfactory and neat picture with them, either. Neither does the depressing dependence of our sensations on our environment, in particular on the physical and chemical state of our body. Nevertheless, the truth of the two kinds of reality seem irrefutable. Will it ever be possible to resolve and understand this desperately unsatisfactory conflict between known phenomena and our expectations? We do not know. However, if it will be possible to "understand" the awakening of the consciousness at birth, and its extinction at death, it will be possible through a study of these phenomena on a broad scale—similar to the study of the properties of materials or of motion.

In our present scientific thinking, either sensations and the consciousness play no role at all, or they are brought in as a *deus ex machina*, as in quantum mechanics. It would be contrary to all our past experience with science, if we had understood the phenomena most deeply affecting the realities of the first kind with as perfunctory an effort as we have made so far. If such an understanding is ever to be obtained, it will be obtained after a careful and detailed study of the awakening and extinction of the consciousness, not only in humans but also in animals. Of the two processes, the awakening may be the more simple to understand because it is not so greatly affected by accidental circumstances.

The Role of Science

The preceding discussion lumps together a great variety of "realities": material, spiritual, and scientific. There are some aspects of the role of scientific recognitions in our set of "realities" which I would like to consider. The discussion will use, to a certain extent, the methods and language of quantum mechanics, not only because I am familiar with that language and not because I believe that the epistemology of quantum mechanics represents the ultimate truth. I will use it because its concepts are undoubtedly more concrete than those of any other language developed for discussing epistemology.

In his remarks on the Future of Science, C. N. Yang quotes Einstein,[9] according to whom the purpose of physics is to find universal laws of nature "from which the cosmos can be built up by pure deduction." This is probably the most ambitious goal of science, reminiscent of the materialistic philosophy. Actually, the validity of this goal could hardly be justified even on the basis of materialistic philosophy because the laws of nature, even if they had the all-pervading force attributed to them by the materialists, would suffice only to predict the future of the cosmos, assuming that its present condition is known. They would not suffice to "build up" the cosmos—this would imply also the explanation of its present condition, or, more precisely, the selection of one state of motion from all the infinitely many states of motion which obey the laws of mechanics.

The concepts of quantum physics are very different from those which created the great expectations concerning the power of the laws of nature which is epitomized by Einstein's words. Rather than justifying

[9] C. N. Yang, address at the Centennial Celebration of M.I.T., 1961. The passage appears in *Essays in Science* (New York: Philosophical Library, 1964).

these expectations, they raise the question whether physics, as we know it, can exist independent of the interpretation of our most usual sensations which are, apparently, born with us and which have no scientific origin.

According to quantum mechanical theory, all of our information about the external world derives from "measurements." These were discussed and described in a general way at the beginning of this article. The relevant point which is important for us now is that no measurement could be interpreted by us if we had no previous knowledge of the properties and structure of the measuring apparatus. This is not the "reading" of the apparatus which was referred to before as a second measurement; it is the knowledge of the type of correlation that is being established between object and apparatus. To put it more crudely, the question is whether the measuring apparatus is a voltmeter or an ammeter, a clock or a balance. In order to obtain any information of the outside world, in order to make any measurement or observation, it is necessary that one already possess a crude knowledge of his surroundings. It is true that this crude knowledge usually comes from other observations, but this only transfers the problem one step further back. Evidently, there is another chain here, similar to that described before, and again the end of the chain—the acquisition of our original and most crude knowledge of the innumerable laws of behavior of our surroundings—is shrouded in mystery. It is probably not only contemporaneous with, but also part of, the awakening of our consciousness, the most mysterious process of all.

Viewing the role of science in this way, one arrives at a much more modest judgment of the role which it plays in our whole body of knowledge. Scientific knowledge always leans on, and is impossible without, the type of knowledge which we acquired in babyhood. Furthermore, this original knowledge was probably not acquired by us in the active sense; most of it must have been given to us in the same mysterious way as, and probably as part of, our consciousness. As to content and usefulness, scientific knowledge is an infinitesimal fraction of the natural knowledge. However, it is a knowledge the structure of which is endowed with beauty because it satisfies abstractions derived from natural knowledge much more clearly than does natural knowledge itself, and we are justly proud of it because we can call it our own creation. It taught us clear thinking and the extent to which clear thinking helps us to order our sensations is a marvel which "fills the mind with ever new and increasing admiration and awe."

Why Are Some Realities of the Second Type So Pervasive?

Another fact that is difficult to understand is the existence of realities of the second type as useful and indispensable as that of material objects. If a little thinking did not make it clear, a recollection of almost every dream would convince us that it could be otherwise, that everything could have a more shadowy existence than do the common objects around us. It is remarkable that, if we look for our glasses, there is no doubt in our mind that they could be found, and one at once stops looking for them on the second floor if one's wife finds them on the first. The fact that this is not so in our dreams shows—if it has to be shown—that reality is a composite empirical fact. Almost equally surprising is the degree of the reality of scientific concepts, facts such as that, if a radio which we put together fails to work, we look for a loose contact and do not suspect the theory.

The question comes up naturally whether it ever will be possible to understand the reality of the second kind, both that given to us by nature, and that acquired by us. Admittedly, the task looks difficult now but so did, only fifty years ago, the task of understanding the properties of materials, such as glass or copper. By understanding, we mean in science the coordination with a larger group of phenomena and the ensuing subordination to a deeper, more general principle. As to the possibility of this, we need not despair even though the problem of the reality of the second kind, the emergence of almost infallible and decisively important concepts, probably lies several layers deeper than could be reached by our present search into the laws of nature.

Perhaps one could find a body of phenomena which would make our concept-building ability less of a single stark fact by studying the concept-forming ability of animals. Perhaps the consciousness of animals is more shadowy than ours and perhaps their perceptions are always dreamlike. On the opposite side, whenever I talked with the sharpest intellect whom I have known—with von Neumann—I always had the impression that only he was fully awake, that I was halfway in a dream.

Let me try to summarize what I have attempted to say here. First,

Two Kinds of Reality   199

that it seems not only possible but rather easy to tell what is real and that there are two kinds of reality. These are so different that they should have different names. The reality of my perceptions, sensations, and consciousness is immediate and absolute. The reality of everything else consists in the usefulness of thinking in terms of it; this reality is relative and changes from object to object, from concept to concept. This is not in conflict with the fact that it would be virtually impossible to live without accepting the reality of some of these objects, such as our surroundings, practically uninterruptedly; in these cases we cannot really avoid thinking in terms of the objects in question. Their reality, although of the second category, is almost complete. It seems to me that these statements follow from a simple analysis of what we call "real." On the contrary, the reason for the existence of any of these things, the consistency and accuracy of our picture of the world, is profoundly baffling. The same is true, perhaps to an even greater extent, of the reality of the concepts of science.

Epistemology of Quantum Mechanics

E. P. Wigner

Contemporary Physics, vol. II. Atomic Energy Agency, Vienna 1969, pp. 431–437
(Reset by Springer-Verlag for this volume)

Abstract. The general consequences of quantum mechanics are discussed and the open questions are outlined, with some of the author's views.

In these Proceedings Bohm, and to some extent Bastin, will discuss the possible structures of possible new theories. The rest of us will discuss the epistemological consequences of the present quantum mechanics. I shall deal first with those consequences which are part of the body of quantum mechanics and should be contained even in textbooks. Second, I shall then outline the open questions and put my own views forward in the last section.

1. If we describe states, that is, physical situations, by a state vector, the temporal change of this state vector is uniquely determined (by the time-dependent Schrödinger equation). Thus, in terms of the state vector, quantum mechanics is a causal theory.

2. We do not experience the state vector directly in the same way as classical theory envisages that we perceive the positions of particles directly. Hence, the temporal change of the state vector has no immediate consequences. Our perceptions, which are what quantum mechanics wants to explain ultimately, are the results of observations which we make. The observations are also called measurements, and the two words are used interchangeably in such discussions as we are carrying out now. The results of the observations are not uniquely determined by the state vector; the same observation on two identical state vectors may give different results, may lead to different perceptions. Several perceptions are possible as a rule; the probabilities with which one or the the other perception will occur can be calculated on the basis of the knowledge of the state vector, but only probabilities can be calculated. The outcome of a particular observation cannot be foreseen unless the probability of a single result happens to be 1.

3. The state vector of the object after the observation has been the subject of some discussion. It has been argued that the state vector after the observation cannot be given, that the object is lost as a result of the observation made thereon. This is unquestionably the case with some observations, as for instance if one observes the presence of a light quantum at a certain silver chloride crystal (contained in a photographic plate) by letting it be absorbed by

that crystal. There are, however, other observations which do not destroy the object; the observation of the direction of the spin of an electron by means of the Stern-Gerlach experiment is of this latter nature. The distinction between the two types of measurements has been made already by Pauli. One avoids many difficult problems if one considers only object-destroying measurements. However, if one disregards the object-preserving measurements, or refuses to make any statement concerning state vector after the measurement, one gives up what appears to most of us the most important objective of quantum mechanics: to furnish probability connections between subsequent observations on the same object. It is true that one can bring objects into definite states, with definite state vectors, by means other than observation. One can then follow the changes of this state vector by means of Schrödinger's equation (or its generalizations). One can then carry out an observation on the system and ascertain that the various possible results of this observation appear with the frequency predicted by quantum mechanics measurement theory. However, if this were the only function of quantum mechanics, it would serve only as its own confirmation and not an image by which one can follow the changes of an object at least by statistical laws.

For this reason, I shall assume the existence of object-preserving observations and in fact will deal only with observations of this nature. This is a somewhat arbitrary restriction, on a somewhat controversial question.

4. There are two attitudes concerning the state vector. One is that it "represents reality" and the observations only serve to ascertain that reality. The other view holds that the state vector is only a tool for calculating the probabilities of the various possible outcomes of measurements. The reality is only the perception, the result of the measurements. Emotionally, surely, the two attitudes are different. However, no observable conclusion depends on which attitude one adopts and it is not necessary to adopt either.

5. The process of measurement is a physical process and physical Theory should be able to describe it. When it does, it regards the measuring instrument, plus the object on which this carries out the measurement, as a single joint system the temporal changes of which describe the measurement. There are two differing opinions on the way the temporal changes should be considered:

(a) According to the first view "measuring instruments must be described classically".
(b) Quantum mechanics is necessary for the description of the measuring instrument also.

The quotation in (a) is from V. Fock but may represent also Bohr's view. von Neumann's discussion of the measuring process reflects the other opinion (b).

von Neumann gave a model for the measuring process. If one denotes the state vector of the initial stae of the apparatus by a_0, the state in which its pointer indicates the result μ of the measurement by a_μ, and the corresponding state of the object by σ_μ, the interaction between instrument and object leads from the state $a_0 \times \sigma_\mu$ to

$$a_0 \times \sigma_\mu \to a_\mu \times \sigma_\mu \qquad (1)$$

σ_μ corresponds to the state of the object for which the measurement gives the result μ with certainty. Equation (1) assumes that, if the object happens to be in such a state, its state σ_μ is not changed by the measurement. This is not a consequence of the quantum mechanical equations of motion. In fact, it is difficult to devise interactions such that

$$a_\mu \times \sigma_\mu \to a_\mu \times \sigma_0 \qquad (2)$$

i.e. that the state of the object becomes after the measurement independent of its initial state μ. However, the rest of the discussion is restricted to such interactions in which this is not the case and only interactions which result in states such as the right side of (1) will be called measurements. These are, then, object-preserving measurements of a special nature: they are measurements which do not change the State σ_μ of the object if this is a state for which the measurement will give a definite result (the result μ) with certainty. It may be well to remark, though, that, as far as can be seen, none of the conclusions which follows is altered if, instead of (1)

$$a_0 \times \sigma_\mu \to a_\mu \times f_\mu \qquad (3)$$

is assumed as long as the f_μ form an orthonormal set.

6. It now follows from the linearity of the equation which describes the temporal development in quantum mechanics that if one starts with the object being in the state $\sum \alpha_\mu \sigma_\mu$ the final state will be

$$a_0 \times \sum \alpha_\mu \sigma_\mu \to \sum \alpha_\mu (a_\mu \times \sigma_\mu). \qquad (4)$$

If the various a_μ correspond to states which have a classical (non-quantum) description, the state on the right side of (4) involves a linear combination of such states and has no classical description. This conclusion remains valid even if one considers interactions of the type of (2) or (3). This simple fact is the reason that I am unable to accept the point of view that the measuring instrument must be described by means of classical mechanics. It seems to me that this assumption is in conflict with the linearity of the quantum theoretical equations of motion. Naturally, if one is willing to alter these equations, the objection becomes invalid.

The preceding discussion is predicated on the assumption that the initial state of the apparatus, a_0, was uniquely determined. It has been recognized already by von Neumann that this is not a necessary assumption and that the apparatus' initial state may not be unique; it may be one of several states, each with a certain probability. However, as d'Espagnat will demonstrate in these Proceedings, this does not alter the conclusion we arrived at.

7. It may be well to observe here that the orthogonality of the σ_μ, in (1) or (2) or (3) can be deduced from the unitary character of the transition represented by the arrow in these equations. I fact, the right sides are orthogonal for $\mu \neq \nu$

$$(a_\mu \times \sigma_\mu, a_\nu \times \sigma_\nu) = (a_\mu, a_\nu)(\sigma_\mu \times \sigma_\nu) \tag{5}$$

and this is 0 because the states a_μ and a_ν are even macroscopically distinguishable, hence orthogonal, $(a_\mu, a_\nu) = 0$. It then follows from the unitary nature of the arrow that

$$(a_0 \times \sigma_\mu, a_0 \times \sigma_\nu) = (a_0, a_0)(\sigma_\mu \times \sigma_\nu) = (\sigma_\mu \times \sigma_\nu) = 0. \tag{6}$$

The same conclusion is arrived at if one uses, instead of (1), either (2) or (3). If one defines the quantity Q which is being measured so that it satisfies

$$Q\sigma_\mu = q_\mu \sigma_\mu \tag{7}$$

and labels the various states σ_μ by real q_μ, (6) guarantees the selfadjoint nature of Q.

8. Naturally, one can escape the preceding conclusions by postulating the quantum mechanics has no validity for macroscopic objects and that all measuring instruments are macroscopic. This point of view was advocated, explicitly, by Ludwig but it may well have been also part of N. Bohr's philosophy. It is not easy to discuss this suggestion, not even if one postulates a definite theory, the classical one, for macroscopic bodies. One reason is that the dividing line between macroscopic and microscopic bodies is unsharp, whereas the separation between quantum and classical theories, with the infinitely greater variety of possible states in quantum theory, is sharp. It is difficult to discuss it also because there are so many quantum phenomena – superconductivity, laser action, among them – which have a quantum origin yet are manifested by macroscopic systems. However, since the present discourse is intended to deal with the epistemology of *quantum mechanics*, it can forego this discussion.

9. If we return to Eq. (4), we see that the quantum mechanical description of observation only describes the establishment of a statistical correlation between apparatus and object. This correlation consists in the appearance of terms $a_\mu \times \sigma_\mu$ only, i.e. the absence of $a_\mu \times \sigma_\nu$ terms in (4), with $\mu \neq \nu$. If we adapt the usual theory of measurement to observations on the joint system of apparatus plus object – the theory which we discussed for the measurements on the object alone – it follows that the renewed observation of the σ_μ system on the object, coupled with an observation of the a_μ system on the apparatus, always gives concordant results: if the apparatus is found in the state a_μ, the object will be found to be in the state σ_μ (and the probability of this is $|\alpha_\mu|^2$) so that we can obtain the state of the object by ascertaining in which of the states a_μ the apparatus is, i.e. by reading the pointer position of the apparatus.

On the other hand, the quantum description implied by the arrow in (4) does not go further than the establishment of a statistical correlation. If a measurement is now undertaken on the apparatus, to read its pointer position

– and one describes this second measurement in a similar way – one merely obtains a state

$$\sum \alpha_\mu (b_\mu \times a_\mu \times \sigma_\mu) \tag{8}$$

where b_μ refers to the second measuring instrument. There is now a correlation between the states of the two apparati and of the original object.

The procedure can be continued by introducing a third apparatus to measure the state of the second, and so on. However, the resulting chain does not alter the fundamental fact: the quantum mechanical equations do not give a description of the completion of the process of observation from which a single one of the terms $a_\mu \times \sigma_\mu$ (or $b_\mu \times a_\mu \times \sigma_\mu$) emerges. This should not be surprising since, as was pointed out at the outset, the temporal development described by quantum mechanics is deterministic, the result of the measuring process subject to statistical laws, i.e. not deterministic. None of this argument is new and all of it is given, for example, in von Neumann's book.

In order to complete the measurement, it is necessary to assume that the observer has, in the words of London and Bauer, an immediate knowledge of the state of the apparatus, that is he directly perceives the state a_μ of the apparatus of (4), or the state b_μ of the second apparatus of (8), or of a subsequent apparatus. If he perceives a_μ, the future behaviour of the state of the object is as if its present state vector where σ_μ, if he perceives b_μ, the future behaviour of object and first apparatus is as if their state vector were $a_\mu \times \sigma_\mu$. These facts are often described more pictorially by the statements that the state vectors "collapse" to σ_μ in the first, to $a_\mu \times \sigma_\mu$ in the second case. Many physicists do not like to accept the postulate that the state vector changes in two ways: in the course of time, continuously, according to the Schrödinger equation, and as a result of observations, abruptly. It is possible to avoid this by considering the state vector to be only a calculational tool, a tool for calculating the probability that an observation which one may undertake will give one result or another (one of the μ of (4) or (8), or another). The two underlying points of view were discussed already in Section 4; there is no difference between their conclusions. The second is, in some ways, more honest and less artifical; it is easier to think in terms of the first one. von Neumann used the first one.

10. We now come to the discussion of the difficulties of the epistemology of quantum mechanics which was presented in the preceding sections. The most obvious of these is that, particularly if the second of the points of view just discussed is accepted, quantum mechanics does not describe reality but only gives statistical correlations between subsequent observations. It does not even commit itself to a "reality" that may be behind these observations in the way classical theory implies this. I am afraid this cannot be helped and only the future will tell us whether our fellow physicists will get used to this as they got used to the idea that there is no absolute time. The jump is much greater.

11. A few more mathematical questions arise. The discussion of these which follows represents, in part, my own views.

The first question which naturally arises is whether it is reasonable to assume an interaction such as implied by (1) or (4). More concretely, for what sets of σ_μ is it reasonable to assume that an instrument can be found, in a proper state a_0, such that the interaction with the object produces what is implied by the arrow? This question was analysed, for most sets σ_μ, by Araki and Yanase, who demonstrated that the instrument must be, in a definite sense fully explained by them, large. Even apart from their point, it appears that the assumption of the existence of an apparatus to implement the arrow in (1) or (4) implies, except for certain sets σ_μ, a good deal of wishful thinking – perhaps as much as exhibited by biologists.

One may inquire, second, whether the picture of observations giving information about the outside world, quantum theory providing a link between these observations, is an adequate picture of our acquiring and using information. The answer must be, surely, in the negative even if we consider observations of a macroscopic nature in the sense that they transmitted by many quanta. One reason for the inadequacy of the picture is that it does not explain how we acquired knowledge concerning the nature of the various apparati we use. The answer of the quantum theorist to the question how we know which apparatus measures what is that we undertook observations in order to ascertain the nature of the apparati we use. However, this only starts another chain because the observation which provided us with information about the nature of one apparatus had to use another apparatus and we had to have information on the nature of this second apparatus. This leads to a chain similar to that encountered in the preceding section, which has an end as little as that chain had, without being resolvable, however, by an epistemological renunciation. More general, the picture of subsequent impressions, with statistical relations between them, is an almost ridiculously simplified picture of our consciousness. It assumes that we are simple automatons – storing and registering information – an oversimplification of a fantastic crudity.

A third question which appears well worth mentioning concerns the reason for all our direct observations to be of macroscopic nature in the sense that they give positions. Quantum mechanics does not give preference in any obvious way to observations of position co-ordinates, not even to those of macroscopic bodies. It is not clear, therefore, why the chain of observations described in Section 9 (the second element of which is illustrated by (8)) has to go through the observation of a position operator. C. Townes has, in a private conversation, revived this problem which is, to my mind, closely connected with the problem last discussed and which probably will not be resolved without resolving that more general problem.

12. Much, if not most, of our knowledge is not acquired directly but is communicated to us by others and many, if not most, results of observations are also acquired second hand. How can such observations be fitted into the general picture sketched in Section 9? The answer of the fully consistent quantum theorist is that a person from whom we obtain information is simply a link in the chain described in Section 9. His observation establishes a statistical

correlation between his own state and the state of the object observed (directly or through a chain) and his communication to us is simply an observation of ours (again presumably through a chain) which informs us of his state. The correlation between this and the state of the object then permits us to infer the state of the latter.

There is no logical flaw in the argument of the "consistent quantum theorist". Nevertheless, his conclusion can hardly be accepted. The observer who has made a direct observation on the system is a person similar to myself, with a consciousness just like my own, and he has made a choice of the various $\ldots \times b_\mu \times a_\mu \times \sigma_\mu$ (which appear in (8) or its generalizations) just as I would make one. The assumption that his state was, before he made his communications to me, a linear superposition of various observation results is not credible. I am therefore led to the view that the state of the one who made the communication to me cannot be described by usual description of quantum mechanics, that this theory is not adequate for the description of life, including consciousness.

This last conclusion should not be a surprising one. Whenever science was extended to new sets of phenomena, it had to introduce new concepts, new descriptions, new ideas. Life and consciousness certainly constitute a new area, divorced from the area of phenomena with which physics is dealing now.

The question naturally arises whether there is a significant difference between the point of view just sketched, which denies the validity of quantum mechanics for the phenomena of life and consciousness, and the point of view mentioned in Section 8, which denies it for all macroscopic bodies. At first sight, the difference appears to be small. In fact, it may be quite large. The point of view mentioned in Section 8 suggests a return to classical theory for all macroscopic bodies, the point of view of the present section suggests the development of a new theory of life. Such a new theory, if it can developed, will have almost surely profound effects on our epistemology. However, as far as the present state of physics, and particularly of quantum mechanics, is concerned, it is my belief that the epistemology sketched in the first part of this discussion expresses their spirit in a natural way.

Epistemological Perspective on Quantum Theory

E. P. Wigner

C. A. Hooker (ed.) Contemporary Research in the Foundations and Philosophy of Quantum Theory. Reidel, Dordrecht 1973, pp. 369–385

One can discuss the epistemology of quantum mechanics from two points of view, and there is some confusion in the literature because the writers (including myself) do not always state clearly on which point of view the discussion is based. The first point of view accepts the observable consequences of quantum mechanics as valid, valid accurately and universally; its objective is the determination of the epistemology on which these consequences can be based. The second point of view from which epistemology can be discussed in the quantum mechanical era is based on the realization of the problems of the epistemology which is based on the acceptance of quantum mechanics as a definitive and final theory. These problems – one of them epitomized by a reference[1] to 'Wigner's friend' – lead one to wonder in what respects quantum mechanics may be modified when the interest of science is extended to a larger set of phenomena characterizing complex living beings. It is justified to speculate in this direction because, clearly, present quantum mechanics is based solely on phenomena involving inanimate objects. It is justified and also interesting to speculate on the extension of our theories to the realm of life and consciousness,[2] even if such speculations do not bear fruit in the form of definite, precise conclusions – as they probably will not.

The confidence in the final nature of quantum theory's epistemology will depend, naturally, on the roundedness and inner coherence of that epistemology. It may be of some relevance for this reason, and also because it is interesting anyway, to mention the blemishes on the otherwise beautiful structure of quantum mechanics. These impair also the roundedness and inner coherence of quantum mechanics' structure and may render us more willing to admit that the epistemology which is based on that structure may be subject to modifications. The weaknesses of the struc= will be, therefore, the first subject of my discussion.

Hooker (ed.), *Contemporary Research in the Foundations and Philosophy of Quantum Theory*, 369–385. All rights reserved

I. RESERVATIONS ABOUT THE STRUCTURE OF
QUANTUM MECHANICAL THEORY

I trust that the remarks on reservations concerning the structure of quantum mechanical theory which follow will not obscure the fact of my profound admiration concerning the practical successes of this theory. I'll group my reservations – there are quite a few of them – into two categories. The problems that arise even in non-relativistic theory form the first, those which render a union between the concepts of quantum mechanics and those of relativity theory difficult form the second category.

(A-1) *Conceptual problems of non-relativistic quantum mechanics – the superposition principle.* The superposition principle – the existence of states $a\psi_1 + b\psi_2$ where a and b are arbitrary constants, ψ_1 and ψ_2 arbitrary states – is such a general principle that one feels there should be some operational procedure to realize it. This would consist in a universal prescription to produce the state $a\psi_1 + b\psi_2$, given in terms of the prescriptions for the realizations of the states ψ_1 and ψ_2. Dr. Gerjuoy has discussed cases in which superpositions can be realized and such realization do constitute some justification of the general principle. However, Dr. Gerjuoy's discussion was confined to highly specialized cases and did not resolve the basic question. It is, in fact, hard to see how the basic question could be approached.

It is surely unnecessary to point out that all the so-called paradoxes of quantum mechanics involve superpositions of classically interpretable states, the superpositions themselves being, however, not interpretable in the naive, classical fashion. This applies to the Einstein-Podolsky-Rosen paradox[3], to Schrödinger's cat[4], and also to the singlet state discussed more recently by J. S. Bell[5]. It is unnecessary to enlarge on this. Let me mention, instead, that the so-called superselection rules[6] do limit the absolute generality of the rule of superposition – they do limit it, however, just enough to impair the mathematical beauty of the general, single and uniform Hilbert space as a frame for the description of all quantum mechanical states. They do not seem to alleviate significantly the conceptual question raised.

(A-2). *The second conceptual problem of non-relativistic quantum mechanics – the problem of measurement.* We learn and teach, respectively, in courses on quantum mechanics that the measurable quantities, or in the words of

Dirac, the observables, are hermitean operators[7]. It can indeed be proved, by means of the theory of measurement, that only hermitean operators can represent measurable quantities[8]. Some books, and some lecturers, go further and claim that *all* hermitean (or more precisely, all self-adjoint) operators can be observables[9]. However, if we ask how the measurement of a given self-adjoint operator should be carried out, the books and the lecturers remain most secretive. One has, of course, no idea how a quantity such as $p + q$ or $pq + qp$ or pqp could be measured – in fact, clearly, most operators cannot. Still, many can. Araki and Yanase[10] furnished the most concise proof that the measurement of almost any self-adjoint operator requires a macroscopic measuring device or, more precisely, a device that is, with comparable probabilities, in a great many of the states of every additive conserved quantity.

There is, however, no rule which would tell us which self-adjoint operators are truly observables, nor is there any prescription known how the measurements are to be carried out, what apparatus to use, etc. In a theory with a positivistic undertone, this is a serious gap.

It may be well if our discussion is a bit more cursory with respect to the second set of difficulties, arising from conflicts with the theory of relativity. I will not mention the difficulties inherent in the relativistic formulation of the quantum mechanics of interacting systems – many of us believe that these are amenable to solution[11], some of us believe that they have been solved already by Isham, Salam, and Strathdee[12] and by Lehmann and Pohlmeyer[1ᵌ], and few of us attribute these difficulties to the conceptual structure of the underlying theory. I will not speak about them when discussing our next subject.

(B-1). *Conceptual problems of relativistic quantum mechanics – the instantaneous nature of the measurement.* Clearly, measurement cannot be an instantaneous process and it is hard to think of one being instantaneous even if one assumes an infinite signal velocity. Much less can it be instantaneous if no signal can travel faster than the velocity of light.[14] Dr. M. Sachs emphasized this point at our meeting.

Actually, the paradoxes which are often found alarming, such as Schrödinger's cat, have little to do with this problem. It is important to realize this point, nevertheless, because it also detracts from the simplicity and hence beauty of the mathematical formulation of the theory.

(B-2). *Conflict between successive measurements.* The relativistic transformation of space-time also leads to a conflict between successive measurements, carried out by observers in different states of motion. The
t=const surfaces of such observers necessarily intersect so that neither
measurement can be considered to occur *after* the other had been carried
out. This is a very serious – though, as we shall see, not insurmountable –
difficulty since what quantum mechanics is supposed to provide us with
are probability connections between subsequent observations. If all these
observations have to be carried out by observers at rest with respect to
each other, the significance of relativistic invariance is much diminished.
The situation is further aggravated by the fact that there is no *operational*
translation of measurements between observers in motion with respect to
each other: a measurement by a moving observer does not appear to be
a measurement for an observer at rest.

It is well known how relativistic field theory overcomes, or tries to overcome, this difficulty.[15] In its original form, it restricted measurements to
the determination of field strengths at space-time points. Such measurements at different points could be compatible (even though, at least in the
simplest case, that of the electromagnetic field, this is questionable). Bohr
and Rosenfeld,[16] in their well known and very ingenious paper, specified
methods for measuring the electromagnetic field at a given point. Their
arguments, I fear, just because of their ingenuity and the equipment they
postulate, lead most readers to the conclusion that the electromagnetic
field *cannot* be measured at points in space-time – not even in arbitrarily
small volumes. The former conclusion is quite in accord with the fact that
an operator which corresponds to the field intensity at a given point does
not exist in the true mathematical sense: only operators which correspond
to integrals of the field intensity over finite domains can be defined in a
mathematically rigorous fashion.[17]

It is for this last reason that modern field theorists postulate the measurement of fields averaged over finite domains.[17] However, if these domains extend only in spatial directions, and are infinitely thin in the
direction of the time axis, at least some of the objections voiced earlier
remain valid. If the domains are bona fide volumes in space-time, one
cannot help feeling that the operator attributed to them can hardly be the
one which is supposed to be measured since, when attributing a field operator to a space-time point in the midst of the domain, it does not take the

effect of the apparatus into account – the apparatus which has interacted with the field at earlier points of the space-time domain.

In order to overcome this last difficulty, and in order to avoid the non-relativistic nature of a domain which is infinitely thin in the time-like direction, one is tempted to use, instead of the traditional constant time cut, the light cone of a point in space-time. This would avoid the difficulties mentioned above but, even though I have tried to implement this, I did not get very far.

(B-3). *Probabilities in a changing world.* Probabilities can be defined only if the process can be repeated many times. This presupposes at least displacement invariance and this is not truly present in a changing world, in an expanding universe.[18]

This point was brought up only for the sake of completeness – few will attribute great significance thereto. The world changes very slowly and the observations take very little time. Nevertheless, it seems to me, the problem does exist.

Our discussion so far concerned the blemishes on the beauty of the conceptual structure of quantum mechanics. Some of these are blemishes because they reduce the mathematical simplicity of the theory – and, as Einstein said, it is easy to believe a theory only if it is truly simple and beautiful. Other blemishes are more closely connected with the present status of the theory, the difficulties encountered in the endeavor to unite quantum and relativistic theories. These are less closely connected with the basic structure of quantum mechanics but are surely not independent therefrom. When concluding this part of our discussion, let me repeat that these are blemishes on a structure which appears to me, in spite of them, truly magnificent and *is*, of course, of immense significance in the whole structure of physics. This should be forgotten even less than the existence of blemishes should – though I believe these should be remembered also.

Let us now turn to the more philosophical problems with which we want to come to grips in spite of the blemishes enumerated.

II. QUESTIONS WITHIN THE FRAMEWORK OF QUANTUM MECHANICS AND TENTATIVE ANSWERS

Before embarking on the truly puzzling questions relating to our subject, let me say a few words on another question which must have bothered many of us. The epistemology of quantum mechanics is supposed to provide the interpretation of the laws of this theory in terms of our observations. It would seem that without such an interpretation the laws of quantum mechanics must be entirely meaningless. How is it possible then that the widespread controversy on the epistemology hardly affects the practical applications of the theory? Most of those working on one of the branches are hardly affected by the controversies and pay very little attention to them. The answer is, in my opinion, that we hardly ever use quantum mechanics in the fashion we use classical mechanics: to predict events. Rather, we use it, as a rule, either to determine material constants, or the possible values of essentially only one observable, the energy. Material constants which we determine by means of the theory are densities, heats of transformation, viscosity, transition probabilities, and some others. These are then inserted into macroscopic equations and we verify the values obtained by means of macroscopic experiments. In this role, quantum mechanics is a servant of macroscopic theories and renders these more definite. The other very common function of quantum mechanics is the determination of the possible energy values of systems, such as hydrogen or helium atoms – and energy is the most important microscopic quantity that we know how to measure. It is perhaps disappointing that so few other conclusions of the general microscopic theory are commonly put to experimental test. However, this is unavoidable since we live in a macroscopic world. It is not surprising, therefore, that the most useful function of quantum mechanics is the providing of material constants for macroscopic equations – a function which at the end of the last century was feared to be the only remaining function of experimental physics.

Let me now take up our subject proper: the epistemology which the complete acceptance of quantum mechanics forces on us. Quantum mechanical theory has two parts: the equations of motion, and the theory of observations. The equations of motion give the change in time of the quantum mechanical determinant of the state of a system: of the state vector. The theory of observations gives the probabilities of the outcomes of observations on systems in terms of the state vector. All is well as long as we do not ask how the observation takes place. However, we do run into difficulties when we try to describe the process of observation by means of the equations of motion, i.e., try to eliminate the second part of the theory by means of the first. Since the equations of motion should be able to describe all events, such an elimination should be possible.[19]

When we try to describe the process of observation by means of the equations of motion, we encounter a contradiction at once: the equations of motion are deterministic, the outcomes of the observation are subject to stochastic laws. Several possible ways can be proposed to eliminate this paradox; they will be discussed next.

(A) The simplest and most natural explanation of the indeterminate outcome of the measurement on an object with completely determined state vector is that the state vector of the measuring apparatus was indeterminate: in one case it had one, in other cases other, directions and the outcome of the measurement is, therefore, different in all these cases.

The possibility just sketched is certainly present. Most measuring apparata, if not all of them, are macroscopic and it would be impossible to determine their state vector. One cc of air can be, at room temperature, with roughly equal probabilities in 10^{20} states, and it would be impossible to ascertain in which one of these it is.

The trouble with this explanation is that, if one discusses this reason for the probabilistic outcome of the measurement – this has been done by von Neumann, myself, and most recently by d'Espagnat[20] – one soon finds that it cannot give the probabilities for the various outcomes of the measurement which are postulated by the second part of quantum theory, by its theory of observations. One therefore has to abandon this explanation for the probabilistic nature of the measurement process, though one may abandon it only reluctantly.

(B) If we could not reduce the theory of observations to the theory of motion, i.e., describe the process of observation in terms of the equations of motion, we can try the opposite: we can eliminate the equations of motion and express all statements of quantum mechanics as correlations between observations. This is indeed possible by attributing the operator[21]

$$Q(t) = e^{iHt/\hbar} \, Q(0) e^{-iHt/\hbar} \tag{1}$$

to the same measurement, carried out at time t, to which we attribute the operator $Q(0)$ if carried out at time 0. We shall denote by P_j the projection operator which leaves the state vectors of outcome j unchanged, annihilates the state vectors of all other measurement outcomes. It is not difficult to see then that an equation similar to the preceding one holds for these P_j

$$P_j(t) = e^{iHt/\hbar} P_j(0) e^{-iHt/\hbar} \tag{1a}$$

Quantum mechanics can be, then, reformulated in terms of the projection operator of the successive measurements. Let us denote the projection operator for the outcome i of the first measurement by P_i, for the outcome of j of the second measurement by P_j'. The probability that the second measurement yields the result j if the first one's outcome was i is then given by

$$\text{Trace } (P_j' P_i)/\text{Trace } P_i = \text{Trace } (P_i P_j' P_i)/\text{Trace } P_i \tag{2}$$

and similar expressions can be given for the probabilities of the different outcomes of several successive measurements.[22] These expressions incorporate the equations of motion if one uses for the operator of a measurement which takes place at time t, the expression in terms of the operator for the measurement at time 0, given by (1).

The preceding reformulation of the equations of quantum mechanics, eliminating explicit reference to the equations of motion and to state vectors, corresponds to a conceptual reformulation thereof. According to this, the function of quantum mechanics is to give statistical correlations between the outcomes of successive observations.[23] From a very positivistic point of view, such a reformulation is quite satisfactory: it refers solely to observations and establishes relations between these directly. From the point of view of everyday experience, in particular our belief in reality and the abstract characterizability of such reality, it appears to be very disturbing. No description of the state of the system is used, by state vector or otherwise. It appears that our theory denies the existence of absolute reality – a denial which is unacceptable to many. It seems to me, however, that it is not necessary to go that far in our conclusions. By referring only to outcomes of observations one does not necessarily deny that there is something real behind the observation – whatever the word 'real' may mean. There may be any amount of old-fashioned reality behind the scenes; it is only that quantum theory does not deal with it but only with probabilities for the outcomes of observations. As to myself, I should admit that the old-fashioned concept of reality seems indeed old-fashioned to me and that I do not know how one could define operationally the reality of anything. My own thoughts, impressions and perceptions surely exist, the existence of all else is inferred only on the basis of these impressions and perceptions.[24]

In my opinion, the restriction of quantum mechanical theory to the determination of the statistical correlations between subsequent observations reproduces most naturally the spirit of that theory; the alternative just discussed, renouncing the definition of reality, is the most natural epistemology of quantum mechanics. It considers the state vector to be only a mathematical tool, useful for carrying out certain calculations, but only a tool and, as (2) shows, a tool the use of which can be avoided.

It is well to admit, however, that the most common applications of quantum mechanics, mentioned at the beginning of this section, do not follow the pattern here described. It is very likely that the calculation of densities, heats of transformation, transition probabilities, etc., could be brought into the form of observations. However, this has not been done – and hence it is not clear what additional assumptions might be needed – nor is it very natural to do so.

III. MODIFICATIONS AND EXTENSIONS OF QUANTUM MECHANICS WHICH MAY AFFECT ITS EPISTEMOLOGY

The formulation of Quantum Mechanical Theory presented last seems to avoid all conceptual difficulties. Surely, it restricts quantum mechanics in the sense that it does not give a picture of 'reality', (whatever this term may mean), but it seems to answer the question which is operationally meaningful: what are the correlations between subsequent observations or perceptions. If one were satisfied with a theory restricted in this way, and if one believed it to extend to all phenomena, one could be satisfied with the theory as formulated above.

There are several reasons for the lack of satisfaction with the theory as described and these reasons have stimulated modifications – unfortunately in every case vaguely formulated modifications – of the theory. Naturally, the modifications suggested depend on the reasons which cause the lack of satisfaction of the proposer of the modification. The rest of this article will consist of a discussion of the proposed modifications.

(A) *Hidden variables*[25]. This is the most widely discussed proposal. It is motivated by the conviction that there is a true reality, hidden from the eyes of the quantum theorist, a reality in terms of which the events, in particular also the observations, have a deterministic character. This reality should be the proper subject of any physical theory. It is not adequately described by the state vector – much less by a theory which is as unconcerned with reality as is the one discussed at the end of the preceding section. It is proposed to call the determinants of the real situation of a system 'hidden variables'. These obey some deterministic equations – at present unknown – which, if known, would, together with the knowledge of the values of the hidden variables, enable one to predict the behavior of the system under all conditions.

It is not clear whether it is possible, or for some fundamental reasons impossible, to ascertain the values of the hidden variables. If it is impossible, their existence has a great deal of similarity with the existence of ghosts which cannot be observed and which cannot influence any events. If it is possible, one can produce states which do not behave according to quantum mechanical theory – a possibility which can, of course, not be denied *a priori*. We shall, for the present, disregard this possibility.

Let us return, therefore, to the concept of hidden variables which will remain hidden, that is, undeterminable. It is neither surprising nor difficult to show that they can be so defined as to reproduce the probabilities for the outcomes of measurements postulated by quantum mechanics.[26] However, as was known already to von Neumann, if this is to be true for an infinite succession of measurements, their number must be infinite.[27] Kochen and Specker[28] proved another disagreeable property which the hidden variables must have, greatly strengthening von Neumann's published work on this subject. The most important contribution to the subject was made, however, by J. S. Bell[5], who showed that in a certain case the hidden variables characterizing one particle must depend on the quantity which is being measured on another particle. This appeared so absurd that the supporters of the hidden variable theories now doubt the conclusions of quantum mechanics which force the acceptance of such hidden variables and are testing them experimentally. As I said before, it is possible that these laws are not valid but, personally, I do not expect that this will be their finding.

(B) *The apparatus must be described by macroscopic theory.* I am not sure it is fair for me to discuss this proposal which was so eloquently articulated by Rosenfeld,[29] Daneri, Loinger, Prosperi,[30] and also by Fock.[31] I cannot bring myself to agree with it.

The proposal is based on the fact that the measuring apparatus is macroscopic. It is, in practice, always described in terms of the concepts of classical, that is macroscopic, physics. The object is microscopic and the interaction between object and apparatus should be described by the equations of quantum mechanics. However, after the interaction, one should revert to a classical description of the apparatus. This is the point where the stochastic element enters: the macroscopic description of a state given microscopically, although on the whole much more crude, does not give a unique set of values to the macroscopic variables. This is clear in the case of the well-known paradoxes, such as that of Einstein-Rosen-Podolsky, or in that of Schrödinger's cat. But one can produce much simpler examples: if the wave function gives finite probabilities to more than one value of a macroscopic variable, the classical translation of what is in quantum mechanics a uniquely defined state, gives finite probabilities to several values of the macroscopic variable in question.

There is little question in my mind but that the proposal which we now discuss is in no way in conflict with the procedure which we are using in practice. The trouble is only that it postulates the miracle which disturbs us: that after the measurement, or if I use my own interpretation, after our observation, the apparatus *is* in a state which has a classical description. Hence, the explanation covers up rather than solves the problem. The transition to a classical description of the apparatus is an arbitrary step which may obscure, but does not eliminate, the basic fact that the true equation of motion is deterministic.

One is also in doubt concerning the dividing line between microscopic and macroscopic systems and on the degree of detail to be used in the macroscopic description. There is also a question what to do about quantum phenomena, such as laser action or superconducting currents, which may take place in the apparatus. The terms 'classical description' and 'macroscopic theory' do not seem to be clearly defined in such cases.

(C) *Quantum mechanics is not valid in the macroscopic domain.* This is a somewhat older suggestion of Ludwig and may be considered as a variant of the preceding one.[32] It admits more clearly our inability to describe the process of observation by means of the equations of motion of quantum mechanics. That these equations are not valid for macroscopic systems may be true, though few physicists are willing to admit this and there is no evidence for such lack of validity. Nevertheless, the clarity of the underlying hypothesis seems to me to be preferable to the somewhat vague nature of the preceding proposal.

(D) *There are no isolated macroscopic systems.* The enormous density of the energy levels of a macroscopic system was pointed out before. It was Zeh[33] who pointed out first that as a result of this enormous density it is practically impossible to isolate a macroscopic object from its surroundings: even a single electron, at a distance of a mile, can cause transitions between the quantum mechanical states of a macroscopic body.

Zeh's observation is unquestionably correct and it is also alarming. We are used to formulating physical theory in terms of equations of motion of isolated systems and Zeh's observation tells us that this is, for macroscopic systems, unrealistic. The same applies, no doubt, for the formulation in terms of outcomes of observations which was given at the end of the preceding section. It is possible, of course, that the conclusions arrived at there remain essentially valid in spite of the interference of accidentally present objects but, at least, the proofs demonstrating that the stochastic element cannot be due to the indeterminate state of the apparatus should be reconsidered.

Zeh's observation may be of very great importance but, in my opinion, it does not convert quantum mechanical theory to the description of reality in the traditional sense. It does not explain how it comes about that looking at the apparatus we get one definite impression. The apparatus plus object may be in a 'mixture' after the measurement, correlated with many other objects – the question how we pick out one of the alternate states in which it may be remains unclear.

(E) *Physical theory should be extended to the phenomena of life and consciousness.*[34] There is little doubt that it would be desirable to follow this proposal. The question is only whether a deeper understanding of the phenomena of life and consciousness will alter our views on the role of quantum mechanics and the meaning of observations. It is my opinion that it is likely to do so.

The basic concepts, in terms of which the laws of nature were formulated, have repeatedly changed in the course of history. They were positions and velocities in Newtonian mechanics, they were field intensities in the theories which followed, and if we accept the conclusion of the preceding section, they are observations in quantum mechanics. A deeper insight into the nature of the observation and the observer is surely desirable since the theory now uses a most stereotyped and crude picture of the observer.

This last point becomes most apparent if one tries to describe the observation of another person.[1] It is true that one could consider, in the spirit of the second alternative of the preceding section, the receipt of the communication of another person concerning his observation, as an observation, on our part, of his state. This observation of ours, of his state of mind, permits the drawing of inferences concerning the state of the object on which he has made his observation, just as one can draw similar inferences concerning the state of the object by looking at a measuring instrument which was in contact with that object. This is formally possible – few would, however, consider such a procedure reasonable. It presupposes that we know the behavior of a person as well as we do that of a measuring instrument and treat him as a measuring instrument. In addition, even though 'reality' in the physical and particularly microphysical world may be a questionable concept, one cannot accept the observations of another person, and the resulting content of his consciousness, to be less real than those of ourselves. This suggests that one treat all observations, undertaken by oneself or another person, on a more nearly equal basis.

It is maintained by some that the laws of physics which we now know, or almost know, suffice for the description of life and the attendant phenomena of consciousness. It is not difficult to adduce evidence to contradict this view, and this has been done also by me.[35] It is, in particular, difficult to accept the possibility that a person's mind is in a superposition of two states in the one of which he has received one, in the other of which he has received another, signal. We ourselves never have felt we were in such superpositions. It seems unlikely, therefore, that the superposition principle applies in full force to beings with consciousness. If it does not, or if the linearity of the equations of motion should be invalid for systems in which life plays a significant role, the determinants of such systems may play the role which proponents of the hidden variable theories attribute to such variables. All proofs of the unreasonable nature of hidden variable theories are based on the linearity of the equations. Actually, it is more likely that the concepts which will emerge when our science encompasses the phenomenon of life will differ as radically from those of present quantum mechanics as for instance the concepts of field theories differ from those of point mechanics or those of quantum mechanics differ from those of earlier theories.

It may be well to recall here the point made at the beginning of the present section. It was emphasized there that there is no logical contradiction in quantum mechanics, particularly not if it is formulated entirely in terms of observations. The proposals dealt with here are based on hopes that physical theory can be modified and reformulated in a way which renders it more attractive and more satisfactory to the proponent. The present proposal is motivated by the desire for a less solipsistic theory which does not deal solely with the observations of a single observer but attributes reality also to the contents of the minds of other observers. In addition, it would be desirable, of course, to know more about mental processes in general. Finally, it is hard to accept as complicated a process as the entering of some cognition into a mind as the fundamental one in terms of which the theory should be formulated. Needless to repeat, none of this points to inner contradictions of present quantum mechanics.

(F) *Only the whole world's wave function is meaningful.* This again is a theory[36] which I find very difficult to accept. It appears to me to be a complete denial of the fact, undeniable in my opinion, that our impressions form the primitive reality. The state vector of my mind, even if it were completely known, would not give its impressions. A translation from state vector to impressions would be necessary; without such a translation the state vector would be meaningless.[37]

Actually, if an outside observer could ascertain the state vector of the world, he would find that I am not in a pure state but in a linear combination of immensely many states, each correlated with the states of immensely many outside objects. The epistemology under discussion postulates that I feel to be in only one of the immensely many states in the linear combination of which I actually am. This means, however, that the other states, and hence the total state vector of the world, are meaningless.

EPISTEMOLOGICAL PERSPECTIVE   **383**

One could postulate, in a similar vein, that the world is homogeneous and isotropic and it is only I who sees differences between different locations and different directions – the real state vector is invariant under all Poincaré transformations. Such ideas appear to me – perhaps wrongly – to be detached from reality.

Princeton University

NOTES

The notes below refer to the specific points made in the body of the article and are, probably, grossly incomplete even concerning those points. Even though they do not refer to points of his article, the writer does want to mention two papers which provided, in his opinion, most of the foundation of our thinking on the subject. These are, first, Heisenberg's 'Über den anschaulichen Inhalt...' (*Zeitschrift für Physik* 43 (1927), 172) which revealed the incompatibility of the classical position – velocity specification of states with the ideas of quantum and wave mechanics. It justified the state vector characterization of such states. The second is N. Bohr's article on complementarity (*Naturwissenschaften* 16 (1928), 245; see also his book *Atomic Physics and Human Knowledge*, John Wiley, New York, 1958, which made clear that the new theory has deep philosophical implications.

An excellent review of the whole literature was given by B. S. De Witt and R. Neill Graham, *American Journal of Physics* 39 (1971), 734.

[1] Ludwik Bass, preprint, to appear in *Hermathena*.
[2] J. H. Greidanus, *Transactions of the Royal Netherlands Academy of Sciences* 23 (1966); to appear in *Foundations of Physics*.
[3] A. Einstein, B. Podolsky, and N. Rosen, *Physical Review* 47 (1935), 777.
[4] E. Schrödinger, *Naturwissenschaften* 23 (1935), 807, 823, 844; *Proceedings of the Cambridge Philosophical Society* 31 (1935), 555.
[5] J. S. Bell, *Physics* 1 (1965), 195. Cf. also E. P. Wigner, *American Journal of Physics* 38 (1970), 1005.
[6] G. C. Wick, A. S. Wightman, and E. P. Wigner, *Physical Review* 88 (1952), 101; D1 (1970), 3267; A. S. Wightman, *Il Nuovo Cimento* 14 (1959), 81; J. M. Jauch, *Helvetica Physica Acta* 33 (1960), 711; E. P. Wigner, *Physikertagung* Vienna, Physik Verlag, Mosbach/Baden, 1962, p. 1; G. C. Hegerfeldt, K. Kraus, and E. P. Wigner, *Journal of Mathematical Physics* 9 (1968), 2029.
[7] P. A. M. Dirac, *The Principles of Quantum Mechanics* (several editions), Chapter II, Clarendon Press, Oxford.
[8] E. P. Wigner, *Zeitschrift für Physik* 133 (1952), 101.
[9] J. von Neumann, *Mathematische Grundlagen der Quantenmechanik*, Julius Springer, Berlin, 1932. (English transl. by R. T. Beyer, Princeton University Press, 1955.) Also W. E. Lamb, *Physics Today* 22 (1969), 23.
[10] H. Araki and M. Yanase, *Physical Review* 120 (1961), 666. An earlier proof was given by E. P. Wigner, Note 8.
[11] I am referring particularly to the axiomatic system proposed by A. S. Wightman. See A. S. Wightman and L. Garding, *Arkiv für Fysik* 28 (1964), 129. Also A. S. Wightman, *Physics Today* 22 (1969), 53.

384   EUGENE P. WIGNER

[12] C. J. Isham, A. Salam, and J. Strathdee, Trieste Preprint, 1970, International Atomic Energy Agency.
[13] H. Lehman and K. Pohlmeyer, DESY Preprint, 1970.
[14] H. Salecker and E. P. Wigner, *Physical Review* 109 (1958), 571; also E. P. Wigner, *Helvetica Physica Acta Supplement* 4 (1956), 210; and S. Schlieder, *Communications in Mathematical Physics* 7 (1968), 305.
[15] A. S. Wightman, Note 11.
[16] N. Bohr and L. Rosenfeld, *Det Koneglige Danske Videnskabernes Selskab, Mathematisk-fysiske Meddelelser* 12 (1933), No. 8; *Physical Review* 78 (1950), 194; E. Corinaldesi, *Il Nuovo Cimento* 8 (1951), 494.
[17] See, A. S. Wightman and L. Garding, Note 11, especially p. 132–4. This article also has references to the older literature.
[18] E. Teller, in *Physical Review* 73 (1948), 801 discusses some of the relevant problems. A more complete discussion of the physical situation was given at the Trieste Symposium of 1968. See *Contemporary Physics*, International Atomic Energy Agency, Vienna, 1969.
[19] J. von Neumann, Note 9. F. London and E. Bauer, *La theorie de l'observation en mecanique quantique*, Hermann et Cie, Paris, 1939.
[20] J. von Neumann, Note 9; E. P. Wigner, *American Journal of Physics* 31 (1963), 6; B. d'Espagnat, *Il Nuovo Cimento* 4 (1966), 828.
[21] It is remarkable how difficult it is to find an explicit statement of this formula, well known to all physicists. Cf., however, Section 6i of S. S. Schweber's *An Introduction to Relativistic Quantum Field Theory*, Row, Peterson and Co., Evanston, 1961 or Section 16.22 of D. Bohm's *Quantum Theory*, Prentice Hall, Englewood Cliffs, 1951.
[22] R. M. F. Houtappel, H. Van Dam, and E. P. Wigner, *Reviews of Modern Physics* 37 (1965), 595, Section 4.4.
[23] This point was made by the present author in his lectures at the 1970 session of the Scuola Internazionale di Fisica E. Fermi – IL Course. It was made, probably, by many others.
[24] Even though this observation appears to be obvious, it also appears to provoke sharp denials. Cf. this writer's article, 'Two Kinds of Reality' in *Essays on Knowledge and Methodology*, collection of papers presented at the Marquette University Conference, Ken Cook and Co., Milwaukee, 1965.
[25] An eloquent discussion of this proposal is given by D. Bohm in his *Causality and Chance in Modern Physics*, Routledge and Kegan Paul, London, 1958. See also D. Bohm, *Physical Review* 85 (1952), 166, 180; D. Bohm and J. Bub, *Reviews of Modern Physics* 38 (1966), 453; L. de Broglie, *Foundations of Physics* 1 (1971), 5.
[26] Cf. e.g., the first section of this author's article, Note 5. The fact itself was surely known to J. von Neumann.
[27] This is the argument which, in this writer's opinion, motivated von Neumann against accepting hidden variables theories. Cf. footnote 1 of the second article in Note 5.
[28] S. B. Kochen and E. Specker, *Journal of Mathematics and Mechanics* 17 (1967), 59.
[29] Cf. e.g., L. Rosenfeld, *Supplement to Progress in Theoretical Physics*, extra No. (1965), 222; *Nuclear Physics* A108 (1968), 241. Dr. Rosenfeld expressed his views also in numerous other articles.
[30] Cf. e.g., A. Daneri, A. Loinger, and G. M. Prosperi, *Nuclear Physics* 33 (1962), 297; *Il Nuovo Cimento* 44 (1966), 119. See also G. M. Prosperi's contribution to the *Scuola Internazionale di Fisica E. Fermi*, IL Course 1970.
[31] V. Fock, see e.g. his paper, 'Classical and Quantum Physics', International Centre for Theoretical Physics, Trieste, 1968.

EPISTEMOLOGICAL PERSPECTIVE   385

[32] G. Ludwig, article in *Werner Heisenberg und die Physik unserer Zeit*, Vieweg u. Sohn, Braunschweig, 1961.

[33] H. D. Zeh, *Foundations of Physics* 1 (1970), 69; see also K. Baumann, *Zeitschrift für Physik* 25a (1970), 1954.

[34] E. P. Wigner, *Foundations of Physics* 1 (1970), 33. The articles of Greidanus, Note 2, and of Zanstra, among others, also support this view.

[35] E. P. Wigner, *Proceedings of the American Philosophical Society* 113 (1969), 95. Also article in *The Scientist Speculates* (ed. by I. J. Good), William Heinemann, London, 1962. A. E. Cochran, *Foundations of Physics* 1 (1971), 235.

[36] H. Everett III, *Reviews of Modern Physics* 29 (1957), 454; J. A. Wheeler, *ibid.* 29 (1957), 463; B. S. De Witt, *Physics Today* 23 (1970), 30, and lectures given at the *Scuola Internazionale di Fisica E. Fermi*, IL Course 1970.

[37] This is articulated also in the articles of Notes 34 and 35.

Realität und Quantenmechanik *

E. P. Wigner

(Reset by Springer-Verlag for this volume)

Als ich hier ankam und das Programm sah, war ich unangenehm überrascht über den Titel meines Vortrages. Doch hat man mir bald gezeigt, daß ich den Gegenstand selber vorgeschlagen habe. Dieser erinnert mich natürlich daran, was Levy-Leblond gesagt hat: Wenn der Wissenschaftler die Reife erreicht hat, so wendet sich sein Interesse mehr und mehr allgemeinen Fragen zu, er wird bald ein Philosoph. Dies ist heute besonders wahr, da sich die Physik so ausgebreitet hat, daß man einen viel kleineren Teil davon beherrscht als man als Junger beherrscht hat. Trotzdem kann und möchte ich Ihnen berichten, daß mir die vergangenen Vorträge großes Vergnügen gebracht haben – ich habe sie sogar verstanden.

Nun zu meinem Gegenstand: der Begriff der Realität. Was bedeutet das? Ich habe mich gefragt was die Existenz eines Magnetfeldes im leeren Raum bedeutet? Wie könnten wir jemandem widersprechen, wenn er behauptete, daß im leeren Raum kein Magnetfeld wirklich existiert? Ich würde sagen, daß ich das zukünftige Benehmen des leeren Raumes viel besser beschreiben kann, viel einfacher, wenn ich die Existenz des Magnetfeldes selbst im leeren Raum zugebe – sonst müßte ich irgendwie begründen, wie das Feld dorthin eingedrungen ist, als ich es später gemessen habe, als natürlich der Raum nicht mehr leer war. Dies bedeutet meiner Ansicht nach, daß Existenz kein völlig definierter Begriff ist, daß sie davon abhängt, ob sie es erleichtert Beobachtungen, in dem erwähnten Falle zukünftige Beobachtungen und Eindrücke, einfach und anschaulich zu beschreiben. Ich kann hinzufügen, daß es möglich ist, die Maxwellsche Theorie der Elektrizität und des Magnetismus umzuformulieren, so daß die Felder in ihr gar nicht erscheinen. Die Wirkung der Felder an Ladungen magnetischen Feldern erscheint in diesen Reformulierungen als eine Fernwirkung der Ladungen und der magnetischen Körper – verzögert in der Zeit. Die Möglichkeit einer solchen Reformulierung der Maxwellschen Gleichungen zeigt am überzeugendsten, daß es möglich ist, die Realität der Felder in Abrede zu stellen – und dies wurde von mehreren Kollegen getan [1]. Man muß zugeben, daß die Realität auch in diesem Falle kein absoluter Begriff ist.

Wenn wir dies völlig akzeptieren, so ist es nicht möglich, die Quantenmechanik in einer scheinbar, aber nur scheinbar, allgemeinen und völlig logischen Weise zu interpretieren. Diese Interpretation sagt, daß die Wellenfunktion nicht

---

* Vortrag gehalten in Lindau, 1.7. und in Tutzing, 7.7.1982.

die Realität beschreibt, sondern nur ein Mittel ist, die statistischen Relationen zwischen meinen aufeinanderfolgenden Beobachtungen zu ermitteln.

Ich möchte aber zugeben, daß ich mit dieser Interpretation nicht zufrieden bin. Erstens ist sie offenbar schrecklich solipsistisch, sie beschreibt Zusammenhänge nur zwischen *meinen* Beobachtungen. Mit anderen Worten, wenn wir diese Interpretation annehmen, haben nur meine Beobachtungen Wirklichkeit, denn die Theorie gibt nur Zusammenhänge zwischen den Resultaten meiner Beobachtungen. Natürlich ist es wahr, daß wenn die Theorie alle meine Beobachtungen, das heißt Eindrücke, wirklich beschreiben würde, wäre sie in gewissem Sinne vollständig – vielleicht vollständig, aber nicht wirklich befriedigend – wir alle fühlen, daß die Realität sich weiter erstreckt als *meine* Eindrücke, wenn es auch möglich ist zu behaupten, daß für mich diese die volle Realität enthalten. Es ist aber schwer, und für mich wirklich unmöglich in dieser Weise zu denken und zum Beispiel den Eindrücken meiner Freunde keine Realität zuzuschreiben. Wenn das so wäre, würde ich nicht zu Euch reden. Außerdem müßten wir natürlich zugeben, daß die meisten meiner Eindrücke nicht durch die Quantenmechanik beschrieben sind. Und endlich, daß ich weiß, daß meine Eindrücke oft einen Irrtum darstellen: manchmal muß mein Augenglas verschoben werden, es täuscht mich. Tatsächlich glaube ich, daß eine zu solipsistische Theorie nicht annehmbar ist.

Ich will hier den krassen Fall beschreiben, der es vielleicht am klarsten zeigt, daß wir nicht wirklich an diese Interpretation der Quantenmechanik glauben, obwohl die Annahme, daß die heutige Quantenmechanik allgemein gültig ist, zu dieser Annahme zwingt. Der Zustand, den ich beschreiben werde, wird oft „Wigner's Friend" genannt [2], obwohl ich sicher weder der einzige noch der erste bin, der diese Schwierigkeit erkannt hat. Der Zustand, den ich beschreiben will, ist, daß eine Messung ausgeführt wird, die mit etwa gleichen Wahrscheinlichkeiten zu zwei verschiedenen Resultaten führt und das erste Resultat führt zu einem Lichtsignal etwas links, das zweite etwas rechts von meinem Freunde. Wenn wir an die Gültigkeit der Quantenmechanik wirklich glauben, so ist der Endzustand der Messung eine Superposition von zwei Wellenfunktionen. In der einen verursachte die Messung einen Lichtstrahl links und meines Freundes Zustand ist der, daß er ein Lichtsignal links gesehen hat. Die andere Wellenfunktion mit der die soeben gegebene superpositioniert ist, beschreibt den anderen Zustand des Objektes an dem mein Freund die Messung ausführt, enthält einen Lichtstrahl rechts von ihm und beschreibt ihn in dem Zustande in dem er das Lichtsignal rechts gesehen hat. Daß die quantenmechanische Theorie zu so einer Superposition führt und nicht *entweder* zu der einen *oder* zu der anderen soeben beschriebenen Wellenfunktion, ist natürlich, denn die Gleichung ist deterministisch und führt nicht einmal zu einer, ein andermal zu einer anderen Wellenfunktion. Nach der Interpretation, die ich beschrieben habe, ist die Messung nur vollendet, wenn ich höre, was mein Freund mir sagt, wo er das Lichtsignal gesehen hat.

Dies glauben wir aber nicht – wir glauben, daß der Freund schon bevor er uns erklärt hat, wo er das Lichtsignal sah, das Lichtsignal dort sah, daß sein Sehen des Lichtsignals Realität hat und daß sein Bewußtsein nie in einer Superposition zweier Eindrücke war – wir können nicht so solipsistisch sein, daß wir seinen Eindrücken keine Realität zuschreiben. Und dies halte ich für eine wirkliche Schwierigkeit der Interpretation, die ich beschrieben habe und nicht nur dieser Interpretation, sondern auch der allgemeinen Gültigkeit der jetzigen Quantenmechanik. Und dies ist ernst. Vielleicht darf ich schon hier erwähnen, daß die Modifikation der Quantenmechanik, die ich erwähnen werde, diese Schwierigkeit eliminiert, obwohl sie, wie ich zugebe, auch keine allgemeine Gültigkeit hat. Aber sie ist nicht deterministisch und führt nicht zu einer Superposition der beiden Wellenfunktionen, die ich beschrieb, sondern führt entweder zu der einen oder zu der anderen Wellenfunktion. Sie gibt Wahrscheinlichkeiten für das linke und das rechte Resultat.

Bevor ich aber meinen Vorschlag beschreibe will ich noch zwei andere Schwierigkeiten der jetzigen Theorie der quantenmechanischen Messung erwähnen. Die erste Schwierigkeit ist, daß, obwohl die Theorie die Möglichkeit unzählbarer verschiedener Messungen fordert, nicht angibt, wie diese Messungen ausgeführt werden können. Die Theorie [3] sagt, daß jeder hermitische Operator meßbar ist, sie sagt aber nicht, wie man diese Messungen ausführen soll. Und ich bin überzeugt, daß sie nicht wie postuliert ausführbar sind. Ich will jetzt nicht auf die Einzelheiten eingehen – manche dieser sind zu offenbar, andere sind in der Literatur diskutiert worden. Die Schwierigkeit existiert und ich habe keine Lösung dafür.

Die zweite Schwierigkeit, die ich erwähnen will, stammt daher, daß zur Zeit als die Meßtheorie geschaffen wurde, in ihrer formalen Form von Johann von Neumann, die Quantenmechanik noch eine nichtrelativistische Theorie war. Deshalb hat man angenommen, daß die Messung momentan über den vollen Raum ausgeführt werden kann, daß sie zu einer Wellenfunktion führt, die den Zustand zu einer *bestimmten Zeit* charakterisiert. Dies ist, so scheint es mir, im Widerspruch zur Relativitätstheorie wonach man keine Signale mit überlichtgeschwindigkeit übermitteln kann. Dies ist, meiner Ansicht nach, auch die Ursache, daß die Messung der Lage nicht nur zu experimentellen, sondern auch zu theoretischen Schwierigkeiten führt:

Die Feststellung, daß das Partikel an einem Punkt des vierdimensionalen Raumes sich befindet, kann man nicht mit einer Lorentz invarianten Wellenfunktion beschreiben [4]. Mir scheint, daß es viel natürlicher wäre zu postulieren, daß die Messung die Wellenfunktion an dem Lichtkegel, dem negativen Lichtkegel des Beobachters bestimmt. Die Spitze des Lichtkegels ist natürlich der Zeitraumpunkt, bei welcher die Messung vom Beobachter vorgenommen wird. Ich glaube selbst dieses ist eine zu scharfe Annahme und vielleicht wird dies nächste Woche in Tutzing diskutiert werden, aber sie eliminiert schon die Schwierigkeit, die ich eben erwähnte habe. Ich sollte vielleicht erwähnen, daß sie es unmöglich macht, daß zwei verschiedene Beobachtungen zu Zeiten an denen ihr Abstand räumlich ist, unabhängige Beobachtungen sind. Ihre negativen Lichtkegel überlappen und die Beobachtung durch einen Beobachter in diesem Gebiet kann die Beobachtung des anderen Beobachters beeinflussen. Ich könnte dies viel weiter diskutieren, aber ich glaube es ist besser, wenn ich zu meinem Gegenstand zurückkehre.

Wenn wir zugeben, daß es eine Wirklichkeit ist, daß mein Freund das Lichtsignal entweder links oder rechts gesehen hat, daß also das Prinzip de Determinismus für seine Beobachtung nicht gilt, so müssen wir die Quantenmechanik etwas modifizieren. Natürlich könnte man anstatt dessen annehmen, daß die Ursache der Unbestimmtheit des Ausganges seiner Beobachtung durch die Unbestimmtheit seines ursprünglichen Zustandes verursacht wurde, doch kann man leicht nachweisen, daß dies schon mit der Linearität der quantenmechanischen Bewegungsgesetze im Widerspruch ist. So sind wir veranlaßt Modifikationen der Bewegungsgleichung vorzuschlagen. Die meisten Vorschläge in dieser Richtung empfahlen nichtlineare Gleichungen für die Zeitabhängigkeit der Wellenfunktion und auch ich habe vorher dies im Auge gehabt. Meine Ansicht in dieser Richtung wurde durch eine Arbeit von Dietrich Zeh [5] – er wies nach, daß makroskopische Systeme in unserer Welt nicht isolierte Systeme sein können – selbst im intergalaktischen Raum werden sie von der Umgebung beeinflußt – sie können keine isolierten Systeme sein und die deterministischen Gleichungen der Physik beziehen sich nur auf solche isolierte Systeme. Ich glaube die Details seiner überlegung müssen für unsere Anwendung etwas modifiziert werden, aber ihr Wesentliches bleibt erhalten und man kann zeigen, daß der innere Zustand eines Makro-Systems von der Umgebung schon in einem hundertsten Teil der Sekunde verändert wird. Dies bezieht sich natürlich nur auf die ungeheuer komplizierte Wellenfunktion des Systems, nicht auf seinen makroskopischen Zustand. Wenn dies die Ursache des nichtkausalen Benehmens des Beobachters ist, so bezieht sich das auch auf makroskopische Meßapparate und es ist natürlich, anzunehmen, daß die Zustände auch dieser bald nicht Superpositionen verschiedener Stellungen des Zeigers sind, sondern mit verschiedenen Wahrscheinlichkeiten verschiedene Stellungen annehmen, wenigstens wenn sie makroskopisch sind. Und im quantenmechanischen Sinne ist das etwas ganz anderes.

In diesem Sinne muß man zugeben, daß sowohl der Beobachter wie auch ein makroskopischer Meßapparat verbunden sind nicht nur mit dem System woran die Messung vorgenommen wird, sondern auch mit ihrer eigenen Umgebung und diese ist auch weiter verbunden. Dies bedeutet, daß ihr Zustand, der Zustand des Beobachters und eines makroskopischen Apparats, in der Quantenmechanik mit einer Dichtematrix, nicht mit einer Wellenfunktion beschrieben werden muß. Die Zeitabhängigkeit der Dichtematrix ist linear, sie hängt aber nicht nur von ihrer inneren Struktur, sondern auch von den äußeren Einflüssen ab. Wenn kein solcher Einfluß gegeben ist, so ist die Zeitabhängigkeit der Dichtematrix ρ durch die alte Gleichung

$$i\hbar\frac{\partial \rho}{\partial t} = H\rho - \rho H \qquad (1)$$

gegeben. Diese wollen wir aber modifizieren, so daß sie dem äußeren Einfluß Rechnung trägt. Wie tun wir das?

Falls die Wellenfunktion eine Superposition von zwei Lagen des Zeigers des Meßapparats enthält, wird diese vielleicht eine Abhängigkeit von der entsprechenden Koordinate x der Form $f(x+a;\xi)+f'(x-a,\xi)$ haben, wo die Funktion

f im wesentlichen verschwindet, wenn sein erstes Argument wesentlich von 0 verschieden ist (das zweite Argument bezieht sich auf alle anderen Koordinaten des Systems und des Meßapparates. Die Wellenfunktion ist dann eine Superposition von zwei Zuständen, die eine von diesen einer Lage bei $-a$, die andere einer Lage des Zeigers bei $+a$ entsprechen. Dasselbe gilt natürlich für jede Wellenfunktion $f(x+a) + wf'(x-a)$, wo w eine komplexe Zahl vom Absolutwert 1 ist. Die Zustände mit verschiedenen w sind aber im Prinzip unterscheidbar, es ist unrichtig zu sagen, daß der Zustand dadurch allein charakterisiert ist, daß der Zeiger entweder nahe zu $-a$ oder nahe zu $+a$ liegt. Die Elemente der Dichtematrix, die der Wellenfunktion $f(x+a, \xi) + f'(x-a, \xi')$ entspricht, sind

$$M(x, \xi; x', \xi') = [f(x+a, \xi) + f'(x-a, \xi)][f(x'+a, \xi') + f'(x'-a, \xi')]^* \quad (2)$$

wo x, ξ die Zeile, x', ξ' die Spalte bezeichnen. Die Elemente der Dichtematrix des Zustandes $\sqrt{2}f(x+a, \xi)$ sind natürlich

$$m_-(x, \xi; x', \xi') = 2f(x+a, \xi)f(x'+a, \xi)^* \quad (3)$$

und ähnliches gilt für die Dichtematrix m_+ des Zustandes $f(x+a, \xi)$. (Der Faktor 2 ist eingeschlossen um der Normierung der Funktion f Rechnung zu tragen.) Der Zustand, der mit einer Wahrscheinlichkeit $1/2$ der Wellenfunktion $f(x+a)$ und derselben Wahrscheinlichkeit der Wellenfunktion $f(x-a)$ entspricht, ist

$$\begin{aligned} M(x, x') &= \tfrac{1}{2}m_-(x, x') + \tfrac{1}{2}m_+(x, x') \\ &= f(x+a)f(x'+a)^* + f(x-a)f(x'-a)^*. \end{aligned} \quad (4)$$

Sie ist von $M(x, \xi; x', \xi)$ dadurch verschieden , daß die Nichtdiagonalelemente in Bezug auf die x Variable, d.h. die Terme $f(x+a, \xi)f(x'-a, \xi')^*$ und $f'(x+a, \xi)f(x'+a, \xi')$ fehlen. Und die Wechselwirkung mit einer äußerlichen Struktur, der Umgebung, führt gerade zu dem Verlust dieser Nichtdiagonalelemente, weil diese Wechselwirkung für die zwei Lagen, bei $-a$ und $+a$, verschieden ist. Es ist natürlich, deshalb die Zeitabhängigkeit so abzuändern, daß die Nichtdiagonalelemente, die makroskopisch unterscheidbare Zustände verbinden, im Laufe der Zeit verschwinden. Ich schlage deshalb einen Zusatzterm zur gewöhnlichen Zeitabhängigkeit der Dichtematrix ρ vor, der Zusatzterm verursacht eine exponentielle Verkleinerung der Nichtdiagonalelemente. Die Verkleinerung ist natürlich umso rascher, je mehr sich die Zustände, die Zeile und die Spalte charakterisieren, makroskopisch voneinander unterscheiden.

Um dies zu formulieren, bezeichne ich die Zeilen und auch die Spalten von ρ mit mehreren Indices. Die ersten drei, X_0, Y_0, Z_0 sind die drei Koordinaten des Systems, die nächsten drei x_1, y_1, z_1, die drei Dipolmomente, die nächsten fünf die Komponenten der Quadrupolmomente, usw. Die ganze Gleichung der Zeitunabhängigkeit wird dann etwa

$$i\hbar \frac{\partial \rho}{\partial t} = H\rho - \rho H - i\hbar \sum_{em} \varepsilon_e \left(\mathcal{L}_{em} - \mathcal{L}'_{em}\right)^2 \rho \quad (5)$$

wo die $\mathcal{L}_{0m}$ für die drei Koordinaten X_0, Y_0, Z_0, die drei $\mathcal{L}_{1m}$ für die drei Dipolmomente, die fünf $\mathcal{L}_{2m}$ für die Quadrupolmomente stehen und so weiter. Die $\mathcal{L}_{em}$ beziehen sich auf die Reihen, die $\mathcal{L}'_{em}$ auf die Spalten von ρ.

Ich nehme an, daß die ε_ℓ die Form

$$\varepsilon_e = \gamma |l^2 a^{2e} \tag{6}$$

haben; γ ist das Reziproke einer charakteristischen Zeit, a ist eine charakteristische Länge. Daß diese nicht explizit gegeben sind, bedeutet schon, daß unsere Gleichung nur näherungsweise gelten kann – in der Tat hängt natürlich die Geschwindigkeit mit der die Größen der Nichtdiagonalelemente abnehmen von der tatsächlichen Umgebung ab. Auch kann man bemerken, daß die vorgeschlagene Gleichung Galilei invariant ist, was nicht zutreffen mag, wenn die Umgebung nicht kugelsymmetrisch ist.

Die vorgeschlagene Gleichung enthält die Tatsache, daß die makroskopischen Koordinaten makroskopischer Systeme keine entfernten Korrelationen behalten und so beschreibt sie das statistische Verhalten der Beobachtungsprozesse. Sie beschreibt in keiner Weise das Eindringen von Information in unser Bewußtsein wie, ganz allgemein, die Existenz und das Verhalten unseres Bewußtseins heute noch ebenso außerhalb des Gebietes der Physik steht wie das zu Newton's Zeiten für das Verhalten der chemischen Kräfte der Fall war. Vielleicht werden die zukünftigen Generationen der Physiker dies ändern können – wir können dies hoffen!

Literatur

1. J.A. Wheeler und R.P. Feynman: Interaction with the Absorber as the Mechanism of Radiation. Rev. Mod. Phys. **17**, 159 (1945)
2. E.P. Wigner: Remarks on the Mind-Body Question. In: I.E. Good (ed.) The Sientist Speculates. Heinemann, London 1962, S. 302
3. J. von Neumann: Mathematische Grundlagen der Quantenmechanik. J. Springer, Berlin 1932
4. T.D. Newton and E.P. Wigner: Rev. Mod. Phys. **21**, 400 (1949). See also M.H.L. Price: Proc. Roy. Soc. **195**A, 62 (1948)
5. D. Zeh: Foundations of Physics **1**, 69 (1970)

Interpretation of Quantum Mechanics

E. P. Wigner

J. A. Wheeler and W. H. Zurek (eds.) Quantum Theory and Measurement.
Princeton University Press, Princeton, New Jersey, 1983, pp. 260–314

§1. Problems Raised by Quantum Theory before the
Advent of Quantum Mechanics   260
§2. The Mathematical Formalism of Quantum Mechanics   267
§3. Direct Product and Quantum Mechanical Description
of the Measuring Process   276
§4. The Physics of the Measurement Description   283
§5. Other Proposed Resolutions of the Measurement
Paradox   288
§6. Experimental Tests of Bell's Inequality   295
§7. Problems of the Standard Interpretation:
Unmeasurable Quantities   297

§1. Problems Raised by Quantum Theory before the Advent of Quantum Mechanics

The conceptual problems generated by the generally accepted interpretation of quantum mechanics overshadow in philosophical depth those generated by the older quantum theory (see for example the collections of basic papers edited by ter Haar, 1967, and Kangro, 1972) to such an extent that one is likely to forget the latter. Nevertheless, these were very real also, though more concrete and lying more within physics proper than those generated by quantum mechanics.

The idea of the quantum emission and absorption of radiation was conceived by M. Planck (1900a,b) in order to explain the finite energy density of the black body radiation. Classical electrodynamic theory gave an infinite density for this radiation. The energy density per unit frequency, as calculated on the basis of this theory, was

$$8\pi v^2 kT/c^3$$

and the integral of this over the frequency v is clearly infinite. Partly on the basis of experimental information, partly on intuition, and partly to maintain conformance with Wien's displacement law, Planck replaced this by

$$\frac{8\pi h v^3/c^3}{e^{hv/kT} - 1},$$

Lectures originally given in the Physics Department of Princeton University during 1976, as revised for publication. 1981. Copyright © 1983, Princeton University Press.

and in order to "explain" it, postulated that both emission and absorption of radiation occur instantaneously and in finite quanta. It may be worth mentioning that, somewhat later, realizing how drastic this assumption was, he modified it somewhat, postulating that the absorption is a continuous process. [An account of the history of Planck's black body theory and of its influence on the development of quantum mechanics can be found in a book by Kuhn (1978) which contains also an extensive list of references. —Eds.]

Bohr's postulate of the quantum condition for the electronic orbits was the second building block of pre-quantum-mechanical quantum theory. This also was highly successful, explaining the spectrum of atomic hydrogen clearly and also, in a qualitative, but only qualitative way, the periodic system and hence some basic properties of all atoms (Bohr. 1913a,b,c; 1914; 1915a,b). The latter considerations were qualitative and it remains true that, in spite of many efforts, no mathematically consistent rules could be formulated specifying the orbits of electrons in systems with more than a single electron. In spite of this, the picture, analogous to the present Hartree-Fock picture, was quite successful. Nevertheless, the absence of mathematically consistent rules on the basis of which the electronic orbits, and hence the energy levels, could be determined was greatly disturbing. In addition, it was, of course, a mystery how the electron jumps from one precisely defined orbit to another. The problems which were most deeply felt were, however, different, and were concerned principally with the flow of the conserved quantities, such as energy.

The large cross section of atoms for the absorption of light, which, for instance, for the Na resonance radiation is around 10^{-9} cm^2, was in agreement with classical electromagnetic theory but difficult to understand if light consisted of quanta with a point-like structure, since the radii of atoms are around 3×10^{-8} cm. On the other hand, the uniform spread of the radiation energy over the area of the light beam, as postulated by classical theory, was in apparent contradiction with the fact that the emission of the photoelectrons was not delayed by a decrease of the intensity of the light beam—not even if the average energy incident on an atom on the surface of the electron emitter was less than a thousandth of the energy needed to liberate the electron. This did indicate a concentration of the energy in point-like quanta, so that the striking of an atom by a quantum could furnish the energy needed to liberate the electron. In order to reconcile the two phenomena, Bohr, Kramers, and Slater (1924) postulated that the conservation law of energy is valid only statistically. However, the experiments of Bothe and Geiger (1924), and of Compton and Simon (1952a,b,c), refuted this assumption.

Another phenomenon which was difficult to explain was the Stern-Gerlach effect (Gerlach and Stern, 1921; 1922). A beam of silver atoms is split, by an inhomogeneous magnetic field, into two beams, such that the angular momentum of the atoms in one of the beams has a definite direction, and that of the atoms in the other beam has the opposite direction. Originally, evidently, the angular momenta were randomly oriented. How could they all assume one of two definite directions? Surely the final state must depend continuously on the intial state. It was also perturbing that the classical picture, developed by Ehrenfest, could not account for the transfer of the angular momentum to those atoms whose angular momentum included, originally, a considerable angle with both of the two final momentum directions (Einstein and Ehrenfest, 1922).

As the last example of the difficulties, chemical association reactions may be mentioned. An illustration is $2NO_2 \rightarrow N_2O_4$. For such a reaction to take place, the two associating molecules must collide with an energy which corresponds to one of the energy levels of the compound molecule, N_2O_4 in the example cited. This has a small probability but since the levels of the compound molecule have a certain width—this was already recognized before the advent of quantum mechanics—it is not entirely impossible. However, the angular momentum of the compound is strictly quantized and the probability of a collision with a sharply defined angular momentum surely has zero probability. In spite of this, association reactions do take place. How this happens was at least as much of a mystery as how the electron from one orbit of the hydrogen atom jumps into another, equally sharply defined, orbit. An explanation was proposed by Polanyi and Wigner (1925): that the angular momentum (as considered from the coordinate system in which the total center of mass is at rest) is increased, as a result of the collision, to the nearest integer multiple of ℏ—an assumption leading to very nearly correct results but surely in contradiction to a basic conservation theorem.

All the phenomena mentioned were very puzzling—so puzzling indeed that many physicists doubted that a rational explanation of quantum phenomena would ever be found. It may be worthwhile, therefore, to mention Einstein's suggestion for overcoming the paradox mentioned in connection with the photoelectric effect. He believed in the concentration of the energy in quanta and that these quanta have structures similar to particles. However, their motion is governed by what he called *Führungsfeld*—that is, "guiding field"—and this obeys the equations of electrodynamics. In this way the existence of interference phenomena could be reconciled with the concentration of energy in very small—perhaps infinitesimally small—volumes. However, the picture could not be reconciled with the conservation laws for energy and momentum, and Einstein firmly believed in these. As a result Einstein never published the *Führungsfeld* idea. Schrödinger's theory (1926) reconciled the two postulates: his *Führungsfeld* the Schrödinger wave, moved not in ordinary space, but in configuration space and referred not to single particles, but to the change of the configuration of the whole system, i.e., the motion of all particles. However, his theory was not a pre-quantum-mechanical theory but a fascinating reformulation and even reinterpretation of the original Heisenberg-Born-Jordan (Heisenberg, 1925; Born and Jordan, 1925; Born, Heisenberg, and Jordan, 1926) quantum mechanics which, originally, attempted only the calculation of energy levels and transition probabilities.

It may be useful, at this point, to recall a few dates:

II.2 INTERPRETATION   263

1900–(Dec. 14) M. Planck announces his quantum theory (Planck, 1900a,b).

1905–Einstein proposes his law of the photoelectric effect (actually, the Nobel prize was awarded to Einstein for this, not for his theory of relativity) (Einstein, 1905).

1913–N. Bohr's theory of the H atom (Bohr, 1913a,b,c).

1914–The experiment of Franck and Hertz (1914).

1916–Einstein's derivation of Planck's black-body-radiation formula (Einstein, 1916a,b; 1917a).

1921–Stern-Gerlach experiment (Gerlach and Stern, 1921).

1923–L. de Broglie suggests that matter also has a wave nature (de Broglie, 1923a,b,c; 1924a,b,c,d,e; 1925; 1926).

1925–W. Heisenberg's article which was to form the basis of matrix mechanics (Heisenberg, 1925).

1925–M. Born and P. Jordan establish matrix mechanics (Born and Jordan, 1925).

1926–E. Schrödinger proposes his equations of wave mechanics (Schrödinger, 1926, 1930).

1927–Davisson and Germer verify the wave nature of matter by their interference experiment (Davisson and Germer, 1927).

1927–W. Heisenberg's article "Über den anschaulichen Inhalt der quantentheoretischen Kinematik und Mechanik"—the uncertainty principle (Heisenberg, 1927).

At first, it was difficult for the community of physicists to accept Planck's quantum idea of absorption and emission of radiation. In fact he himself proposed a modification of it which, he expected, would make it more palatable: he suggested that only the emission process is instantaneous, while the absorption of light is a continuous process (Kuhn, 1978). Einstein was the first to accept Planck's original idea at face value and his proposal of the law of photoelectric emission was a result of this. As to the interpretation of the Schrödinger waves as *Führungsfeld*, this became generally accepted as a result of Heisenberg's 1927 paper. This paper also convinced most physicists that it is not meaningful to attribute a definite orbit, or a definite path, to particles—that the concepts in terms of which classical mechanics characterizes the states of a particle are not applicable in the microscopic domain.

Actually, Heisenberg's argument was not entirely rigorous—it was based on the analysis of measurements by means of a γ-ray microscope, but the analysis was not complete in all details. Nevertheless, the article had a profound influence. One can substitute for Heisenberg's γ-ray microscope the measurement of position by means of a light quantum sent out toward the object at time t_1, reflected by it, and received at the position of the emission at time t_2 (Figure 1). The collision of the light quantum with the object occurred at time

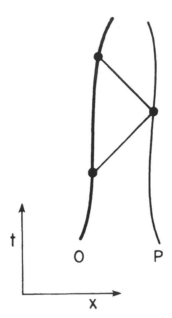

FIGURE 1. The uncertainty principle, illustrated for the particle, P. The observer, O, sends out a pulse of light at t_1 and receives this pulse, reflected, at t_2. The distance of the particle is determined to be $(t_2 - t_1)/2$, but only with the accuracy of the length of the pulse. The momentum of the particles is uncertain to the extent of the kick the photon has given to it—i.e., at least $\sim h/[$the length of the pulse$]$.

$$t_{collision} = \tfrac{1}{2}(t_1 + t_2) \tag{1}$$

and the object's position at that time was

$$x_{collision} = \tfrac{1}{2}c(t_2 - t_1). \tag{1a}$$

However, the times t_1 and t_2 cannot be measured exactly because the light quantum's field had to have a finite extension, to be denoted by Δx. Its frequency, hence, was indeterminate to the extent $c/\Delta x$, its energy had an uncertainty $hc/\Delta x$, and its momentum $h/\Delta x$. It imparted twice its momentum to the particle: thus even if the momentum of this was accurately known before the measurement, after the measurement it was uncertain to about $\Delta p = 2h/\Delta x$. Hence the measurement described permitted the determination of the position only with an accuracy of the order $\Delta x \approx h/\Delta p$—the conclusion arrived at by Heisenberg. The point is that the measurement described could be made accurate only if one had a light wave of sharply defined frequency which was at the same time accurately localized in space. The two requirements cannot be met simultaneously.

Let us now give a mathematical proof of the uncertainty relation. We will give two different proofs, the second one of which is due to H. P. Robertson (1929).

II.2 INTERPRETATION 265

We consider a system of units in which $\hbar \ (\equiv h/2\pi) = 1$. Then we have

$$\bar{x} = \int |\psi(x)|^2 x \, dx, \qquad \bar{p} = \int |\phi(p)|^2 p \, dp, \tag{2a}$$

with

$$\phi(p) = \frac{1}{\sqrt{2\pi}} \int \psi(x) e^{-ipx} \, dx. \tag{2b}$$

Now it will simplify the calculation a bit to set $\bar{x} = 0$ and $\bar{p} = 0$. This represents no loss of generality as it merely amounts to a change of the integration variable as far as x is concerned and a multiplication of ψ by $\exp(-i\bar{p}x)$ to annul the new $\bar{p}$. The uncertainties Δx, Δp are then defined by

$$(\Delta x)^2 = \int |\psi(x)|^2 x^2 \, dx, \tag{3a}$$

and

$$(\Delta p)^2 = \int |\phi(p)|^2 p^2 \, dp = \int |\partial \psi / \partial x|^2 \, dx. \tag{3b}$$

We want to prove that for every state

$$(\Delta x)^2 (\Delta p)^2 \geq \tfrac{1}{4}.$$

We can restrict ourselves without loss of generality to the case where the two uncertainties are equal. This is so because the substitution of $\psi_{new}(x) = \sqrt{\alpha}\psi(\alpha x)$ (the $\sqrt{\alpha}$ is inserted to keep ψ_{new} normalized) decreases Δx by α and increases Δp by α. It does not change $\Delta x \, \Delta p$, but makes it possible to adjust them so that Δx and Δp become equal. Hence, we can assume that this is the case and we can set

$$\Delta x \, \Delta p = \tfrac{1}{2} \left[(\Delta x)^2 + (\Delta p)^2 \right]$$
$$= \tfrac{1}{2} \int \left[x^2 |\psi(x)|^2 + |\partial \psi / \partial x|^2 \right] dx. \tag{4}$$

This is the functional that gives the energy of the harmonic oscillator. It is minimized by $\psi = c \exp(-x^2/2)$ (c is the normalization constant, $\pi^{-1/4}$). The two terms inside the integral of (4) are then indeed equal, and equal to $\tfrac{1}{2}$, and add to unity. Hence we find, in this case,

$$\Delta x \, \Delta p \geq \tfrac{1}{2} \tag{5}$$

and, by virtue of the remark before (4), this holds also generally, as we wanted to show.

The second proof goes as follows. One writes

$$\int \psi^* \left(-x \frac{\partial}{\partial x} \right) \psi \, dx = a, \qquad \int \psi^* \left(\frac{\partial}{\partial x} x \right) \psi \, dx = b, \tag{6}$$

the star denoting transition to the conjugate complex. Then, because of the normalization of ψ, we know that

$$\int \psi^* \left(-x \frac{\partial}{\partial x} + \frac{\partial}{\partial x} x \right) \psi \, dx = \int \psi^* \psi \, dx = \int |\psi|^2 \, dx = 1,$$

or

$$a + b = 1. \tag{6a}$$

Now, by Schwartz's inequality, we have

$$|a|^2 = \left| \int \psi^* x \frac{\partial \psi}{\partial x} \, dx \right|^2 \leq \int |x \psi(x)|^2 \, dx \int \left| \frac{\partial \psi}{\partial x} \right|^2 \, dx, \tag{7}$$

so (again assuming for simplicity $\bar{x} = \bar{p} = 0$)

$$|a|^2 \leq (\Delta x)^2 (\Delta p)^2. \tag{8}$$

Next one recognizes by partial integration that

$$b = a^*. \tag{9}$$

This means that if we set

$$a = u + iv \tag{9a}$$

we must have, by (6a),

$$2u = 1. \tag{9b}$$

This result in turn implies

$$|a|^2 = u^2 + v^2 \geq \tfrac{1}{4}; \tag{10}$$

and this together with (7) again proves (5).

It is worthwhile to remark here that the above "proof" for the uncertainty relation is fundamentally different from Heisenberg's. Like equations (3a) and (3b), it is based on the statistical interpretation of (nonrelativistic) quantum mechanics. Those equations do not refer to the actual ways x and p could be measured—the subjects of Heisenberg's article. Thus, they avoid the basic question: how is a measurement to be carried out, how do we prepare the object on which the measurement is taking place? They assume that the probabilities of the various outcomes of the measurement of x and p are correctly given by the present quantum mechanics. The conclusion we obtained is of a statistical nature: There is no state of any system such that the results of the measurements of x and p would be predictable with greater accuracies Δx and Δp than are indicated by (5). Evidently, the verification of this statement requires, first, the repeated production of the same state many times so that the statistical distribution of the outcomes of the measurements of x and of p can be ascertained. This is in itself a difficult problem; one can never be absolutely sure that one has produced the same state of the system. At the first preparation somebody may have looked on—can we be sure that this did not affect the state of the system produced? Similar remarks apply to the measurements. There is also the question: When is the measurement completed? We will see that, if we adhere strictly to the principles of quantum mechanics, the measurement is completed only when we have observed its outcome, i.e., have read the recording of the measuring apparatus. Apparently, this is not a precisely repeatable process. Could the fluctuations of the outcome of the measurements of x and p come from these sources? Evidently we believe that this is not the case; but it would be difficult to explain this belief convincingly to a fully detached person. In spite of this, we are all fully convinced of all this and do not believe that any refinement of the preparation of the system the x and p of which is to be measured, or any improvement of the measuring technique, would lead to a violation of (5).

Before going into a more detailed discussion of these questions, we will review the mathematical structure of quantum mechanics, its description of states, and its calculation of the probabilities of the outcomes of measurements. The equations used to derive (5) will naturally appear as special cases of the general theory.

§2. The Mathematical Formalism of Quantum Mechanics

Hilbert Space

In quantum mechanics, as in classical physics, we postulate the existence of *isolated systems*. In both theories, if a complete description of an isolated system is given at one time, a complete description for any other time is uniquely determined as long as the system remains isolated—i.e., is not influenced by any other system. In this sense, both theories are deterministic. The means of description of the state of the system have, however, undergone drastic changes over the course of the history of physics. In the original Newtonian mechanics, the state of the system was described by the positions and velocities of its constituents. In the later field theories, it was described by the field strengths at all points of space—that is, by one or more—in the case of electrodynamics six—functions of three real variables, the latter characterizing the points of ordinary space. The usual quantum-mechanical description of the state of a system is much more abstract; it is given by a vector (or, more accurately, by a ray) in an abstract complex Hilbert space. This vector is usually called the state vector. Since the Hilbert space has infinitely many dimensions, this amounts to the specification of the state by an infinite set of complex numbers, $a_1, a_2, a_3, \ldots$, the components of the state vector in Hilbert space, which however, satisfy the condition that the sum

$$\sum |a_i|^2 < \infty \qquad (1)$$

is finite. This means that the *length* of the vectors in Hilbert space is finite. Furthermore, it is postulated that two state vectors which have the same direction, i.e., the components of which differ only by a common factor, characterize the same state. The set of vectors the components of which differ only by a common factor, e.g., the vectors with the components $a_1, a_2, a_3 \ldots$ and the vectors $ca_1, ca_2, ca_3 \ldots$ for any $c \neq 0$, are said to form a *ray*. We can choose, therefore, one of these vectors to describe the state, and we usually choose one that is normalized, i.e., one for which

$$\sum |a_i|^2 = 1. \qquad (1a)$$

The different normalized vectors describing the same state differ only by a factor ω of modulus 1.

Schrödinger's original formulation of his "wave mechanics" was, of course, not in terms of the Hilbert space. Wave mechanics characterized the states of systems in terms of complex valued functions, actually functions in configuration space. However, if one introduces an orthonormal set of functions $u_n(x_1, x_2 \ldots)$ in any space,

$$\int \cdots \int u_n(x_1, x_2 \ldots)^* u_m(x_1, x_2 \ldots) dx_1 \, dx_2 \cdots = \delta_{nm}, \qquad (2)$$

one can expand the wave function ψ of any state in terms of this set. Thus,

$$\psi(x_1, x_2 \ldots) = \sum a_n u_n(x_1, x_2 \ldots), \qquad (2a)$$

$$a_n = \int\!\!\int u_n(x_1, x_2 \ldots)^* \psi(x_1, x_2 \ldots) dx_1 \, dx_2 \ldots; \qquad (2b)$$

and the numbers a_n can be considered to be the components of a vector in Hilbert space as long as

$$\int \cdots \int |\psi(x_1, x_2 \ldots)|^2 \, dx_1 \, dx_2 \cdots = \sum |a_n|^2 \qquad (2c)$$

is finite—which we assume to be the case for Schrödinger's wave functions.

It may be worthwhile to observe at this point that the correspondence between functions ψ and the corresponding vector a in Hilbert space is not one to one. Two functions ψ and ψ' which differ only on a set of measure 0 (for instance only on a denumerable set of points) correspond to the same vector in Hilbert space. However, this observation rarely plays an important role in actual calculations.

The most important derivative concept in Hilbert space is that of the scalar product. The scalar product of two vectors a and b is defined as

$$(a, b) = \sum a_n^* b_n. \tag{3}$$

In terms of the wave functions this is

$$(\psi, \phi) = \int \cdots \int \psi(x_1, x_2 \ldots)^* \phi(x_1, x_2 \ldots) dx_1 \, dx_2 \cdots$$
$$= \sum a_n^* b_n. \tag{3a}$$

the last part of the equation being valid if the a are, according to (2a), the expansion coefficients of ψ and the b are, similarly, the expansion coefficients of ϕ. That the sum in (3) is finite follows from (1), and the similar restriction follows for b, by means of Schwartz's inequality. It follows from (3) or (3a) also that

$$(a, b) = (b, a)^*, \tag{3b}$$

the star denoting, as before, the conjugate complex.

If a and b are normalized according to (1a), the absolute square $|(a, b)|^2$ of the scalar product (a, b) is called, for reasons which will appear soon, the transition probability from the state a into the state b—or conversely, because of (3b). Otherwise, the transition probability is

$$\frac{|(a, b)|^2}{(a, a)(b, b)}. \tag{4}$$

It may be worth remarking that (a, a) is real and also positive unless $a = 0$, and that the scalar product is linear in the second factor,

$$(a, \beta b + \beta' b') = \beta(a, b) + \beta'(a \ b'), \tag{5}$$

and antilinear in the first factor,

$$(\alpha a + \alpha' a', b) = \alpha^*(a, b) + \alpha'^*(a', b), \tag{5a}$$

where α, α', β, and β' are arbitrary complex numbers. These equations are immediate consequences of the definition (3) of the scalar product.

Linear Operators in Hilbert Space

An operator in Hilbert space transforms a vector in Hilbert space into another (or the same) vector in the same space. One should really write $A(\phi)$ for the vector into which the operator A transforms the vector ϕ, but, at least for linear operators A, one writes $A\phi$ instead. An operator A is called linear if for any two vectors ϕ, ψ and any two numbers α, β

$$A(\alpha\phi + \beta\psi) = \alpha A\phi + \beta A\psi \qquad (6)$$

holds. More generally, one can write

$$A(\sum a_n\psi_n) = \sum a_n A\psi_n. \qquad (6a)$$

Hence if we assume that the ψ_n in (6a) form a complete orthonormal set of vectors, and if we expand the $A\psi_n$ in terms of these vectors, we can write

$$A\psi_n = \sum_m A_{mn}\psi_m. \qquad (7)$$

Then (6a) gives for the transformation properties of the coefficients a_n,

$$a_n \rightarrow \sum_m A_{nm}a_m. \qquad (7a)$$

The operation A acts on the expansion coefficients as the matrix (A_{nm}), the matrix elements of which are, because of (7)

$$A_{mn} = (\psi_m, A\psi_n). \qquad (7b)$$

The last two equations are valid if the ψ_m form a complete orthonormal set; but (6a) is valid as a consequence of (6) for any set of vectors ψ_m.

Two special kinds of linear operators play particularly important roles in quantum mechanics. The invariance transformations, such as the time-displacement transformation, are mediated by unitary transformations U which leave the scalar products, and hence the transition probabilities (4), between any two vectors unchanged:

$$(U\phi, U\psi) = (\phi, \psi). \qquad (8)$$

They will play a lesser role in the considerations which follows. The other special kind of linear operators of basic significance for what follows are the self-adjoint operators. In order to define them, one first defines the adjoint $A^\dagger$ to an operator A. This is so defined that for any two vectors ϕ and ψ,

$$(\phi, A\psi) = (A^\dagger \phi, \psi) \tag{9}$$

is valid. One concludes from (9) and (3b) that

$$(A\psi, \phi) = (\psi, A^\dagger \phi), \tag{9a}$$

so that A is the adjoint of $A^\dagger$ if $A^\dagger$ is the adjoint of A. The matrix elements of $A^\dagger$ are, as one can infer from (7b),

$$(A^\dagger)_{nm} = A^*_{mn}, \tag{9b}$$

i.e., the matrices of A and $A^\dagger$ are what we call hermitian adjoints. The second type of operators of basic significance can now be defined. They are those which are equal to their adjoints,

$$A = A^\dagger. \tag{10}$$

In the corresponding matrices the matrix elements which lie symmetrically with respect to the main diagonal are the conjugate complexes of each other.

For the sake of accuracy it should be pointed out that the existence of the adjoint $A^\dagger$ to an arbitrary linear operator is by no means obvious, and requires, if all the questions of convergence are to be treated rigorously, a reasonably elaborate proof (von Neumann, 1932). Once its existence is established, i.e., once it is shown that there is a ϕ' independent of ψ satisfying the equation

$$(\phi, A\psi) = (\phi', \psi) \tag{11}$$

(ϕ' depending only on A and ϕ), the linear dependence of this $\phi' = A^\dagger \phi$ on ϕ is easily proved. We have on the one hand,

$$(\alpha\phi_1 + \beta\phi_2, A\psi) = [A^\dagger(\alpha\phi_1 + \beta\phi_2), \psi], \tag{12}$$

and on the other,

$$\begin{aligned}
(\alpha\phi_1 + \beta\phi_2, A\psi) &= \alpha^*(\phi_1, A\psi) + \beta^*(\phi_2, A\psi) \\
&= \alpha^*(A^\dagger \phi_1, \psi) + \beta^*(A^\dagger \phi_2, \psi) \\
&= (\alpha A^\dagger \phi_1 + \beta A^\dagger \phi_2, \psi).
\end{aligned} \tag{12a}$$

Hence the right sides of (12) and (12a) are equal and, since this is true for every ψ,

$$A^\dagger(\alpha\phi_1 + \beta\phi_2) = \alpha A^\dagger\phi_1 + \beta A^\dagger\phi_2, \tag{12b}$$

i.e., $A^\dagger$ is linear.

Normal Form of Self-Adjoint Operators

Except for the proof of the existence of the hermitian adjoint $A^\dagger$, the preceding discussion is straightforward and easy. This is not the case for the following discussion, particularly not if the concept of the operator is extended so that it encompasses also unbounded operators. The following discussion gives only the results. The detailed proofs can be found, for instance, in the book of J. von Neumann (1932) or in the book of G. W. Mackey (1963).

Let us consider first bounded self-adjoint operators, i.e., operators such that $(A\psi, A\psi)$ has an upper limit. The usual procedure to obtain the normal form is to look for the characteristic vectors ψ_ν of A

$$A\psi_\nu = \lambda_\nu\psi_\nu. \tag{13}$$

One easily sees that the characteristic values λ_ν have to be real and the characteristic vectors ψ_ν, which are solutions for different λ_ν, are orthogonal. We assume that they are also normalized, i.e., that $(\psi_\nu, \psi_\nu) = 1$. If the characteristic vectors ψ_ν form a complete orthonormal set, we say that A has only a point spectrum, this consisting of the λ_ν. The effect of A on an arbitrary vector ϕ

$$\phi = \sum (\psi_\nu, \phi)\psi_\nu \tag{14}$$

is then, because of its linear character,

$$A\phi = A \sum (\psi_\nu, \phi)\psi_\nu = \sum (\psi_\nu, \phi)A\psi_\nu$$
$$= \sum (\psi_\nu, \phi)\lambda_\nu\psi_\nu. \tag{14a}$$

The measurement theory then postulates that the measurement of A on a system in the state ϕ gives one of the values λ and that it gives the value λ_ν with the probability

$$p_\nu = |(\psi_\nu, \phi)|^2. \tag{15}$$

It is assumed here that ϕ and the ψ_ν are normalized,

$$(\phi, \phi) = (\psi_\nu, \psi_\nu) = 1. \tag{15a}$$

It is also reasonable to assume that the system which was originally in the state ϕ is, after the measurement—if the result of that is λ_ν—in the state ψ_ν. Naturally it would be good to justify this postulate of measurement theory by means of the equations of motion of quantum mechanics—that is, to describe the measurement process. Such an analysis would have to treat in quantum-mechanical language the apparatus used for the measurement of A. We will discuss later the extent to which this has so far proved to be possible, and the examples which show that for certain A it is impossible.

The preceding remarks apply to a self-adjoint operator which has a point spectrum only. One can well say that most operators do not have such a spectrum. A self-adjoint operator may have a continuous spectrum also—or only a continuous spectrum. In other words the solutions ψ_v of (13)—with a finite length so that (15a) can be made valid—may not form a complete set. In fact, there may be no such solution at all; that is, there may be no vector ψ_v in Hilbert space which satisfies (13) for some λ_v. In the general case the preceding equations have to be replaced by much more complicated ones. In order to illustrate this point, it is helpful to rewrite (14) and (15) somewhat by decomposing A into projection operators P_v. A projection operator is defined by

$$P_v\phi = (\psi_v, \phi)\psi_v. \tag{16}$$

It then follows from (14) that

$$\sum P_v = 1, \tag{17}$$

and from (14a),

$$\sum \lambda_v P_v = A. \tag{17a}$$

One easily verifies that the P_v are self-adjoint, identical with their squares, and more generally satisfy the equations,

$$P_v P_\mu = \delta_{v\mu} P_v. \tag{17b}$$

In fact,

$$
\begin{aligned}
P_v P_\mu \phi &\equiv P_v(P_\mu \phi) = P_v(\psi_\mu, \phi)\psi_\mu = (\psi_\mu, \phi)P_v\psi_\mu \\
&= (\psi_\mu, \phi)(\psi_v, \psi_\mu)\psi_v = \delta_{v\mu}(\psi_\mu, \phi)\psi_v = \delta_{v\mu}P_v\phi.
\end{aligned} \tag{17c}
$$

The expression for the transition probability into ψ_v—which is also the probability that the outcome of the measurement of A on ϕ will be λ_v—becomes, in terms of P_v,

$$p_v = (\phi, P_v\phi). \tag{18}$$

One can also write this in the form

$$p_v = (P_v\phi, P_v\phi) \tag{18a}$$

as, because of the self-adjoint nature of P_v and (17),

$$(P_v\phi, P_v\phi) = (\phi, P_v^2\phi) = (\phi, P_v\phi). \tag{18b}$$

The derivation so far has supposed that A has only a discrete spectrum, a condition for the validity of equations (17).

We now proceed to the more general case that A may also have a continuous spectrum. In this case one has to admit that the measurement will not yield a mathematically precise value—no one asks whether the outcome of this measurement is a rational or irrational number. It is more reasonable to ask, for instance, whether the outcome is smaller than a number λ. The probability for a "yes" answer to this question can be written, in the case of a discrete spectrum, as

$$p(\lambda) = [\phi, P(\lambda)\phi]. \tag{19}$$

Here

$$P(\lambda) = \sum_{\lambda_v < \lambda} P_v. \tag{19a}$$

It then follows from (18) that the probability for an outcome of the measurement between λ' and $\lambda > \lambda'$ is

$$p(\lambda) - p(\lambda') = \{\phi, [P(\lambda) - P(\lambda')]\phi\}. \tag{20}$$

It follows from the theory of self-adjoint operators in Hilbert space that the operators $P(\lambda)$ can also be defined if the spectrum is not exclusively discrete. In other words, the $P(\lambda)$ can be defined for every self-adjoint operator, for any real λ. One clearly has

$$P(\lambda) \to 0 \quad \text{for} \quad \lambda \to -\infty; \tag{21a}$$

and the analogs of (17) and (17b) are

$$P(\lambda) \to 1 \quad \text{for} \quad \lambda \to \infty \tag{21b}$$

and

$$P(\lambda)P(\lambda') = P(\lambda')P(\lambda) = P(\lambda') \quad \text{for} \quad \lambda' \leq \lambda. \tag{21c}$$

These equations, of course, do not determine the $P(\lambda)$ since they do not involve A. In order fully to define the $P(\lambda)$, one has to write the analog of (17a). This becomes, in the general case, a somewhat complicated expression—a limit of a sum with increasingly many terms. One has to form series of increasingly many λ values which cover the real line with increasing density. If the density is $1/N$, one has the series

II.2 INTERPRETATION   275

$$A = \sum_{n=-\infty}^{\infty} \sum_{m=1}^{N} (n + m/N)[P(n + m/N) - P(n + (m - 1)/N)], \qquad (22)$$

and this is valid in the limit $N = \infty$. Naturally, this is a special form for A in which the distance between successive λ values is uniformly $1/N$. There are infinitely many other ways to increase the density of the λ. The mathematician writes a Stieltjes integral for (22) and, in fact, (22) is the definition of the Stieltjes integral. The expression for A is written in the elementary form (22) because it does not presuppose familiarity with that integral. The point is that (22), together with the equations (21), fully defines the λ and hence determines the mathematical expression (20) for the probability of the outcome of the measurement of A lying between λ' and λ.

A few remarks are needed to complete the discussion. First, the possibility of finding a mathematical expression for the outcome of the measurement of an operator obviously does not guarantee the possibility of such a measurement. Second, the theory postulates measurements corresponding to operators, not to classical quantities. Originally, the attempt was made to define operators to correspond to classical quantities. The present theory does not strive for such a coordination. Thus, one does not try to choose, for instance, between $(px^2 + x^2p)/2$ and xpx as operators to correspond to the classical quantity, "product of the square of the coordinate and first power of the momentum." Instead, one asks if an experimental device can be constructed that will realize a measurement of the operator $(px^2 + x^2p)/2$ and whether another can be built that will measure xpx.

On the mathematical side: if A is bounded, $P(\lambda)$ remains 0 up to the lower bound of A and becomes 1 at the upper bound. Hence that part of the sum in (22) vanishes for which $n + m/N$ is below the lower bound; and so does that which is above the upper bound. However, and this theorem is due to von Neumann (1932), the preceding equations as written are valid also for general self-adjoint operators, not only for bounded ones. It is worthwhile, nevertheless, to add a few words relating to the former, since most common operators are not bounded. There are vectors in Hilbert space to which an unbounded operator cannot be applied. Both multiplication by x and $i\partial/\partial x$ are unbounded operators. For instance, multiplication of the state vector $1/(i + x)$ by x does not produce a state vector, since $\psi(x) = x/(i + x)$ is not square integrable; for it the expression (2c) is infinite. For an unbounded operator it is postulated only that there be an everywhere dense set of vectors to which it can be applied—i.e., that in the neighborhood of any chosen vector there be vectors arbitrarily close to it to which the operator can be applied. In the case of $1/(i + x)$ such vectors are, for instance $e^{-\varepsilon x^2}/(i + x)$ with decreasing values of ε. In fact, the square of the difference vector,

$$\int \left| \frac{1}{i + x} - \frac{e^{-\varepsilon x^2}}{i + x} \right|^2 dx. \qquad (23)$$

goes to zero as ε goes to zero. Moreover, multiplication with x leaves $e^{-\varepsilon x^2}/(i + x)$ within the Hilbert space no matter how small ε is, as long as it remains positive.

It is remarkable that the probability of the outcome of measurements, given by (20), is well defined even for unbounded operators, even in the case of state vectors ϕ to which the unbounded operator cannot be applied. The intrinsic reason for this is that the measurement hardly refers to the operator and its spectrum—what it really purports to define is the transition probability into the state $[P(\lambda) - P(\lambda')]\phi$ if the original state was ϕ. The projection operators $P(\lambda)$ could be given other labels, such as $\lambda/(\lambda^2 + 1)^{1/2}$, and the measurement would still be the same, only its outcome would be called differently. To repeat, however, all this is theory, and there is no mathematical guarantee that even a transition probability into an arbitrary state can be measured. In order to guarantee the possibility of such a measurement one would have to describe a way to do it. The mathematical theory of the measurement, as formulated first by von Neumann but now generally accepted, ingenious as it is, does not do that.

It may be worthwhile to observe, nevertheless, that the self-adjoint nature of an unbounded operator imposes rather rigorous criteria. First, the operator and its adjoint must give the same vector if applied to a vector which lies in the intersection of the domains of definition of the two operators. Second, the two domains of definition must be identical (von Neumann called them hypermaximal). Fortunately, this criterion apparently plays no practical role in the theory.

§3. Direct Product and Quantum Mechanical Description of the Measuring Process

The Direct Product of Hilbert Spaces

Starting from two Hilbert spaces, one can construct a single larger Hilbert space, a sort of union of the two. If the axes of the first Hilbert space are specified by Greek indices v. and of the second by Latin indices such as n, then the axes of the direct product are specified by a double index, such as vn. Thus vn specifies a single axis of the Hilbert space which is the direct product of the original two Hilbert spaces. The number of axes vn is still denumerable even though it appears to be greater than the number of axes of the factors: one can order pairs of ordered numbers into a single series such as 11; 12, 21; 13, 22, 31; 14, 23, and so on. The components Ψ_{vn} of a vector Ψ in the space of the direct product have two indices, and the scalar product of two vectors Ψ and Φ in that space is

$$(\Psi, \Phi) = \sum_{vn} \Psi_{vn}^* \Phi_{vn}. \tag{24}$$

The Hilbert space which is the direct product of two Hilbert spaces H_1 and H_2 is usually denoted by $H_1 \otimes H_2$.

One also defines the direct product of two vectors, one of which, ϕ, is in H_1, and the other, ψ, in H_2. This direct product, denoted by $\phi \otimes \psi$, is in the space $H_1 \otimes H_2$. Its vn component is

$$(\phi \otimes \psi)_{vn} = \phi_v \psi_n. \tag{25}$$

The direct product is linear in its two factors. Thus

$$(\alpha\phi + \alpha'\phi' \otimes \psi) = \alpha(\phi \otimes \psi) + \alpha'(\phi' \otimes \psi), \tag{25a}$$

and a similar equation applies if ψ is replaced by $\beta\psi + \beta\psi'$. In order to verify (25a), one compares the vn components of the two sides. These are $(\alpha\phi + \alpha'\phi')_v \psi_n$ and $\alpha\phi_v\psi_n + \alpha'\phi'_v\psi_n$ and the two are equal.

The square length of $\phi \otimes \psi$ is

$$(\phi \otimes \psi, \phi \otimes \psi) = \sum_{vn} |(\phi \otimes \psi)_{vn}|^2 = \sum_{vn} |\phi_v \psi_n|^2$$
$$= \sum_v |\phi_v|^2 \sum_n |\psi_n|^2, \tag{25a}$$

and is the product of the squares of the lengths of the two factors ϕ and ψ. The scalar product of two direct product vectors $\phi \otimes \psi$ and $\phi' \otimes \psi'$ in the new Hilbert space becomes, similarly,

$$(\phi \otimes \psi, \phi' \otimes \psi') = \sum_{vn} (\phi \otimes \psi)_{vn}^*(\phi' \otimes \psi')_{vn} = \sum_{vn} (\phi_v\psi_n)^*\phi'_v\psi'_n$$
$$= \sum_v \phi_v^*\phi'_v \sum_n \psi_n^*\psi'_n = (\phi, \phi')(\psi, \psi'), \tag{25b}$$

that is, the product of the scalar products of the two factors.

It is important to remark, finally, that the direct product of two factors ϕ and ψ is the same vector in the new Hilbert space no matter which coordinate systems are used for its definition in (25). If the unit vectors in the direction of the coordinate axes used in (25) for the two Hilbert spaces H_1 and H_2 are denoted by e_v and f_n, the vn component of $\phi \otimes \psi$ is, by (25),

$$(\phi \otimes \psi)_{vn} = \phi_v\psi_n = (e_v, \phi)(f_n, \psi) = (e_v \otimes f_n, \phi \otimes \psi), \tag{26}$$

This equation remains valid for any other choice of the coordinate axes in the original Hilbert spaces. One can verify this statement by explicit calculation. It follows from (26) that

$$\phi \otimes \psi = \sum_{vn} (e_v, \phi)(f_n, \psi)e_v \otimes f_n. \tag{26a}$$

Hence, the component of the state vector in the new $e'_\mu \otimes f'_m$ direction becomes

$$(e'_\mu \otimes f'_m, \phi \otimes \psi) = \sum_{vn} (e_v, \phi)(f_n, \psi)(e'_\mu \otimes f'_m, e_v \otimes f_n)$$

$$= \sum_{vn} (e_v, \phi)(f_n, \psi)(e'_\mu, e_v)(f'_m, f_n). \qquad (26b)$$

But we have the equality,

$$\sum_v (e'_\mu, e_v)(e_v, \phi) = \sum_v ((e_v, e'_\mu)e_v, \phi) = (e'_\mu, \phi); \qquad (26c)$$

and a similar equation applies for the f. Therefore, we have verified explicitly that

$$(e'_\mu \otimes f'_m, \phi \otimes \psi) = (e'_\mu, \phi)(f'_m, \psi). \qquad (26d)$$

This result proves that the equation (25) defining the direct product of the two vectors is independent of the coordinate systems used in the original Hilbert spaces.

The direct product of two Hilbert spaces, and the direct product of vectors in them, is introduced in order to describe the joining of two systems into a single system. This is important if one wants to describe the interaction of two systems which were originally separated—in our case the interaction between the measuring apparatus and the system on which the measurement is undertaken. If it is possible to describe the two systems in separate Hilbert spaces, their union can indeed be most easily characterized in the direct product of these Hilbert spaces. In this "product space" the state vector of the union is the direct product of the state vectors of the components. Indeed, if one wants to calculate the probability that the first system is in state ϕ' and the second is in state ψ' when their actual state vectors are ϕ and ψ, one obtains by (25b),

$$|(\phi' \otimes \psi', \phi \otimes \psi)|^2 = |(\phi', \phi)|^2|(\psi', \psi)|^2, \qquad (27)$$

which is the product of the transition probabilities from ϕ into ϕ' and from ψ into ψ'. Since the two system are assumed to be independent, this is the expected result.

Direct Products of Operators

In order to obtain the more general expressions corresponding to (18), etc. for the joint system, it is useful to introduce the concept of the direct product of two operators A and B, acting in the two original Hilbert spaces H_1 and H_2. The action of A is described by the equation

$$(A\phi)_v = \sum_{v'} A_{v'v}\phi_{v'};$$ (28a)

that of B by

$$(B\psi)_n = \sum_{n'} (B_{n'n}\psi_{n'}.$$ (28b)

Their direct product, to be denoted by $A \otimes B$, will then transform Ψ with the components Ψ_{vn} into $A \otimes B\Psi$, the components of which are

$$((A \otimes B)\Psi)_{vn} = \sum_{v'n'} A_{v'v}B_{n'n}\Psi_{v'n}.$$ (28)

This is the definition of $A \otimes B$. Clearly, if Ψ is a direct product, $\Psi = \phi \otimes \psi$. The action of $A \otimes B$ on this will give

$$(A \otimes B)\phi \otimes \psi = A\phi \otimes B\psi,$$ (29)

the direct product of the results of the actions of A and of B in their respective Hilbert spaces.

The concept of the direct product of two operators enables us to generalize the expression for the transition probability (27) as the similar expression (15) for a single system was generalized, first in (18) for the case of a discrete spectrum, and then in (19) and (20) to the case where a continuous spectrum may also be present. First, if both operators A and B, to be measured on the two systems, have only a discrete spectrum, then let us denote the projection operators which correspond to the characteristic values α_v of A and b_n of B with P_v and Q_n, respectively. The probability that the outcome is a_v for the measurement of A on ϕ is then $(\phi, P_v\phi)$. The probability that the outcome is b_n for the measurement of B on ψ is $(\psi, Q_n\psi)$. The probability that the two outcomes on the joint system be a_v and b_n is then given by

$$P_{vn} = (\phi \otimes \psi, (P_v \otimes Q_n)(\phi \otimes \psi)) = (\phi, P_v\phi)(\psi, Q_n\psi),$$ (30)

as follows from (29). We conclude that $P_v \otimes Q_n$ is the projection operator for the outcome a_v of A and b_n of B. Similarly, we can generalize (19): the probability for A giving a result smaller than λ *and* B giving a result smaller than l is

$$p(\lambda, l) = \{\phi \otimes \psi, [P(\lambda) \otimes Q(l)]\phi \otimes \psi\} = [\phi, P(\lambda)\phi][\psi, Q(l)\psi].$$ (31)

Here $P(\lambda)$ is defined for operator A as in (19a) and $Q(l)$ is defined similarly for B. Equation (31) is valid even when there is a continuous spectrum. The formula for the probability that the outcomes will fall in the intervals λ, λ' and l, l', respectively, is obtained equally easily as

$$[p(\lambda', l') - p(\lambda, l')] - [p(\lambda', l) - p(\lambda, l)]$$
$$= (\phi \otimes \psi, [P(\lambda') - P(\lambda)][Q(l') - Q(l)]\phi \otimes \psi), \qquad (32)$$

in generalization of (20). The first two terms provide the probability that B will fall below l' and A between λ and λ'. The square bracket gives the probability for the same interval of A and that B will fall below l. Hence (32), the difference between the two, gives the probability that A will fall between λ and λ' and B between l and l'. The right side is a symmetric expression for this probability. Actually, these formulae are given only for the sake of completeness; they will not be used. The same applies to the general formula,

$$(A \otimes B)(A' \otimes B') = (AA' \otimes BB'). \qquad (33)$$

It is worth remembering, though, that the projection operator in the direct product space for the probability of the outcome a_v for A without specification of the outcome for any measurement on the second system is $P_v \otimes 1$; and conversely, it is $1 \otimes Q_n$ for the outcome b_n of B as measured on the second system if no measurement on the first system is undertaken. This concludes our mathematical discussion of the direct product concept.

It should be admitted, however, that the exclusion principle's requirement of the antisymmetrization of the state vector puts a certain restriction on the postulate that the states of the two noninteracting systems be described in separate Hilbert spaces. As a result of the antisymmetrization, this separation is not actually directly possible. A similar remark applies for particles obeying Bose statistics and demanding a symmetrized state vector. It does not seem, though, that this restriction is truly relevant in the case to be considered: the two Hilbert spaces may describe not the two different objects but the conditions in two distinct parts of space, in our case that of the object and that of the apparatus. It seems that field theories might not encounter this difficulty, but since actually no one seems to believe that a more precise discussion, taking the symmetric or antisymmetric nature of the Schrödinger wave functions into account, would alter any of the conclusions, such a precise discussion of this point does not appear in the literature. It may be true that it would be good to provide such a discussion for the sake of accuracy.

The Quantum Mechanical Description of Measurement

The concept of the direct product greatly facilitates the quantum mechanical description of the measurement process. This consists of a temporary interaction between the object on which the measurement is undertaken and the apparatus which performs the measurement. Let us consider, first, a measurement on an object which is in a state for which the outcome is surely determined—i.e., a state, the state vector σ_κ of which is a characteristic vector of the quantity to be measured.

II.2 INTERPRETATION   281

If we denote the initial state vector of the apparatus by a_0, then the initial state of apparatus plus object is $a_0 \otimes \sigma_\kappa$. After the measurement, the apparatus has assumed a state which shows the outcome of the measurement; we denote its state vector by a_κ. Hence, if the object did not change its state as a result of the measurement—this assumption will be discussed further later—the interaction between apparatus and object transforms $a_0 \otimes \sigma_\kappa$ into

$$a_0 \otimes \sigma_\kappa \to a_\kappa \otimes \sigma_\kappa, \tag{34}$$

and this is assumed to be valid for all κ. Thus, the art of measuring the quantity with the characteristic vectors σ_κ consists in producing an apparatus the interaction of which with the object has the result indicated by (34).

It now follows from the linear nature of the quantum-mechanical equations of motion that if the initial state of the object is a linear combination of the σ_κ, say $\sum \alpha_\kappa \sigma_\kappa$, the final state of object plus apparatus will be given by

$$a_0 \otimes \sum \alpha_\kappa \sigma_\kappa = \sum \alpha_\kappa (a_0 \otimes \sigma_\kappa) \to \sum \alpha_\kappa (a_\kappa \otimes \sigma_\kappa). \tag{35}$$

The second member of (35) follows from the linearity of the direct product in terms of its factors, an immediate consequence of (25); the last member, from the linear nature of the time-development operator.

Do the processes postulated in (34) and (35) fully describe the measurement? In the case of (34) this is true: the apparatus assumes a definite state which indicates the state of the object. In the general case, described by (35), this is not the case: the apparatus is not, with the desired probability $|\alpha_\kappa|^2$, in the state a_κ. In fact, the joint state of object and apparatus appears quite complicated. This could not have been expected to be otherwise: the transformation indicated by the arrow is a consequence of the quantum-mechanical equation of motion and this is determin- istic. The outcome of the measurement in the general case of (35) has a probabilistic nature. What the final state of (35) does show is that a statistical correlation between the state of the apparatus and that of the object has been established: the prob- ability of finding both in the state v is, according to the standard postulate of quan- tum mechanics,

$$\left| a_v \otimes \sigma_v, \sum \alpha_\kappa (a_\kappa \otimes \sigma_\kappa) \right|^2 = \left| \sum_\kappa \alpha_\kappa (a_v, a_\kappa)(\sigma_v, \sigma_\kappa) \right|^2$$

$$= \left| \sum_\kappa \alpha_\kappa \delta_{v\kappa} \right|^2 = |\alpha_v|^2. \tag{36}$$

Both (a_v, a_κ) and $(\sigma_v, \sigma_\kappa)$ have been set equal to $\delta_{\kappa v}$—the former because the states a_v and a_κ are supposed to be distinguishable even at the macroscopic level, the latter because the σ are normalized characteristic vectors of the self-adjoint operator which is being measured. A calculation entirely similar to (36) then shows also that the probability is zero of finding the apparatus in state a_λ and the object in the state σ_κ with $\kappa \neq \lambda$—i.e., that indeed the right side of (35) represents a joint state of object and apparatus with the statistical correlation between the states as indicated.

It is evident, therefore, that interesting and suggestive as (35) may be (it is due, essentially, to von Neumann, 1932) it does not completely describe the quantum-mechanical measurement. In order to give (35) the meaning just outlined, we must assume that a measurement on the state of the right side of (35) is possible and that it gives the different possible values with the probabilities postulated by the usual interpretation. Actually, in order to obtain the state of the object, it is necessary only to measure the state of the apparatus. However, the quantum-mechanical description of this measurement suffices no more than before to pick out one definite value of κ from many. The equations, being deterministic, lead always from a superposition to a superposition. Nevertheless, if (1) one measures the state of the apparatus a by a second apparatus b, and if (2) the states a_κ of a definitely put the apparatus b into the state b_κ; in other words, if

$$b_0 \otimes a_\kappa \to b_\kappa \otimes a_\kappa, \tag{37}$$

then (3) the interaction of this apparatus with the state (34) resulting from the measurement on σ_κ by a, will give

$$b_0 \otimes (a_\kappa \otimes \sigma_\kappa) \to b_\kappa \otimes a_\kappa \otimes \sigma_\kappa. \tag{37a}$$

This is true, at least, if b interacts only with a, not with the object. Hence, in this case, the interaction of b_0 with the result of the measurement on the general state (cf. [35]) will give

$$b_0 \otimes \sum_\kappa \alpha_\kappa (a_\kappa \otimes \sigma_\kappa) \to \sum_\kappa \alpha_\kappa (b_\kappa \otimes a_\kappa \otimes \sigma_\kappa). \tag{37b}$$

Thus, a correlation between the states of all three systems—object, apparatus a, and apparatus b—is established; but, naturally, no choice between the different states σ_κ is made. A similar statement applies if a fourth apparatus is used to "measure" the state of b, and so on. The measurement process, as far as it can be described by standard quantum mechanics, only establishes statistical correlations between the states of the apparata and those of the object. No choice for a definite state emerges, except if the object was, to begin with, in a state in which the outcome of the measurement is unique, as it is in the case considered in (34).

Before discussing some further problems arising from the assumption embodied in (34) and (35), and before proposing a possible resolution of the problem here encountered, it may be worth pointing out the obvious fact that while it may be possible to envisage an interaction of the form (34) between apparatus and object, this in no way guarantees that an apparatus with this interaction with the object can be found. This also will be discussed further later.

§4. THE PHYSICS OF THE MEASUREMENT DESCRIPTION (35)

The description of the measurement indicated by (34) and (35) has one very happy consequence: it shows that only self-adjoint operators are measurable. This follows from the unitary nature of the transformation indicated by the arrow. If one writes (34) for two different indices, say κ and λ, the final states of the corresponding transformations are orthogonal.

We have

$$(a_\kappa \otimes \sigma_\kappa, a_\lambda \otimes \sigma_\lambda) = (a_\kappa, a_\lambda)(\sigma_\kappa, \sigma_\lambda). \tag{38}$$

Moreover, a_κ and a_λ are two clearly distinguishable states—distinguishable, as a rule, even macroscopically. Therefore $(a_\kappa, a_\lambda) = 0$. Hence, it follows from the unitary nature of the transformation indicated by the arrow that the initial states were also orthogonal:

$$(a_0 \otimes \sigma_\kappa)(a_0 \otimes \sigma_\lambda) = (a_0, a_0)(\sigma_\kappa, \sigma_\lambda) = 0. \tag{38a}$$

In addition we know that $(a_0, a_0) = 1$. We conclude that $(\sigma_\kappa, \sigma_\lambda) = 0$ if a_κ and a_λ indicate two different outcomes of the measurement. If we further postulate that the outcomes are described by real numbers, i.e., that the characteristic values of the operator Q of which the a_κ are characteristic vectors are real,

$$Q a_\kappa = q_\kappa a_\kappa, \quad \text{with} \quad q_\kappa = q_\kappa^*. \tag{38b}$$

then the operator Q is necessarily self-adjoint. This means that for any two functions $\sum \alpha_\kappa a_\kappa$ and $\sum \beta_\lambda a_\lambda$ we have

$$\left(\sum \alpha_\kappa a_\kappa, Q \sum \beta_\lambda a_\lambda\right) = \sum_\kappa \alpha_\kappa^* (a_\kappa, \sum \beta_\lambda Q a_\lambda) = \sum_{\kappa\lambda} \alpha_\kappa^* \beta_\lambda (a_\kappa, q_\lambda a_\lambda)$$

$$= \sum_{\kappa\lambda} \alpha_\kappa^* \beta_\lambda q_\lambda \delta_{\kappa\lambda} = \sum_\kappa \alpha_\kappa^* \beta_\kappa q_\kappa \tag{38c}$$

and one obtains the same expression for $(Q \sum \alpha_\kappa a_\kappa, \sum \beta_\lambda a_\lambda)$. This is satisfactory. It may be worth observing that all the preceding calculations seem to imply a discrete set of outcomes of the measurement—the last proof implies that the measured quantity Q has a discrete spectrum. This is both realistic and essentially unavoidable: the outcome of a measurement is restricted to a discrete set; in the case of the measurement of a continuous quantity, such as the momentum, the finite accuracy of the measuring device still guarantees that the different outcomes realized form only a discrete set, which in practice is even finite. If one does not want to accept this argument, it is ncessary to rewrite the preceding discussion, essentially in terms of Stieltjes integrals, and this is quite possible, though, in the opinion of this writer, unnecessary.

This is on the favorable side. One has to admit, on the other hand, that (35) is a highly idealized description of the measurement. It does not specify the duration of the measuring process. In fact, most writers, including von Neumann, at least imply that the transition indicated by the arrow in (35) is instantaneous. Of course, even if one accepts this idealization, unless the quantity to be measured commutes with the Hamiltonian of the system, its value will change after the measurement, and this applies also for the microscopic, that is, quantum-mechanical, description of the apparatus. However, these changes in the system and in the apparatus do not introduce basic problems. Thus, if the σ_κ are orthogonal to each other immediately after the measurement, orthogonality will continue also after the measurement as long as the system remains isolated.

The fact that the measurement is of finite duration introduces a more serious problem. If the operator of the quantity which is being measured does not commute with the Hamiltonian of the system, as is the case, for instance, when position is measured, it will change in the course of the measurement. To which position at which time does the measurement then refer? This issue is unclear and is rarely discussed. The existence of this issue reemphasizes that the quantum-mechanical description of the measurement, embodied in (34) and (35), is a highly idealized description—unless, as was mentioned before, the quantity to be measured is stationary.

In view of all these reservations, it is worthwhile to give a practical example of a measurement (Wigner, 1963). The example most often given is the measurement of the operator s_z—that is, the component of the spin of a particle in a fixed direction. This is a quantity which, for the free particle, is stationary; it commutes with the Hamiltonian. The measurement is called the Stern-Gerlach experiment and was discussed above in a cursory fashion. A particle with spin is passed through an inhomogeneous magnetic field. It is easiest to discuss the case of total spin $\frac{1}{2}$. When the magnetic moment of the particle points in the direction of increase of the field, the particle experiences a force—and is deflected—in the direction of decrease of the field, and conversely. The incoming beam is split into two beams. In one, the spin variable in the field direction has one value, and in the other beam, the other value. Hence, a statistical correlation is established between the position of the particle and its spin direction. The "a" of (34) or (35) is, in this case, the position of the particle. The two σ_κ are the two states of the spin, one parallel, the other antiparallel to the magnetic field. The right side of (35) resembles an expression such as

$$\alpha_+ e^{-x^2-y^2-(z-c)^2}\delta(s_z, \tfrac{1}{2}) + \alpha_- e^{-x^2-y^2-(z+c)^2}\delta(s_z, -\tfrac{1}{2}). \tag{39}$$

Of course, even if the beam is split, one does not yet know what the spin of the particle is: (39) is still a consequence of the quantum-mechanical equation of motion. In order to have a particle with a definite spin direction one must make an additional measurement, determining whether it is in one beam or the other—whether its z coordinate is positive or negative. This is the process indicated in equations (37a) and (37b)—and it is much more difficult to describe quantum mechanically because the quantum-mechanical description of the position measurement is not at all unique. Nevertheless, this example is instructive, and the expression for the state vector after the original measurements, indicated in (39), is a relatively easy consequence of the quantum-mechanical equations of motion. This is not the case for the measurements of most quantities—including the position of a particle.

Conceptual Problems of the Measurement Description

It was pointed out at the end of §3—and also sometime ago, particularly by von Neumann—that the description given is incomplete. Even if we accept the validity of the equations postulated, (34) and (35), we can account only for the establishment of a statistical correlation between the states of the object and those of the apparatus, not for the fact that the measurement gives one result once, another result another time—i.e., not for the statistical nature of the outcome of the measurement process. As was also pointed out earlier, this is hardly surprising because the deterministic nature of the equations of motion prevents them from accounting for a probabilistic result. The present section will deal with the problem resulting from this circumstance.

Naturally, one way out of this difficulty would be to postulate that the equations are not fully correct, or, at least, that they do not fully describe the actual situation. This has often been suggested, by various writers, and this possibility will be discussed in some detail later—on the whole with a rather negative, but not completely negative result. A more natural explanation would be that the statistical nature of the measurement outcome arises because the initial state of the apparatus—of a macroscopic apparatus—is not unique. Some initial states of the apparatus give one, others another result, and the statistical nature of the measurement outcome is due to the statistical nature of the initial state of the apparatus. Of course this proposed explanation would not apply to all processes of measurement. It does not apply to the Stern-Gerlach experiment insofar as it leads to a state like that given by (39). However, the final process of observation always involves some macroscopic apparatus, and the statistical explantion could be imagined to apply to that phase of the measurement. We shall see, however, that this proposed way out is also unacceptable. If the outcome of the measurement obeys the statistics postulated by quantum-mechanical theory, this correctness of prediction cannot be explained by the statistical nature of the initial state of the apparatus if we continue to believe in the unlimited validity of the present quantum-mechanical equations.

Let us consider, for the sake of simplicity, a measurement which can have, as in the Stern-Gerlach experiment discussed before, only two outcomes. For σ_+ it will have the outcome 1, for σ_- the outcome -1, for $\alpha\sigma_+ + \beta\sigma_-$ (with $|\alpha|^2 + |\beta|^2 = 1$) the outcome 1 with probability $|\alpha|^2$, the outcome -1 with probability $|\beta|^2$. The possible initial states of the apparatus will be denoted by a_0', a_0'', a_0''', and so on. If the object's initial state is, for instance, $(\sigma_+ + \sigma_-)/\sqrt{2}$, we would expect on this view that half of the states a_0', a_0'', $a_0''' \cdots$ will give the result 1, half of them -1.

This expectation, however, is at odds with the linear nature of the equations. We are supposing that each of the apparatus states a_0 gives, with σ_+, the result 1, with σ_- the result -1. This supposition means that

$$a_0 \otimes \sigma_+ \rightarrow a_+ \otimes \sigma_+ \quad \text{and} \quad a_0 \otimes \sigma_- \rightarrow a_- \otimes \sigma_- \tag{40}$$

or, in other words, that a_+ shows a result $+1$, and a_- the result -1. It now follows from the linear nature of the interaction operator that

$$a_0 \otimes (\sigma_+ + \sigma_-)/\sqrt{2} \rightarrow a_+ \otimes \sigma_+/\sqrt{2} + a_- \otimes \sigma_-/\sqrt{2}; \tag{40a}$$

that is, that none of the a_0', a_0'', $a_0''' \ldots$ can give definitely $a = +1$ or definitely $a = -1$ as a result. We believe that each apparatus will give the correct result whenever the state of the object is definitely σ_+, and also whenever the state of the object is definitely σ_-. But we see from (40a) that as long as we maintain this belief, we cannot blame the statistical nature of the outcome of the measurement for the state $(\sigma_+ + \sigma_-)/\sqrt{2}$ on the uncertain initial state of the apparatus. Thus, if (40) holds, none of the states a_0 can give either purely $a_+ \otimes \sigma_+$ or purely $a_- \otimes \sigma_-$. All give the linear combination (40a) of these two states.

This situation suggests a drastic reformulation of the basic concepts of quantum mechanics. It appears that the statistical nature of the outcome of a measurement is a basic postulate, that the function of quantum mechanics is not to describe some "reality," whatever this term means, but only to furnish statistical correlations between subsequent observations. This assessment reduces the state vector to a calculational tool, an important and useful tool, but not a representation of "reality." The statistical correlations can be calculated with the aid of the state vector. If the first observation tells us that the state vector is σ_κ, we can calculate the probability of the outcome λ of an experiment the characteristic functions of the operator of which are $b_1, b_2 \ldots$. This probability is $|(a_\kappa, b_\lambda)|^2$. The probability that the next measurement of the quantity the characteristic functions of which are

$c_1, c_2 \ldots$ will give the result μ is then $|(b_\lambda, c_\mu)|^2$ and the probability of both outcomes b_λ and c_μ is

$$|(a_\kappa, b_\lambda)(b_\lambda, c_\mu)|^2 = (a_\kappa, b_\lambda)(b_\lambda, c_\mu)(c_\mu, b_\lambda)(b_\lambda, a_\kappa). \tag{41}$$

The generalization to more measurements is obvious (Houtappel, Van Dam, and Wigner, 1963).

It is of some interest that (41) can be given a more concise form. If the state vectors in (41) can be written as functions of a variable x, then (41) can be given the form,

$$\int \cdots \int a_\kappa(x)^* b_\lambda(x) b_\lambda(x')^* c_\mu(x') c_\mu(x'')^* \cdot b_\lambda(x'') b_\lambda(x''')^* a_\kappa(x''') dx \cdots dx'''. \tag{41a}$$

However, $b_\lambda(x) b_\lambda(x'')^*$ is the kernel of the projection operator P defined in (16) and equations (17) for the operator A. Hence, if we make $a_\kappa(x''')$ of (41a) the first factor, the factors $a_\kappa(x''') a_\kappa(x)^*$, $b_\lambda(x) b_\lambda(x'')^*$, and so on, are the kernels of the projection operators P_κ for the first quantity measured, P'_λ for the second one, P''_μ for the third, and again P'_λ for the second. The integrations over x, x', x'' give the product of these projection operators, and the integration over x''' the trace of the product. Hence, the first measurement having given the outcome κ, the probability that the second one gives λ, the third one μ etc. becomes

$$\frac{Trace\ P_\kappa P'_\lambda P''_\mu P'''_\nu \cdots P'''_\nu P''_\mu P'_\lambda}{Trace\ P_\kappa}, \tag{42}$$

where P, P', $P'' \ldots$ are the proper projection operators for the first, second, third, etc. measurement, respectively. The *Trace* P_κ in the denominator appears because the characteristic value κ of the operator of the first measurement may not be simple—in which case one has to divide with the multiplicity of this characteristic value. It may be worth remarking that one can add a last factor P_κ in the numerator of (42) to make it more symmetric. This does not change its value because, with $P_\kappa^2 = P_\kappa$, one can insert another factor P_κ on the left side of the expression the trace of which appears in the numerator. This extra factor P_κ can then be shifted to the right end of the numerator since, quite generally, *Trace* $AB = Trace\ BA$ or, in our case, *Trace* $P_\kappa Q = Trace\ QP_\kappa$ where Q is the expression in the numerator of (42). The insertion of the P_κ factor on the right therefore does not change the value of (42); it only makes it appear more symmetric.

The preceding derivation of (42) is incomplete, because the possibility of multiple characteristic values of the measured operators is not explicitly taken into account. This has little significance for (35)—it does not matter whether or not the same "pointer position" corresponds to several a_κ. As far as the primed P are concerned (that is, as far as the measurements after the initial one are concerned), the possibility of multiple characteristic values can easily be taken into account and the result is correctly represented by (42). However, (42) also contains the assumption that, if P_κ corresponds to a multiple characteristic value, that is, if $Trace\ P_\kappa \geq 2$, the state produced by this initial measurement contains the different states of this characteristic value with equal probabilities. This is an assumption which cannot be fully justified.

It should be observed also that if the later measurements, those to which the primed P correspond, take place at later times, the corresponding P must be modified so that they are in the "Heisenberg picture." This means that if the measurement to which $P_v^{(n)}$ corresponds takes place at time t_n after the first measurement, the $P_v^{(n)}$ in (42) can be obtained from the $P_v^{(n)}$ applicable if the measurement n had taken place immediately after the first one by the formula,

$$P_v^{(n)}(t_n) = \exp(-iHt_n/\hbar)P_v^{(n)}(0)\exp(iHt_n/\hbar). \tag{43}$$

(Naturally $t_1 \leq t_2 \leq t_3 \ldots$.)

The preceding interpretation of the quantum-mechanical formalism reflects its concern with observations, and with giving probabilities for the outcomes of these observations. It does not eliminate the difficulty which the finite length of the measurement time creates, the difficulty mentioned at the end of the preceding section. Nor does it alter the fact that we have not specified how the measurements are to be carried out. A discussion of limitations on the possibility of measuring certain operators will be given in the next chapter. In spite of all these reservations, it remains essentially correct to say that the basic statement of quantum mechanics can be given in a formula as simple as (42).

§5. OTHER PROPOSED RESOLUTIONS OF THE MEASUREMENT PARADOX

The measurement paradox referred to in the title of this section is the contradiction between the deterministic nature of the quantum-mechanical equations of motion and the probabilistic outcome of the measurements—processes which should be describable by the quantum-mechanical equations of motion. Section 4 above proposed a resolution of the paradox: the quantum-mechanical equations of motion do not describe the measurement process; they only help in the calculation of the probabilities of the different outcomes. These probabilities form the real content of quantum-mechanical theory. The formalism of state vectors, equations of motion, etc., are only means to calculate these probabilities. The observation results are the true "reality" which underlie quantum mechanics. The state vector does not represent "reality." It is a calculational tool. It should be mentioned that von Neumann's idea was not truly different from this. He postulated that the state vector varies in two different ways. As long as the system is isolated, its state vector is subject to the quantum-mechanical equation of motion and its behavior is deterministic. When an observation takes place, there is a second type of change of the state vector. Its change then has a probabilistic nature. It jumps discontinuously. It becomes one of the characteristic vectors of the operator which is being measured. If the initial state vector (normalized) was ϕ, it jumps with the probability $|(\psi_v, \phi)|^2$ into the state ψ_v which is one of the (normalized) characteristic vectors of the operator which is being measured. Naturally, the sum of these probabilities must be 1 and this is the consequence of the normalized nature of ϕ. The second type of change of the state vector, the jump from ϕ into one of the ψ_v, according to von Neumann's picture, is not described by the quantum-mechanical equations of motion.

The other resolutions of the measurement paradox propose more "physical" pictures. The most popular of these is the picture of hidden parameters—embraced particularly by D. Bohm (1952a,b), Y. Aharonov (Bohm and Aharonov, 1957), J. Bub (1969), (see also Bohm and Bub, 1966a,b), J. P. Vigier (1951; 1956), and Bohm and Vigier (1954), but also by many others. The story of this picture is given in some detail by F. Belinfante (1973) as well as by M. Jammer in his book, *The Philosophy of Quantum Mechanics* (Jammer, 1974) which is worth reading for other reasons—it encompasses an amazing amount of information on the history of our general subject. The other attempt at the resolution of the paradox is due to H. Everett (1957), B. S. DeWitt (1970), and J. A. Wheeler (1957) (see also DeWitt and Graham, 1973). This is the "relative state" theory, postulating that, as a result of an observation, the world splits into several new worlds, existing independently of each other. Both these pictures will be discussed in more detail, and arguments against them will be presented. It should perhaps be added now that J. A. Wheeler (1977) no longer supports this view.

Theory of "Hidden Variables"

The idea of "hidden variables" postulates that the description of states, by the quantum-mechanical state vector, is incomplete; that there is a more detailed description, by means of variables now "hidden," which would be complete and the knowledge of which would permit one to foresee the actual outcomes of observations—observations about whose outcomes present-day quantum mechanics makes only probabilistic statements. The relation of the postulated theory of hidden variables to present quantum mechanics would be similar to the relation of classical microscopic physics to macroscopic physics. The former uses the positions and velocities of the atoms as variables, while macroscopic physics, such as, for instance, hydrodynamics, describes only the average velocities of the atoms situated in volumes which are large on the microscopic scale and contain many atoms.

There is no clear specification of the nature of the "hidden variables"—they are hidden. The best known replacement for Schrödinger's equation is Bohm's

(1952a,b), in which he reverses Schrödinger's analysis. Schrödinger derived the wave equation named after him by viewing the classical Hamilton-Jacobi equation as giving an incomplete and approximate description of this wave: incomplete, in the sense that it deals only with the phase, $S(x, y, z, t)/\hbar$, of this wave, $\psi(x\ y, z, t)$; approximate, in the sense that the equation for the Hamilton-Jacobi function S is nonlinear, whereas by demanding linearity Schrödinger got the right equation for ψ. Bohm turns this reasoning around. He assumes that Schrödinger's equation for

$$\psi = Re^{-iS/\hbar} \tag{44}$$

is valid (R and S being real and functions of the position coordinates) and obtains equations for R and S. The equation for S,

$$\frac{\partial S}{\partial t} = \sum_j \frac{1}{2m_j}\left[\left(\frac{\partial S}{\partial x_j}\right)^2 + \left(\frac{\partial S}{\partial y_j}\right)^2 + \left(\frac{\partial S}{\partial z_j}\right)^2\right] + V + Q, \tag{44a}$$

differs from the classical Hamilton-Jacobi equation by containing an extra "quantum term" Q. All the other terms are obtained from the Hamilton equation for the energy, $E = H(x, p)$, by replacing E by $-\partial S/\partial t$ and p_{x_j} by $\partial S/\partial x_j$, etc. The additional term, Q, is the "quantum potential,"

$$Q = -\sum_j \frac{\hbar^2}{2m_jR}\left(\frac{\partial^2}{\partial x_j^2} + \frac{\partial^2}{\partial y_j^2} + \frac{\partial^2}{\partial z_j^2}\right)R, \tag{44b}$$

in the equation for S. The equation for R is the classical one:

$$\frac{\partial R}{\partial t} = \sum_j \frac{1}{m_j}\left(\frac{\partial R}{\partial x_j}\frac{\partial S}{\partial x_j} + \frac{\partial R}{\partial y_j}\frac{\partial S}{\partial y_j} + \frac{\partial R}{\partial z_j}\frac{\partial S}{\partial z_j}\right) + \frac{R}{2m_j}\left(\frac{\partial^2 S}{\partial x_j^2} + \frac{\partial^2 S}{\partial y_j^2} + \frac{\partial^2 S}{\partial z_j^2}\right). \tag{44c}$$

One can obtain these equations by introducing (44) into Schrödinger's time-dependent equation for ψ and separating real and imaginary parts.

The interpretation of these equations from the point of view of hidden variables theory is not so simple. It seems to be agreed that R^2 gives, as in the quantum interpretation, the probability of the configuration indicated by its variables, i.e., that it refers to an ensemble. For S, the classical interpretation is postulated, but it is difficult to understand then how the properties of the ensemble, described by R^2, can influence the motion of an individual system, the system which S is supposed to describe. Yet Q depends on R and R describes the *ensemble* containing the system the motion of which should be described by S. Is it that a system's behavior is different depending on the set of systems of which it may be a part?

This dependence of the individual on the ensemble is an objection against a specific theory of hidden variables, not a general objection against all such theories; i.e., it is not an argument against the existence of a deterministic theory of the motion of atomic objects and constituents. Von Neumann was convinced that no such deterministic theory is compatible with quantum mechanics. The reason is easily illustrated by the Stern-Gerlach experiment, or rather, by an indefinite number of repetitions of that experiment. One may consider the measurement of the spin component first in the z direction, then in the x direction, then again in the z direction, then again in the x direction, and so on. If the total spin of the particle on which the measurements are undertaken is $\frac{1}{2}$, all measurements succeeding the first one give the two possible results with a probability $\frac{1}{2}$. If these results are, fundamentally, all determined by the initial values of the hidden parameters, the outcome of each measurement should give some information on the initial values of these parameters. Eventually, it would seem, the values of the "hidden parameters" which determine the outcomes of the first N measurements would be in such a narrow range that they would determine, if N is large enough, the outcomes of all later measurements. Yet this is in contradiction to the quantum-mechanical prediction.

The preceding argument can be made more convincing by substituting other measurements, such as a position and momentum measurements, for the measurements of spin direction. Yet this argument apparently cannot be made mathematically rigorous and it was not published by von Neumann. The proof he published (see p. 173 of von Neumann, 1932; pp. 326–28 in English translation), though it was made much more convincing later on by Kochen and Specker (1967), still uses assumptions which, in my opinion, can quite reasonably be questioned.

Bell's Argument

In my opinion, the most convincing argument against the theory of hidden variables was presented by J. S. Bell (1964). His argument, in its simplest form, starts with a system of two particles with spins $\frac{1}{2}$, these spins being antiparallel, i.e., forming a singlet state. The spin part of the state vector is

$$\sigma_+(1)\sigma_-(2) - \sigma_-(1)\sigma_+(2). \tag{45}$$

Here $\sigma_+(1)$ is the state vector of particle 1, with positive component in the z direction. The meaning of the other symbols should then be obvious. Actually, the state (45) is spherically symmetric. It is the only antisymmetric combination possible for the two spin functions. Thus the z direction mentioned above can be replaced by any other direction without changing the value of (45) (as can be verified, of course, by actual calculation). The state (45) is called the spin singlet state. The spin parts of the state vectors of the two electrons of the He atom, or of the H_2 molecule, are actually in that state. In fact, the spin state remains essentially unchanged even if the electrons are torn away from the atom or molecule by dipole radiation, so that their spins remain, to a very good approximation, in the state (45) even when they are widely separated from each other, as is demanded for some applications of Bell's argument. Even for that case, (45) represents a state of the two spins which is not only theoretically possible but also, to a good approximation, experimentally realizable.

For the state (45), there is, according to quantum mechanics, a statistical correlation for the measurement of the spin components in two directions, e_1 and e_2, which enclose an angle θ_{12}. The probability that both components are positive is

$$P_{++} = (\tfrac{1}{2}) \sin^2 (\theta_{12}/2) = P_{--} \tag{46}$$

and this is also the probability that the measurement of the spin components of the two particles, in the e_1 and e_2 directions respectively, will have a negative outcome for both. The probability that one component will be positive the other negative, or for the converse, is

$$P_{+-} = P_{-+} = (\tfrac{1}{2}) \cos^2 (\theta_{12}/2). \tag{46a}$$

If $\theta_{12} = 0$, i.e., if the two directions are parallel, the probability (46) of both having the same component is zero. This vanishing of P_{++} is what makes the aggregate spin 0, i.e., makes the total state a singlet. The expressions (46) and (46a) for the probabilities will not be proved explicitly. They are contained, at least implicitly, in the usual textbooks on quantum mechanics.

Bell's inequality, based on the assumption that the "hidden variables" of the two particles uniquely determine the outcomes of the measurements of spin components in all directions, will be shown to be in conflict with equations (46) and (46a). Actually, it suffices to consider three directions, e_1, e_2, e_3, and measurements of the spin components in these directions. Let us denote $(+ - + ; - - +)$ the probability-weighted integral over the values of the hidden variables, an integral taken over that domain of these variables which ensures that the first particle's spin has a positive component in the e_1 and e_3 directions, and a negative component in the e_2 direction. while the second particle's spin has a negative component in the e_1 and e_2 directions, and a positive component in the e_3 direction. The meaning of the other combinations of six + and − signs is similar; thus $(+ + + ; - - -)$ gives the probability that the spin component of the first particle is positive in all three directions, that of the second negative. Since the hidden variables are supposed to determine completely the properties of the objects to which they refer, all these quantities are fully defined and are, surely, non-negative. It appears that there are $2^6 = 64$ of them.

It can be noted, next, that most of the 64 symbols are 0. Thus $(+ + - ; + - +)$ must vanish because it corresponds to such values of the hidden variables as give, for both particles, a positive component of the spin as measured in the e_1 direction. Because of (46), the probability for this spin configuration vanishes. That is, no value of the hidden variables produces a positive spin component for both particles in the e_1 direction, at least not if quantum mechanics, and hence (46), properly gives the probabilities of the outcomes of measurements. Similarly, if quantum mechanics is correct, all symbols vanish in which the same sign appears at the same position both before and after the semicolon. This then leaves 8 symbols such as $(+--;-++)$ which can have non-zero values.

Next, we write down the probabilities for those outcomes of the measurements which indicate a positive spin component for both particles if these are measured in two different directions, and equate these with the corresponding quantum-mechanical expressions. First, if the spin components are measured in the e_1 and e_2 directions, respectively, the probability of finding positive components is, according to the theory of hidden variables (the first sign before, and the second sign after the semicolon must be $+$),

$$(+-+;-+-) + (+--;-++) = (\tfrac{1}{2})\sin^2(\theta_{12}/2). \tag{47}$$

All other bracket symbols vanish in which the first sign before the semicolon and the second sign after the semicolon are $+$. Similarly, if the measurements are made in the e_2 and e_3 directions, respectively, we have

$$(++-;--+) + (-+-;+-+) = (\tfrac{1}{2})\sin^2(\theta_{23}/2). \tag{47a}$$

Likewise we have

$$(++-;--+) + (+--;-++) = (\tfrac{1}{2})\sin^2(\theta_{13}/2). \tag{47b}$$

It follows from equations (47) that

$$\sin^2(\theta_{12}/2) + \sin^2(\theta_{23}/2) = \sin^2(\theta_{13}/2) + 2(+-+;-+-) + 2(-+-;+-+)$$
$$\geq \sin^2(\theta_{13}/2). \tag{48}$$

This is the Bell inequality. It should be fulfilled if the theory of hidden variables is correct, and if, in addition, quantum mechanics correctly predicts—as it apparently does—the probabilities of all conceivable outcomes of a measurement.

It is easy to find directions e_1, e_2, e_3, however, for which (48) is not valid. It is not valid, in particular, if the three directions are in the same plane and e_2 is between e_1 and e_3. Specifically, if $\theta_{12} = \theta_{23} = \pi/3$, $\theta_{13} = 2\pi/3$, then the left side of (48) is $\tfrac{1}{4} + \tfrac{1}{4} = \tfrac{1}{2}$, the right side $\tfrac{3}{4}$. Hence, the hidden variable theory, as applied here, leads to a contradiction.

There are then two questions. First, is (46) correct, i.e., can it be confirmed experimentally? This will be discussed later. Second, can one modify the theory of hidden variables in such a way that Bell's inequality (48) does not follow? The answer to this second question is "yes"; but the modified theory implies the use of hidden variables which specify not only the state of the two particles but also the states of the measuring devices, and postulates correlations between the state of the particles and the directions in which the spin is going to be measured. The theory that assumes that there are no such correlations is often called "local" hidden variable theory. The preceding argument shows that any theory of hidden variables conforming with the postulate of locality is in conflict with quantum mechanics. On the other hand, if one admits hidden variables which establish correlations between the measuring devices and the objects, i.e., describe the states of both together, one cannot see the limit of the complex that has to be described jointly. It seems highly questionable whether a theory which does not permit the specification of the states of isolated objects can be in any way useful.

Of course, if future experiments were to demonstrate that Bell's inequalities, in particular (48), are correct and the quantum-mechanical equations, in particular (46) and (46a), incorrect, this finding would constitute strong evidence in favor of some theory of hidden variables. As was mentioned before, the present status of the experimental research will be discussed later. As of 1981 it appears to be definitely established that what is called "Bell's inequality" is strongly violated, whereas the quantum-mechanical predictions appear to be well supported.

Many-World Theories

"Many-world" theories are much more difficult to discuss than theories of hidden variables. They postulate—as mentioned earlier—that if a measurement with a probabilistic outcome is undertaken, the world splits into several worlds, and each possible outcome of the observations appears in the fraction of the new worlds given by the quantum-mechanical probability of that outcome. This re-establishes the determinism which the laws of nature are expected to exhibit.

It is, of course, difficult to see the meaning of the statement that there are other worlds with which we never will have any contact, which have no influence on us, and which we cannot influence or perceive in any way. From a positivistic point of view, the statement that there are such worlds, and that they are constantly created in large numbers, is entirely meaningless. It can be neither confirmed nor refuted.

Let us admit, however, in conclusion, that the weakness of the theories which have been proposed to replace the standard interpretation of quantum mechanics, and which form the subject of most of these notes, does not establish the full validity of quantum theory. The weaknesses of quantum theory, though not as marked as the deficiencies of the theories discussed in the present chapter, are real nevertheless. They will be discussed subsequently.

§6. Experimental Tests of Bell's Inequality

The conflict between quantum theory and the theory of hidden variables, as manifested by Bell's inequality, was discussed in the last section. The original version of the present section was kindly provided by S. J. Freedman, then at Princeton University. It dealt, principally, with the experimental work then available and aimed at deciding which of the two formulations—Bell's inequality or the conclusions of quantum mechanics—was correct. Since that time (1976) the experimental information has been greatly extended, and also some errors in the earlier work have been detected. As a result, it would not be reasonable to repeat Freedman's analysis fully. Instead we refer the reader to Pipkin's (1978) summary of the experimental findings. These findings disagree with Bell's inequality and seem to agree with the consequences of quantum theory. This is a very important point.

In spite of the fact that the more recent experimental results are of great relevance and are, naturally, not contained in Freedman's original analysis, his discussion will be presented below, though in much abbreviated form, as being of some historical interest. It is also good to realize that experimental results can be in error. It is even better to realize that these errors can be corrected.

Freedman also presented some remarks on three general, largely theoretical, questions on our subject. (1) He characterized the structure of a general hidden variable model. One of his "hidden variables" was the wave function (and, in many cases, this also is hidden). But he introduced a set of other variables which *always* remain hidden and which are supposed to give a complete description of the state of the system. (2) He presented a special model in order to prove that it is conceivable that hidden variables could be found which would determine the future of the system completely. He admitted, though, at this point, that in order to eliminate the difficulty (for hidden variables!) of the invalidity of Bell's inequality one has to abandon the postulate of the local character of the hidden variables. These must be so constituted, he said, that they establish statistical correlations between the state of the object on which the measurement will be undertaken, and the state of the apparatus which will carry out the measurement. These correlations, he supposed, exist before the measurement takes place. (3) Freedman also presented a modification of Bell's inequality which is more easily and more directly subject to an experimental test than is the original inequality.

Perhaps the present writer will be permitted to voice his opinion on theories of hidden variables of the type which were crudely described under (2) and which are often proposed. In my opinion, it is not very meaningful to introduce variables, the magnitude of which cannot be determined. These variables could be defined as giving the state of the system for the entire future. Such a description would require the introduction of one additional variable, most naturally denoted by t.

to describe the system. However, it is the purpose of physics to give information on the future based on facts ascertainable at present. The unqualified existence of "hidden variables" would nullify this purpose.

Let us go over to a short description of the experimental information available at the time Freedman made his remarks. Actually, the first two experiments were not carried out with spin-$\frac{1}{2}$ particles in the singlet state of (45) but with two photons emitted in succession by an atom. Their polarizations then give correlations similar to those of the two spin-$\frac{1}{2}$ particles discussed in the preceding chapter.

The first set of experiments was carried out on Ca by Freedman in collaboration with Clauser (1972). The result was a correlation factor of 0.300 ± 0.009. This agrees with the quantum-mechanical value of 0.301 but contradicts Bell's inequality which postulates a value smaller than 0.25.

The next set of experiments was done by Holt and Pipkin (1974) with isotopically pure Hg^{198} (zero nuclear spin). Their result was 0.216 ± 0.013. This agrees with Bell's inequality but differs grossly from the quantum-mechanical value of 0.301. Perhaps fortunately, it was demonstrated not long after this section was originally written that an experimental error had crept into the original value (see Clauser, 1976).

Still other experiments deal with the decay of positronium, in a ground state of spin zero, into two photons. Experiments on the correlation of the polarization of these two photons were originally suggested by Wheeler (1946). The early experiments, carried out before 1970, gave excellent agreement with quantum mechanics (Wu and Shaknov, 1950; Kasday, Ullman, and Wu, 1970). Their results did not really contradict Bell's inequality, but it is hard to believe that the laws of quantum mechanics are violated in any area under consideration if they are so well obeyed in the region investigated. But more recent work disagrees with these results, referring to conditions in which quantum mechanics and Bell's inequality are in contradiction (Faraci, Gutkowski, Notarrigo, and Pennisi, 1974). This creates a confusing situation.

The last experiment referred to by Freedman related to the spin-spin correlation when a low energy proton is scattered by a proton at rest. Low energy scattering is dominated by proton pairs of zero orbital momentum. The space part of the wave function is therefore symmetric on interchange of the two particles. Consequently the spin part must be antisymmetric, of the form (45), corresponding to zero total spin. The original experiments did not seem to confirm quantum mechanics, but the later, improved ones, did, and were in contradiction with Bell's inequality (Lamehi-Rachti and Mittig, 1976).

This will conclude the somewhat abbreviated review of the experimental material available in 1975 on the validity of Bell's inequality and hence on the possibility of explaining the statistical nature of the outcomes of measurements with hidden variables of a local and reasonable nature. As was mentioned before, the subject was treated in more detail in the original version of these lecture notes as given by Freedman. As was also mentioned, more recent experimental data definitely deny the possibility that Bell's inequality is generally valid, and hence rule out any possibility of eliminating the statistical nature of the measurement process by the introduction of local hidden variables.

The next section, the last one, will deal with problems of the standard interpretation. That interpretation assumes that the principle of determinism does not apply to the measurement process. It also shows weaknesses. Nevertheless, no current experimental information contradicts it. Its principal weakness is that it gives no clear and simple rule for the way the measurement can be carried out nor for the limits of the accuracy of the measurement process. The preceding discussion was confined to measuring the spin component or the state of polarization, both of which clearly appear to be determinable with high accuracy.

§7. Problems of the Standard Interpretation: Unmeasurable Quantities

Most physicists working on or with quantum mechanics were quite surprised that the experimental confirmation of some of its simple consequences, such as the violation of Bell's inequality, was as ambiguous as described in the preceding sections. Most of us are convinced, nevertheless, that the consequences of quantum theory, which were tested in the experiments described, will be unambiguously confirmed by further and perhaps more precise experiments.

The present section will be devoted to a discussion of internal problems of the standard interpretation as discussed in §3. Let us admit that these problems do mar the mathematical beauty of the theory by demonstrating the difficulty of making measurements and the unavoidable limitations on the accuracy of most of them. These difficulties and limitations indicate, as do the remarks of §4, that quantum mechanics shares a degree of incompleteness with all other theories of physics.

It is a reassuring feature of quantum theory, however, that the earliest problem of measurement—a problem which generated a great deal of discussion—was not really a problem. The apparent problem was to construct the proper quantum-mechanical operator to correspond to a given classical expression. For example, what is the operator which corresponds to the classical expression xp? Surely xp, where p stands for $(\hbar/i)\partial/\partial x$, has to be rejected because it is not self-adjoint. It is natural to choose, instead,

$$xp \rightarrow \frac{1}{2}(xp + px) = \frac{1}{2}\frac{\hbar}{i}\left(x\frac{\partial}{\partial x} + \frac{\partial}{\partial x}x\right) = \frac{\hbar}{i}\left(x\frac{\partial}{\partial x} + \frac{1}{2}\right). \qquad (49)$$

However, in somewhat more complicated cases, such as x^2p^2, the choice is not unique. Two possible choices are $\frac{1}{2}(x^2p^2 + p^2x^2)$ and xp^2x (p here stands again for $(\hbar/i)\partial/\partial x$), both of which are self-adjoint but not equal to each other. It should perhaps be mentioned that Weyl (1927) has proposed a definite operator to correspond to any classical function of position and momentum. But as long as no experimental prescription is provided for the measurement of that operator, the meaning of the Weyl prescription is not clear.

The quandary just described has been resolved by considering the operator to be defined by the quantity which is measured. Thus one does not ask for the operator which corresponds to the classical expression x^2p^2 but asks for ways in which either $\frac{1}{2}(x^2p^2 + p^2x^2)$ or xp^2x ($= px^2p$) can be measured. We will be concerned henceforth with questions of this nature. Unfortunately, as we shall see, there are serious limitations on the measurability of an arbitrary quantity. They blur the mathematical elegance of von Neumann's original postulate that all self-adjoint operators are measurable. Von Neumann does produce an expression for the interaction between object and apparatus which would lead to equation (34) of our §3. However, the ensuing considerations show that for many if not most operators, this expression—or any other expression which might lead to that equation—contradicts some of the basic principles of quantum theory. What then are the limitations of measurability?

Only Quantities Which Commute with All Additive Conserved Quantities Are Precisely Measurable

This theorem, dating back to 1952 (Wigner, 1952; see also Araki and Yanase, 1960; Yanase, 1961), will not be proved generally. Only a characteristic example will be given. The quantity to be measured is the component of the spin in the x direction. The "additive conserved quantity" will be the angular momentum in the z direction. By "additive" we mean that the magnitude of the quantity for object-plus-apparatus is the sum of this quantity for object and for apparatus, both before and after the measurement.

In the case considered, the argument is very simple. Let us denote the spin states with positive and negative z-component by α and β respectively. The state vector associated with the positive x-component spin is then $2^{-1/2} (\alpha + \beta)$, and that of the negative one, $2^{-1/2} (\alpha - \beta)$. Hence equation (34) of §3 reads in this case

$$a \otimes (\alpha + \beta) \to a^+ \otimes (\alpha + \beta), \tag{50a}$$

$$a \otimes (\alpha - \beta) \to a^- \otimes (\alpha - \beta). \tag{50b}$$

Let us decompose the apparatus states a, a^+, and a^- into state vectors each having a definite angular momentum in the z direction,

$$a = \sum a_m; \qquad a^+ = \sum a_m^+; \qquad a^- = \sum a_m^-. \tag{51}$$

It then follows from the addition and subtraction of equations (51) that

II.2 INTERPRETATION   299

$$2 \sum a_m \otimes \alpha \rightarrow \sum a_m^+ \otimes (\alpha + \beta) + \sum a_m^- \otimes (\alpha - \beta), \tag{51a}$$

$$2 \sum a_m \otimes \beta \rightarrow \sum a_m^+ \otimes (\alpha + \beta) - \sum a_m^- \otimes (\alpha - \beta). \tag{51b}$$

It simplifies these equations to introduce the symbols

$$2b_m = a_m^+ + a_m^- \quad \text{and} \quad 2c_m = a_m^+ - a_m^-. \tag{52}$$

Hence (51a) and (51b) become

$$\sum a_m \otimes \alpha \rightarrow \sum (b_m \otimes \alpha + c_m \otimes \beta), \tag{52a}$$

$$\sum a_m \otimes \beta \rightarrow \sum (b_m \otimes \beta + c_m \otimes \alpha). \tag{52b}$$

The angular momentum of $a_m \otimes \alpha$ in the z direction is $m + \frac{1}{2}$, that of $a_m \otimes \beta$ is $m - \frac{1}{2}$. Since the angular momentum in the z direction (or any other direction) does not change as a result of the interaction, it follows that the interaction of the state a_m alone with α will yield

$$a_m \otimes \alpha \rightarrow b_m \otimes \alpha + c_{m+1} \otimes \beta, \tag{53a}$$

and of a_m alone with β will give

$$a_m \otimes \beta \rightarrow b_m \otimes \beta + c_{m-1} \otimes \alpha. \tag{53b}$$

Now the lengths of the vectors on the left and right sides of (53a) must be equal. Moreover, the two terms on the right side are orthogonal. Therefore we have

$$(a_m, a_m) = (b_m, b_m) + (c_{m+1}, c_{m+1}); \tag{54a}$$

and similarly, from (53b)

$$(a_m, a_m) = (b_m, b_m) + (c_{m-1}, c_{m-1}). \tag{54b}$$

Hence, we have

$$(c_{m+1}, c_{m+1}) = (c_{m-1}, c_{m-1}), \tag{54}$$

which is a contradiction. Thus (54) means either (1) that the c_m are all 0—in which case the $a_m^+ = a_m^-$ so that $a^+ = a^-$ whereas they should be orthogonal—or (2) that

$$(c, c) = \sum (c_m, c_m) \tag{55}$$

is infinite unless all (c_m, c_m) vanish. But this is impossible. Thus from (52) $(c, c) = \frac{1}{2}$ because a^+ and a^- are of the same unit length and are orthogonal. However, (55) already indicates how this difficulty will be overcome. We will recognize that (52a) and (52b) can be only approximately valid (see equations 56a and 56b). In other words, all (c_m, c_m) will be very small and they will not be absolutely equal (Gaussian dependence on m, with a very large spread in m-values). This means that the measurement is not absolutely accurate, as (52a) and (52b) would imply. However, the magnitude of the inaccuracy, i.e., the magnitude of the terms η in (56a) and (56b), will be minimized.

The conclusion thus arrived at seems to deny the possibility of measuring the spin component precisely in an arbitrary direction, as described in §3. However, two points have to be remembered. First, the measuring equipment uses an external magnetic field. If that field is considered to be "external," it invalidates the law of conservation of angular momentum. The only way to remedy this difficulty is to consider the magnet as part of the apparatus—as it is. The thus enlarged apparatus becomes quite macroscopic. Second, as equation (39) indicates, the measurement is not perfect. The two beams, with positive and negative s_z, overlap to a certain extent. These two points suggest first, that the measurement of a quantity not commuting with additive conserved quantities requires a large apparatus and, second, that the limitations on the possible accuracy of the measurement can decrease with increasing "size" of the apparatus. "Size," we shall see, means the amount of the additive conserved quantity that is contained in the apparatus.

Instead of (50a) and (50b) we now write

$$a \otimes (\alpha + \beta) \to a^+ \otimes (\alpha + \beta) + \eta^+ \otimes (\alpha - \beta), \qquad (56a)$$

$$a \otimes (\alpha - \beta) \to a^- \otimes (\alpha - \beta) + \eta^- \otimes (\alpha + \beta). \qquad (56b)$$

Here the terms η^+ and η^- express the error in the measurement. We will try to make η^+ and η^- as small as possible. We write, similar to (51),

$$\eta^+ = \sum \eta_m^+, \qquad \eta^- = \sum \eta_m^- \qquad (57)$$

and, in analogy to (52), we set

$$\eta_m^+ + \eta_m^- = 2\sigma_m \quad \text{and} \quad \eta_m^+ - \eta_m^- = 2\delta_m. \qquad (58)$$

Hence, (56a) and (56b) give as a result of the conservation of the z angular momentum the relations

$$a_m \otimes \alpha \to (b_m + \sigma_m) \otimes \alpha + (c_{m+1} - \delta_{m+1}) \otimes \beta, \qquad (58a)$$

$$a_m \otimes \beta \to (b_m - \sigma_m) \otimes \beta + (c_{m-1} + \delta_{m-1}) \otimes \alpha. \qquad (58b)$$

These are the analogs of (52a) and (52b). The arrow here, as there, denotes a unitary transformation. Can this requirement be satisfied? Previously, using only part of the unitary restrictions, we arrived at the contradiction (54). Now we have to take into account all the consequences of unitarity.

The left sides of (58a) and (58b) are orthogonal to each other and orthogonal to all similar expressions with other m-values. Their unitary transforms must likewise be orthogonal. This they are automatically when they have different values of the conserved quantity, "total z angular momentum." When the left sides, say $a_m \otimes \alpha$ and $a_{m+1} \otimes \beta$, have the same value for this quantity, orthogonality of the right sides gives

$$(b_m + \sigma_m, c_m + \delta_m) + (c_{m+1} - \delta_{m+1}, b_{m+1} - \sigma_{m+1}) = 0. \qquad (59)$$

Since the lengths of the a_m vectors are free, the only further relation imposed by the unitarity condition is that the lengths of the vectors on the right sides of (58a) and (58b) be equal. This gives, again for all m,

$$(a_m, a_m) = (b_m + \sigma_m, b_m + \sigma_m) + (c_{m+1} - \delta_{m+1}, c_{m+1} - \delta_{m+1}), \qquad (60a)$$

$$(a_m, a_m) = (b_m - \sigma_m, b_m - \sigma_m) + (c_{m-1} + \delta_{m-1}, c_{m-1} + \delta_{m-1}). \qquad (60b)$$

As when (54) was derived, so here we take the difference of these two equations. The (b_m, b_m) term drops out to give

$$\begin{aligned}
(c_{m+1}, c_{m+1}) &- (c_{m-1}, c_{m-1}) \\
&= 2Re[(c_{m+1}, \delta_{m+1}) + (c_{m-1}, \delta_{m-1}) - 2(b_m, \sigma_m) \\
&\quad + (\delta_{m-1}, \delta_{m-1}) - (\delta_{m+1}, \delta_{m+1})].
\end{aligned} \qquad (60)$$

In the case of an ideal measurement, the left side vanished; we had $\sigma = \delta = 0$; and we encountered a contradiction. The sums of (60a) and (60b) for definite m-values need not be considered; they only give the probabilities (a_m, a_m) of the specified angular momenta of the measuring device in terms of the other quantities that appear in the state vector of the system after the measurement has taken place. The sum of these probabilities for all m-values, however, does give the normalization condition for the apparatus. Because $(\alpha, \alpha) = (\beta, \beta) = 1$, this condition becomes

$$\begin{aligned}
\sum (a_m, a_m) = \sum \{ &(b_m, b_m) + (\sigma_m, \sigma_m) + \tfrac{1}{2}(\delta_{m-1}, \delta_{m-1}) + \tfrac{1}{2}(\delta_{m+1}, \delta_{m+1}) \\
&+ \tfrac{1}{2}(c_{m+1}, c_{m+1}) + \tfrac{1}{2}(c_{m-1}, c_{m-1}) + Re[(c_{m-1}, \delta_{m-1}) - (c_{m+1}, \delta_{m+1})] \} = 1.
\end{aligned} \qquad (61)$$

The second member of (61) gives (a_m, a_m) as the average of the two expressions given by (60). We have a final condition. We require that at least the first terms on the right sides of (50a) and (50b) should represent different states—i.e., that $(a^+, a^-) = 0$. In terms of the b_m and c_m this condition says,

$$\sum_m (b_m + c_m, b_m - c_m) = 0. \tag{62}$$

Equations (59), (60) and (61), (62) represent the conditions which b, c, σ, and δ must satisfy in order to guarantee the unitary nature of the transformation indicated by (56a) and (56b). In equations (61) and (62) there is summation over m. They give the normalization condition and the requirement that the transformations (56a) and (56b) represent a measurement at least in the approximate sense.

Next, we define the error ε implicit in the measurement process indicated by these expressions,

$$\varepsilon = \sum \left[(\eta_m^+, \eta_m^+) + (\eta_m^-, \eta_m^-) \right] = 2 \sum \left[(\sigma_m, \sigma_m) + (\delta_m, \delta_m) \right]. \tag{63}$$

The "size," M, of the apparatus which permits an error so small will be defined by

$$\begin{aligned} M^2 &= \sum m^2 (a_m, a_m) \\ &= \tfrac{1}{2} \sum m^2 [2(b_m, b_m) + (c_{m+1}, c_{m+1}) + (c_{m-1}, c_{m-1}) \\ &\quad + 2Re(c_{m-1}, \delta_{m-1}) - 2Re(c_{m+1}, \delta_{m+1}) + (\delta_{m+1}, \delta_{m+1}) + (\delta_{m-1}, \delta_{m-1})]. \end{aligned} \tag{64}$$

Again, the average of the right sides of (60a) and (60b) was substituted for (a_m, a_m). The problem then is to find expressions for b_m, c_m, σ_m, and δ_m satisfying equations (59) to (62) which, for a given error ε, make M as small as possible; or, equivalently, for a given "size" of the apparatus, M, make the error ε as small as possible.

This minimum problem will not be solved exactly but only under the assumption that the error ε is small as compared with 1—i.e., that the σ_m and δ_m are small as compared with b_m and c_m. This means that (b_m, b_m) and (c_m, c_m) are appreciable for a rather wide range of m-values. Therefore they will be assumed to depend continuously on m. Even though this idealization is not mathematically rigorous, we adopt it. We rewrite in terms of this continuum approximation all of the equations from (59) to (64) except for (60), which only gives (a_m, a_m) in terms of the other quantities:

$$(b_m, c_m) = 0; \tag{59a}$$

$$2 \frac{d}{dm} (c_m, c_m) = 2Re[(c_m, \delta_m) - (b_m, \sigma_m)]; \tag{60c}$$

$$\sum [(b_m, b_m) + (c_m, c_m)] = 1;$$ (61a)

$$\sum (b_m, b_m) = \sum (c_m, c_m), \qquad Im \sum (b_m, c_m) = 0;$$ (62a)

$$\varepsilon = 2 \sum [(\sigma_m, \sigma_m) + (\delta_m, \delta_m)];$$ (63a)

$$M^2 = \sum m^2 [(b_m, b_m) + (c_m, c_m)].$$ (64a)

It now follows that the Hilbert space vectors δ_m and σ_m are best assumed to be parallel to c_m and b_m, respectively, and indeed the proportionality constant between them is real. A part of δ_m which would be orthogonal to c_m would not change any of the equations, except by adding to the error ε. The same applies to σ_m with respect to b_m. The same argument shows that the proportionality factor between δ_m and c_m, and between σ_m and b_m, has to be real. Next, we assume that the lengths of b_m and c_m, and of δ_m and σ_m, are equal, but of course, wherever (c_m, δ_m) is positive, (b_m, σ_m) is negative, and conversely. Otherwise the derivative of (c_m, c_m) would become 0 and we would face the same difficulty as in the case of the idealized measurement (cf. equation 54). The equalities assumed can be justified, but this will not be done here. Capitalizing on these observations, and writing $c(m)$ for the length of c_m and $c(m)\delta(m)$ for (c_m, δ_m), we can restate the preceding set of equations:

$$4c(m) \frac{dc(m)}{dm} = 4c(m)\delta(m);$$ (60d)

$$\int c(m)^2 \, dm = \tfrac{1}{2};$$ (61b)

$$\varepsilon = 4 \int \sigma(m) \, dm;$$ (63b)

$$M^2 = 2 \int m^2 c(m)^2 \, dm.$$ (64b)

The other equations are automatically satisfied—except that the vectors b_m and c_m must be assumed to be orthogonal.

It follows from (60a) that $\delta(m)$ is the derivative of $c(m)$, so that the error becomes

$$\varepsilon = 4 \int \left(\frac{dc(m)}{dm} \right)^2 dm.$$ (63c)

The problem, therefore, reduces either to minimizing the size M (or its square) given by (64b) while fixing the error ε of (63c) and the normalization (61b)—or to minimizing the error but fixing the size and taking the normalization into account. The Lagrange equation of the minimum problem reduces in both cases to a linear relation between $c(m)$, $m^2 c(m)$, and $d^2 c(m)/dm^2$—i.e., the quantum-mechanical equation of the oscillator. The solution is, therefore, that $c(m)$ is a Gauss error curve

$$c(m) = \alpha e^{-(\beta/2)m^2}, \tag{65}$$

the constants of which are determined by (61b) and either (63c) or (64b). The relation between error ε and size naturally is the same no matter which procedure is used:

$$\varepsilon = 1/(2M^2). \tag{66}$$

This is an interesting result, due in this form to Araki and Yanase (1960). It shows that most measurements cannot be made mathematically precise. Moreover, the more accuracy we demand, the greater must be the "size" (equation 64b) of the measuring apparatus.

Are there any quantities where the preceding size-accuracy correlation does not apply, quantities which, as far as this argument goes, may be measured precisely with a small apparatus? For elementary particles, only the mass and the magnitude of the spin are such precisely measurable quantities; none of the quantities specifying the actual state is measurable with arbitrary precision. If one has a more complex system, the preceding quantities do depend on the state because the degree of excitation is determined by the Minkowski length of the four-dimensional momentum vector (the rest mass). In addition, if the system contains several particles, the scalar product of the momentum vectors and many other similar quantities do commute with all additive conserved quantities. Nevertheless, it is clear that the limitation on the measurability of many quantities with a small apparatus is very severe (Yanase, 1961).

The preceding discussion, leading to equation (66), is incomplete—and this not only from the point of view of mathematical rigor. We considered only a system of spin $\frac{1}{2}$ and only one conservation law, conservation of the angular momentum in one of the directions perpendicular to that in which the spin component is to be measured. It is true that the components of the linear momentum do commute with the spin components; but the other components of the angular momentum do not. It would be interesting to generalize the result (66) both as to the system considered, and as to the inclusion of all additive conservation laws. It should also be repeated that (66) represents only a necessary condition for the apparatus to measure the spin component, and even such a generalization as is suggested above would only give a *necessary* condition. To demonstrate the actual measurability, one would have to describe the measuring apparatus and its functioning—as was done to some degree in §3 with respect to the measurement here considered, but not generally. In fact we shall derive in what follows several other constraints on measurability.

It has been observed that, in order to define a position coordinate, one has to have a coordinate system with a definite position in space and equipped with a clock with well-defined zero of time. The former requires that the apparatus which defines the coordinate system have a spread in momentum which corresponds to the position-momentum uncertainty relation, and a similar remark, to be discussed later, applies to the clock. Hence, in the case of the measurement of the position coordinate, the need to have an apparatus with a large momentum uncertainty, that is, momentum spread, is obvious. The same applies for many other measurements, but not with the generality established in the preceding discussion. In particular, the argument in this section also applies to the conservation laws for electric charge, for baryon number, etc., and already shows the difficulty of measuring any operator not commuting with these quantities. This last point will be taken up again when the superselection rules are discussed.

The Boson-Fermion Superselection Rule

The preceding section dealt with difficulties inherent in the measurement of a great many, if not most, operators, but the difficulties considered there could be overcome to an increasing extent by an increase in the "size" of the measuring equipment. The superselection rules absolutely exclude the measurability of certain operators. This exclusion is demonstrated in the first case, the boson-fermion superselection rule, on the basis of the postulate of the rotational invariance of the theory. The demonstration in the second case, the charge, baryon-number, etc. superselection rules, is more intricate and touches on a deeper philosophical problem.

The boson-fermion superselection rule tells us that no operator is measurable which has a finite matrix element between two states, one of which has an integer, the other a half-integer, angular momentum. The former states are associated with bosons, the latter with fermions but, of course, in the sense here considered, a hydrogen atom is a boson—the half-integer spins of the electron and of the proton combine to an integer spin. If an operator with a finite matrix element is measured, and the measurement results in characteristic vectors of the operator being produced, some of these characteristic vectors will have components with both integer and half integer spin: $b + f$.

From here on, the proof of the inconsistency with rotational invariance can proceed in many ways. A rather natural way to carry out the demonstration starts from the decompositions of b and f in terms of (1) total angular momentum and (2) z-component of the angular momentum, as viewed from some arbitrary coordinate system:

$$b = \sum_{Jm} b_m^{Jm}, \tag{67}$$

$$f = \sum_{Jm} f_m^{Jm}. \tag{67a}$$

Naturally, the J and m in (67) are integers, in (67a) half integers—this is the definition of b and f. The index m appears in (67) and (67a) twice in order to remind us that the terms in (67) or (67a) are not partners in the sense of invariance theory. Rather, the effect of a rotation R on these is given by the formulae

306   WIGNER

$$O_R b_m^{Jm} = \sum_{m'} D^{(J)}(R)_{m'm} b_{m'}^{Jm} \qquad (68)$$

and the same formula applies to the f_m^{Jm}, the b_m^{Jm} being the "partners" of b_m^{Jm}, the f_m^{Jm} partners of f_m^{Jm}. If the rotation R is applied on $b_{m'}^{Jm}$, we have

$$O_R b_{m'}^{Jm} = \sum_{m''} D^{(J)}(R)_{m''m'} b_{m''}^{Jm}, \qquad (69)$$

and a similar formula applies to the rotation of the f_m^{Jm}. The $D^{(J)}(R)$ are the matrices of the representations of the rotation group in our three-dimensional space (Wigner, 1959).

It now follows that the state vector of the state which looks like $b + f$ from the point of view of a coordinate system obtained from the original one by a rotation R is

$$O_R(b + f) = \sum_{Jm} O_R b_m^{Jm} + O_R f_m^{Jm}$$

$$= \sum_{Jm} \sum_{m'} D^{(J)}(R)_{m'm} b_{m'}^{Jm} + \omega_R \sum_{Jm} \sum_{m'} D^{(J)}(R)_{m'm} f_{m'}^{Jm}. \qquad (70)$$

The ω_R, with absolute value 1, had to be introduced in (70) because the effect of a transformation is always indefinite within a factor of absolute value 1— the state determines the state vector only up to such a factor. Actually, the first term of (70) should also be provided with such a factor. However, one factor of this nature can be eliminated from all such equations. One has only to postulate that the state vector be one of the state vectors representing the state in question. The others can be obtained from it by multiplication with an arbitrary phase factor, i.e., a factor of modulus 1. Actually, the omission of this phase factor simplifies the calculation relatively little.

It may be well to observe at this point that if the operator projecting into the state $b + f$ is observable—and we have assumed it is—then the operator projecting into $O_R(b + f)$ is likewise, by means of the same apparatus rotated by R. We shall take for R, first a rotation by π about z, to be denoted by Z. We then have, using the usual formulae for $D^{(J)}$ (remembering, though, that the m of b is an integer, that of f a half integer),

$$O_Z(b + f) = \sum_{Jm} i^{2m}(b_m^{Jm} + \omega_Z f_m^{Jm}) \qquad (71)$$

and

$$O_Z^2(b + f) = \sum_{Jm} i^{4m}[b_m^{Jm} + \omega_Z(\omega_Z f_m^{Jm})]. \qquad (71a)$$

We recall that O_Z^2 is the unit operation. Therefore (71a) should differ only by a factor from $b + f$. In the first term m is an integer and the factor in question indeed has the value unity. Hence $i^{4m}\omega_Z^2$ must also be 1 for the half integer m of the second term. It follows that $\omega_Z^2 = -1$, or

$$\omega_Z = \pm i. \tag{71b}$$

Let us observe that, in order to derive this result, it was necessary to assume that neither b nor f is a null-vector.

We consider next a rotation by π about the y axis. The standard formulae give

$$O_Y(b + f) = \sum_{Jm} i^{2J - 2m}(b_{-m}^{Jm} + \omega_Y f_{-m}^{Jm}). \tag{72}$$

Calculation of $O_Y^2(b + f)$ yields this time:

$$O_Y^2(b + f) = \sum_{Jm} i^{2J - 2m} O_Y(b_{-m}^{Jm} + \omega_Y f_{-m}^{Jm})$$

$$= \sum_{Jm} i^{2J - 2m} i^{2J + 2m}(b_m^{Jm} + \omega_Y^2 f_m^{Jm}). \tag{72a}$$

The fact that the second ω_Z factor in (71a) and the second ω_Y factor in (72a) must be the same as the first follows from the unitary nature of the transformations O_Z and O_Y. Otherwise the scalar products $[b + f, O_Z(b + f)]$ and $[O_Z(b + f), O_Z^2(b + f)]$ would not be equal. A similar remark applies to O_Y. Now $O_Y^2(b + f)$ can again differ from $b + f$ only in a factor. Moreover, the first term shows that this factor is 1. Also we have $i^{4J} = -1$ for the J of the second term. Therefore we arrive at a result similar to (71b),

$$\omega_Y = \pm i. \tag{72b}$$

A similar calculation yields

$$\omega_X = \pm i, \tag{73}$$

as was to be expected. However, the product of rotations by π about z and by π about y is a rotation by π about x. Therefore the vector $O_Y O_Z(b + f)$ can differ from $O_X(b + f)$ only by a factor. Contrary to this requirement we find

$$O_Y O_Z(b + f) = O_Y \sum_{Jm} i^{2m}(b_m^{Jm} + \omega_Z f_m^{Jm})$$

$$= \sum_{Jm} i^{2m} i^{2J - 2m}(b_{-m}^{Jm} + \omega_Y \omega_Z f_{-m}^{Jm}). \tag{74}$$

Comparison of this expression with that for $O_X(b + f)$ shows that the factor in the first term is 1. Comparison of the second terms then shows that

$$\omega_X = \omega_Y\omega_Z, \tag{74a}$$

which is in contradiction to (71b), (72b), and (73). This disagreement shows that the existence of a state $b + f$, with finite b and finite f, is incompatible with the principle of rotational invariance. As was mentioned above, this "boson-fermion superselection rule" can be established in many different ways, each of which, however, relies to some degree on the theory of rotational invariance, in particular on the difference between the $D^{(J)}$ which apply to bosons and to fermions.

Superselection Rules for Charge, etc.

The creation of a state $b + f$ with both b (boson integer spin) and f (fermion, half-integer spin) finite, or, in other words, "any violation of the boson-fermion superselection rule," would conflict, as we have just seen, with the postulate of the rotational invariance of the theory, provided that the basic ideas of quantum mechanics are valid. This deduction may be a blemish on the very general principles of quantum theory, but it does not affect any of its practical applications or conceivable experimental conclusions.

According to the "charge superselection rule," it is also impossible to produce states

$$\sum c_n, \tag{75}$$

with c_n representing a state with electric charge number n and more than one of the vectors c_n finite. According to the "baryon superselection rule," the same prohibition applies to the baryon number. In contrast to the situation with the state vectors $b + f$, no conflict with any of the fundamental principles has been derived from the possibility of producing a state with the state vector (75). Rather, the validity of the somewhat controversial superselection rules for charge, etc. is based on the physical impossibility of producing states such as (75). We do not know of any macroscopic object that would be in a state such as (75). Even if we do not know the exact electric charge of an object—and in the case of a macroscopic object this is difficult to know—we cannot distinguish between (75) and other states, such as

$$\sum e^{i\phi_n}c_n, \tag{75a}$$

the exponentials in (75a) being phase factors of modulus 1. In the mathematical terminology of quantum mechanics this ambiguity is expressed by the statement that the actual state is not a superposition of the states c_n but a mixture of them. The state then cannot be characterized by a state vector. The system can be in any of the states c_n. The probability that a measurement of the charge will give the value n is (c_n, c_n). The theory of mixtures has been given an elegant mathematical form, but it will not be elaborated here.

State vectors with different charges do not interact with each other. Thus the time development of every c_n of (75) is the same as if it were present alone. Therefore the mixture character of the state remains preserved in time. Similarly, if two systems are united to form a joint system, and if each of the two is a mixture of states with definite charge numbers, the same will be true of the united system. It follows from this reasoning that unless nature supplies some superposition of different charge states, such as (75), no such superposition state will ever be found. Then the superselection rule for charge will be valid. The same applies to the superselection rules for baryon number, etc.

The law of conservation of electric charge, the independence of the time development of the states with different charges, and our observation on the union of two systems, remain equally valid if "linear momentum" is substituted for "electric charge." It is natural to ask, therefore, why no superselection rule applies for the components of linear momentum or angular momentum. How is it that one can produce a state which is a superposition of states with different linear momenta? And how is it that one can ascertain the coefficients $e^{i\phi_n}$ in the expression corresponding to (75a) with n now referring to a component of linear momentum (n a continuous variable) or to a component of the angular momentum? The answer to this question, it must be admitted, cannot be read out of the basic equations of quantum mechanics. We really do not know how we acquired the ability to see light signals and to feel objects, and why there are no similar phenomena in connection with electric charges. Superconductivity is often claimed to provide such signals. Indeed, the usual theory does use a description of the superconducting state which is a definite superposition of states with different electric charges. However, it is evident that the conclusions of the theory, describing currents, magnetic fields, and similar observable phenomena, would not change in any way if the different charge states were multiplied with different phase factors as they are in (75a). None of the observables used has matrix elements connecting states with different charges. Only the finiteness of such matrix elements could permit one to distinguish (75a) from (75). As far as linear momentum is concerned, the situation is entirely different. Any position operator, for instance, has major matrix elements between states with different momenta.

The question naturally arises whether and how phase relations between different charge states could manifest themselves. Surely the fact that the observed phenomena of superconductivity do not prove the existence of such phase relations does not prove their absence. Naturally, the existence of such phase relations, that is, the breakdown of the charge superselection rule, would be very surprising. Present quantum mechanics would do as little to explain such a breakdown as to explain how the states of different energy in a thermodynamic ensemble could manifest phase relations. Nevertheless, it may be interesting to find an experimental criterion to tell whether such phase relations exist—that is, to tell whether superpositions such as (75) or (75a) with definite ψ_n are distinguishable.

The simplest experiment would consist in the uniting of two systems. Let each system be imagined to be in a superposition of states with the charges n and m. If these are superpositions such as (75) or (75a), their state vectors could be denoted by

$$\psi_n + e^{i\phi}\psi_m \quad \text{and} \quad \psi'_n + e^{i\phi'}\psi'_m, \tag{76}$$

with definite ϕ and ϕ'. Their union would then have the state vector

$$\psi_n \otimes \psi'_n + \left[e^{i\phi'}\psi_n \otimes \psi'_m + e^{i\phi}\psi_m \otimes \psi'_n\right] + e^{i\phi + i\phi'}\psi_m \otimes \psi'_m. \tag{76a}$$

This formula indicates that the component with charge $n + m$, given in the brackets of (76a), would have a definite structure and might well be distinguished by conceivable experiments from a mixture of the two states $\psi_n \otimes \psi'_m$ and $\psi_m \otimes \psi'_n$. The factor $\exp[i(\phi - \phi')]$ measuring the difference in phase between the two states of equal charge would be definite. There is no reason to believe that that factor would be unobservable. Needless to say, (76) and hence (76a) could be greatly generalized and, equally obvious, the experiment alluded to here is far from concrete. Moreover, it may be impossible to carry it out.

The Measurement of Position

As was mentioned at the beginning of this section, the early interest in the correspondence of a quantum-mechanical operator to classical expressions, such as x^2p^2, has largely ceased. Nevertheless, there are classical quantities, such as linear and angular momentum, energy, and position, to which one would like to coordinate an operator. It appears, at least superficially, that these quantities have a simple meaning, and can be measured, and that it should be possible to find the proper operator for them.

Whether easily measurable or not, the coordination of operators to energy, linear and angular momentum components (and the three other quantities associated with them in relativistic theory) can easily be made. The operators are the infinitesimal operators of the (special) relativistic invariance group, also called the Poincaré group. Position is different. Quite apart from the difficulties of measuring position soon to be discussed, we find great problems in coordinating an operator to position. These difficulties have been most elegantly demonstrated by Hegerfeldt (1974) (see also Pryce, 1948; Møller, 1972; Wightman, 1962; and Fleming, 1965). The following discussion is based on his remark. It is more technical than the rest of these notes, being based on the theory of the Poincaré invariance of quantum mechanics, and may be, for some, difficult to follow.

In order to simplify the demonstration, we shall use a spacetime with only one spacelike dimension. The generalization to three spacelike dimensions is not difficult but would make the discussion a good deal longer. The spirit of the demonstration is not impaired by the restriction to one spacelike dimension.

Let us assume that there is a position operator and consider its projection operators $P(x)$ as defined, for an arbitrary operator, by equations (21) of §2 ("all values up to x"). Let us then produce out of some initial state ϕ, by the operator $P(x_2) - P(x_1)$ (with $x_2 > x_1$), a state ψ for which the position is confined to the x_1, x_2 interval. We express the state vector of such a state in the momentum representation, as is usual in the quantum-mechanical invariance theory. Thus we write $\psi = \psi(p, \sigma)$, where p is the momentum vector (one-dimensional for our world with one spacelike dimension) and σ is a discrete variable, characterizing the other variables, such as spin component in one direction, etc. Because of the confinement of the position to the x_1, x_2 interval, that is, because of

$$\psi(p, \sigma) = [P(x_2) - P(x_1)]\phi,$$

we have

$$P(x)\psi(p, \sigma) = 0 \qquad \text{for all} \quad x \leq x_1, \tag{77}$$

and

$$P(x)\psi(p, \sigma) = \psi(p, \sigma) \qquad \text{for all} \quad x \geq x_2. \tag{77a}$$

The scalar product of two vectors ϕ and ϕ', expressed in the (p, σ)-representation, is

$$(\phi, \phi') = \sum_\sigma \int \phi(p, \sigma)^* \phi'(p, \sigma) \, dp/p_0. \tag{78}$$

Here cp_0 is the energy associated with the momentum p,

$$p_0 = (m^2 c^2 + p^2)^{1/2}. \tag{78a}$$

The p_0 in the denominator of (78) will play no significant role. It is introduced in the usual definition for the scalar product because it simplifies the expressions for Lorentz transformations. We shall not make use of general transformations but only of displacements in space and time. The operator for a displacement in space by a and displacement in time by t is simply multiplication by

$$e^{-ip_0t + ipa}. \tag{79}$$

If $|a| > x_2 - x_1$, a space displacement by a moves the original confinement into a new, nonoverlapping one so that $e^{ipa}\psi(p, \sigma)$ will be orthogonal to $\psi(p, \sigma)$,

$$(e^{ipa}\psi(p, \sigma), \psi(p, \sigma)) = \int e^{-ipa} \sum_{\sigma} |\psi(p, \sigma)|^2 \, dp/p_0 = 0, \tag{80}$$

for $|a| > x_2 - x_1$. This can be proved also more formally by noting that

$$P(x - a)e^{ipa} = e^{ipa}P(x) \tag{79a}$$

and making use of (21c) of §2. From the validity of (79), and the confinement of x for $\psi(p, \sigma)$ to an interval of the width $x_2 - x_1$, (80) should be evident anyway.

Equation (80) expresses the fact that the Fourier transform of

$$\sum_{\sigma} \psi(p, \sigma)^*\psi(p, \sigma)/p_0 \tag{80a}$$

is confined to a region of width $2(x_2 - x_1)$. It then follows from the expression of (80a) in terms of its Fourier transform—an integral over a region of width $2(x_2 - x_1)$—that the expression (80a) is a holomorphic function of p; that is, it has no singularity in the finite complex plane. The scalar product of a space displaced $\psi(p, \sigma)$ and a time displaced $\psi(p, \sigma)$,

$$(e^{ipa}\psi(p, \sigma), e^{-ip_0t}\psi(p, \sigma)) = \int e^{-ipa-ip_0t} \sum_{\sigma} |\psi(p, \sigma)|^2 \, dp/p_0, \tag{81}$$

evidently is the Fourier coefficient of the function

$$e^{-ip_0t} \sum_{\sigma} \psi(p, \sigma)^*\psi(p, \sigma)/p_0. \tag{81a}$$

This function, for finite t, is not a holomorphic function of p. To be sure, (80a) *is* such a function; but the p_0 in the exponent has singularities at $\pm imc$. Hence, there is no finite interval of a outside of which its Fourier transform, (81), will vanish. This means that the time displacement spreads the position, originally confined to the interval x_1, x_2, over all space. Otherwise there could be no finite transition probability to functions $e^{ipa}\psi(p, \sigma)$ for arbitrarily large a, since the latter state is confined to the interval $x_1 - a$, $x_2 - a$. No matter how one defines the position, one has to conclude that the velocity, defined as the ratio of two subsequent position measurements divided by the time interval between them, has a finite probability of assuming an arbitrarily large value, exceeding c. One either has to accept this, or deny the possibility of measuring the position precisely or even giving significance to this concept: a very difficult choice! And the author of these notes will admit that since writing them, he has devoted a good deal of attention to the problem (Ahmad and Wigner, 1975; O'Connell and Wigner, 1977, 1978).

II.2 INTERPRETATION 313

Summary

The present section starts with the observation that the measured quantities correspond to (self-adjoint) operators, not to the quantities of classical physics. It then points to the demonstrable difficulties of certain measurements, i.e., where it can be shown that it is difficult to find an apparatus the interaction of which with the object has the result indicated by the basic equation (34) of §3.

It is shown, first, that no precise measurement of any quantity is possible unless this commutes with all conserved additive quantities. In some cases this conclusion can be derived, at least qualitatively, with ease. In other cases, such as operators which do not commute with the operator of electric charge, the situation is less obvious. It is shown then that a measurement which has a small probability of giving an incorrect result surely needs a large apparatus, i.e., one in a state in which the additive conserved quantities, not commuting with the quantity to be measured, have probabilities for values spread over a wide spectrum.

The second type of case discussed refers to quantities which cannot be measured at all. Examples are, first, operators which have a finite matrix element between a state with integer and another state with half-integer angular momentum. The so-called boson-fermion superselection rule shows that it is impossible to produce states which are superpositions of states with integer and half-integer angular momenta. A similar statement is then made for the superposition of states with different electric charges and also with different baryon numbers, and perhaps other similar descriptors.

Finally, we had to recognize, every attempt to provide a precise definition of a position coordinate stands in direct contradiction with special relativity.

All these are concrete and clearly demonstrated limitations on the measurability of operators. They should not obscure the other, perhaps even more fundamental weakness of the standard theory, that it postulates the measurability of operators but does not give directions as to how the measurement should be carried out. This problem has already been emphasized at the end of §3, the description of the quantum theory of measurement.

All the preceding discussion is based on the usual quantum-mechanical theory, as taught in classes, not on the more advanced axiomatic quantum field theory. The question therefore arises whether quantum field theory avoids the difficulties mentioned. Surely, as far as the superselection rules are concerned, it does not. As to the rest, my opinion is negative also. The best known discussion of the measurement of field strengths, that given by Bohr and Rosenfeld, postulates an electric test charge with arbitrarily large charge and arbitrarily small size (Bohr and Rosenfeld, 1933, 1950). Naturally, the fact that the quantum field theory does not resolve our problems, though regrettable, should not be considered as an argument against that theory.

314   WIGNER

Acknowledgments

The author would like to express his appreciation to all who helped him to write up his lectures, and this includes not only S. J. Freedman but several others who attended the lectures. He also wishes to record his gratitude to W. H. Zurek for reviewing the material and, even more, for assembling the references.

The Limitations of Determinism

E. P. Wigner

Absolute Values and the Creation of the New World (Proc. 11th ICUS, 1982).
International Cultural Foundation Press, New York 1983, pp. 1365–1370

It is a very great pleasure to talk to you this morning and I very much appreciate the introduction of Dr. Fukuda, even though, as you will see, in some minor points I disagree with him.

My problem is that physics is very much more technical than the discussions which we had yesterday and which I enjoyed very much. This is not surprising because physics is much better founded, it is much older, and I am afraid much more successful so far than psychology, or physiology, or sociology. As a result, I am afraid that some of the things I will tell you about will not be as easily understood as I could understand what we heard and enjoyed yesterday.

I am also afraid that much of what I'll say will be a little boring for the physicists among you, because I try to be on a level that even the non-physicists can understand. Let me start by saying that, in my opinion, present-day physics started about three hundred years ago and was best incorporated into the book of Isaac Newton, Principia Mathematica Philosophiae Naturalis. I would like to say, however, something that is not usually emphasized: that Isaac Newton's greatest accomplishment was, at least in my opinion, not the law of gravitational attraction, which is of course a wonderful thing, but a very fundamental and semiphilosophical proposition: the separation of initial conditions from the laws of nature. The laws of nature are, as Einstein put it, of great simplicity and mathematical beauty. In other words you can formulate them with a few words if you know a bit of mathematics, as you all know. Consider, for instance, the law of gravitation, that the gravitational attraction is proportional to the product of the two masses attracting each other and inversely proportional to the square of their distance. This law of nature can be very simply formulated and has, as Einstein said, mathematical beauty. The initial conditions, on the other hand, are as confused and arbitrary as possible. I don't think any law of physics will ever postulate the fact that there is one lady in the first row here, as this is, so to say, an initial condition. The actual initial conditions are, one can say, outside the area of physics. Perhaps I should also mention that the increase of entropy is just based on the fact that the initial conditions are as irregular as possible. The first question which I wish to mention is: the division of irregularity of initial conditions and great simplicity of the laws of nature, will this sharp separation of these two concepts be always with us? In the past I believe we all thought so and we didn't even really value this accomplishment of Newton because we took it for granted. But there is an Austrian philosopher, Ernst Mach, who said that all laws of physics are approximate and this has been, as I learned here recently, endorsed by Professor Geoffrey Chew at University of California.

Let me tell you now why I became interested in this subject. Quantum mechanics does not entirely adhere to this principle. The equations of quantum mechanics are deterministic, but the outcome of observations is subject to probability laws. The example I usually quote is the so-called Stern-Gerlach experiment in which a beam of electrons is split into two beams and into which one of the beams the electrons is to enter is not determined. In fact, if we believe quantum mechanics—and this has been confirmed experimentally— it is not right to claim that it enters into one of the two beams, because if the two beams are united they interfere with each other, even if there is only a single particle in the beam. But if I observe in which one it is, if I see a flash of light, either above or below on a suitably placed plate, then the outcome of the observation has a probabilistic nature—I see it either above or below and it is impossible to predict where I'll see it. Now of course, if I took quantum mechanics seriously, I would say that the person who observed the flash is also in a superposition of two states, having seen it above and having seen it below. But if I think about it really and if a friend of mine looks at the flash, then surely he sees it either above or below, so that it is clear that quantum mechanics does not apply to the observation of my friend, does not apply to life and, as I will point out, it has also limitations as far as all macroscopic objects are concerned. In fact this will be my principal point: the validity of quantum mechanics has limitations, it is valid only to microscopic systems and is not valid for macroscopic objects.

One more remark: The laws of physics, and the laws of quantum mechanics in particular, but in fact all the laws of physics, apply only to isolated systems. And this is very obvious. If I have a system and apply the equations of quantum mechanics or physics to it and suddenly there comes some interference from the outside, then my equations, which assume that the system remains isolated, will not take the effect of the outside system into account, and their result will not be valid.

The question then comes up, and this was very interestingly brought up by a German physicist D. Zeh, whether a macroscopic system in our particular world can be isolated from outside effects, whether it is possible to keep isolated. And he found that it cannot. Let me give an example for this, not the one which D. Zeh gave because he talks about a cubic centimeter of gas, but of course a cubic centimeter of gas cannot be isolated, because it has to be in a container and the container interacts with it. What I was considering instead was a cubic centimeter of tungsten in intergalactic space where we can *hope* it to be isolated. However, it is not isolated because there is cosmic radiation with a temperature of about three degrees absolute, and to my surprise, when I calculated how many light quanta there are in a cubic centimeter, it turned out that there are about five hundred. This means that the cubic centimeter is struck per second by about 10^{13} light quanta. Not all of these light quanta influence its state because there are all sorts of rules and all sorts of types of interactions. But if one calculates how long will its microscopic state, that is its internal vibrations, remain uninfluenced by the light quanta which strike the cube, it turns out that it is about a thousandth of a second. This means that microscopically, in the sense of quantum mechanics, the time dependent equation which gives its future microscopic state is valid only for a thousandth of a second. Of course, I should emphasize that this refers to its microscopic state which tells us which vibrations are excited, which vibrations are not excited, and how strongly they hang together. Macroscopically this is not observable, and of course, the whole idea of determinism comes from the observation of the macroscopic state. The macroscopic description is not influenced by a tiny vibration but the microscopic state is influenced. Well, this shows that the probabilistic phenomenon enters not only when a living being observes, as I believed some time ago, but already if any macroscopic system plays a role.

The question which now arises naturally is: can we write equations which describe this situation. One would say at first: no, all equations are deterministic, the time derivative of the description is given by an equation. But this is not so. There are quantities which obey definite equations but which describe probabilistic outcomes. These are equations which apply, and I am sorry to admit that this is quite technical, to the so-called density matrix. The density matrix on the whole describes a situation in which we do not know what the exact quantum mechanical description is, but it can be decomposed into probabilities of different states. Well perhaps I do deviate a little from what I should say and I describe a bit roughly, the equation which I am proposing. The density matrix is, of course, a square set of numbers and I propose to describe the rows and columns by several indices: the position of the system, its dipole moment, its quadrupole moment, and so on down the line. If this were a classical description, it would have only diagonal elements because in a classical description, the position for instance is determined and there is no interference between different positions. The same applies to the dipole moment, to the quadrupole moment, and so on. In classical description the silver atom is either up here or down there. In quantum description there can be a superposition of the two states so that if I bring them together, as I described before, it becomes evident that it was neither really up here nor really down there. But as a result of the interference with other objects, its state would approach the one given by the classical description. Hence the equations which I propose eliminate the off-diagonal elements, they eliminate them slowly. It eliminates fastest the off-diagonal elements of the position, a little less slowly the off-diagonal element of the dipole moment, even less fast the off-diagonal elements of the quadrupole moment and so on.

The equation which I propose in this way is clearly an approximate equation because the situation is certainly very different if I am down in intergalactic space where it takes a full thousandth of a second to make a change in the state and very different here where the influence of you gentlemen and of the materials around us is very much greater. So the equation really depends on the environment but just the same it is an equation.

All this has an important consequence for the problem which I mentioned to begin with and which induced my interest in the subject: the probabilistic outcome of the measurement. It shows that there is an indeterminism if the observing apparatus or the observing person, is macroscopic. And people are macroscopic. Hence, it is clear that, for the observation, or the so-called measurement process, the probabilistic nature of the equation plays an important role. Let me admit therefore that present-day physics, and in particular quantum mechanics, has a limited validity and it should be formulated in a somewhat probabilistic way, particularly so for a very macroscopic body and this is what I tried to do. I should admit now that my equation still applies only to non-living bodies.

But the equation may describe the behavior of the measuring apparatus. According to the present equations of quantum mechanics, if the outcome of the measurement can assume, for instance, two values, the position of the pointer will become indeterminate—it will be a superposition of the two positions which correspond to the two possible outcomes of the measurement. Such a superposition is not equivalent with the situation in which one of the two positions may be assumed by the pointer with the corresponding probabilities— just as, as was discussed before, the passage of a light quantum through two holes does not lead to the situation in which it would be either behind the first or behind the second hole—the two parts of the light quantum interfere with each other, it is, in a sense, behind both holes. But if the probability equation discussed above is valid, the position of the pointer will assume a definite value—after about a millionth of a second—so that the two positions cannot be brought to interference with each other—not even theoretically.

Having praised the probabilistic equation which I proposed, let me admit again that, as to consciousness, it does not give a description thereof. Consciousness is still terribly outside the area of present-day physics and also the ideas which were formulated even before quantum mechanics, such as dialetical materialism, have, in my opinion, no sense. And let me admit also that my idea is valid only if we accept an extremely positivistic point of view. But if we look back at the history of physics, the positivistic point of view appears to have been very successful and very generally accepted. You know probably that Lorentz who created the basic Lorentz transformations, the basis of the special relativity theory, didn't believe in the fact that the simultaneity of two events depends on the observer, which is a positivistic concept. Well, now every physicist accepts the special theory of relativity. In fact, most of us accept also the general theory and this also is an advocate of the positivistic philosophy. Let me say that positivistic philosophy means that we attribute reality only to what can be observed. And as I said, relativity theory was at one time also considered to be very positivistic.

What I am proposing is a further step, and possibly a dangerous step, in the direction of positivism by denying full causality, particularly for macroscopic bodies, because they cannot be isolated from outside influences. Of course, the whole universe is an "isolated system," it is under no outside influence, but we surely cannot know its state at any definite time so that its alleged causal behavior is meaningless. This means that absolute causality is meaningless and it is reasonable to consider a probabilistic description of the events around us.

The Nonrelativistic Nature of the Present Quantum Mechanical Measurement Theory

E. P. Wigner

Annals of the New York Academy of Sciences *480*, 1–5 (1986)

SOME PROBLEMS OF EARLIER THEORIES

I wish to recall first some problems of earlier theories of physics that are similar to the present ones of quantum mechanics. (These will be discussed later.) There was, after each great discovery in physics, some puzzlement and realization of incompleteness in the definition of some of the basic concepts in terms of which the "laws" were formulated. Let me begin with the first great discovery of Copernicus, Kepler, Tycho Brahe, and others, which took place about 450 years ago. Copernicus (1473–1543), afraid of the church's disapproval, said that it is "easier to describe" the motion of the planets by assuming that the sun is at rest, while the planets, including the Earth, are moving. However, even then, it was not clear what "at rest" and what "moving" meant—it was only clear that the description of their relative positions was easier to describe in a coordinate system attached to the sun. Of course, the motion of the planets was communicated to us only by the light they reflected, and the basic properties of light were not described by the physics of that time.

Their description of the planets' motion led to the wonderful theory of Newton. His description of the forces by which the sun attracts the planets was based on Galileo's observation on the laws of free-fall here on the Earth, and on the improved determination of the moon's distance from it, leading to an improved value of the moon's centripetal acceleration. However, the transmittal of the information about the motion of the planets, that is, light, was not part of that time's physics, and the specification of the basic coordinate system in which the postulated laws of motion are valid was left to the distant future—in fact to Einstein's general relativity. Still, the establishment of laws of nature and the separation of initial conditions from them were truly miraculous accomplishments that cannot be forgotten.

The next fundamental discovery, describing a new set of phenomena, was initiated by James C. Maxwell. His laws of nature abandoned the idea of the "action at a distance" of Newton gravitational forces and introduced a field theory. (For the gravitational forces, this was accomplished only much later by Einstein's general relativity theory.) His laws of nature, apparently designed to describe electromagnetic phenomena, also described light as an electromagnetic process (1873). His idea of fields, quantities depending on three space coordinates and one time coordinate, introduced a much more complicated description of nature than that of Newton, whose description of the state of his systems contained only a set of numbers—six times more than were objects contained in the system (three position coordinates and three velocity components for each). The idea of fields was also a wonderful idea, but its union with the idea of discrete objects—particles—led to difficulties (two very important ones, in fact). The existence of particles, in particular of electrons, could not be denied.

The first of these difficulties—much more at the center of interest in the early part of this century than it is now—concerned the energy of the field connected with a pointlike particle. The electric field strength near to an electron is inversely proportional to the square of the distance from the electron, with the energy density then proportional to the inverse fourth power. The total energy of the electric field next to a charged pointlike particle would then be infinite. There were attempts to overcome this difficulty (leading to an infinite mass of the particle) by attributing a finite size thereto, but no such proposal turned out to be simple and attractive. The other difficulty stemmed from the consideration of a thermal equilibrium between particles and the field. It turned out that the heat capacity of the field was infinite so that, in a true equilibrium, the field would deprive the particles of all their kinetic energy. This was evidently totally unreasonable.

It was this second difficulty that the next terribly big step intended to eliminate. Max Planck (1858–1947) in 1900 proposed a new law of nature: that the particles, in particular, the atoms (and molecules), can absorb and emit only definite amounts of light energy, the amount being $h\nu$, with ν giving the frequency of the light absorbed or emitted and h being a new natural constant (Planck's constant). This was, of course, a terribly new idea, but considering the generally realized need for a new idea, it was quite soon favorably considered. It was evident that a new natural constant was needed also for an expression of the thermal equilibrium energy density of the radiation. If we disregard the possible use of the electron's electric charge, e, the earlier natural constants, c and k, cannot produce an expression of the dimension of energy per unit volume as a function of the temperature, T. With h, there is such an expression, which has, naturally, a factor of T^4 (it is $k^4 T^4/h^3 c^3$ and has a factor 48π).

Even though Planck's ideas originally sounded somewhat unreasonable, they were soon (very soon) generally accepted. Einstein's observation (1905) on the dependence of the maximum energy of the electrons produced by the photoelectric effect on the frequency of the light producing them helped a great deal, as did observations by M. von Laue, W. H. Nernst, and many, many others.

THE DEVELOPMENT OF QUANTUM THEORY

A new breakthrough came with Niels Bohr's explanation of the hydrogen spectrum. He introduced the postulate that the orbits of electrons in atoms obey certain rules; thus, no orbit violating these rules is possible. He formulated the rules for the possible orbits of a single electron, that is, the electron of the hydrogen atom or that of ionized helium, and his rules explained the spectra of these elements with amazing accuracy. Unfortunately, his postulates could not be really applied to systems with more than one electron. However, Bohr also established a wonderful school for young physicists, and many of the later contributions to the development of the physics of atoms and molecules greatly benefited from that school. Unfortunately, no theory for systems with more than one electron could be developed by them. Naturally, therefore, the whole theory was very puzzling: how does the electron jump from one orbit to another? The light emission that results from such a jump takes of the order of 10^{-8} seconds of time, whereas the electron runs around the orbit in about 10^{-16} seconds or a little more.

Even though I was studying for a degree in chemical engineering at the "Technical High School" in Berlin, I attended the physics colloquia at the University. They were very interesting and I learned a lot there. However, what I want to tell you in this connection is that I had a subconscious impression that those in the first row, including Einstein, Planck, von Laue, and some others, were afraid that man is not bright enough to formulate a consistent theory for microscopic (that is, atomic) phenomena, of elementary systems involving small numbers of electrons, protons, atoms, etc. I also believed that many others had the same impression. My impression, though, was changed suddenly after two years (which I spent as a chemical engineer) when I read an article by M. Born and P. Jordan that was based on a 1925 article by Werner Heisenberg. This was then followed (in 1926) by an article of the three of them. In these articles, they abandoned the idea of describing the positions and motions of the electrons in the atom (or molecule) and restricted the problem of microscopic physics to the determination of the energy levels and the probabilities of the possible transitions between these by the emission or absorption of light. The latter process's probability is, of course, proportional to the intensity of light of the proper frequency. This was a wonderful suggestion and, in fact, led to the establishment of a theory that permits the calculation of the quantities just mentioned—absorption and emission probabilities. It was the beginning of quantum mechanics and it showed that the classical description of atomic systems must be greatly modified.

The wonderful Heisenberg-Born-Jordan theory, though, has some weaknesses. The most fundamental of these is that it described only light absorption and emission, and as it turned out later, even these were not described completely. What is even more fundamental, though, is that it did not describe at all collision phenomena, which play a very important role in physics. In fact, in a sense, light absorption is also a collision phenomenon. What it did show, persuasively, was that ordinary physics, based, as far as particles are concerned, on the description of the states of these by giving positions and velocities, must be fundamentally modified. This was a very useful and important observation.

The next fundamental change in our description of basic (that is, microscopic) phenomena was brought about by Schrödinger's introduction of wave mechanics. This is, in a way, a return to the concepts of the earlier theories of physics inasmuch as it postulates the existence of a description of the state of the system and not only probabilities for the various possible observable changes. In fact, it was terribly effective—the quantities that the earlier theory gave only for one-electron systems were now given quite generally. It also described other changes, including the effects of collisions, which was a very important accomplishment. Actually, eventually Heisenberg claimed that the collision matrix (calculable from the wave function) is the most fundamental concept of quantum theory. However, the fact that there is an equation describing the time-dependence of the "wave function" is perhaps an equally valid evidence of its effectiveness.

In the whole discussion so far, relativity theory was not mentioned; it will play an

4   ANNALS NEW YORK ACADEMY OF SCIENCES

important role in the discussion of the weaknesses of the theory that follows. In addition, the special theory of relativity, though immensely important, did not introduce new concepts; it only changed some basic ones, in particular, the meaning of simultaneity. General relativity, though, did give a new theory of gravitation and a wonderful one of that. It also eliminated, as was mentioned before, the problem of defining the coordinate systems in which the common equations of physics are valid. However, it has, also in this last regard, difficulties when used for microscopic (in particular, for atomic) systems. This last point will be discussed in the next section.

THE PROBLEMS AND DIFFICULTIES OF QUANTUM MECHANICS

The pre-quantum-mechanical laws of physics told us how to calculate the future state of a physical system (which is "isolated," that is, not influenced by any other material or radiation) if its initial state (that is, its state at a definite earlier time) is given. It must be admitted that the processes to determine either its initial or its final states were rarely discussed (as was mentioned in the discussion of the motion of the planets and Newton's law of gravitation). However, it was assumed that this is possible, and indeed it was, with sufficient accuracy, for the physical systems considered, which were macroscopic.

The big change, though, came as a result of the description of the states of microscopic systems by Schrödinger's wave function or a mathematical equivalent of this—a vector in an infinite dimensional space. Because the systems considered are microscopic, their state is influenced by the process of observation. According to the usually propagated theory, the observation, in general, changes the state of the system; its state after the observation is one of the characteristic functions of the "observed quantity." The same holds for the determination of the final state; that is, the equations of quantum mechanics give us only the probabilities for the different possible results of the second observation. One can, therefore, say that quantum mechanics gives us the probabilities of the possible outcomes of the second observation and, of course, the state of the system is then again known. Thus, it becomes the characteristic function of the quantity observed, which corresponds to the characteristic value that the observation gave.

I described the basic idea of the verification of the quantum mechanical laws probably in much more detail than is necessary. However, I wish to give now the weaknesses of this theory, which was, incidentally, most clearly formulated by John von Neumann.

First, just as in classical theory, the method of observation is not generally described. In fact, it has been demonstrated that the variety of quantities (functions of position and momentum) that can be observed by a finite "measuring apparatus" is very limited; the corresponding operator must commute with all additive conserved quantities. These involve electric charge, number of protons plus neutrons, total linear momentum, and total angular momentum in any direction. Most operators, of course, do not commute with all these quantities and they can be measured with high (but finite) accuracy only with large measuring apparatuses. Incidentally, the measuring apparatuses must clearly be large because they show the outcome of the observation to the observer who can perceive only macroscopic pointers or other indicators of the measuring apparatuses.

This brings us to two other problems of the quantum mechanical measurement theory. The first and perhaps most obvious is that because the wave function of the system on which the measurement is undertaken has a finite size in space, it takes some time for the information from its distant parts to reach the apparatus, even if it comes with light velocity. Hence, the information that the apparatus can obtain does not refer to an instantaneous state of the system (the state of which is measured), but, at best, to its state on the negative light cone. Similarly, the change of its state vector, which the observation provides, can influence the state of the system at best on the positive light cone. Because quantum mechanics deals, as a rule, only with microscopic systems, these are not terrible limitations; they are substantial ones, though, if the state of the system is changing fast, which is the rule for systems in the description of which relativity theory plays a significant role. The original theory of measurement, as formulated by von Neumann, disregards this point. This is natural, though, because quantum mechanics was a nonrelativistic theory in the days of his contributions to the measurement problem.

The second and, in my opinion, equally important limitation of quantum theory is the fact that the measurement theory implicitly assumes that the outcome of the measurement (for instance, the position of the indicator of the measuring apparatus) has a definite value. If the linear nature of the quantum mechanical equations were valid, this could not be the case. Therefore, the only thing that the equations permit us to postulate for the measurement results is a state of object plus apparatus, which shows a correlation between the states of both. This is essentially a wave function of the two, which is of the form,

$$\Sigma\, a_\kappa \psi_\kappa \phi_\kappa.$$

In this, the a's are constants, the ψ_κ is the state of the object in which the measured quantity surely has the value κ, and ϕ_κ is a state of the apparatus showing that result. However, the equations of quantum mechanics are deterministic and the outcome of the measurement, which gives *one* of the possible values for κ, would be in contradiction to linearity.

This shows that the process of measurements cannot be described by the equations of quantum mechanics because their existence is in contradiction to its principles. It is important to realize this fact. I originally thought that the limitation of quantum mechanics' validity excludes only living beings. However, an important article of J. S. Bell convinced me that the limitation excludes macroscopic objects because he found evidence for this fact and proposed some superficial modification of the equations of quantum mechanics. (I did also.) This contradicts its linear character for the state vector, but preserves it for the density matrix. However, that equation, though taking care of the problem here discussed, surely does not have universal validity.

PART II

Quantum-Mechanical Measuring Process

Die Messung quantenmechanischer Operatoren

E. P. Wigner

Zeitschrift für Physik *133*, 101–108 (1952)

Eingegangen am 24. Mai 1952

Die übliche Annahme der statistischen Deutung der Quantenmechanik, daß alle hermiteschen Operatoren meßbare Größen darstellen, wird wohl allgemein als eine bequeme mathematische Idealisierung und nicht als ein Ausdruck eines Tatbestandes anerkannt. Es wird hier gezeigt, daß schon die Gültigkeit von Erhaltungssätzen für gequantelte Größen (wie der Drehimpulssatz oder der Satz für die Erhaltung der elektrischen Ladung), die die Wechselwirkung von Meßobjekt und Meßapparat beherrschen, die Messung der meisten Operatoren nur als einen Grenzfall gestattet. Insbesondere sind die Bedingungen für die Messung von Operatoren, die mit der Gesamtladung unvertauschbar sind, wahrscheinlich unerfüllbar. Dasselbe dürfte für Operatoren gelten, die mit der Anzahl der schweren Teilchen unvertauschbar sind.

1. Die Grundidee der statistischen Deutung der Quantenmechanik wurde zuerst von BORN ausgesprochen[1]. Seine Gedanken wurden durch die Untersuchungen von HEISENBERG und BOHR[2] u. a. in der anschaulich-physikalischen Richtung vertieft und weiter entwickelt. Die mathematische Formalisierung der Theorie verdankt man besonders den Untersuchungen von NEUMANNs[3]. Ein Grundstein der Theorie besteht in der Annahme, daß jedem selbstadjungierten Operator Q eine meßbare physikalische Größe entspricht. Das Meßergebnis ist immer ein Eigenwert des Operators Q; gleichzeitig führt die Messung das System in den Zustand über, der durch die Eigenfunktion des Meßergebnisses beschrieben wird. Sei etwa φ die ursprüngliche Zustandsfunktion des Systems und bezeichnen wir die Eigenwerte und Eigenfunktionen von Q mit $q_1, q_2, \ldots$ bzw. mit $\psi_1, \psi_2, \ldots$. Dann liefert die Messung mit der Wahrscheinlichkeit $|(\psi_\nu, \varphi)|^2$ das Resultat q_ν und das System befindet sich nach der Messung[4] im Zustand ψ_ν. Mit (ψ_ν, φ) wurde das hermitesche

---

[1] BORN, M.: Z. Physik **37**, 803 (1926). „Die Bewegung der Partikeln folgt Wahrscheinlichkeitsgesetzen, die Wahrscheinlichkeit selbst aber breitet sich im Einklang mit dem Kausalgesetz aus."

[2] HEISENBERG, W.: Z. Physik **43**, 172 (1927). — Die PhysikalischenPrinzipien der Quantenmechanik. Leipzig 1930. — BOHR, N.: Nature, Lond. **121**, 580 (1928). — Naturwiss. **17**, 483 (1929) und weitere Artikel in der Max Planck-Nummer der Naturwissenschaften. Vgl. auch MOTT, N. F.: Proc. Roy. Soc. Lond. **126**, 79 (1929) und BOHR, N., u. L. ROSENFELD: Phys. Rev. **78**, 794 (1950).

[3] NEUMANN, J. v.: Mathematische Grundlagen der Quantenmechanik, bes. Kap. VI. Berlin 1932.

[4] Falls der Eigenwert q_ν entartet ist und mehrere Eigenzustände $\psi_{\nu 1}, \psi_{\nu 2}, \ldots$ umfaßt, so ist die Wahrscheinlichkeit von q_ν gleich $w_\nu = \sum_\varkappa |(\psi_{\nu\varkappa}, \varphi)|^2$ und der Zustand nach der Messung $\sum_\varkappa w_\nu^{-\frac{1}{2}} (\psi_{\nu\varkappa}, \varphi) \psi_{\nu\varkappa}$.

102   E. P. WIGNER:

skalare Produkt von ψ_ν und φ bezeichnet. Es möge hier noch bemerkt werden, daß ein Operator I, der sowohl die Übergangswahrscheinlichkeiten wie auch die zeitliche Änderung des Systems invariant läßt, auch den Zustand selber nicht beeinflussen kann. Konkreter ausgedrückt, wenn für alle φ und ψ sowohl $|(\psi, \varphi)|^2 = |(\psi, I\varphi)|^2$ wie auch $(I\varphi)_t = I(\varphi_t)$ (wo der Index t die zeitliche Änderung des Systems beschreibt), so sind die Zustände φ und $I\varphi$ überhaupt ununterscheidbar. In der orthodoxen Formulierung der Theorie sind die I komplexe Zahlen vom Absolutwert 1.

Die große Schwäche des oben skizzierten Formalismus ist, daß er keine Vorschrift enthält, wie die Messung des Operators Q ausgeführt werden kann. Schematisch läßt sich zwar eine solche Vorschrift leicht angeben[1]: Man vereinige das Meßobjekt φ mit einem Meßinstrument, dessen Zustandsfunktion mit ξ bezeichnet werden möge. Das Meßinstrument ist so beschaffen, daß die Zustandsfunktion des Gesamtsystems $\varphi\xi$, bestehend aus Meßobjekt und Meßinstrument, nach einer gewissen Zeit in

$$\varphi\xi \to \sum_\nu (\psi_\nu, \varphi)\, \psi_\nu \chi_\nu \qquad (1)$$

übergeht, worin die χ_ν makroskopisch unterscheidbare Zustände des Meßinstruments sind: der Zustand χ_ν kündigt das Meßresultat q_ν an[2]. Diese Vorschrift bleibt aber rein formal, solange nicht angegeben wird, wie man das Meßinstrument im Zustand ξ konstruieren soll. Eine weitere, erkenntnistheoretisch noch tieferliegende Schwierigkeit der Theorie ist mit den Worten „markroskopisch unterscheidbare Zustände" überdeckt. Dieser Punkt wurde schon von HEISENBERG ausführlich, und soweit das zur Zeit möglich ist, vollkommen diskutiert. Er soll hier nicht wieder aufgenommen werden. Die Frage, die diskutiert werden soll, bezieht sich vielmehr auf die Möglichkeit einer Wechselwirkung zwischen Meßobjekt und Meßinstrument, wie sie durch (1) symbolisiert wird. Daher werden wir lediglich notwendige Bedingungen für die Meßbarkeit einer Größe erhalten. Selbst wenn eine Wechselwirkung, die (1) entspricht, keinem hier erkannten Prinzip widerspricht, ist es sehr wohl möglich, daß entweder ξ prinzipiell unrealisierbar ist, oder daß die χ_ν einer direkten oder auch indirekten makroskopischen Unterscheidung unzugänglich

---

[1] NEUMANN, J. v.: Mathematische Grundlagen der Quantenmechanik, bes. Kap. VI. Berlin 1932.

[2] Man erkennt aus (1) auch die Ursache für den hermiteschen Charakter der Operatoren, die beobachtbaren Größen entsprechen. Der Übergang von der linken Seite von (1) zur rechten wird durch einen unitären Operator bewirkt. Er führt $\psi_\nu \xi$ bzw. $\psi_\mu \xi$ in $\psi_\nu \chi_\nu$ bzw. in $\psi_\mu \chi_\mu$ über. Die letzten Funktionen sind aber orthogonal zueinander, weil die χ, als makroskopisch unterscheidbar, zueinander orthogonal sein müssen. Wegen der Unitarität des Überganges folgt dies dann auch für die $\psi_\nu \xi$, d.h. auch für die ψ_ν, die das Eigenfunktionensystem von Q bilden. Da zudem die q_ν als Meßresultate reell sind, folgt der selbstadjungierte Charakter von Q.

Die Messung quantenmechanischer Operatoren.   **103**

sind. In der Tat[1] verschiebt (1) nur die Frage der Unterscheidbarkeit der ψ_ν auf die Unterscheidbarkeit der χ_ν, und es werden hier nur die Bedingungen und die Möglichkeit einer solchen Verschiebung untersucht.

2. Solange man im Rahmen der allgemeinen, durch (1) ausgedrückten Theorie verbleibt, kann man über die Meßbarkeit keine konkreten Aussagen machen, die über die Fußnote 2 der vorigen Seite hinausgehen. In der Tat könnte das Meßinstrument von (1) im allgemeinen ein sehr einfaches und elementares System sein, in einem Beispiel von HEISENBERG besteht es aus einem einzigen Lichtquant[2]. Wenn man aber das Postulat der relativistischen Invarianz mit heranzieht, so sollte es wohl weitgehend bekannt sein, daß wenigstens ein Operator I_1 mit allen beobachtbaren Größen Q vertauschbar ist. Der Operator I_1 läßt alle Zustände mit ganzzahligem Drehimpuls ungeändert, multipliziert aber alle Zustände mit halbzahligem Drehimpuls mit -1. Die Beobachtung einer Größe, deren Operator mit I_1 unvertauschbar ist [wie etwa die der gequantelten Amplituden $\psi(x, y, z) + \psi(x, y, z)^*$], würde es ermöglichen, zwischen Zuständen zu unterscheiden, die nach der Relativitätstheorie ununterscheidbar bleiben müssen[3]. Diese Beschränkung der Beobachtbarkeit ist unabhängig von der Theorie der Messung, wie sie in (1) ausgedrückt ist. Wir wollen hier aber eine andere Art Beschränkung besprechen, die ihren Ursprung in den Erhaltungssätzen gequantelter Größen hat und die auf einer Diskussion der Möglichkeit der Abbildung (1) beruht. Diese Beschränkung wird nicht so streng sein, wie die vorerwähnte und wird lediglich als Folge haben, daß das Meßinstrument sehr groß sein muß in dem Sinne, daß es, mit beträchtlicher Wahrscheinlichkeit, einen sehr großen Betrag der gequantelten Erhaltungsgröße enthalten muß, mit deren Operator sein Operator unvertauschbar ist.

3. Gequantelte Erhaltungsgrößen der obenerwähnten Art sind etwa die Komponente des Drehimpulses in einer gewissen Richtung, die gesamte elektrische Ladung des Systems, die Anzahl der „schweren Teilchen" darin. Von nun an soll der untere Index einer Zustandsfunktion die Anzahl der Quanten ($\hbar$, e, usw.) angeben, die der durch die Zustandsfunktion beschriebene Zustand enthält. Um gebrochene Indizes zu vermeiden, wird aber, wenn nötig, der Index um $\frac{1}{2}$ erhöht werden. Weiterhin wird zunächst (1) in ihrer ursprünglichen Gestalt angenommen, andere Definitionen der Messung werden am Ende dieses Aufsatzes diskutiert werden.

---

[1] Vgl. HEISENBERG, W.: l. c. und NEUMANN, J. v.: l. c., bes. S. 223, 224.

[2] HEISENBERG, W.: l. c., Kapitel II, 2, Beispiel b.

[3] Dieser Punkt wird in einer bald zu erscheinenden Arbeit von WICK, WIGHTMAN und WIGNER in etwas populärer Weise weiter erörtert werden. Der vorliegende Aufsatz verdankt seinen Ursprung einer Fragestellung, die im Laufe der Abfassung der obenerwähnten Arbeit auftauchte.

104   E. P. Wigner:

Im einfachsten Fall haben die Eigenfunktionen eines typischen Operators, der mit der Erhaltungsgröße nicht vertauschbar ist, die Form $(\psi_0 + \psi_1)/\sqrt{2}$ und $(\psi_0 - \psi_1)/\sqrt{2}$. Wenn z.B. die Erhaltungsgröße des Drehimpuls in der Z-Richtung ist, sind $\psi_0 + \psi_1$ und $\psi_0 - \psi_1$ Eigenfunktionen der X-Komponente des Spins eines Teilchens. Der Operator, der dieser Komponente zugeordnet ist, ist offenbar unvertauschbar mit dem Drehimpuls in der Z-Richtung. Die Gl. (1) besagt also

$$\left.\begin{array}{l} (\psi_0 + \psi_1)\,\xi \to (\psi_0 + \psi_1)\,\chi \\ (\psi_0 - \psi_1)\,\xi \to (\psi_0 - \psi_1)\,\chi', \end{array}\right\} \tag{2}$$

worin $(\chi, \chi') = 0$ und der Pfeil eine lineare unitäre Transformation darstellt, die mit dem Operator der Erhaltungsgröße vertauschbar ist. Addieren und subtrahieren wir die beiden Gln. (2) und zerlegen wir gleichzeitig $\chi + \chi'$ und $\chi - \chi'$ in die Teile σ_ν bzw. τ_ν, die Zustände mit einem scharf bestimmten Wert der Erhaltungsgröße darstellen

$$(\chi + \chi')/\sqrt{2} = \sum \sigma_\nu , \qquad (\chi - \chi')/\sqrt{2} = \sum \tau_\nu , \tag{3}$$

so erhalten wir

$$\psi_0\,\xi \to (\psi_0 \sum \sigma_\nu + \psi_1 \sum \tau_\nu)/\sqrt{2} , \tag{4a}$$

$$\psi_1\,\xi \to (\psi_0 \sum \tau_\nu + \psi_1 \sum \sigma_\nu)/\sqrt{2} . \tag{4b}$$

Daß der Meßapparat einen unendlichen Betrag der Erhaltungsgröße enthalten muß, erhellt schon aus (4). Gemäß (4) ist der Erwartungswert der Erhaltungsgröße gleich für die beiden Zustände, die durch die rechten Seiten von (4) dargestellt werden. Ihr Erwartungswert für das Meßobjekt ist in beiden Fällen $\frac{1}{2}$, ihr Erwartungswert für das Meßinstrument ist in beiden Fällen das arithmetische Mittel der Erwartungswerte für $\sum \sigma_\nu$ und $\sum \tau_\nu$. Nach der Messung sind ja Meßobjekt und Meßinstrument wieder getrennt und der Gesamtinhalt des Systems an der Erhaltungsgröße setzt sich additiv zusammen aus den Inhalten von Meßobjekt und Meßinstrument. Der Erwartungwert der Erhaltungsgröße ist aber offenbar um 1 größer für die linke Seite von (4b) als (4a).

Man kann diesen Widerspruch noch verschärfen, wenn man beachtet, daß aus (4) und dem Erhaltungssatz die Gleichungen

$$\left.\begin{array}{l} \psi_0\,\xi_\nu \to (\psi_0\,\sigma_\nu + \psi_1\,\tau_{\nu-1})/\sqrt{2} \\ \psi_1\,\xi_{\nu-1} \to (\psi_0\,\tau_\nu + \psi_1\,\sigma_{\nu-1})/\sqrt{2} \end{array}\right\} \tag{5}$$

folgen, worin die ξ_ν die Bestandteile von ξ an Eigenfunktionen der Erhaltungsgröße sind

$$\xi = \sum \xi_\nu . \tag{6}$$

Bezeichnen wir nun

$$(\xi_\nu, \xi_\nu) = x_\nu; \qquad (\sigma_\nu, \sigma_\nu) = s_\nu; \qquad (\tau_\nu, \tau_\nu) = t_\nu; \qquad (\sigma_\nu, \tau_\nu) = a_\nu + i\,b_\nu \tag{7}$$

Die Messung quantenmechanischer Operatoren.   **105**

$(x_\nu, s_\nu, t_\nu, a_\nu, b_\nu$ reell), so drücken die Gleichungen

$$x_\nu = \tfrac{1}{2} s_\nu + \tfrac{1}{2} t_{\nu-1}, \qquad x_{\nu-1} = \tfrac{1}{2} t_\nu + \tfrac{1}{2} s_{\nu-1}, \tag{8a}$$

$$0 = a_\nu - i b_\nu + a_{\nu-1} + i b_{\nu-1} \tag{8b}$$

den unitären Charakter des durch den Pfeil an (5) angedeuteten Überganges,

$$\sum x_\nu = \sum s_\nu = \sum t_\nu = 1, \qquad \sum a_\nu = \sum b_\nu = 0 \tag{9}$$

die Normalisierung von ξ, χ, χ' und die Orthogonalität von χ und χ' aus. Aus (8b) und (9) folgt aber unmittelbar $a_\nu = b_\nu = 0$, aus (8a) folgt

$$x_{\nu+1} - \tfrac{1}{2} s_{\nu+1} = \tfrac{1}{2} t_\nu = x_{\nu-1} - \tfrac{1}{2} s_{\nu-1},$$

d.h. daß sowohl $x_{2\nu+1} - \tfrac{1}{2} s_{2\nu+1}$ wie auch $x_{2\nu} - \tfrac{1}{2} s_{2\nu}$ von ν unabhängig sind. Dies gilt dann auch für t_ν, was aber mit (9) unverträglich ist. Streng genommen ist also eine Messung, die zur Separation von $\psi_0 + \psi_1$ und $\psi_0 - \psi_1$ führt, unmöglich. Dasselbe läßt sich mit Hilfe einer etwas umständlicheren Algebra, die aber nicht wesentlich von der obigen verschieden ist, auch für die Zustände $\alpha \psi_0 + \beta \psi_1$ und $-\bar\beta \psi_0 + \bar\alpha \psi_1$ zeigen, wo α und β beliebige komplexe Zahlen sind.

Da eine Messung der ·Spinkomponenten praktisch möglich ist, muß es auch möglich sein, die vorangehende Überlegung so zu modifizieren, daß sie die Möglichkeit einer solchen Messung mit beliebiger Genauigkeit demonstriert. Bezeichnen wir zu diesem Zweck die Zustände, in die $(\psi_0 + \psi_1)\xi$ bzw. $(\psi_0 - \psi_1)\xi$ durch den Meßprozeß übergeführt werden mit

$$\left.\begin{array}{l} (\psi_0 + \psi_1)\, \xi \to (\psi_0 + \psi_1)\, \chi + (\psi_0 - \psi_1)\, \eta \\ (\psi_0 - \psi_1)\, \xi \to (\psi_0 - \psi_1)\, \chi' + (\psi_0 + \psi_1)\, \eta'. \end{array}\right\} \tag{10}$$

Wenn dann $(\chi, \chi') = 0$ verbleibt, (η, η) und (η', η') beliebig klein gemacht werden können, so kann man durch Feststellung des Zustandes χ bzw. χ' des Meßinstrumentes in fast allen Fällen auf den Zustand des Meßobjekts schließen.

Wir werden es so einrichten, daß $\eta = -\eta'$ und auch $(\eta, \chi) = (\eta, \chi') = (\chi, \chi') = 0$. Dies bedeutet, daß die Messung drei Resultate haben kann: der Zustand ist $(\psi_0 + \psi_1)/\sqrt{2}$, der Zustand ist $(\psi_0 - \psi_1)/\sqrt{2}$, der Zustand ist unbestimmt. Doch ist die Wahrscheinlichkeit dafür, daß man das letzte Resultat erhält, (η, η) und dies kann, wie wir sogleich sehen werden, beliebig klein gemacht werden. Um dies zu erreichen, wird allerdings die Zerlegung von ξ nach (6) sehr viele Komponenten haben müssen.

Wir nehmen an, daß diese Zahl n ist und daß das Meßinstrument nicht weniger als ein und nicht mehr als n Einheiten der Erhaltungsgröße enthalten kann. Dann verschwinden die ξ_ν außer für $0 < \nu \leq n$.

106   E. P. WIGNER:

Weiterhin führen wir zur Abkürzung

$$
\left.
\begin{aligned}
2\chi &= 2\sigma + \varrho + \tau \\
2\chi' &= 2\sigma - \varrho - \tau \\
2\eta &= -2\eta' = \tau - \varrho
\end{aligned}
\right\} \tag{11}
$$

ein. Dies ergibt aus (10)

$$
\left.
\begin{aligned}
\psi_0 \xi &\to \psi_0 \sigma + \psi_1 \varrho \\
\psi_1 \xi &\to \psi_0 \tau + \psi_1 \sigma.
\end{aligned}
\right\} \tag{12}
$$

Die σ, τ, ϱ können dann, ähnlich zu (6), als eine Summe von Eigenfunktionen der Erhaltungsgröße geschrieben werden. Unter den σ_ν sind nur jene mit $0 < \nu \leq n$ endlich, dagegen bleibt ϱ_0 endlich, während ϱ_n schon verschwindet. Umgekehrt ist $\tau_1 = 0$, während τ_{n+1} endlich ist.

Die Orthogonalität der rechten Seiten von (12) führt zu

$$
(\sigma_\nu, \tau_\nu) + (\varrho_{\nu-1}, \sigma_{\nu-1}) = 0 , \tag{13}
$$

die Normalisierungsbedingung ist

$$
(\xi_\nu, \xi_\nu) = (\sigma_\nu, \sigma_\nu) + (\varrho_{\nu-1}, \varrho_{\nu-1}) = (\sigma_\nu, \sigma_\nu) + (\tau_{\nu+1}, \tau_{\nu+1}) . \tag{13a}
$$

Hierzu kommen die Bedingungen

$$
(\xi, \xi) = \sum (\xi_\nu, \xi_\nu) = 1 , \tag{14a}
$$

$$
(\chi, \chi') = 4 \sum (\sigma_\nu, \sigma_\nu) - \sum (\varrho_\nu + \tau_\nu, \varrho_\nu + \tau_\nu) = 0 , \tag{14b}
$$

und wegen $(\chi, \eta) = (\chi', \eta) = 0$

$$
\sum (\sigma_\nu, \tau_\nu - \varrho_\nu) = 0 \tag{14c}
$$

$$
\sum (\tau_\nu + \varrho_\nu, \tau_\nu - \varrho_\nu) = 0 . \tag{14d}
$$

Man kann diese Gleichungen in mannigfacher Weise befriedigen. Die einfachste Wahl — die aber nicht zu dem kleinstmöglichen Wert von (η, η) führt — ist wohl die, bei der alle

$$
(\sigma_\nu, \tau_\nu) = (\sigma_\nu, \varrho_\nu) = 0 \tag{15}
$$

verschwinden. Damit hat man (13) und (14c) erfüllt. Sodann kann man für jene ν, für die sowohl ϱ_ν wie auch τ_ν endlich sein können,

$$
\varrho_\nu = \tau_\nu \qquad (1 < \nu \leq n-1) \tag{15a}
$$

annehmen, und allen nicht verschwindenden ϱ, τ

$$
(\varrho_\nu, \varrho_\nu) = (\tau_\nu, \tau_\nu) = c' \tag{15b}
$$

die gleiche Norm geben. Damit ist (14d) befriedigt und auch (13a), wenn man es zur Bestimmung der (ξ_ν, ξ_ν) benutzt.

Schließlich kann man auch

$$
(\sigma_\nu, \sigma_\nu) = c \tag{15c}
$$

unabhängig von ν annehmen (für $0 < \nu \leq n$). Es folgt dann auch $(\xi_\nu, \xi_\nu) = c + c'$ und wegen (14a)

$$n(c + c') = 1. \tag{16a}$$

Es bleibt nur noch (14b) übrig. Dies gibt

$$4nc = (\varrho_0, \varrho_0) + (\varrho_1, \varrho_1) + (\tau_n, \tau_n) + (\tau_{n+1}, \tau_{n+1}) + \left. + \sum_{\nu=2}^{n-1} (2\varrho_\nu, 2\varrho_\nu) = 4c' + 4(n-2)c' = 4(n-1)c'. \right\} \tag{16b}$$

Aus (16a) und (16b) berechnet sich $c' = 1/(2n-1)$. Wenn man schließlich (η, η) berechnet, so fallen wegen (15a) die Glieder mit $\nu = 2, 3, \ldots, n-1$ weg und man erhält

$$(\eta, \eta) = c' = 1/(2n-1). \tag{17}$$

Dies geht tatsächlich zu Null, wenn n sehr groß wird. Durch eine vorteilhaftere Wahl der σ, τ, ϱ hätte man erreichen können, daß (η, η) wie $1/n^2$ zu Null geht. Trotzdem wird ξ eine sehr große Anzahl von Komponenten, also der Meßapparat einen sehr großen Bestand an der Erhaltungsgröße, haben müssen, wenn man eine große Sicherheit haben will, daß die Wechselwirkung zwischen Meßobjekt und Meßapparat zu einer Messung führt. Insbesondere, wenn man den Phasenunterschied zwischen Teilen der Zustandsfunktion messen will, die verschiedenen Gesamtladungen entsprechen[1], muß die elektrische Ladung des Meßapparats — wenn eine solche Messung überhaupt möglich ist — weitgehend unbestimmt sein.

4. Es fragt sich noch, ob die Beschreibung der Messung, die in (1) oder (2) enthalten ist, nicht eine zu anspruchsvolle ist. Es wird sich aber zeigen, daß, obwohl dies wahrscheinlich der Fall ist, selbst eine erheblich lockerere Definition des Meßprozesses zu ähnlichen Ergebnissen führt.

Die wichtigste Verallgemeinerung von (2) besteht wohl darin, daß man eine Änderung des Zustandes des Meßobjekts zuläßt, selbst wenn es ursprünglich in einem der beiden Zustände $\psi_0 + \psi_1$ oder $\psi_0 - \psi_1$ war. Falls die Messung lediglich diese Zustände voneinander unterscheiden soll, bleibt ja der Endzustand des Meßobjekts belanglos (vgl. auch mehrere der Beispiele in Fußnote 2 am Anfang). Es wurde weiter oben schon erwähnt, daß selbst die Feststellung eines Unterschiedes zwischen $\varphi + \varphi'$ und $\varphi - \varphi'$ unstatthaft ist, wenn φ einen Zustand mit ganzzahligem, φ' einen Zustand mit halbzahligem Drehimpuls beschreibt.

Wenn wir (2) so abändern, daß wir in der rechten Seite lediglich ψ_0 und ψ_1 durch ψ_0' und ψ_1' ersetzen, ändert sich an den vorangehenden

---

[1] Dieser Punkt wird in einer bald zu erscheinenden Arbeit von WICK, WIGHTMAN und WIGNER in etwas populärer Weise weiter erörtert werden. Der vorliegende Aufsatz verdankt seinen Ursprung einer Fragestellung, die im Laufe der Abfassung der obenerwähnten Arbeit auftauchte.

108   E. P. WIGNER: Die Messung quantenmechanischer Operatoren.

Überlegungen gar nichts. In der Tat würde die Erfüllbarkeit der so gewonnenen Gleichungen auch die Erfüllbarkeit von (2) in ihrer ursprünglichen Gestalt mit sich ziehen. Wir wollen daher gleich allgemein

$$
\begin{aligned}
(\psi_0 + \psi_1)\,\xi &\to (\textstyle\sum \psi'_\mu)\,(\textstyle\sum \chi'_\lambda) \\
(\psi_0 - \psi_1)\,\xi &\to (\textstyle\sum \psi''_\mu)\,(\textstyle\sum \chi''_\lambda)
\end{aligned} \Bigg\}
\tag{18}
$$

annehmen. Dagegen beschränken wir uns auf den Fall, daß die Anzahl der Quanten der Erhaltungsgröße in ξ eine bestimmte ist. Diese Anzahl kann dann, ohne die Allgemeinheit der Betrachtung zu beeinträchtigen, gleich Null angenommen werden.

Da die linken Seiten von (18) entweder kein, oder ein Quantum der Erhaltungsgröße besitzen, folgt

$$
\sum_\mu \psi'_\mu \chi'_{\nu-\mu} = 0 , \qquad \nu \neq 0, 1 . \tag{19}
$$

Wegen der Orthogonalität der Glieder der Summe in (19) müssen sie alle schon einzeln verschwinden. Es kann wieder angenommen werden, daß ψ'_0 und χ'_0 endlich sind und es bleiben demnach nur die folgenden zwei Fälle möglich:

1. $\psi'_0, \psi'_1, \chi'_0$ endlich, alle anderen verschwinden;
2. $\psi'_0, \chi'_0, \chi'_1$ endlich, alle anderen verschwinden.

Auch für die doppeltgestrichenen Größen gilt, daß entweder nur zwei ψ'' und ein χ'', oder nur ein ψ'' und zwei χ'' endlich sein können. Zudem folgt im Fall 1 aus

$$
2\psi_0\,\xi \to (\psi'_0 + \psi'_1)\,\chi'_0 + \sum \psi''_\mu \chi''_\lambda ,
$$

daß $\psi''_1 \chi''_0$ endlich, und zwar gleich $-\psi'_1 \chi'_0$ sein muß. Im Fall 2 folgt ebenso $\psi'_0 \chi'_1 = -\psi''_0 \chi''_1$. Eine ganz einfache Diskussion führt nun zum Ergebnis, daß Fall 1 zu der Modifikation von (2) führt, die schon im vorangehenden Absatz besprochen wurde. Fall 2 führt dagegen im wesentlichen zu

$$
\begin{aligned}
(\psi_0 + \psi_1)\,\xi &\to \psi'_0\,(\chi_0 + \chi_1) \\
(\psi_0 - \psi_1)\,\xi &\to \psi'_0\,(\chi_0 - \chi_1)
\end{aligned} \Bigg\}
\tag{20}
$$

anstatt (2). In diesem Fall führt der Meßprozeß zu einem Austausch der Erhaltungsgröße zwischen Meßobjekt und Meßinstrument. Insbesondere wird das Problem der Unterscheidung von $\psi_0 + \psi_1$ und $\psi_0 - \psi_1$ durch das fast gleichbedeutende Problem der Unterscheidung von $\chi_0 + \chi_1$ und $\chi_0 - \chi_1$ ersetzt. Falls dieser Unterschied nicht unmittelbar apperzipierbar ist, bleibt also das Resultat des vorangehenden Abschnittes ungeändert bestehen.

Princeton (N. J.), Palmer Physical Laboratory, University.

Theorie der quantenmechanischen Messung

E. P. Wigner

Physikertagung Wien 1961. Physik Verlag, Mosbach/Baden 1962, pp. 1–8

Einleitung

Heute wissen wir, daß *Planck*, als er die Idee der quantenhaften Emission und Absorption der Strahlung geschaffen hat, sich sehr wohl der Bedeutung des Schrittes bewußt war, den er getan hatte. Ich zweifle aber, ob er die volle Tragweite seiner Entdeckung erkannt hatte, ob er voraussehen konnte, daß seine Entdeckung die Forschung über die Grundbegriffe der Physik durch wenigstens 50 Jahre ebenso stark beherrschen werde, wie *Newtons* Gesetze sie seinerzeit beherrscht hatten. Heute will ich nur über einen sehr kleinen Teil der Auswirkungen der Quantentheorie sprechen, einen Teil, der auf mich immer eine besondere Anziehung ausgeübt hat, weil er so nahe an allgemeine Probleme der menschlichen Erkenntnis führt.

Die Entwicklung der Quantentheorie aus der Erkenntnis der quantenhaften Natur der Emission und Absorption ist ein fast ebenso großes Wunder, wie die Entwicklung des biblischen Senfsamens. Über den langen Weg der, ausgehend von *Plancks* Idee, zum Gebäude der Quantenmechanik führte, will ich heute nicht sprechen. Doch will ich einige Worte über das Auftauchen erkenntnistheoretischer Probleme sagen.

Für die meisten von uns war *Heisenbergs* Arbeit über die Unbestimmtheitsrelation [1] die erste Andeutung, daß die — damals noch junge — Quantenmechanik nicht nur eine Quantelungsvorschrift ist, sondern ein neues Begriffssystem zur Beschreibung der Regelmäßigkeiten der unbelebten Welt. Wenn ich sage, daß *Heisenbergs* Arbeit eine Andeutung war, so sage ich zu wenig. Sie war vielmehr eine Ankündigung, die man zwar in ihren Einzelheiten nicht leicht verstehen konnte, die aber den Ton der Autorität hatte. Nur wenige, die sie gelesen hatten, bezweifelten weiterhin, daß die Quantenmechanik die Lösung der damals brennenden prinzipiellen Fragen der Physik bringen würde.

Die Ankündigung erwies sich auch als richtig. Zunächst wurden *Heisenbergs* Gedanken besonders von *Niels Bohr* — aber auch von anderen — vertieft und in ihrem physikalischen Inhalt verallgemeinert. Die mathematische Formulierung, die sich als die knappste und prägnanteste Darstellung erwiesen hat, verdanken wir in ihrer endgültigen Fassung *v. Neumann*. Bekanntlich wird nach *v. Neumann* der Zustand eines quantenmechanischen Systems durch einen Zustandsvektor, d. h. einen Vektor im *Hilbert*-Raum beschrieben. Dieser Vektor ist aber nicht direkt durch eine Beobachtung feststellbar. Vielmehr wird durch jede Beobachtung der Wert einer Größe, wie etwa Impuls, Ortskoordinate oder Energie, gemessen. Im mathematischen Rahmen der Quantenmechanik ist einer solchen Größe ein selbstadjungierter Operator zugeordnet. Die möglichen Meßresultate sind die Eigenwerte dieses Operators. Die Entwicklungs-

1

koeffizienten des Zustandsvektors in bezug auf die Eigenvektoren des Operators bestimmen die Wahrscheinlichkeiten für das Auftreten der zugehörigen Eigenwerte als Meßresultate [2].

An der vorangehenden Deutung der quanten- und wellenmechanischen Begriffe hat sich durch die Jahre — es sind inzwischen etwa 35 vergangen — nichts geändert. Hinzugekommen ist lediglich eine eingehendere Analyse der eben beschriebenen Vorschriften und ihrer Grundlagen, eine weitere Verfolgung der zugrunde liegenden Ideen, und schließlich eine höhere Einschätzung der Bedeutung der damals neu eingeführten Begriffe. Ich will mich zunächst der eingehenderen Analyse zuwenden, die uns auch auf natürlichem Wege zu den erkenntnistheoretischen Folgerungen führen wird.

Orthodoxe Auffassung des Meßprozesses

Die erste Frage, die sich beim Betrachten des *v. Neumann*schen Meßprozesses aufwirft, betrifft den Zustand des Systems nach der Messung. *v. Neumann* hatte eine eindeutige Antwort auf die Frage: Nach der Messung ist der Zustandsvektor gleich dem Eigenvektor, der zu dem Eigenwert gehört, der sich als Meßresultat ergeben hat. Daher ändert sich nach *v. Neumann* der Zustandsvektor in zwei grundverschiedenen Weisen: Die Änderung zwischen zwei Messungen erfolgt nach den quantenmechanischen Bewegungsgleichungen kontinuierlich und kausal. Die Änderung während der Messung — oder besser gesagt infolge der Messung — ist diskontinuierlich und nur durch Wahrscheinlichkeitsgesetze beschreibbar.

Wenn die Messung vollzogen wurde, das Meßresultat aber nicht notiert ist, so ist der Zustand des Objektes im allgemeinen ein Gemisch der Eigenvektoren des gemessenen Operators. Daß durch die Messung ein Gemisch entsteht, ist nicht überraschend, da die Messung als ein Stoßprozeß zwischen Apparat und Objekt betrachtet werden kann, und da es wohl bekannt ist, daß nach einem Stoß zwar das Gesamtsystem der beiden Stoßpartner durch einen Zustandsvektor charakterisiert werden kann, nicht aber jeder Einzelne der beiden Partner. Vielmehr kann sich jeder nach dem Stoß in verschiedenen Zuständen befinden; der Meßprozeß ist eben ein Stoß solcher Art, daß die Komponenten des auftretenden Gemisches die Eigenvektoren des gemessenen Operators sind.

Das Merkwürdige und Neuartige tritt nur auf, wenn das Meßresultat notiert wird: Durch unsere Kenntnisnahme des Meßresultates verwandelt sich das Gemisch in den Eigenvektor, der dem Meßresultat entspricht. Man sagt, daß das Gemisch auf einen einzigen Eigenvektor „zusammenschrumpft".

Daß der Akt der Kenntnisnahme den Zustandsvektor beeinflußt, scheint zuerst eine absurde Behauptung. Man fragt sich, was geschehen würde, wenn das Meßinstrument, ein unbelebtes Wesen, das Meßresultat nur registrieren würde. Die Antwort ist nach der orthodoxen Theorie, daß das Meßinstrument nach der Messung auch nicht in einem bestimmten Zustand ist, sondern mit entsprechenden Wahrscheinlichkeiten in verschiedenen Zuständen, die etwa verschiedenen Zeigerstellungen des Instruments entsprechen. Betrachten wir zwei miteinander verträgliche Messungen am System Objekt-Apparat. Die erste betrifft den betrachteten Operator des Objekts; die zweite die Stellung

2

des Zeigers des Apparates. Für eine gegebene Stellung des Zeigers gibt es genau einen korrespondierenden Zustand des Objekts. Die Wahrscheinlichkeiten für nicht entsprechende Kombinationen von Zeigerstellung und Zustand sind Null. Die Wechselwirkung zwischen Objekt und Apparat hat einen solchen Zustand des vereinten Systems geschaffen, daß die Feststellung des Zustandes des Apparates auch den Zustand des Objektes festlegt. Es ist etwa so, als zöge man einen Geldschein aus einer Urne: Sieht man sich eine Seite des Scheines an, so weiß man auch wie die andere aussieht. Diese statistische Kopplung zwischen den Zuständen des Meßinstruments und des Objekts bildet die Erklärung für das Schrumpfen des Wellenpaketes, das wir so seltsam fanden. Jetzt, auf den zweiten Blick, erscheint diese Erscheinung nicht mehr so seltsam. Wenn wir den Zustand des Meßinstruments nach der Messung kennen, so wissen wir ebenso genau, was der Zustand des Objekts ist, wie wir die Rückseite des Scheines kennen, wenn wir wissen, was auf der Vorderseite steht. Unser zweiter Eindruck ist, daß das Schrumpfen des Wellenpaketes ganz selbstverständlich ist.

Wenn wir uns aber den Vorgang zum dritten Mal überlegen, so sehen wir, daß die Lage trotz allem recht sonderbar, wenn auch nicht unvernünftig ist. Der Zustandsvektor hängt ja schließlich doch von unserer Kenntnisnahme des Meßresultates ab. Es ist also keine objektive Charakterisierung des Zustandes möglich. Anders ausgedrückt: Die Quantenmechanik liefert lediglich Wahrscheinlichkeitszusammenhänge zwischen Eindrücken, die wir vom Objekt erhalten, wenn wir wiederholt damit in Wechselwirkung treten, indem wir Messungen daran vornehmen. Der Zustandsvektor ist nur eine Hilfsgröße mit deren Hilfe wir die Wahrscheinlichkeiten berechnen können, mit denen wir bei der nächsten Messung die verschiedenen möglichen Eindrücke erhalten werden.

Im vorangehenden habe ich die orthodoxe, v. Neumannsche Auffassung [2], wiedergegeben. Die Messung, die die beschriebenen Eigenschaften hat, wurde von *Pauli* Messung erster Art genannt [3]. Der Ausdruck „Messung" wurde von *Pauli* und von *Landau* [4] auch in einem weiteren Sinne verwendet. Die Art der Wechselwirkungen, die man als Messungen bezeichnen will, ist offenbar einer gewissen Willkür unterworfen. Man sollte aber nicht vergessen, daß Wechselwirkungen der Art wie sie v. *Neumann* betrachtet hat, existieren, sie existieren auch wenn man sie nicht als Messungen bezeichnen wollte. Daher existiert auch das voran besprochene „Schrumpfen" und die erkenntnistheoretische Folgerung wurde zu Recht gezogen.

Nichts, was ich bisher gesagt habe ist neu. v. *Neumann*s Buch [2] enthält schon eine, wenn auch sehr knappe, Schilderung der Zusammenhänge, die ich darzulegen versucht habe. Das Buch von *London* und *Bauer* [5] beschreibt die Lage eingehender und mit großer Klarheit. Es gibt auch andere vorzügliche Darstellungen.

Versuche die Interpretation zu ändern

Wenn es auch durchaus nicht unvernünftig ist zuzugeben, daß die Quantenmechanik eine objektive Beschreibung der Welt nicht zuläßt, so ist es auch natürlich, sich gegen diese Folgerung zu sträuben. Vor allem will man sich

3

nicht zu einem erkenntnistheoretischen Standpunkt von einer Theorie drängen lassen, die wahrscheinlich doch nicht endgültig ist. Daher fehlt es auch nicht an Versuchen, das vorstehend entwickelte Bild zu mildern.

Von den mir bekannten, in der Literatur vorhandenen oder bald erscheinenden, Versuchen erscheint mir nur die von *Ludwig* [6] logisch einwandfrei. Die Erscheinung des Schrumpfens des Wellenpakets kann nicht durch Neudefinitionen eliminiert werden; sie folgt aus den üblichen Prinzipien der Quantenmechanik. Wenn man sie nicht haben will, so muß man die Zuständigkeit der Quantenmechanik leugnen. Daher nimmt *Ludwig* an, daß die Quantenmechanik nur im Grenzfalle mikroskopisch kleiner Objekte gilt, daß sie für den Meßapparat, den er immer als makroskopisch betrachtet, nicht anwendbar ist. Makroskopische Objekte können mit Hilfe von verhältnismäßig wenigen Parametern voll beschrieben werden, und diese Parameter haben immer bestimmte Werte. Die quantenmechanische Auffassung, daß verschiedene Werte der Zustandsparameter endliche Wahrscheinlichkeitsamplituden haben, trifft nach *Ludwig* für makroskopische Körper nicht zu.

Die starken wie auch die schwachen Punkte der *Ludwig*schen Auffassung sind zu offenbar, und ich will sie nicht im einzelnen anführen. Ungefähr gleichzeitig mit *Ludwig* habe auch ich einen Vorschlag in derselben Richtung gemacht [7], der aber nicht so radikal ist wie *Ludwig*s. Ich will auch auf diesen Vorschlag hier nicht näher eingehen und erwähne ihn nur, weil er mich auf die Begriffe geführt hat, die ich am Ende der Diskussion besprechen will.

Physikalische Probleme des Meßprozesses

Die vorangehende Diskussion beschäftigt sich mit den erkenntnistheoretischen Folgen des Quantentheoretischen Meßbegriffs. Für uns Physiker ist es aber auch wichtig, die Art und Mannigfaltigkeit der möglichen Meßprozesse kennenzulernen. Je zahlreicher sie sind, um so reicher werden die Aussagen der Quantenmechanik. Umgekehrt wird man es als den größten Mangel der großen Allgemeinheit der *v. Neumann*schen Gedanken ansehen müssen, daß Meßprozesse nur allgemein postuliert werden, ohne Vorschriften zu geben, wie sie ausgeführt werden können. Deshalb erhebt sich die Frage, ob man mit Hilfe allgemeiner Prinzipien etwas über die Meßbarkeit von Operatoren aussagen kann.

Es läßt sich leicht zeigen, daß im strengen orthodoxen Sinne und mit einem Apparat von endlicher Größe nur Operatoren gemessen werden können, die mit allen additiven Erhaltungsgrößen, wie Impuls, Energie, elektrische Ladung, vertauschbar sind. Wenn Operatoren mit diesen Erhaltungsgrößen nicht vertauschbar sind, so ist ihre Messung mit einem endlichen Apparat stets mit einem Fehler behaftet; d.h. die Messung kann fehlschlagen. Die Wahrscheinlichkeit für das Fehlschlagen kann um so kleiner gemacht werden, je größer der benutzte Apparat ist. Unter ,,Größe" verstehen wir hierbei seinen Gehalt am Quadrat der additiven Erhaltungsgröße[1]. Wie groß dieser aber auch sei, man kann einen mit einer Erhaltungsgröße unvertauschbaren

---

[1] Für gewöhnlich wird nicht das Quadrat der Erhaltungsgröße selbst, sondern deren mittleres Schwankungsquadrat benutzt; aber mit der ersten macht man auch die zweite Größe zu einem Minimum.

4

Operator nicht mit Sicherheit messen. Das folgt aus einer mathematischen Analyse des Meßprozesses, deren einfachste Formulierung wir *Araki* und *Yanase* [8] verdanken. Ich würde den mathematischen Beweis gern vorführen, weil er so einfach ist. Er beruht auf der Anwendung der Erhaltungssätze auf den „Zusammenstoß" von Objekt und Apparat. Ich will statt dessen nur am einfachsten physikalischen Fall den Ursprung des Satzes besprechen.

Betrachten wir als Beispiel die Messung einer Lagenkoordinate, die ja bekanntlich mit dem Impuls nicht vertauschbar ist. Will man die Ortskoordinate mit einer gewissen Genauigkeit messen, dann muß die Lage der Apparatur wenigstens mit der verlangten Genauigkeit bekannt sein. Das bedeutet aber, daß der Impuls des Apparates entsprechend unbestimmt ist, und das Quadrat des Impulses hat einen beträchtlichen Erwartungswert. Es ist auch klar, daß eine *genaue* Messung des Ortes jedenfalls unmöglich ist. Beim Drehimpuls sind die Verhältnisse nicht ganz so einfach und für elektrische und Baryonenladung muß man sich auf den mathematischen Beweis berufen. Es wird sich zeigen, daß dies nicht ohne Ursache ist.

Welche Größen kann man demnach genau messen? Am wichtigsten sind die absoluten Invarianten, wie Ruhemasse, Spin, elektrische und Baryonenladung. Ich glaube, daß *Jauch* diesen den Namen „essential variables" gegeben hat. Für sehr einfache Systeme sind diese auch die einzigen Invarianten, für kompliziertere Systeme gibt es noch andere.

Aus der vorstehenden Überlegung folgt zunächst, daß nur die Kenntnis über Invarianten leicht und ohne Verlust von einem Meßapparat zum anderen übertragen werden kann. Eine wiederholte Übertragung des Meßresultates, bis es unser Bewußtsein erreicht, spielt aber bei jeder Messung eine wesentliche Rolle. *Yanase* [9] hat deshalb vorgeschlagen, daß man nur solche Wechselwirkungen als Messung bezeichnen soll, bei denen die Zustände des Apparates, die zu den verschiedenen möglichen Zuständen des Objekts gekoppelt sind, durch Werte von *Invarianten* unterschieden werden können. Das wollen wir auch annehmen. Diese Forderung ist durchaus erfüllt bei den üblichen Messungen mit einem makroskopischen Meßinstrument: Die Markierung des Punktes, der mit dem Zeiger zusammenfällt, ist sicher eine Invariante. Die Forderung, die wir eben annahmen, geht etwas weiter, indem sie die Registrierung durch eine Invariante auch im Falle eines mikroskopischen Apparates fordert.

Nicht genau meßbare Größen: Die schiefe Information

Betrachtet man die Lagenmessung vom Standpunkt der Erhaltungsgrößen, so ist z. B. die Messung der x-Koordinate eine Messung der Phasenbeziehungen zwischen den *de Broglie*-Wellen mit verschiedenen Werten der Impulskomponente in der x-Richtung. Entspricht die Lage z. B. dem Punkte $x = 0$, so sind alle *de Broglie*-Wellen in Phase. Die Phase der *de Broglie*-Welle mit der Impulskomponente p_x ist $\exp(-i\,p_x\,a)$, wenn x den Mittelwert a hat. Die Lage mag keine exakt meßbare Größe sein, sie ist aber eine der Größen, die am häufigsten gemessen werden. Das bedeutet vom invariantentheoretischen Gesichtspunkt, daß die Messung von Phasenbeziehungen zwischen Zuständen, für die die Erhaltungsgröße p_x verschiedene Werte hat, zu den üblichsten Messungen gehört.

5

Dagegen hat noch niemand eine Phasenbeziehung zwischen Zuständen mit verschiedenen Werten der Erhaltungsgröße „elektrische Ladung" gemessen und es wird allgemein angenommen, daß eine solche Messung unmöglich ist. Woher stammt dieser Unterschied? Gibt es Übergänge zwischen diesen extremen Fällen? Ich glaube nicht, daß man die Theorie der Messung völlig versteht, solange man keine Antwort auf diese Fragen geben kann.

Wir haben gesehen, daß es viel leichter ist eine Invariante zu messen, als eine Größe, deren Eigenvektoren nicht parallel zu den Eigenvektoren der Invarianten liegen. Um das Maß unserer Kenntnis über die weniger leicht zugänglichen Eigenschaften des Zustandes zu charakterisieren, empfiehlt es sich eine Größe zu definieren, die wir „schiefe Information" nennen wollen. Sie soll unsere Kenntnis über nicht genau meßbare Größen durch eine Zahl charakterisieren, ähnlich wie die Entropie unsere gesamte Unkenntnis durch eine Zahl beschreibt. Die Bezeichnung „schiefe Information" mag nicht glücklich gewählt sein, aber vielleicht ist auch der mathematische Ausdruck, den ich vorschlagen will, nicht glücklich gewählt, in welchem Fall der Name nicht so wichtig ist.

Wir bezeichnen die Komponenten des Zustandsvektors mit zwei Indizes: k gibt den Wert der Erhaltungsgröße an und a steht für die übrigen Variablen. In einem Zustand, indem k nur einen Wert haben kann (wie man oft sagt „scharf" ist), soll die schiefe Information Null sein. Dies legt nahe, für die schiefe Information das mittlere Quadrat der Abweichung von k von seinem Mittelwert, also den Ausdruck

$$I = \sum_{ka} |\psi(k, a)|^2 k^2 - |(\sum_{ka} |\psi(k, a)|^2 k)^2 \tag{1}$$

zu wählen. Ist der Zustand eine statistische Gesamtheit, mit der statistischen Matrix $\varrho(k, a; k', a')$ so empfiehlt sich

$$I = \frac{\sum_{ka} \sum_{k'a'} |\varrho(k, a; k', a')|^2 (k - k')^2}{\sum_{ka} \sum_{k'a'} |\varrho(k, a; k', a')|^2} \tag{2}$$

zu setzen. Das ist nicht der einfachste Ausdruck, der im Falle, daß die Gesamtheit nur einen Zustandsvektor hat, in Gl. (1) übergeht. In diesem Fall ist z.B. der Nenner von Gl. (2) gleich 1, so daß er weggelassen werden könnte. Der Ausdruck (2) hat aber verschiedene Vorteile gegenüber allen anderen, die ich betrachtet habe. Im folgenden sollen die Eigenschaften von I, die mir wesentlich erscheinen, aufgezählt werden.

I ändert sich durch vier Prozesse: Durch den Ablauf der Zeit, da sich ϱ ändert; wenn man zwei Systeme vereinigt; wenn man ein System in zwei Systeme trennt; und schließlich, wenn man eine Messung am System vornimmt. Im Gegensatz zur Entropie sollte I im allgemeinen fallen, weil ja Information verloren gehen kann. I kann aber nicht immer fallen oder konstant sein, weil unsere Kenntnis über gewisse schiefe Größen, wie die Lagenkoordinate, gelegentlich wächst. Diese Möglichkeit des Wachsens von I

6

im Falle der Lagenkoordinaten und das ständige Null-sein im Falle von Größen, die schief zur elektrischen Ladung stehen, ist das, was wir erklären wollten.

Um die Änderung von I in den vier Fällen zu beschreiben, können wir damit anfangen, daß I konstant bleibt, wenn sich ψ oder ϱ gemäß der *Schrödinger*-Gleichung ändern. Das entspricht dem *Liouville*schen Satz im Falle des Ausdrucks für die Entropie. 2) Wenn man zwei Systeme zu einem Gesamtsystem vereinigt, ist das I des Gesamtsystems gleich der Summe der I der Teile. 3) Wenn man eine Messung am System vornimmt, so fällt I im Mittel (die Mittelung ist auszuführen über die verschiedenen möglichen Meßresultate mit den entsprechenden Wahrscheinlichkeiten). I kann aber steigen, obwohl es zumeist fällt, wenn man ein System in zwei Teile trennt. Das scheint zunächst im Widerspruch damit zu sein, daß I sich bei der Vereinigung von zwei Systemen additiv verhält. Man muß aber bedenken, daß die Systeme, die man vereinigt, statistisch unabhängig voneinander sind, während dies im allgemeinen nicht zutrifft für die Teile, in die man ein System spaltet.

Das Steigen von I kann also nur bei einer Trennung erfolgen, und auch dann ist es ein Ausnahmefall. Verschwindet I für das Gesamtsystem, so sind auch die Werte von I für die beiden Teilsysteme nach der Trennung Null. In diesem Fall kann I also nicht anwachsen und wird immer Null bleiben. Dies ist offenbar der Fall für die Information, die schief in bezug auf die elektrische Ladung steht. Wegen irgendeiner Ursache, die ich nicht angeben kann, wurde uns nie Information gegeben, die schief zur elektrischen und Baryonenladung usw. steht, und wir können daher keine solche Information erwerben. Dagegen haben wir schiefe Information in bezug auf den Impuls, und wir können diese durch zweckmäßiges Vereinen und späteres Trennen von Systemen vermehren. Wie wir die ursprüngliche schiefe Information erworben haben, kann auch nicht gesagt werden, sie ist uns fast angeboren.

Zusammenfassung

Die quantenmechanische Theorie der Messung zeigt, daß die Aussagen der Quantenmechanik als Wahrscheinlichkeitszusammenhänge zwischen aufeinanderfolgenden Beobachtungen gekennzeichnet werden können. Will man diesen Schluß vermeiden, so muß man, wie es z. B. *Ludwig* vorgeschlagen hat, die allgemeine Gültigkeit der Quantenmechanik in Abrede stellen.

Die übliche Theorie der quantenmechanischen Messung unterscheidet nicht zwischen Messungen verschiedener Größen. Es zeigt sich aber, daß es hinsichtlich Meßbarkeit drei Arten von Größen gibt. Die mit den additiven Erhaltungsgrößen vertauschbaren Operatoren können im Prinzip auch mit einem mikroskopischen Apparat gemessen werden. Die mit den additiven Erhaltungsgrößen nicht vertauschbaren Operatoren können nur mit makroskopischen Apparaten gemessen werden, und auch nur dann, wenn wenigstens etwas schiefe Information in bezug auf die Erhaltungsgrößen vorhanden ist, mit denen der Operator nicht vertauschbar ist. Diese schiefe Information kann dann vermehrt werden. Wenn keine solche schiefe Information vorhanden ist, wie z. B. in bezug auf die elektrische Ladung, so kann ein damit unvertauschbarer Operator überhaupt nicht gemessen werden. Dies ist die Ursache der entsprechenden Überauswahlregeln (superselection rules).

7

Literatur

[1] *W. Heisenberg*, Über den anschaulichen Inhalt der quantentheoretischen Kinematik und Mechanik. Z. Phys. **43**, 172 (1927).

[2] *J. v. Neumann*, Mathematische Grundlagen der Quantenmechanik (Berlin 1932, englische Übersetzung Princeton University Press 1955). Vgl. besonders Kapitel VI.

[3] *W. Pauli* Artikel im Handbuch der Phisyk (Berlin 1933) Band XXIV 1, Seite 83. Vgl. insbesondere Seiten 143—154.

[4] *L. D. Landau* und *E. M. Lifshitz*. Quantum Mechanics (Pergamon Press, London 1958. Vgl. besonders Kapitel I.

[5] *F. London* und *E. Bauer*, La Theorie de l'Observation en Mécanique Quantique (Paris 1939). Auch *H. Margenau* hat sich in verschiedenen Veröffentlichungen, ähnlichen Ansichten angeschlossen.

[6] *G. Ludwig*, Gelöste und ungelöste Probleme des Meßprozesses in der Quantenmechanik. In Sammelband: *Werner Heisenberg* und die Physik unserer Zeit, Braunschweig 1961.

[7] In der bald zu erscheinenden Sammlung The Scientist Speculates (*J. Good*, Editor).

[8] *G. Araki* und *M. Yanase*, Phys. Rev. **120**, 662 (1960).

[9] *M. Yanase*, Phys. Rev. **123**, 666 (1961).

(Eingegangen am 6. 2. 1962)

The Problem of Measurement

E. P. Wigner

Symmetries and Reflections.
Indiana University Press, Bloomington, Indiana, 1967, pp. 153–170

Introduction

The last few years have seen a revival of interest in the conceptual foundations of quantum mechanics.[1] This revival was stimulated by the attempts to alter the probabilistic interpretation of quantum mechanics. However, even when these attempts turned out to be less fruitful than its protagonists had hoped,[2] the interest continued. Hence, after the subject had been dormant for more than two decades, we again hear discussions on the basic principles of quantum theory and the epistemologies that are compatible with it. As is often the case under similar

Reprinted by permission from the *American Journal of Physics*, Vol. 31, No. 1 (January, 1963).

[1] Some of the more recent papers on the subject are: Y. Aharonov and D. Bohm, *Phys. Rev.*, 122, 1649 (1961); *Nuovo Cimento*, 17, 964 (1960); B. Bertotti, *Nuovo Cimento Suppl.*, 17, 1 (1960); L. de Broglie, *J. Phys. Radium*, 20, 963 (1959); J. A. de Silva, *Ann. Inst. Henri Poincaré*, 16, 289 (1960); A. Datzeff, *Compt. Rend.*, 251, 1462 (1960); *J. Phys. Radium*, 21, 201 (1960); 22, 101 (1961); J. M. Jauch, *Helv. Phys. Acta*, 33, 711 (1960); A. Landé, *Z. Physik*, 162, 410 (1961); 164, 558 (1961); *Am. J. Phys.*, 29, 503 (1961); H. Margenau and R. N. Hill, *Progr. Theoret. Phys.*, 26, 727 (1961); A. Peres and P. Singer, *Nuovo Cimento*, 15, 907 (1960); H. Putnam, *Phil. Sci.*, 28, 234 (1961); M. Renninger, *Z. Physik*, 158, 417 (1960); L. Rosenfeld, *Nature*, 190, 384 (1961); F. Schlögl, *Z. Physik*, 159, 411 (1960); J. Schwinger, *Proc. Natl. Acad. Sci. U.S.*, 46, 570 (1960); J. Tharrats, *Compt. Rend.*, 250, 3786 (1960); H. Wakita, *Progr. Theoret. Phys.*, 23, 32 (1960); 27, 139 (1962); W. Weidlich, *Z. Naturforsch.*, 15a, 651 (1960); J. P. Wesley, *Phys. Rev.*, 122, 1932 (1961). See also the articles of E. Teller, M. Born, A. Landé, F. Bopp, and G. Ludwig in *Werner Heisenberg und die Physik unserer Zeit* (Braunschweig: Friedrich Vieweg und Sohn, 1961).

[2] See the comments of V. Fock in the *Max Planck Festschrift* (Berlin: Deutscher Verlag der Wissenschaften, 1958), p. 177, particularly Sec. II.

circumstances, some of the early thinking had been forgotten; in fact, a small fraction of it remains as yet unrediscovered in the modern literature. Equally naturally, some of the language has changed but, above all, new ideas and new attempts have been introduced. Having spoken to many friends on the subject which will be discussed here, it became clear to me that it is useful to review the standard view of the late "Twenties," and this will be the first task of this article. The standard view is an outgrowth of Heisenberg's paper in which the uncertainty relation was first formulated.[3] The far-reaching implications of the consequences of Heisenberg's ideas were first fully appreciated, I believe, by von Neumann,[4] but many others arrived independently at conclusions similar to his. There is a very nice little book, by London and Bauer,[5] which summarizes quite completely what I shall call the orthodox view.

The orthodox view is very specific in its epistemological implications. This makes it desirable to scrutinize the orthodox view very carefully and to look for loopholes which would make it possible to avoid the conclusions to which the orthodox view leads. A large group of physicists finds it difficult to accept these conclusions and, even though this does not apply to the present writer, he admits that the far-reaching nature of the epistemological conclusions makes one uneasy. The misgivings, which are surely shared by many others who adhere to the orthodox view, stem from a suspicion that one cannot arrive at valid epistemological conclusions without a careful analysis of the *process of the acquisition of knowledge*. What will be analyzed, instead, is only the type of information which we can acquire and possess concerning the external inanimate world, according to quantum-mechanical theory.

We are facing here the perennial question whether we physicists do not go beyond our competence when searching for philosophical truth.

[3] W. Heisenberg, Z. *Physik*, 43, 172 (1927); also his article in *Niels Bohr and the Development of Physics* (London: Pergamon Press, 1955); N. Bohr, *Nature*, 121, 580 (1928); *Naturwissen.*, 17, 483 (1929) and particularly *Atomic Physics and Human Knowledge* (New York: John Wiley & Sons, Inc., 1958).

[4] See J. von Neumann, *Mathematische Grundlagen der Quantenmechanik* (Berlin: Verlag Julius Springer, 1932), English translation (Princeton, N.J.: Princeton University Press, 1955). See also P. Jordan, *Anschauliche Quantentheorie* (Berlin: Julius Springer, 1936), Chapter V.

[5] F. London and E. Bauer, *La Théorie de l'observation en mécanique quantique* (Paris: Hermann et Cie., 1939); or E. Schrödinger, *Naturwissen.*, 23, 807 ff. (1935); *Proc. Cambridge Phil. Soc.*, 31, 555 (1935).

The Problem of Measurement   155

I believe that we probably do.[6] Nevertheless, the ultimate implications of quantum theory's formulation of the laws of physics appear interesting even if one admits that the conclusions to be arrived at may not be the ultimate truth.

The Orthodox View

The possible states of a system can be characterized, according to quantum-mechanical theory, by state vectors. These state vectors—and this is an almost verbatim quotation of von Neumann—change in two ways. As a result of the passage of time, they change continuously, according to Schrödinger's time-dependent equation—this equation will be called the equation of motion of quantum mechanics. The state vector also changes discontinuously, according to probability laws, if a measurement is carried out on the system. This second type of change is often called the reduction of the wavefunction. It is this reduction of the state vector which is unacceptable to many of our colleagues.

The assumption of two types of changes of the state vector is a strange dualism. It is good to emphasize at this point that the dualism in question has little to do with the oft-discussed wave-versus-particle dualism. This latter dualism is only part of a more general pluralism or even "infinitesilism" which refers to the infinity of noncommuting measurable quantities. One can measure the position of the particles, or one can measure their velocity, or, in fact, an infinity of other observables. The dualism here discussed is a true dualism and refers to the *two* ways in which the state vector changes. It is also worth noting, though only parenthetically, that the probabilistic aspect of the theory is almost diametrically opposite to what ordinary experience would lead one to expect. The place where one expects probability laws to prevail is the change of the system with time. The interaction of the particles, their collisions, are the events which are ordinarily expected to be governed by statistical laws. This is not at all the case here: the uncertainty in the behavior of a system does not increase in time if the system is left alone, that is, if it is not subjected to measurements. In this case, the properties of the system, as described by its state vector,

---

[6] This point is particularly well expressed by H. Margenau, in the first two sections of the article in *Phil. Sci.*, 25, 23 (1958).

change causally, no matter what the period of time is during which it is left alone. On the contrary, the phenomenon of chance enters when a measurement is carried out on the system, when we try to check whether its properties did change in the way our causal equations told us they would change. However, the extent to which the results of all possible measurements on the system can be predicted does not decrease, according to quantum-mechanical theory, with the time during which the system was left alone; it is as great right after an observation as it is a long time thereafter. The uncertainty of the result, so to say, increases with time for some measurements just as much as it decreases for others. The Liouville theorem is the analog for this in classical mechanics. It tells us that, if the point which represents the system in phase space is known to be in a finite volume element at one given time, an equally large volume element can be specified for a given later time which will then contain the point representing the state of the system. Similarly, the uncertainty in the result of the measurement of Q, at time 0, is exactly equal to the uncertainty of the measurement of $Q_t = \exp(-iHt/\hbar)Q_0 \exp(iHt/\hbar)$ at time t. The information which is available at a later time may be less valuable than the information which was available on an earlier state of the system (this is the cause of the increase of the entropy); in principle, the amount of information does not change in time.

Consistency of the Orthodox View

The simplest way that one may try to reduce the two kinds of changes of the state vector to a single kind is to describe the whole process of measurement as an event in time, governed by the quantum-mechanical equations of motion. One might think that, if such a description is possible, there is no need to assume a second kind of change of the state vector; if it is impossible, one might conclude, the postulate of the measurement is incompatible with the rest of quantum mechanics. Unfortunately, the situation will turn out not to be this simple.

If one wants to describe the process of measurement by the equations of quantum mechanics, one will have to analyze the interaction between object and measuring apparatus. Let us consider a measurement from the point of view of which the "sharp" states are $\sigma^{(1)}, \sigma^{(2)}, \cdots$. For these states of the object the measurement will surely yield the values $\lambda_1, \lambda_2,$

The Problem of Measurement   **157**

$\cdots$, respectively. Let us further denote the initial state of the apparatus by a; then, if the initial state of the system was $\sigma^{(\nu)}$, the total system—apparatus plus object—will be characterized, before they come into interaction, by $a \times \sigma^{(\nu)}$. The interaction should not change the state of the object in this case and hence will lead to

$$a \times \sigma^{(\nu)} \rightarrow a^{(\nu)} \times \sigma^{(\nu)}. \tag{1}$$

The state of the object has not changed, but the state of the apparatus has and will depend on the original state of the object. The different states $a^{(\nu)}$ may correspond to states of the apparatus in which the pointer has different positions, which indicate the state of the object. The state $a^{(\nu)}$ of the apparatus will therefore be called also "pointer position ν." The state vectors $a^{(1)}, a^{(2)}, \cdots$ are orthogonal to each other —usually the corresponding states can be distinguished even macroscopically. Since we have considered, so far, only "sharp" states, for each of which the measurement in question surely yields one definite value, no statistical element has yet entered into our considerations.[7]

Let us now see what happens if the initial state of the object is not sharp, but an arbitrary linear combination $\alpha_1 \sigma^{(1)} + \alpha_2 \sigma^{(2)} + \cdots$. It then *follows* from the linear character of the quantum-mechanical equation of motion (as a result of the so-called superposition principle) that the state vector of object-plus-apparatus after the measurement becomes the right side of

$$a \times [\sum \alpha_\nu \sigma^{(\nu)}] \rightarrow \sum \alpha_\nu [a^{(\nu)} \times \sigma^{(\nu)}]. \tag{2}$$

Naturally, there is no statistical element in this result, as there cannot be. However, in the state (2), obtained by the measurement, there is a statistical correlation between the state of the object and that of the apparatus: the simultaneous measurement on the system—object-plus-apparatus—of the two quantities, one of which is the originally measured quantity of the object and the second the position of the pointer of the apparatus, always leads to concordant results. As a result, one of these measurements is unnecessary: The state of the object can be ascertained by an observation on the apparatus. This is a consequence of the special form of the state vector (2), of not containing any $a^{(\nu)} \times \sigma^{(\mu)}$ term with $\nu \neq \mu$.

---

[7] The self-adjoint (Hermitean) character of every observable can be derived from Eq. (1) and the unitary nature of the transformation indicated by the arrow. Cf. E. Wigner, Z. *Physik*, **133**, 101 (1952), footnote 2 on p. 102.

It is well known that statistical correlations of the nature just described play a most important role in the structure of quantum mechanics. One of the earliest observations in this direction is Mott's explanation of the straight track left by the spherical wave of outgoing α particles.[8] In fact, the principal conceptual difference between quantum mechanics and the earlier Bohr-Kramers-Slater theory is that the former, by its use of configuration space rather than ordinary space for its waves, allows for such statistical correlations.

Returning to the problem of measurement, we see that we have not arrived either at a conflict between the theory of measurement and the equations of motion, nor have we obtained an explanation of that theory in terms of the equations of motion. The equations of motion permit the description of the process whereby the state of the object is mirrored by the state of an apparatus. The problem of a measurement on the object is thereby transformed into the problem of an observation on the apparatus. Clearly, further transfers can be made by introducing a second apparatus to ascertain the state of the first, and so on. However, the fundamental point remains unchanged and a full description of an observation must remain impossible since the quantum-mechanical equations of motion are causal and contain no statistical element, whereas the measurement does.

It should be admitted that when the quantum theorist discusses measurements, he makes many idealizations. He assumes, for instance, that the measuring apparatus will yield some result, no matter what the initial state of the object was. This is clearly unrealistic since the object may move away from the apparatus and never come into contact with it. More importantly, he has appropriated the word "measurement" and used it to characterize a special type of interaction by means of which information can be obtained on the state of a definite object. Thus, the measurement of a physical constant, such as cross section, does not fall into the category called "measurement" by the theorist. His measurements answer only questions relating to the ephemeral state of a physical system, such as, "What is the x component of the momentum of this atom?" On the other hand, since he is unable to follow the path of the information until it enters his, or the observer's, mind, he considers the measurement completed as soon as a statistical relation has been established between the quantity to be measured and the state of some

[8] N. F. Mott, *Proc. Roy. Soc.* (London), 126, 79 (1929).

The Problem of Measurement   *159*

idealized apparatus. He would do well to emphasize his rather specialized use of the word "measurement."

This will conclude the review of the orthodox theory of measurement. As was mentioned before, practically all the foregoing is contained, for instance, in the book of London and Bauer.

Critiques of the Orthodox Theory

There are attempts to modify the orthodox theory of measurement by a complete departure from the picture epitomized by Eqs. (1) and (2). The only attempts of this nature which will be discussed here presuppose that the result of the measurement is not a state vector, such as (2), but a so-called mixture, namely, *one* of the state vectors

$$a^{(\mu)} \times \sigma^{(\mu)}, \tag{3}$$

and that this particular state will emerge from the interaction between object and apparatus with the probability $|\alpha_\mu|^2$. If this were so, the state of the system would not be changed when one ascertains—in some unspecified way—which of the state vectors (3) corresponds to the actual state of the system; one would merely "ascertain which of various possibilities has occurred." In other words, the final observation only increases our knowledge of the system; it does not change anything. This is not true if the state vector, after the interaction between object and apparatus, is given by (2) because *the state represented by the vector (2) has properties which neither of the states (3) has.* It may be worthwhile to illustrate this point, which is fundamental though often disregarded, by an example.

The example is the Stern-Gerlach experiment,[9] in which the projection of the spin of an incident beam of particles, into the direction which is perpendicular to the plane of the drawing, is measured. (See Fig. 1.) The index ν has two values in this case; they correspond to the two possible orientations of the spin. The "apparatus" is that positional coordinate of the particle which is also perpendicular to the plane of the drawing. If this coordinate becomes, in the experiment illustrated, positive, the spin is directed toward us; if it is negative, the spin is directed away from us. The experiment illustrates the statistical correla-

[9] The same experiment was discussed recently from another point of view by H. Wakita, *Progr. Theoret. Phys.*, 27, 139 (1962).

tion between the state of the "apparatus" (the position coordinate) and the state of the object (the spin) which we have discussed. The ordinary use of the experiment is to obtain the spin direction, by observing the position, i.e., the location of the beam. The measurement is, therefore, as far as the establishment of a statistical correlation is concerned, complete when the particle reaches the place where the horizontal spin arrows are located.

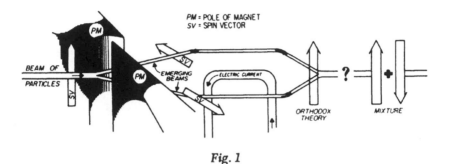

Fig. 1

What is important for us, however, is the right side of the drawing. This shows that the state of the system—object-plus-apparatus (spin and positional coordinates of the particle, i.e., the whole state of the particle)—shows characteristics which neither of the separated beams alone would have. If the two beams are brought together by the magnetic field due to the current in the cable indicated, the two beams will interfere and the spin will be vertical again. This could be verified by letting the united beam pass through a second magnet which is, however, not shown on the figure. If the state of the system corresponded to the beam toward us, its passage through the second magnet would show that it has equal probabilities to assume its initial and the opposite directions. The same is true of the second beam which was deflected away from us. Even though the experiment indicated would be difficult to perform, there is little doubt that the behavior of particles and of their spins conforms to the equations of motion of quantum mechanics under the conditions considered. Hence, the properties of the system, object-plus-apparatus, are surely correctly represented by an expression of the form (2) which gives, *in this case,* properties which are different from those of *either* alternative (3).

In the case of the Stern-Gerlach experiment, one can thus point to a specific and probably experimentally realizable way to distinguish between the state vector (2), furnished by the orthodox theory, and the more easily visualizable mixture of the states (3) which one would offhand expect. There is little doubt that in this case the orthodox theory is correct. It remains remarkable how difficult it is, even in this very simple case, to distinguish between the two, and this raises two questions. The first of these is whether there is, in more complicated cases, a principle which makes the distinction between the state vector (2), and the mixture of the states (3), impossible. As far as is known to the present writer, this question has not ever been posed seriously heretofore, and it will be considered in the present discussion also only obliquely. The second question is whether there is a continuous transition between (2) and the mixture of states (3) so that in simpler cases (2) is the result of the interaction between object and measuring apparatus, but in more complicated and more realistic cases the actual state of object-plus-apparatus more nearly resembles a mixture of the states (3). Again, this question can be investigated within the framework of quantum mechanics, or one can postulate deviations from the quantum-mechanical equations of motion, in particular from the superposition principle.

"More complicated" and "more realistic" mean in the present context that the measuring apparatus, the state of which is to be correlated with the quantity to be measured, is of such a nature that it is easy to measure *its* states, i.e., correlate it with the state of another "apparatus." If this is done, the state of that second "apparatus" will be correlated also to the state of the object. The case of establishing correlations between the state of the apparatus which came into direct contact with the object and another "apparatus" is usually greatest if the first one is of macroscopic nature, i.e., complicated from the quantum-mechanical point of view. The ease with which the secondary correlations can be established is a direct measure of how realistically one can say that the measurement has been completed. Clearly, if the state of the apparatus which carried out the primary measurement is just as difficult to ascertain as the state of the object, it is not very realistic to say that the establishment of a correlation between its and the object's state is a fully completed measurement. Nevertheless, it is so regarded by the orthodox theory. The question which we pose is, therefore, whether it is consistent with the principles of quantum mechanics to assume that at the end of a realistic measurement the state of object-plus-apparatus is not a wavefunction, as given by (2), but a mixture of the states (3). We shall see that the answer is negative. Hence, the modification of the orthodox theory of measurement mentioned at the beginning of this section is not consistent with the principles of quantum mechanics.

Let us now proceed with the calculation. Even though this point is not usually emphasized, it is clear that, in order to obtain a mixture of states as a result of the interaction, the initial state must have been a mixture already.[10] This follows from the general theorem that the characteristic values of the density matrix are constants of motion. The assumption that the initial state of the system, object-plus-apparatus, is a mixture, is indeed a very natural one because the state vector of the apparatus, which is under the conditions now considered usually a macroscopic object, is hardly ever known. Let us assume, therefore, that the initial state of the apparatus is a mixture of the states $A^{(1)}$, $A^{(2)}$, $\cdots$, the probability of $A^{(\rho)}$ being p_ρ. The vectors $A^{(\rho)}$ can be assumed to be mutually orthogonal. The equations of motion will yield, for the state $A^{(\rho)}$ of the apparatus and the state $\sigma^{(\nu)}$ of the object, a final state

$$A^{(\rho)} \times \sigma^{(\nu)} \to A^{(\rho\nu)} \times \sigma^{(\nu)}. \tag{4}$$

Every state $A^{(1\nu)}$, $A^{(2\nu)}$, $\cdots$ will indicate the same state $\sigma^{(\nu)}$ of the object; the position of the pointer is ν for all of these. For different ν, however, the position of the pointer is also different. It follows that the $A^{(\rho\nu)}$, for different ν, are orthogonal, even if the ρ are also different. On the other hand, $A^{(\rho\nu)}$ and $A^{(\sigma\nu)}$, for $\rho \neq \sigma$, are also orthogonal because

---

[10] This point is disregarded by several authors who have rediscovered von Neumann's description of the measurement, as given by (1) and (2). These authors assume that it follows from the macroscopic nature of the measuring apparatus that if several values of the "pointer position" have finite probabilities [as is the case if the state vector is (2)], the state is necessarily a *mixture* (rather than a linear combination) of the states (3)—that is, of states in each of which the pointer position is definite (sharp). The argument given is that classical mechanics applies to macroscopic objects, and states such as (2) have no counterpart in classical theory. *This argument is contrary to present quantum-mechanical theory.* It is true that the motion of a macroscopic body can be adequately described by the classical equations of motion if its state has a classical description. That this last premise is, according to present theory, not always fulfilled, is clearly, though in an extreme fashion, demonstrated by Schrödinger's cat-paradox (cf. reference 5). Further, the discussion of the Stern-Gerlach experiment, given in the text, illustrates the fact that there are, in principle, observable differences between the state vector given by the right side of (2), and the *mixture* of the states (3), each of which has a definite position. Proposals to modify the quantum-mechanical equations of motion so as to permit a mixture of the states (3) to be the result of the measurement even though the initial state was a state vector, will be touched upon later.

The Problem of Measurement   163

$A^{(\rho\nu)} \times \sigma^{(\nu)}$ and $A^{(\sigma\nu)} \times \sigma^{(\nu)}$ are obtained by a unitary transformation from two orthogonal states, $A^{(\rho)} \times \sigma^{(\nu)}$ and $A^{(\sigma)} \times \sigma^{(\nu)}$ and the scalar product of $A^{(\rho\nu)} \times \sigma^{(\nu)}$ with $A^{(\sigma\nu)} \times \sigma^{(\nu)}$ is $(A^{(\rho\nu)}, A^{(\sigma\nu)})$. Hence, the $A^{(\rho\nu)}$ form an orthonormal (though probably not complete) system

$$(A^{(\rho\nu)}, A^{(\sigma\mu)}) = \delta_{\rho\sigma}\delta_{\nu\mu}. \tag{5}$$

It again follows from the linear character of the equation of motion that, if the initial state of the object is the linear combination $\sum\alpha_\nu\sigma^{(\nu)}$, the state of object-plus-apparatus will be, after the measurement, a mixture of the states

$$A^{(\rho)} \times \sum \alpha_\nu\sigma^{(\nu)} \to \sum_\nu \alpha_\nu[A^{(\rho\nu)} \times \sigma^{(\nu)}] = \Phi^{(\rho)}, \tag{6}$$

with probabilities p_ρ. This same mixture should then be, according to the postulate in question, equivalent to a mixture of orthogonal states

$$\Psi^{(\mu k)} = \sum_\rho x_\rho^{(\mu k)}[A^{(\rho\mu)} \times \sigma^{(\mu)}]. \tag{7}$$

These are the most general states for which the originally measured quantity has a definite value, namely λ_μ, and in which this state is coupled with some state (one of the states $\sum_\rho x_\rho^{(\mu\kappa)}A^{(\rho\mu)}$) with a pointer position μ. Further, if the probability of the state $\Psi^{(\mu k)}$ is denoted by $P_{\mu k}$, we must have

$$\sum_k P_{\mu k} = |\alpha_\mu|^2. \tag{7a}$$

The $x_\rho^{(\mu k)}$ will naturally depend on the α.

It turns out, however, that a mixture of the states $\Phi^{(\rho)}$ cannot be, at the same time, a mixture of the states $\Psi^{(\mu k)}$ (unless only one of the α is different from zero). A necessary condition for this would be that the $\Psi^{(\mu k)}$ are linear combinations of the $\Phi^{(\rho)}$, so that one should be able to find coefficients u so that

$$\sum_\rho x_\rho^{(\mu k)}[A^{(\rho\mu)} \times \sigma^{(\mu)}] = \Psi^{(\mu k)} = \sum_\rho u_\rho\Phi^{(\rho)}$$
$$= \sum_{\rho\nu} u_\rho\alpha_\nu[A^{(\rho\nu)} \times \sigma^{(\nu)}]. \tag{8}$$

From the linear independence of the $A^{(\rho\nu)}$ it then follows that

$$u_\rho\alpha_\nu = \delta_{\nu\mu}x_\rho^{(\mu k)}, \tag{8a}$$

which cannot be fulfilled if more than one α is finite. It follows that it is not compatible with the equations of motion of quantum mechanics to assume that the state of object-plus-apparatus is, after a measurement, a mixture of states each with one definite position of the pointer.

It must be concluded that *measurements which leave the system ob-ject-plus-apparatus in one of the states with a definite position of the pointer cannot be described by the linear laws of quantum mechanics.* Hence, if there are such measurements, quantum mechanics has only limited validity. This conclusion must have been familiar to many even though the detailed argument just given was not put forward before. Ludwig, in Germany, and the present writer have independently suggested that the equations of motion of quantum mechanics must be modified so as to permit measurements of the aforementioned type.[11] These suggestions will not be discussed in detail because they are suggestions and do not have convincing power at present. Even though either may well be valid, one must conclude that the only known theory of measurement which has a solid foundation is the orthodox one and that this implies the dualistic theory concerning the changes of the state vector. It implies, in particular, the so-called reduction of the state vector. However, to answer the question posed earlier: yes, there is a continuous transition between the state vector (2), furnished by orthodox theory, and the requisite mixture of the states (3), postulated by a more visualizable theory of measurement.[11]

What Is the State Vector?

The state vector concept plays such an important part in the formulation of quantum-mechanical theory that it is desirable to discuss its role and the ways to determine it. Since, according to quantum mechanics, all information is obtained in the form of the results of measurements, the standard way to obtain the state vector is also by carrying out measurements on the system.[12]

In order to answer the question proposed, we shall first obtain a for-

[11] See G. Ludwig's article "Solved and Unsolved Problems in the Quantum Mechanics of Measurement" (reference 1) and the present author's article "Remarks on the Mind-Body Question" in *The Scientist Speculates*, edited by I. J. Good (London: William Heinemann, 1962), p. 284, reprinted in this volume.

[12] There are, nevertheless, other procedures to bring a system into a definite state. These are based on the fact that a small system, if it interacts with a large system in a definite and well-known state, may assume itself a definite state with almost absolute certainty. Thus, a hydrogen atom, in some state of excitation, if placed into a large container with no radiation in it, will almost surely transfer all its energy to the radiation field of the container and go over into its normal state. This method of preparing a state has been particularly stressed by H. Margenau.

The Problem of Measurement   165

mula for the probability that successive measurements carried out on a system will give certain specified results. This formula will be given both in the Schrödinger and in the Heisenberg picture. Let us assume that n successive measurements are carried out on the system, at times $t_1, t_2, \cdots, t_n$. The operators of the quantities which are measured are, in the Schrödinger picture, $Q_1, Q_2, \cdots, Q_n$. The characteristic vectors of these will all be denoted by ψ with suitable upper indices. Similarly, the characteristic values will be denoted by q so that

$$Q_j \psi_\kappa{}^{(j)} = q_\kappa{}^{(j)} \psi_\kappa{}^{(j)} \tag{9}$$

The Heisenberg operators which correspond to these quantities, if measured at the corresponding times, are

$$Q_j{}^H = e^{iHt_j} Q_j e^{-iHt_j} \tag{10}$$

and the characteristic vectors of these will be denoted by $\varphi_\kappa{}^{(j)}$, where

$$\varphi_\kappa{}^{(j)} = e^{iHt_j} \psi_\kappa{}^{(j)} \qquad Q_j{}^H \varphi_\kappa{}^{(j)} = q_\kappa{}^{(j)} \varphi_\kappa{}^{(j)}. \tag{10a}$$

If the state vector is originally Φ, the probability for the sequence $q_\alpha{}^{(1)}$, $q_\beta{}^{(2)}, \ldots, q_\mu{}^{(n)}$ of measurement-results is the absolute square of

$$\left(e^{-iHt_1}\Phi, \psi_\alpha{}^{(1)}\right) \left(e^{-iH(t_2-t_1)} \psi_\alpha{}^{(1)}, \psi_\beta{}^{(2)}\right) \cdots$$
$$\left(e^{-iH(t_n-t_{n-1})}\psi_\lambda{}^{(n-1)}, \psi_\mu{}^{(n)}\right). \tag{11}$$

The same expression in terms of the characteristic vectors of the Heisenberg operators is simpler,

$$\left(\Phi, \varphi_\alpha{}^{(1)}\right)\left(\varphi_\alpha{}^{(1)}, \varphi_\beta{}^{(2)}\right) \cdots \left(\varphi_\lambda{}^{(n-1)}, \varphi_\mu{}^{(n)}\right). \tag{11a}$$

It should be noted that the probability is not determined by the n Heisenberg operators $Q_j{}^H$ and their characteristic vectors: the *time order* in which the measurements are carried out enters into the result essentially. Von Neumann already derived these expressions as well as their generalizations for the case in which the characteristic values $q_\alpha{}^{(1)}$, $q_\beta{}^{(2)}, \cdots$ have several characteristic vectors. In this case, it is more appropriate to introduce projection operators for every characteristic value $q^{(j)}$ of every Heisenberg operator $Q_j{}^H$. If the projection operator in question is denoted by $P_{j\kappa}$, the probability for the sequence $q_\alpha{}^{(1)}, q_\beta{}^{(2)}, \cdots$, $q_\mu{}^{(n)}$ of measurement-results is

$$(P_{n\mu} \cdots P_{2\beta} P_{1\alpha}\Phi,\, P_{n\mu} \cdots P_{2\beta} P_{1\alpha}\Phi). \tag{12}$$

The expressions (11) or (11a) can be obtained also by postulating that the state vector became $\psi_\kappa^{(j)}$ when the measurement of $Q^{(j)}$ gave the result $q_\kappa^{(j)}$. Indeed, the statement that the state vector is $\psi_\kappa^{(j)}$ is only a short expression for the fact that the last measurement on the system, of the quantity $Q^{(j)}$, just carried out, gave the result $q_\kappa^{(j)}$. In the case of simple characteristic values the state vector depends only on the result of the last measurement and the future behavior of the system is independent of the more distant past history thereof. This is not the case if the characteristic value $q^{(j)}$ is multiple.

The most simple expression for the Heisenberg state vector, when the *j*th measurement gave the value $q_\kappa^{(j)}$, is, in this case,

$$P_{j\kappa} \cdots P_{2\beta} P_{1\alpha} \Phi, \tag{12a}$$

properly normalized. If, after normalization, the expression (12a) is independent of the original state vector Φ, the number of measurements has sufficed to determine the state of the system completely and a pure state has been produced. If the vector (12a) still depends on the original state vector Φ, and if this was not known to begin with, the state of the system is a mixture, a mixture of all the states (12a), with all possible Φ. Evidently, the measurement of a single quantity Q, the characteristic values of which are all nondegenerate, suffices to bring the system into a pure state though it is not in general foreseeable which pure state will result.

We recognize, from the preceding discussion, that the state vector is only a shorthand expression of that part of our information concerning the past of the system which is relevant for predicting (as far as possible) the future behavior thereof. The density matrix, incidentally, plays a similar role except that it does not predict the future behavior as completely as does the state vector. We also recognize that *the laws of quantum mechanics only furnish probability connections between results of subsequent observations carried out on a system*. It is true, of course, that the laws of classical mechanics can also be formulated in terms of such probability connections. However, they can be formulated also in terms of objective reality. The important point is that the laws of quantum mechanics can be expressed only in terms of probability connections.

The Problem of Measurement   167

Problems of the Orthodox View

The incompatibility of a more visualizable interpretation of the laws of quantum mechanics with the equations of motion, in particular the superposition principle, may mean that the orthodox interpretation is here to stay; it may also mean that the superposition principle will have to be abandoned. This may be done in the sense indicated by Ludwig, in the sense proposed by me, or in some third, as yet unfathomed sense. The dilemma which we are facing in this regard makes it desirable to review any possible conceptual weaknesses of the orthodox interpretation and the present, last, section will be devoted to such a review.

The principal conceptual weakness of the orthodox view is, in my opinion, that it merely abstractly postulates interactions which have the effect of the arrows in (1) or (4). For some observables, in fact for the majority of them (such as xyp_z), nobody seriously believes that a measuring apparatus exists. It can even be shown that no observable which does not commute with the additive conserved quantities (such as linear or angular momentum or electric charge) can be measured precisely, and in order to increase the accuracy of the measurement one has to use a very large measuring apparatus. The simplest form of the proof heretofore was given by Araki and Yanase.[13] On the other hand, most quantities which we believe to be able to measure, and surely all the very important quantities such as position, momentum, fail to commute with all the conserved quantities, so that their measurement cannot be possible with a microscopic apparatus. This raises the suspicion that the macroscopic nature of the apparatus is necessary in principle and reminds us that our doubts concerning the validity of the superposition principle for the measurement process were connected with the macroscopic nature of the apparatus. The joint state vector (2), resulting from a measurement with a very large apparatus, surely *cannot be distinguished* as *simply from a mixture* as was the state vector obtained in the Stern-Gerlach experiment which we discussed.[14]

A second, though probably less serious, difficulty arises if one tries to

[13] H. Araki and M. Yanase, *Phys. Rev.*, 120, 666 (1961); cf. also E. P. Wigner, Z. *Physik*, 131, 101 (1952).
[14] This point was recognized already by D. Bohm. See Section 22.11 of his *Quantum Theory* (Englewood Cliffs, New Jersey: Prentice-Hall, Inc., 1951).

calculate the probability that the interaction between object and apparatus be of such nature that there exist states $\sigma^{(\nu)}$ for which (1) is valid. We recall that an interaction leading to this equation was simply postulated as the type of interaction which leads to a measurement. When I talk about the probability of a certain interaction, I mean this in the sense specified by Rosenzweig or by Dyson, who have considered ensembles of possible interactions and defined probabilities for definite interactions.[15] If one adopts their definition (or any similar definition) the probability becomes zero for the interaction to be such that there are states $\sigma^{(\nu)}$ satisfying (1). The proof for this is very similar to that[16] which shows that the probability is zero for finding reproducing systems—in fact, according to (1), each $\sigma^{(\nu)}$ is a reproducing system. The resolution of this difficulty is presumably that if the system with the state vector a—that is, the apparatus—is very large, (1) can be satisfied with a very small error. Again, the large size of the apparatus appears to be essential for the possibility of a measurement.

The simplest and least technical summary of the conclusions which we arrived at when discussing the orthodox interpretation of the quantum laws is that these laws merely provide probability connections between the results of several consecutive observations on a system. This is not at all unreasonable and, in fact, this is what one would naturally strive for once it is established that there remains some inescapable element of chance in our measurements. However, there is a certain weakness in the word "consecutive," as this is not a relativistic concept. Most observations are not local and one will assume, similarly, that they have an irreducible extension in time, that is, duration. However, the "observables" of the present theory are instantaneous, and hence unrelativistic, quantities. The only exceptions from this are the local field operators and we know, from the discussion of Bohr and Rosenfeld, how many extreme abstractions have to be made in order to describe their measurement.[17] This is not a reassuring state of affairs.

[15] C. E. Porter and N. Rosenzweig, *Suomalaisen Tiedeakatemian Toimotuksia*, VI, No. 44 (1960); *Phys. Rev.*, 120, 1698 (1960); F. Dyson, *J. Math. Phys.*, 3, 140, 157, 166 (1962). See also E. P. Wigner, *Proceedings of the Fourth Canadian Mathematics Congress* (Toronto: University of Toronto Press, 1959), p. 174, reprinted in this volume.

[16] Cf. the writer's article in *The Logic of Personal Knowledge* (London: Routledge and Kegan Paul, 1961), p. 231, reprinted in this volume.

[17] N. Bohr and L. Rosenfeld, *Kgl. Danske Videnskab. Selskab, Mat.-fys. Medd.*, 12, No. 8 (1933); *Phys. Rev.*, 78, 194 (1950); E. Corinaldesi, *Nuovo Cimento*, 8, 494 (1951); B. Ferretti, *ibid.*, 12, 558 (1954).

The Problem of Measurement   169

The three problems just discussed—or at least two of them—are real. It may be useful, therefore, to re-emphasize that they are problems of the formal mathematical theory of measurement, and of the description of measurements by macroscopic apparatus. They do not affect the conclusion that a "reduction of the wave packet" (however bad this terminology may be) takes place in some cases. Let us consider, for instance, the collision of a proton and a neutron and let us imagine that we view this phenomenon from the coordinate system in which the center of mass of the colliding pair is at rest. The state vector is then, if we disregard the unscattered beam, in very good approximation (since there is only S-scattering present),

$$\psi(r_p, r_n) = r^{-1} e^{ikr} w(r), \tag{13}$$

where $r = |r_p - r_n|$ is the distance of the two particles and $w(r)$ some very slowly varying damping function which vanishes for $r < r_0 - \frac{1}{2}c$ and $r > r_0 + \frac{1}{2}c$, where r_0 is the mean distance of the two particles at the time in question and c the coherence length of the beam. If a measurement of the momentum of one of the particles is carried out—the possibility of this is never questioned—and gives the result **p**, the state vector of the other particle suddenly becomes a (slightly damped) plane wave with the momentum − **p**. This statement is synonymous with the statement that a measurement of the momentum of the second particle would give the result − **p**, as follows from the conservation law for linear momentum. The same conclusion can be arrived at also by a formal calculation of the possible results of a joint measurement of the momenta of the two particles.

One can go even further[18]: instead of measuring the linear momentum of one particle, one can measure its angular momentum about a fixed axis. If this measurement yields the value $m\hbar$, the state vector of the other particle suddenly becomes a cylindrical wave for which the same component of the angular momentum is − $m\hbar$. This statement is again synonymous with the statement that a measurement of the said component of the angular momentum of the second particle certainly would give the value − $m\hbar$. This can be inferred again from the conservation law of the angular momentum (which is zero for the two

---

[18] See, in this connection, the rather similar situation discussed by A. Einstein, B. Podolsky, and N. Rosen, *Phys. Rev.*, 47, 777 (1935).

particles together) or by means of a formal analysis. Hence, a "contraction of the wave packet" took place again.

It is also clear that it would be wrong, in the preceding example, to say that even before any measurement, the state was a mixture of plane waves of the two particles, traveling in opposite directions. For no such pair of plane waves would one expect the angular momenta to show the correlation just described. This is natural since plane waves are not cylindrical waves, or since (13) is a state vector with properties different from those of any mixture. The statistical correlations which are clearly postulated by quantum mechanics (and which can be shown also experimentally, for instance in the Bothe-Geiger experiment) demand in certain cases a "reduction of the state vector." The only possible question which can yet be asked is whether such a reduction must be postulated also when a measurement with a macroscopic apparatus is carried out. The considerations around Eq. (8) show that even this is true *if* the validity of quantum mechanics is admitted for all systems.

Some Comments
Concerning Measurements in Quantum Mechanics

J. M. Jauch, E. P. Wigner, and M. M. Yanase

Il Nuovo Cimento *48B*, 144–151 (1967)
(Reset by Springer-Verlag for this volume)

Ricevuto il 1° Dicembre 1966

Summary. The paradoxical aspects of the quantum-mechanical measuring process continue to provoke numerous discussions, indicating a growing interest in the basic problems of quantum mechanics [1,2]. The purpose of the present note is to clarify our views on this important problem and to contrast them with the views expressed in some recent publications on the same subject [1h,i].

1. Discussion Under the Assumption
That Quantum Mechanics Has Unrestricted Validity

We shall start the discussion on the basis of the assumption that quantum mechanics has unrestricted validity. We do not mean by this that any particular set of equations is absolutely and accurately valid – in the present state of the theory we would not even know to which equation to attribute such validity. Nor do we wish to imply here that the physical content of quantum mechanics excludes necessarily a causal infrastructure. Rather, we mean only that the general framework of quantum mechanics has unrestricted validity, from which it follows that the outcome of all possible future observations on a system can be predicted only statistically, and that the limits of predictability are definite. Thus, to repeat von Neumann's example, no matter how often we ascertained whether the spin of an atom is in one or the opposite direction, we will still be unable to foresee whether it will adjust into a new direction or into the opposite one unless the direction is parallel or antiparallel to the last direction

---

[1] a) A. Daneri, A. Loinger and G. M. Prosperi: Nucl. Phys. **33**, 297 (1962); b) E. P. Wigner: Am. Journ. Phys. **31**, 6 (1963); c) A. Shimony: Am. Journ. Phys. **31**, 755 (1963); d) P. A. Moldauer: Am. Journ. Phys. **32**, 172 (1964): e) M. M. Yanase: Am. Journ. Phys. **32**, 208 (1964); f) J. M. Jauch: Helv. Phys. Acta **37**, 193 (1964); g) W. C. Davidon and H. Ekstein: Journ. Math. Phys. **5**, 1588 (1964); h) L. Rosenfeld: Suppl. Progr. Theor. Phys., extra number, p. 222 (1965); i) A. Daneri, A. Loinger and G. M. Prosperi: Nuovo Cimento **44** B, 119 (1966); j) J. S. Bell: Rev. Mod. Phys. **38**, 447 (1966).

[2] B. d'Espagnat: Conceptions de la Physique Contemporaine, in Actualités Scientifiques et Industrielles (Paris, 1965). This book has a complete bibliography of the older references on the problem of measurement in quantum mechanics.

which it assumed. Furthermore, the probability for a direction and its opposite depends only on the outcome of the last observation we made on its direction. In addition, the linear character of the equations of motion will also be assumed – it is a very general and simple principle.

Let us state at once that we continue to believe that von Neumann's description of the measurement process is basically correct and that it is in fact in agreement with the principles of quantum theory in the sense just described. As has been pointed out many times before, von Neumann's theory, if followed to its ultimate consequences, leads to an epistemological dilemma.

The first alternative is to assume that quantum mechanics refrains from making any definite statements about "physical reality" just as relativity theory has taught us to refrain from defining "absolute rest". Rather quantum mechanics only furnishes us with correlations between subsequent observations and these correlations are, in general, only statistical.

The second alternative is to say that the wave function, or more generally the state vector, is a description of the "physical reality". In that case one has to admit that the state vector may change in two fundamentally different ways: continously and causally, as a result of the lapse of time, according to the time-dependent Schrödinger equation, and discontinuously and erratically, as a result of observations. This second change is often called "collapse of the wave function".

According to the accepted principles of scientific epistemology neither of these two points of view seems satisfactory. According to the first the state vector is reduced to the status of a mere mathematical tool expressing the part of earlier observations which have relevance for predicting results of later ones. It thus no longer represents the "state" of an individual system but describes only some properties of ensembles of such systems prepared under identical relevant conditions.

The second point of view elevates the state vector into a description of "physical reality" (whatever that may mean). The price one pays for this amounts to a breakdown of the unity of physical science: all physical systems behave according to the principles of quantum mechanics under all circumstances except one, *viz.* when they are under observation.

Actually, it is not clear whether the dilemma is a real one. The concept of "physical reality" as far as inanimate objects are concerned, may itself lack of "physical reality" just as the concept of absolute rest does. What we can do is to foresee, to some degree, what we are going to experience and all other questions concerning "reality" may constitute only an unnecessary superstructure. In the terminology of K.R. Popper, the reality may not be "falsifiable". This does not mean that we have to abandon the concept of "reality" altogether. It only means that this concept does not seem to be necessary for the formulation of the conclusions to which physical theory leads us.

2. Attempts to Describe the Observation by the Equations of Quantum Mechanics

It has been proposed to escape the dilemma described in the preceding Section by substituting an apparatus for the observer, the behavior of which, together with the behavior of the object, can be described by the time-dependent Schrödinger equation, *i.e.* without the postulate of the "collapse of the state vector" (without the "sudden and erratic" that is the second kind of change in the state vector). This is, indeed, possible and the process was, in fact, described in this manner already by von Neumann. However, the interaction between object and apparatus leads only to a joint state of the two in which there is a correlation between them in the sense that an observation of the state of the apparatus also provides information on the state of the object. This is satisfactory but does not resolve the aforementioned dilemma. It only substitutes the problem of the observation on the object by the problem of the observation of the apparatus. It is, in particular, not true that the final state of the apparatus is one which has a complete classical description. Thus, if one subjects the joint state of object plus apparatus to a time inversion (reverses the directions of all velocities), it will return to the initial state which is a different state from those obtained by time inversion of any of the final macroscopic states.

It may be well to illustrate the preceding point on an example. The example will be the measurement of the direction of the spin. Let us denote the wave functions for the states in which the spin is parallel and antiparallel to a specified direction (usually called the z-axis) by α and β, respectively. The initial state of the apparatus will be called a, the final state which indicates "parallel" as the result of the observation shall be called a_+, the state indicating "antiparallel" by a_-. The proper working of the apparatus then means that the interaction between apparatus and spin results in the transformation of the wave function αa into αa_+ and βa into βa_-:

$$(1) \qquad \alpha a \to \alpha a_+ , \qquad \beta a \to \beta a_- .$$

It follows from the linearity of the equations of motion that, if we start with an arbitrary state of the spin, such as $c\alpha + c'\beta$, the final state will be

$$(2) \qquad (c\alpha + c'\beta)a \to c\alpha a_+ + c'\beta a_- .$$

In this final state there is a correlation between the direction of the spin and the state of the apparatus. This is apparent from (2): the only two states of the joint system, spin plus apparatus, which appear in (2) are αa_- and βa_- and there is no αa_- or βa_-. These would represent states in which the spin is parallel to the preferred direction but the apparatus indicates "antiparallel", and conversely. This is an intuitive interpretation of (2) which can be fully confirmed by a formal application of the theory of measurement on the joint system. What we emphasized before is that the transition indicated by (2) does not conclude the measurement: it only establishes a correlation between the states of spin and apparatus. In other words, the + is not an "or" but addition of two vectors in Hilbert space. In fact, the state $c\alpha a_+ + c'\beta a_-$ of the joint system has properties which neither the state αa_+ nor the state βa_- has. In particular, under time inversion (*i.e.* if one reverses all the velocities) it will return into $(c\alpha + c'\beta)a$, which is different from both αa and βa. Thus, if $c = c' = 2^{-\frac{1}{2}}$, the measurement of the spin in the x-direction, perpendicular to the z-axis considered before, certainly will give the result "parallel" (for $c = -c' = 2^{-\frac{1}{2}}$ it would have given the result "antiparallel"). This is not true for either αa nor for βa: both of these measurements of the spin in the x-direction give "antiparallel" with a probability $\frac{1}{2}$.

It follows that the transition from $(c\alpha + c'\beta)a$ into $c\alpha a_- - c'\beta a_-$ does not conclude the measurement and does not result either in the state αa_+, or in the state βa_-, but in the state different from both. This is, as Daneri, Loinger and Prosperi say, "quite correct and obvious". It does not, however, represent the most general situation which may be invoked to give a complete quantum-mechanical description of the measurement.

It is possible, and in fact natural, to go further and to assume that the initial state of the apparatus – which may be of macroscopic nature – is not known, that it may be in any of the several states $a_1, \ldots, a_n$. The probabilistic nature of the outcome of the observation then could be blamed on our ignorance of the initial state of the apparatus. Let us consider, as an example, the state of a spin which is a linear combination of the parallel and antiparallel orientations in such a way that a measurement of its direction gives, with the probability w the "parallel" result (and with probability $1 - w$ the "antiparallel" result), this apparently acausal situation would not have to contradict the principle of causality. One could assume that the wn among the states $a_1, a_2, \ldots, a_n$ give, with certainty, the answer "parallel", the remaining $(1 - w)n$ of the a the answer "antiparallel". This is an attractive possibility which would blame the acausal nature of the observation process on the accident of starting with one or another of the states of the apparatus. Unfortunately, it can be shown that this explanation of the statistical nature of the observation is in conflict with the linearity of the equation of motion. This was demonstrated by one of the present writers and we have no explanation for the contrary assertion of Daneri, Loinger and Prosperi[3].

There seems, thus, no escape from the conclusions of the preceding Section and the resulting epistemological dilemma as long as one assumes the validity of the basic principles of quantum mechanics, in particular of the superposition principle and the linearity of the equations of motion.

---

[3] See ref. 1b) and 1i) respectively.

3. The Rôle of the Classical Observer

The dilemma mentioned in the proceding Sections has been discussed many times before. In common with many of these previous discussions we have omitted to mention an important qualification for a suitable measuring device: *viz.* that it must be a "classical" system. This point was especially stressed by Bohr in all of his discussions of this problem. What does this mean and why is this necessary for a measuring device?

The origin of this requirement is deeply rooted in the very essence of scientific methodology. A measurement, in order to qualify for the raw material of a scientific theory, must permit us to verify facts which are "unambiguous" and "objective", terms which appear indeed frequently in Bohr's writings, without, however, being defined more explicitly by him.

There would be no need to do so if we were concerned with measurements on classical systems. Indeed in such systems every measurement is unambiguous (as long as it is not an observational error) and objective. However, in a strict sense of the word there do not exist any classical systems. Every system, no matter how closely it may approximate the "classical" ideal, is composed of atoms and it is therefore subject to the laws of quantum mechanics.

Once this difficulty is recognized, it is easy to overcome it, and here again we need only formalize the ideas of Bohr on this question: "A system is said to show classical behaviour when in this particular behaviour the specifically quantal effects (that is any effect which disappears with $\hbar \to 0$) can be neglected." This does not mean that these effects do necessarily vanish, but rather that they are negligibly small in comparison to the effect of interest.

This means two things, first of all that the concept "classical" has only approximate significance for real physical systems and secondly that classical systems can be described as systems with a restricted class of observables (*viz.* only those observables which cannot see $\hbar$, we might say).

The precise mathematical analysis of these concepts which we shall not give here shows that "classical", "unambiguous" and "objective" are characterized by one and the same condition, *viz.* that the restricted class of observables of a classical system must all be compatible with one another. Such a system of observables is represented in quantum mechanics by a commuting set of self-adjoint operators.

When this idealized notion of "classical" is adopted as an additional requirement of a suitable measuring device, the solution of the dilemma is easy, as was shown by one of us [1f]. It is based on the theory of equivalence classes of states. A measurement determines not individual microstates but only equivalence classes of such which we might call macrostates. It is then easy to show that the two states $c\alpha a_+ + c'\beta a_-$ and the mixture of αa_+ and βa_- with probabilities $|c|^2$ and $|c'|^2$ are always in the same equivalence class with respect to that specific measurement. Thus there is no possibility at all to "collapse the state vector". This means that certain operations, such as the reversal of the velocities discussed before – and many others – are not only impractical but impossible also in principle. This impossibility is a new principle outside and contrary to the present doctrine of quantum mechanics. If it is accepted there is no need to assume a "collapse of the state vector". The latter operation, which was such a disturbing aspect in von Neumann's theory, disappears from the scene. It is for this reason that we have come to view the classical property of measuring devices as the origin of this "collapse" prescribed by the usual formulation of quantum mechanics.

In the light of this explanation we may now better assess the discussion of Daneri et al. [1i] as well as that of Rosenfeld [1h]. The former dismiss all the recent publications as giving no "new substantial contributions to the subject". By contrast these authors consider their own paper as the "natural crowning of the basic structure of present-day quantum mechanics" and they are convinced that further progress in this field of research will consist essentially in refinements of their approach. Rosenfeld on the other hand believes that the recent discussions of these questions are based on "misunderstandings, which go back to deficiencies in von Neumann's axiomatic treatment" and that only the work of Daneri et al. have completely removed these deficiencies.

Lest these exaggerated claims might repel a sensitive reader, we want to stress here that the work of Daneri et al., as far as its essential is concerned, is a useful contribution to the theory of measurement. What they show is that certain macroscopic systems (characterized by them by certain ergodic properties) do in fact show a behavior which makes it virtually impossible to measure with them other than compatible properties. Such systems must necessarily contain many atoms, that is, they are macroscopic compared to the measured effects on the atomic scale. In their view the microscopic part of the measuring acts merely as a triggering device, while the essential macroscopic part of the measuring process, that part which wipes out the phase relations, is "related to a process taking place in the latter apparatus after all interaction with the atomic system has ceased" (Rosenfeld [1c]). The unfortunate feature of this last statement is that it is quite easy to give counter-examples of measurements which do not proceed according to the scheme of a triggering device followed by an ergodic amplification in a macroscopic system. The most startling examples of this kind are for instance the so-called "negative-result measurements" discussed by Renninger [4]. It follows from these examples that the macroscopic and ergodic systems are useful (and practically indispensable) devices to raise the events to the level of data (cf. ref. 1f), but that they do not touch the basic aspects of the dilemma.

[4] M. Renninger: Zeits. f. Phys. **158**, 417 (1960).

4. Discussion Under the Assumption
That the Basic Principles of Quantum Mechanics
Have Only Limited Validity

The preceding discussion uses only the basic principles of quantum mechanics, such as the superposition principle, but uses these under conditions under which they were never tested. Thus, the demonstration of the difference between superposition and mixture, discussed in Sect. 2, involves an operation (reversal of velocities on the microscopic scale) which cannot be carried out in practice on a macroscopic system, such as a measuring apparatus. It is true that reversal of velocities is only one of many means to demonstrate the difference between the two states. However, it appears very likely that none of these means can be put to work in practice. It is tempting, therefore, to conclude that the basic principles of quantum mechanics have only limited validity and they cannot fully describe the measuring process.

The assumption that quantum mechanics is inapplicable to the description of the process of observation gives a great deal of leeway, but not an explanation. Proposals that the principles of quantum mechanics have only limited validity were made repeatedly, in particular by Ludwig[5], but also by one of the present authors[6]. Ludwig postulates that classical mechanics is valid for macroscopic bodies and these can be only in states which have a macroscopic description. This excludes, in particular, also linear combinations of macroscopic states, and there is no way to arrive at the dilemma of Sect. 1. However, there is also no complete theory which could be consistently applied. Naturally, the question of the validity of quantum mechanics for macroscopic bodies should be subject to experimental test. Even though it is, evidently, not easy to design valid experiments to decide the issue, it would be most desirable to do so.

In a certain sense both the work of one of us concerning the introduction of macrostates and the work of Daneri *et al.* are steps in the direction of a generalization of the quantum-mechanical description of physical systems. Neither macrostates nor the classical states of the macroscopic apparatus of Daneri *et al.* do evolve according to a Schrödinger equation. In fact the latter give an explicit form of the evolution of such states in the form of a master equation which, as is well known, cannot be derived from a Schrödinger equation without additional assumptions.

None of these generalizations can, however, be considered as basic modifications which would allow an experimental test of the validity of conventional quantum mechanics, and our present understanding of the measuring process does not give us any clue for such a test.

---

[5] G. Ludwig: Gelöste und Ungelöste Probleme des Messprozesses in der Quantenmechanik, article in Werner Heisenberg und die Physik unserer Zeit (Braunschweig 1961), p. 150. Cf. particularly Sect. 3a.
[6] E. P. Wigner: Remarks on the mind-body question, article in *The Scientist Speculates* (London 1961), p. 284.

Riassunto*

Gli aspetti paradossali del procedimento di misura in meccanica quantistica continuano a provocare numerose discussioni, che indicano un crescente interessamento ai problemi fondamentali della meccanica quantistica ([1, 2]). Scopo del presente scritto è di chiarire le nostre vedute su questo importante problema e di contrapporle ai punti di vista espressi in alcune recenti pubblicazioni sulla questione [1h,i].

---

* Traduzione a cara della Redazione.

Некоторые замечания относительно процесса измерения в квантовой механике.

Резюме (*). — Парадоксальные аспекты процесса измерения в квантовой механике продолжают провоцировать многочисленные дискуссии, что указывает на растущий интерес к основным проблемам квантовой механики ([1.2]). Цель настоящей работы — разъяснить наши взгляды на эту важную проблему и сравнить их с взглядами, выраженными в некоторых недавних публикациях на эту тему ([1h.i]).

(*) *Переведено редакцией.*

On the Change of the Skew Information in the Process of Quantum Mechanical Measurements

A. Frenkel, E. Wigner, and M. Yanase

Mimeographed notes, ca. 1970
(Reset by Springer-Verlag for this volume)

Abstract. The amount of information on the values of observables which do not commute with the additive conserved quantities may be measured by the skew information I defined in Eq. (1). In Section 1 we enumerate the basic properties of this expression. In Section 2 the change of the skew information is investigated in the process of quantum mechanical measurements in which the pointer position of the apparatus uniquely determines its content of the conserved quantity. In Section 3 the consequences of this limitation on the types of measurements considered are discussed and some open problems concerning the skew information are mentioned.

1. Introduction

A few years ago a quantity called "skew information" was defined as a measure of the information content of a quantum mechanical system with respect to observables which do not commute with (are skew to) some of the additive conserved quantities of the system [1]. The special attention to these skew observables is motivated by the circumstance that it has been shown [2] that they are more difficult to measure than those which commute with all the additive conserved quantities. Actually, the skew information has been defined in [1] only for systems with a single additive conserved quantity. The same restriction will be adopted in the present note. Let us denote the conserved quantity in question by k and the density matrix of the system by ρ. The mathematical expression for the skew information proposed for this case in [1] reads

$$I = -\tfrac{1}{2}\operatorname{Tr}\left[\sqrt{\rho},k\right]^2 = \operatorname{Tr}\rho k^2 - \operatorname{Tr}\left(\sqrt{\rho}\,k\right)^2 . \tag{1}$$

In (1) the brackets denote the commutator, and $\sqrt{\rho}$ stands for the positive semidefinite square root of the positive semidefinite ρ. The density matrix has also well known properties, $U^\dagger$ denotes the hermitean adjoint of U. The symbol "Tr" refers to the trace over the full system. The trace over a subsystem "a" will be denoted by Tr_a. Since both $\sqrt{\rho}$ and k are self-adjoint, their commutator is skew hermitean and its square a negative semidefinite hermitean operator. It follows that I is positive unless ρ commutes with k, in which case it vanishes. Let the system be in a pure state

$$|\psi\rangle = \sum_{k,\alpha} \psi(k,\alpha)|k,\alpha\rangle \tag{3}$$

so that

$$\rho = |\psi\rangle\langle\psi| = \sqrt{\rho} \qquad (3')$$

where the degeneracy index α labels the states belonging to a given eigenvalue of k. In this case the skew information equals the mean square deviation of k from its average value squared:

$$I = \langle\psi|k^2|\psi\rangle - \langle\psi|k|\psi\rangle^2 = \sum_{k,\alpha}|\psi(k,\alpha)|^2 k^2 - \left(\sum_{k,\alpha}|\psi(k,\alpha)|^2 k\right)^2. \qquad (4)$$

In particular, if the system is in an eigenstate of k (e.g.,in an eigenstate of the x component of the linear momentum), then $I = 0$, i.e., the skew information on the observables commuting with k (e.g., the x-coordinate of the system) is zero. In this case, indeed, x is uniformly distributed in space. On the other hand, if $|\psi\rangle$ is a superposition of states with different eigenvalues of k, then $I > 0$, and our amount of information concerning the position of the system is indeed increased, it is now more sharply localized than in the previous case.

The expression (1) is not the only one which goes over into (4) for pure states, but it is probably the simplest among those exhibiting the following basic properties which a quantity measuring an information content of a sytem might be expected to have:

(a) If the representative ensemble of a sytem arises as a result of the union of two other ensembles representing this system, then the information content of the united ensemble should not be larger than the average information content of the two original ensembles. Indeed, by uniting two ensembles one "forgets" from which of them a particular sample stems, i.e., some amount of information is lost. If the density matrices of the two original ensembles are denoted by ρ_1 and ρ_2, and if their weights in the united ensemble are w_1 and $w_2 = 1 - w$, the density matrix of the united ensemble becomes

$$\rho = w_1\rho_1 + w_2\rho_2 \qquad (5)$$

and the skew information has to satisfy the inequality:

$$I = (w_1\rho_1 + w_2\rho_2) \le w_1 I(\rho_1) + w_2 I(\rho_2) \qquad (6)$$

the right-hand side being the average information content of the two ensembles.

(b) If two uncorrelated systems with density matrices ρ_1 and ρ_2 are considered as a single system, the information content of the united system must be the sum of the information contents of the components. Hence, if $\rho = \rho_1 \times \rho_2$, we expect

$$I(\rho_1 \times \rho_2) = I(\rho_1) + I(\rho_2). \qquad (7)$$

$\rho_1 \times \rho_2$ stands for the direct (Kronecker) product of the density matrices ρ_1 and ρ_2.

(c) The information content of an isolated system should be independent of time, because in quantum (and also in classical) mechanics the evolution of an isolated system is fully determined by the initial state and by the equation of motion.

(d) If a system with density matrix ρ is decomposed into subsystems a and b with density matrices $\rho^a = \mathrm{Tr}_b \, \rho$ and $\rho^b = \mathrm{Tr}_a \, \rho$ respectively, then the inequality

$$I(\rho) \geq I(\rho^a) + I(\rho^b) \tag{8}$$

is expected to hold. Indeed, in the process of decomposition the information concerning the statistical correlations between the two subsystems is in general lost.

It has been shown in [1] that the skew information (1) satisfies the properties (a)–(c). However, as far as (d) is concerned, the proof has been given only for the special case in which ρ corresponds to a pure state of the system. Moreover, the question of the change of the skew information in the process of a quantum mechanical measurement was not discussed in [1]. In the present note some aspects of this problem are clarified. In Section 2 we investigate the change of the skew information of the system apparatus + object (A + O) caused by the measurement. We shall be particularly concerned with the second part of the measurement, the so-called "reduction (or collapse) of the wave packet" which takes place when the apparatus is observed by an observer. We shall show that, if one of the observed quantities of the apparatus is the conserved quantity entering (1), the measurement cannot cause an increase of the skew information of the joint system, consisting of both apparatus and object. The demonstration will be given first for the simple case in which the eigenvalues of the conserved quantity k are simple (non-degenerate), and then for the general, more realistic situation. In Section 3 we give a short discussion of our results and also point to further, as yet unsolved, problems.

2. Change of the Skew Information by Measurement

The process of quantum mechanical measurement is conveniently divided into two parts, called the first and the second parts of the measurement. The first part of the measurement is a collision of the object O to be measured with the measuring apparatus A. Both the object and the apparatus are described as quantum mechanical systems and their interation takes place according to the laws of quantum mechanics. The process of measurement is completed by the second part of the measurement, in which the apparatus is observed by an observer. This leads to the so-called reduction (or collapse) of the wave packet or state vector.

Let us calculate the change of the skew information of the joint system A+O by the measurement. We denote by $I^{(0)}$, $I^{(1)}$ and $I^{(2)}$ the skew information before the first, after the first and after the second part of the measurement, respectively. We infer from point (c) in Section 1 that

$$I^{(0)} = I^{(1)} \tag{9}$$

since the first part of the measurement is a quantum mechanical collision process between parts of the isolated system consisting of both apparatus and object, A + O. Such a process can be described by an unitary transformation $\rho_{(1)} = U\rho_{(0)}U^\dagger$, and the operator U must commute with k since k is conserved. Then (9) easily follows from the definition (1).

Let us now turn to the second part of the measurement. First we deal with the simple case when the eigenvalues k_r and k_l of the conserved quantities, k^A of the apparatus and k^0 of the object are simple so that the corresponding amplitudes $\psi(r, l)$ fully describe the state of the system A + O. If before the first measurement both the apparatus and the object were described by a state vector (not a mixture) in their respective Hilbert spaces H_A and H_O, then after the first measurement the system A + O will be also described by a state vector $|\psi^{(1)}\rangle$ in the Hilbert space $H_A \otimes H_O$. The most general form of this state obviously reads

$$|\psi^{(1)}\rangle = \sum_{rl} \psi(r, l)|k_r\rangle \times |k_l\rangle \tag{10}$$

where

$$\sum_{r,l} |\psi(r, l)|^2 = 1. \tag{11}$$

Taking into account that k is additive

$$k = k^A \times 1^0 + 1^A \times k^0 \tag{12}$$

and that

$$\rho_{(1)} = |\psi^{(1)}\rangle\langle\psi^{(1)}| = \sqrt{\rho_{(1)}} \tag{13}$$

we find from (1)

$$I^{(1)} = \sum_{r,l} w(r, l)(k_r + k_l)^2 - \left(\sum_{r,l} w(r, l)(k_r + k_l)\right)^2 \tag{14}$$

where

$$w(r, l) = |\psi(r, l)|^2 \tag{15}$$

Let us now calculate $I^{(2)}$. If the second part of the measurement consists in observing the conserved quantity k^A of the apparatus, i.e., if k^A represents the "pointer position", then according to the theory of measurement of von Neumann [3], the probability for finding the system A + O in the (normalized) state

$$|\psi^{(2)}(r)\rangle = |k_r\rangle \times \frac{1}{\sqrt{w(r)}} \sum_l \psi(r, l)|k_l\rangle \tag{16}$$

which corresponds to the pointer position k_r becomes

$$w(r) = \sum_l w(r, l). \tag{17}$$

Notice that (11) and (15) imply

$$\sum_r w(r) = 1. \tag{18}$$

Since, as a result of the second part of the measurement, the samples belonging to the different states $|\psi^{(2)}(r)\rangle$ are separated from each other, the skew information of the system A + O after the second measurement is the average skew information content of these samples. Hence,

$$I^{(2)} = \sum w(r)I(\rho_2(r)) \tag{19}$$

where, since $\rho_2(r)$ represents a pure state,

$$\sqrt{\rho_2(r)} = \rho_2(r) = |\psi^{(2)}(r)\rangle\langle\psi^{(2)}(r)|$$
$$= w(r)^{-1} \sum_{ll'} \psi(r,l)\psi(r,l')^* |k_r\rangle\langle k_r|x|k_l\rangle\langle k_{l'}| \tag{19a}$$

and we have for its skew information

$$I(\rho_2(r)) = \frac{1}{2}w(r)^{-2} \sum_{ll'} w(r,l)w(r,l')(k_l - k_{l'})^2 . \tag{19b}$$

With (19) this gives finally

$$I^{(2)} = \sum_{r,l} w(r,l)k_l^2 - \sum_r \frac{1}{w(r)}\left(\sum_l w(r,l)k_l\right)^2 . \tag{20}$$

In a lengthy but straightforward calculation the reader can verify that the difference

$$I^{(1)} - I^{(2)} = \sum_{r,l} w(r,l)(k_r + k_l)^2 - \left(\sum_{r,l} w(r,l)(k_r + k_e ll)\right)^2$$
$$- \sum_{r,l} w(r,l)k_l^2 + \sum_r \frac{1}{w(r)}\left(\sum_l w(r,l)k_l\right)^2 \tag{21}$$

can also be written in the form

$$I^{(1)} - I^{(2)} = \sum_r \left\{ \sum_l \left[\frac{w(r,l)}{\sqrt{w(r)}} - \sqrt{w(r)}\sum_{r'} w(r',l) \right] k_l \right.$$
$$\left. + \sqrt{w(r)}\left[k_r - \sum_{r'} w(r')k_{r'} \right] \right\}^2 . \tag{22}$$

The proof of (22) essentially consists in writing the squares in (21) and (22) and using the relations $\sum_l w(r,l) = w(r), \sum_r w(r) = \sum_{rl} w(r,l) = 1$.

The right-hand side of (22) being a sum of positive semidefinite terms, we see that

$$I^{(1)} - I^{(2)} \geq 0 . \tag{23}$$

Thus, for the case that both k^A and k^0 are non-degenerate and, if in the second part of the measurement k^A itself is observed, the skew information cannot increase in the process of measurement:

$$I^{(0)} = I^{(1)} \geq I^{(2)} . \tag{24}$$

Let us now consider the case when the states of the apparatus and of the object are not completely specified by the eigenvalues k_r and k_l of the conserved quantities k^A and k^0. To make the description complete, we introduce the degeneracy index α for the apparatus and β for the object. Then the most general state $|\psi^{(1)}\rangle$ of the system A + O after the first measurement reads

$$|\psi^{(1)}\rangle = \sum_{\substack{r\alpha \\ l\beta}} \psi(r,\alpha,l,\beta)|k_r\alpha\rangle \times |k_l,\beta\rangle \tag{25}$$

where

$$\sum_{\substack{r\alpha \\ l\beta}} |\psi(r,\alpha,l,\beta)|^2 = 1. \tag{26}$$

(We again assumed that, before the first part of the measurement, the apparatus and the object were described by state vectors in their respective Hilbert spaces, not by a mixture.) According to (1) the skew information $I^{(1)}$ of the system in the state (25) turns out to be

$$I^{(1)} = \sum_{\substack{r\alpha \\ l\beta}} |\psi(r,\alpha,l,\beta)|^2(k_r + k_l)^2 - \left(\sum_{\substack{r\alpha \\ l\beta}} |\psi(r,\alpha,l,\beta)|^2(k_r + k_l) \right)^2. \tag{27}$$

Introducing the notations

$$|\psi(r,\alpha,l,\beta)|^2 = w(r,\alpha,l,\beta), \tag{28}$$

$$\sum_{\alpha\beta} w(r,\alpha,l,\beta) = w(r,l) \tag{29}$$

we see that $I^{(1)}$ takes the form (14) found for the simple case. In particular, $I^{(1)}$ can be written in terms of the $w(r,l)$ alone without using the $w(r,\alpha,l,\beta)$. Naturally, we have again

$$\sum_{r,l} w(r,l) = 1. \tag{29a}$$

We now calculate $I^{(2)}$ assuming that the quantity k^A and the quantity corresponding to α are observed on the apparatus, i.e., the pointer position is specified by the values of k^A and α. The probability that the composite system A + O will be found in the (normalized) state

$$|\psi^{(2)}(k_r,\alpha)\rangle = |k_r,\alpha\rangle \times \frac{1}{\sqrt{w(r,\alpha)}} \sum_{l\beta} \psi(r,\alpha,l,\beta)|k_l,\beta\rangle \tag{30}$$

now equals

$$w(r,\alpha) = \sum_{l\beta} \psi(r,\alpha,l,\beta). \tag{31}$$

The average skew information content of the ensembles created in the second part of the measurement then reads

$$I^{(2)} = \sum_{r,\alpha} w(r,\alpha) I(\rho_2(r,\alpha)) \tag{32}$$

where

$$\rho_2(r,\alpha) = |\psi^{(2)}(r,\alpha)\rangle\langle\psi^{(2)}(r,\alpha)| = \sqrt{\rho_2(r,\alpha)}. \tag{33}$$

The calculation of $I(\rho_2(r,\alpha))$ according to (1) and of $I^{(2)}$ according to (32) leads to

$$I^{(2)} = \sum_{rl} w(r,l) k_l^2 - \sum_{r\alpha} \frac{1}{w(r\alpha)} \left(\sum_{l\beta} w(r,\alpha,l,\beta) k_l \right)^2. \tag{34}$$

The first sum in (34) is identical with the first sum in the expression(20) for $I^{(2)}$ in the simple case. Let us now compare a term for fixed r of the second sum in (34) with the corresponding term of the second sum in (20). The term in (34) is

$$\sum_{r\alpha} w(r\alpha)^{-1} \left(\sum_{l\beta} w(r,\alpha,l,\beta) k_l \right)^2. \tag{35a}$$

The corresponding term in (20) is

$$\frac{\left(\sum_l w(r,l) k_l \right)^2}{w(r)} = \frac{\left(\sum_{l\alpha\beta} w(r,\alpha,l,\beta) k_l \right)^2}{\sum_\alpha w(r,\alpha)}. \tag{35b}$$

In order to see that (35a) is larger than (35b), we introduce the quantity

$$L_{r\alpha} = \frac{\sum_{l\beta} w(r,\alpha,l,\beta) k_l}{\sqrt{w(r,\alpha)}}. \tag{36}$$

In terms of this, (35a) becomes simply $\sum_\alpha L_{r\alpha}^2$. The corresponding term in (20), that is (35b), becomes

$$\frac{\left(\sum_\alpha L_{r\alpha} \sqrt{w(r,\alpha)} \right)^2}{\sum_\alpha \left(\sqrt{w(r,\alpha)} \right)^2}. \tag{37}$$

Multiplying both (37) and $\sum L_{r\alpha}^2$ with the denominator of (37) we see, by Schwarz' inequality that $\sum L_{r\alpha}^2$ is indeed larger than (37) and, hence, that (35a) is larger than (35b). Thus, while the value of $I^{(1)}$ is the same as in the simple case, that is (14), $I^{(2)}$ will be in general smaller than the corresponding expression (20) in the simple case and thus a fortiori in general smaller than $I^{(1)}$. It cannot exceed the latter and hence, by the observation (c) of the first section, cannot exceed $I^{(0)} = I^{(1)}$.

We can conclude, therefore, that the average skew information does not increase as a result of a quantum mechanical measurement, at least not as long as the different pointer positions of the apparatus, i.e., the directly observed quantities, correspond to definite values (and not to linear combinations) of the conserved quantity. The argument presented assumed only one conserved quantity so that the expression (1) for the skew information could be used.

3. Discussion

When deriving the relation $I^{(0)} = I^{(1)} \geq I^{(2)}$, we left the amplitudes $\psi(r, l)$ and $\psi(r, \alpha, l, \beta)$ completely arbitrary, except for the normalization conditions (11) and (26). The process of the first part of the measurement (the quantum mechanical interaction between object and apparatus) was also treated only in general terms and it was postulated only that initial and final states are connected by a unitary operator which commutes with the conserved quantity. On the other hand, as far as the second part of the measurement is concerned (the collapse of the state vector as a result of the reading of the pointer position), the very stringent assumption was made that the pointer position also determines the value of the conserved quantity. This means that the apparatus is, after this reading in an eigenstate of the operator representing that quantity. It is a result of this assumption that the skew information of the apparatus vanishes when the second part of the measurement has been carried out. On the other hand, the assumption in question is quite unrealistic, casting doubt on the general validity of the decrease of the skew information in the second part of the measurement process. These doubts have not been resolved to date. Neither have our doubts been resolved concerning the desirability of the universal validity of the postulate of this decrease.

We wish to add a few remarks on the limitations on the first part of the measurement, brought about by the stipulation that the distinguishable states of the apparatus be eigenstates of the conserved quantity. Two types of measurements will be considered for this purpose. Pauli's first kind of measurement [4] (we speak here only about the first part of the measurement, that which can be described by the quantum mechanical equations of motion) can be characterized by the transition

$$|a^0 \times \sigma^\mu\rangle \rightarrow |a^\mu \times \sigma^\mu\rangle. \tag{38f}$$

In (38f), a^0 is the initial state of the apparatus, a^μ the state corresponding to the pointer position μ. Hence, σ^μ is the state of the object which surely yields this pointer position, it is an eigenstate of the quantity which is being measured. The distinguishing mark of the first type of measurement is that the state of the object, if it was one of these eigenstates before the measurement, will remain unchanged by the process of interaction with the apparatus. In contrast, the second type of measurement (again the first part there of) is characterized by

$$|a^0 \times \sigma^\mu\rangle \rightarrow |a^\mu \times \tau^\mu\rangle. \tag{38s}$$

i.e., the state of the object is changed by such a measurement even if it was, before it, in an eigenstate of the quantity which is being measured. Since, for (38s), no assumption is made about the τ, the first kind of measurement is a special case of the second kind.

Since the arrows in (38f) and (38s) represent unitary operators, the scalar products of two expressions of the type given by the left sides must be equal to the scalar products of the corresponding expressions on the right sides. In the case of (38s) this gives

$$\langle a^0 \times \sigma^\mu | a^0 \times \sigma^\lambda \rangle = \langle a^\mu \times \tau^\mu | a^\lambda \times \tau^\lambda \rangle \tag{39}$$

or

$$\langle a^0 | a^0 \rangle \langle \sigma^\mu | \sigma^\lambda \rangle = \langle a^\mu | a^\lambda \rangle \langle \tau^\mu | \tau^\lambda \rangle . \tag{40}$$

Since the pointer positions for different μ and λ should not overlap and since all state vectors are normalized, we have

$$\langle a^\mu | a^\lambda \rangle = \delta_{\lambda\mu} \qquad \langle a^0 | a^0 \rangle = 1 \tag{40a}$$

so that we obtain the well known result

$$\langle \sigma^\mu | \sigma^\lambda \rangle = \delta_{\lambda\mu} \tag{40b}$$

that the states of the object which lead to different pointer positions of the apparatus are orthogonal. This applies also to the first type of measurement.

Making use of the fact that the conserved quantity k commutes with the unitary operator represented by the arrows in (38) we find from (38s)

$$\langle a^0 \times \sigma^\mu | k | a^0 \times \sigma^\lambda \rangle = \langle a^\mu \times \tau^\mu | k | a^\lambda \times \tau^\lambda \rangle \tag{41}$$

Because of the additive nature of the conserved quantity k,

$$k = (k^A \times 1^0) + (1^A \times k^0) . \tag{42}$$

(41) can be simplified to give, for $\lambda \neq \mu$

$$\langle \sigma^\mu | k^0 | \sigma^\lambda \rangle = \langle a^\mu | k^A | a^\lambda \rangle \langle \tau^\mu | \tau^\lambda \rangle \tag{43a}$$

and for $\lambda = \mu$ the rather obvious relation

$$\langle a^0 | k^A | a^0 \rangle + \langle \sigma^\mu | k^0 | \sigma^\mu \rangle = \langle a^\mu | k^A | a^\mu \rangle + \langle \tau^\mu | k^0 | \tau^\mu \rangle . \tag{43b}$$

If the τ^μ form an orthogonal system $\langle \tau^\mu | \tau^\lambda \rangle = \delta_{\mu\lambda}$ – as (40b) shows this is always the case for first type of measurement for which the τ and σ are identical – it follows from (43a) that

$$\langle \sigma^\mu | k^0 | \sigma^\lambda \rangle = 0 \qquad \text{for } \mu \neq \lambda . \tag{44f}$$

If the σ form, furthermore, a complete set of vectors, it follows that k^0 is diagonal in the σ, i.e., that the measured quantity commutes with the conserved one. This is the result of (2). All this applies whether or not the measurement is of the kind considered in Section 2, i.e., whether or not the a^μ are eigenvectors of k^A.

In case of the measurements considered in Section 2, one can go a step further. We shall show that, if the a^μ are eigenvectors of k^A, the corresponding eigenvalues are all the same if the measurement is of the first kind and that this holds also for a^0. This means that, as far as pointer positions in first kind of measurements are concerned, the variable r is not a valid variable: it assumes a single value throughout. In order to show this we note that, if the τ and σ are equal, the last terms on both sides of (43b) cancel and one sees that the expectation values of k^A for all a^μ are the same. Since these are characteristic vectors of k^A, and since the σ are, because of (44f), characteristic vectors of k^0, the $a^\mu \times \sigma^\mu$ is also a characteristic vector of k. Since this is true also for σ^μ alone, it is true for a^0 too. The corresponding characteristics value is then the same as that of the a^μ. This shows that the arguments of Section 2 apply for measurements of the first kind at best in the degenerate case $|a^\mu\rangle = |k_r, \alpha\rangle$. Even in this case, the observation of the conserved quantity is useless – it always gives the same reading. None of the preceding applies for measurements of the second type, not even if the τ are orthogonal so that (44f) still holds. In such measurements an exchange of the conserved quantity between object and apparatus is possible, hence the pointer position is no longer restricted to a single value k_r.

The discussion so far was concerned with the limitations due to the assumption that the observed states of the apparatus $|a^\mu\rangle = |k_r, \alpha\rangle$ were eigenstates of the conserved quantity. A further limitation of the preceding considerations stems from the fact, mentioned before, that the skew information has been defined for only one additive conserved quantity. No expression for the skew information is available which would apply if several additive conserved quantities are present. Furthermore, if there is only one conserved quantity, there might exist expressions, different from (1), which also satisfy the conditions (a), (b), (c) and (d) of section 1.

The authors are indebted to the International School of Physics "Enrico Fermi" which brought them together and revived their interest in the problems of the skew information.

References

1. E.P. Wigner and M.M. Yanase: Proc. N.A.S. **49**, 910 (1963), Can. Jour. Math. **16**, 397 (1964). See also E.P. Wigner: Physikertagung Wien (Mosbach/Baden: Physik Verlag, 1962) p. 1
2. E.P. Wigner: Z. Physik **133**, 101 (1952); H. Araki and M.M. Yanase: Phys. Rev. **120**, 622 (1960); M.M. Yanase: Phys. Rev. **123**, 666 (1961)
3. J. von Neumann: Mathematical Foundations of Quantum Mechanics. Princeton University Press, Princeton 1955. – See also E.P. Wigner: Epistemology of Quantum Mechanics, in: Contemporary Physics, Vol. II, p. 431. Trieste Symposium 1968. (IAEA, Vienna 1969)
4. W. Pauli: Die allgemeinen Prinzipien der Wellenmechanik. In: Handbuch der Physik, Bd. V, Teil 1, 1

The Subject of Our Discussions

E. P. Wigner

B. d'Espagnat (ed.) Foundations of Quantum Mechanics. International School of Physics
"Enrico Fermi" 1970. Academic Press, New York 1971, pp. 1–19

1. – Physics and philosophy.

Our task during the present session of the Enrico Fermi Institute will not
be an easy one. We'll discuss a subject on the borderline between physics
and philosophy and, since most of us are not philosophers, we may say things
which appear dilettantish to true philosophers. Also, few of us are truly familiar
with the ideas and accomplishments of philosophers, earlier and contemporaneous. We may unnecessarily invent a new terminology, flaunting well established custom and neglecting to establish connection with past thinking. It
is good that there are some true philosophers among us who, I hope, will
correct our errors.

Since we shall be thinking about questions on the borderline between physics and philosophy, it may be well to say a few words on the essence of these
disciplines and how they differ from each other. Let us begin with physics.
Physics, it is often said, explains the phenomena of inanimate nature. I find
this definition a bit hollow and more than a bit boastful. The great progress
of physics was initiated by Newton's division of the determinants of our surroundings into two categories: initial conditions and laws of nature. Physics,
as we know it, deals only with the second category and Newton gave up Kepler's
idea to derive the sizes of the planetary orbits from simple rules. What he
gave, rather, were rules on how to obtain the position of a planet at any given
time, using as input its positions at two earlier times. More generally, we can
say that physics establishes regularities in the behavior of inanimate objects.
That these regularities can be described by means of beautiful and conceptually
simple mathematical models has often been commented on as a near miracle.
But we need not be concerned with this now.

Whereas it is easy to characterize the objective of physics as the search
for, and description of, regularities in the behavior of inanimate objects, philosophy is more difficult to define. It is the search for an encompassing view

of nature and life, for an elevated picture of the world and our role therein. Naturally, the interface between physics and philosophy covers only a small part of all this and consists, essentially, of an analysis of the character and significance of the regularities established by physicists. One of the tasks; of philosophy is the assimilation of the information which other sciences produce and one of the tasks of physics is to provide such information by exploring the unifying principles underlying its many detailed results.

All of this shows that physics is, fundamentally, a much more modest discipline than philosophy.. This also has rewards. Physical theories are often superseded and replaced by more accurate, by more general and deeper theories. However, the superseded theory still retains validity as an approximation, applicable in a perhaps restricted but still multifaceted set of circumstances. As far as the original domain of the theory is concerned, the superseding theory does, as a rule, little more than to delineate more sharply the boundaries of that domain. Also, the new theory owes, almost invariably, a great deal to the older one—it usually could not have been invented without a knowledge thereof.

On the contrary, the different philosophical pictures seem to represent *alternatives* and do not have a space in each other as do successive physical theories. Different philosophies represent different images of the world, conflicting images. As a result, philosophers often dislike each others' theories. It will be good to keep in mind this difference between the relations of physical theories, and of philosophical theories, to each other. In this regard, the ideas on the epistemological implications of quantum theory resemble more the philosophical than the physical theories.

2. – Common applications *vs.* fundamental problems of quantum mechanics.

As was implied before, and as is well known, there is no full unanimity among physicists concerning the fundamental principles which underlie quantum mechanics. This is very surprising at first, and remains surprising even after it is explained away to a certain extent. One of the reasons that the basic principles affect the day-to-day work of the physicist rather little is that quantum mechanics is used very rarely to establish the regularities between events directly. It is used more frequently in conjunction with classical, that is macroscopic, theories by providing material constants for these, such as cohesive strength, viscosity, chemical affinity, etc.—quantities which are arbitrary material constants as far as classical, macroscopic theory is concerned. The most important exception to this rule is collision theory and we shall speak a good deal about idealized collisions, measurements, or observations. However, all collisions, not only the idealized ones, are in this category—they are events rather than properties and their description uses the equations of motion of quantum mechanics rather than calculate characteristic values. Collision theory is closely tied to the fundamental, epistemological problems of quantum mechanics and the work on the foundations of collision theory is more dependent on, and supports more directly, the epistemology of quantum mechanics than other applications of that theory. This will be, no doubt, discussed further by others in the course of our study here.

3. – Epistemology and everyday life.

There is a very instructive joke in the Introduction to Boltzmann's *Kinetic Theory of Gases*. He tells us, in his story, that he was, as a young man, very critical of the logical rigor of the books on physics with which he became familiar. He was quite elated, therefore, when he heard about a physics book which was, he was told, strictly logical. He rushed to the library—only to find, first, that the book was out and, second, that it was all in English. Boltzmann spoke no English at that time. He went home, quite down-hearted and complained to his brother. His brother, however, told him that, if the book was all that good it was surely worth waiting for it until it would be returned to the library. As to its being in English, Boltzmann's brother said, that surely will be immaterial. If the book is entirely logical, the authors won't use any term before having defined and explained it carefully.

Incidentally, HEISENBERG, in his *The Whole and Its Parts*, also points to the impossibility of strict logical rigor in scientific work.

What follows from all this? Simply, that the statements and conclusions of science are, and have to be, expressed in common language, that science cannot be independent from everyday experience, the concepts and information we acquired in babyhood. The homo scientificus who bases his actions and knowledge on science alone does not, and cannot, exist. In fact, we feel that the more primitive the notions are which one uses to express the regularities observed, the more fundamental can the theory be. This does not mean that science accepts or needs all notions which we acquired as children, or that it accepts them uncritically—relativity theory showed that this is not the case— but that it cannot exist without the notions and has to use some of them as a fundament. Boltzmann's anecdote brings this point home most clearly.

4. – The basic quantities of various physical theories.

What, then, are the concepts which physics, in its different stages of development, uses to describe the regularities which are its subject? Newtonian mechanics' primitive concepts were the positions of objects at different times;

there is, for instance, a connection between any three positions which a planet occupies in the course of its motion. They all lie on the same ellipse, one of the foci of which is the sun. The position of an object seems to be a very primitive concept—we know, though, relativity theory's critique of the primitivity thereof. In Maxwell's theory the components of the electromagnetic field, as functions of position and time, are the primitive concepts. The manifold and mathematical sophistication of these is much greater than those of the concepts of Newtonian mechanics, and elaborate preparations are necessary for the verification of the regularities between Maxwell's quantities, postulated by his theory. Again, relativity theory criticized the *a priori* nature of these concepts, but not very severely.

The problems become critical, however, when we come to quantum mechanics, in particular to those conclusions of the theory in which the theory stands on its feet and is not merely supplying material constants for classical theory. The primitive concepts here are the observations, between the outcomes of which quantum mechanics postulates probabilistic connections. The problematical element is not so much that the regularities of quantum mechanics do not have an absolute but only a probabilistic or statistical nature, although it is good to keep this point in mind. Thus, a typical postulate is that if an energy measurement on the electron of a hydrogen atom shows that it is in the normal state, a position measurement theoreon will give, with a probability 0.67, that it is further from its proton than a Bohr radius, and with a probability 0.33 that it is closer thereto. No definite statement as to the exact position of the electron is possible.

As I said, the principal difficulty is not the statistical nature of the regularities postulated by quantum mechanics. The principal difficulty is, rather, that it elevates the measurement, that is the observation of a quantity, to the basic concept of the theory. This introduces two difficulties. First, surely, the observation of the position of an electron is not something primitive, and the observation of most other quantities is even less primitive. They require elaborate apparatus, have severe limitations, some of which have been derived from fundamental conservation laws by participants of this meeting.

Second, however, and this will be argued and elaborated from various and even opposing points of view, it seems dangerous to consider the act of observation, a human act, as the basic one for a theory of inanimate objects. It is, nevertheless, at least in my opinion, an unavoidable conclusion. If it is accepted, we have considered the act of observation, a mental act, as the primitive concept of physics, in terms of which the regularities and correlations of quantum mechanics are formulated. The mental acts may be primitive in the sense in which we demanded primitiveness for the basic concepts of physics. They are, however, little known to us in the detail in which we assume them to have certain properties. It may well be said that we explain a riddle by a mystery.

The preceding discussion deals with the interface of physics and philosophy in a general way and touches on our concrete problem only peripherally. It anticipates the fact that quantum mechanical theory, if followed through consistently, leads to difficult epistemological and philosophical questions. The following discussion of the basic principles of quantum mechanics will trace the path leading to those difficulties by trying to describe the measurement or observation process—the two are synonyms in the language of the quantum theorist—in terms of physical concepts. We shall see that there is no complete unanimity concerning the significance and the solution of the problems which this path encounters but it will be attempted to represent all points of view fairly.

5. – The quantum mechanics of the measurement process.

5˙1. *A preliminary remark.* – Quantum mechanics, as far as the change of the state vector is concerned, is a causal or deterministic theory. In this regard it differs from the theories which appear a-causal only because the input, the initial conditions, are inadequately specified. It has been shown variously, in particular also by von Neumann, that, in classical mechanics, an arbitrarily small uncertaintity in the initial conditions results, after a sufficiently long time, in an arbitrarily large difference in the final co-ordinates as long as this difference is compatible with the conservation laws and the restrictions on the system, such as limitations on the values which the co-ordinates can assume. This is not so in quantum mechanics as far as the theory-given variation of the state vector is concerned. If the initial state vector is known, the final one is completely determined so that the theory is causal (as is also classical mechanics). Furthermore, in contrast to classical mechanics, if there is an initial uncertainty in the state vector, this uncertaintity is preserved at later times but does not grow. This is a consequence of the unitary nature of the operator of time development. The acausal nature of the theory is due to the possibility of an observation giving various possible results even on a system with a well defined and completely known state vector but the variety and spread of these results does not systematically increase in time. In fact, to every observation at any time there corresponds some observation at any later time which gives the same results with the same probabilities and hence also with the same spread. The acausality of the theory manifests itself only at the observations undertaken.

5˙2. *The role of the state vector.* – The preceding remark dealt with the state vector. There are two epistemological attitudes toward this. The first attitude considers the state vector to represent reality, the second attitude regards it to be a mathematical tool to be used to calculate the probabilities for the various possible outcomes of observations. It is not easy to give an operational meaning to the difference of opinion which is involved because, fundamentally, the realities of objects and concepts are ill defined. One can adopt the compromise attitude according to which there is a reality to objects but quantum mechanics is not concerned therewith. It only furnishes the probabilities for the various possible outcomes of observations or measurements—in quantum mechanics these two words are used synonimously.

5·3. *Two kinds of measurements.* – The effect of the measurement on the state of the system on which the measurement was undertaken has been discussed a good deal. Pauli already distinguished two types of measurements. The second of these either annihilates the system or else changes its state arbitrarily. The first type of measurement brings (or leaves) the system into the state in which the quantity which is measured surely has the value which is the outcome of the measurement. We shall be concerned, almost exclusively, with this type of measurement; this can be used also to « prepare a state », *i.e.* to bring the system into a known state.

Let us denote by A_0 a state vector of the measuring apparatus which describes its state in which it is ready for carrying out the measurement. Let us denote by σ the characteristic vectors of the operator Q which is being measured by the apparatus. We then have

$$(1) \qquad\qquad Q\sigma_\mu = q_\mu \sigma_\mu$$

q_μ being « the outcome » of the measurement for the state σ_μ of the system. It will be assumed that the Hilbert space of measuring apparatus plus the system on which the measurement is undertaken is the direct product of the Hilbert spaces of the apparatus and that of the system. If the state vector of the system before the measurement was undertaken was σ_μ then the state vector of apparatus plus system was $A_0 \times \sigma_\mu$. In the first type of measurement, this goes over into

$$(2) \qquad\qquad A_0 \times \sigma_\mu \to A_\mu \times \sigma_\mu,$$

A_μ being a state vector of the apparatus in which it indicates the outcome q_μ of the measurement. In the second type of measurement, which will not be discussed in detail, either the Hilbert space undergoes a change, or the system on which the measurement was undertaken goes over into a neutral or unknown state

$$(3) \qquad\qquad A_0 \times \sigma_\mu \to A_\mu \times \sigma_0 .$$

If, after a measurement of the first type, the measurement is repeated before the system's state vector had a chance to change, the outcome of the measurement will be q_μ again. This is not the case for the second type of measurement; this gives information only on the past state of the system and does not permit the preparation of the states σ_μ thereof. As was mentioned before, we shall deal with the first type of measurements, corresponding to (2), which permits the preparation of the states σ_μ.

5˙4. *Is the measuring apparatus macroscopic?* – In the preceding discussion the measuring apparatus was described by state vectors: A_0 before and one of the A_μ after the measurement if the system was in one of the states σ_μ, a characteristic vector of the operator Q which is beeing measured. This is not the only way which is advocated for the description of the measuring apparatus. « Measuring instruments must be described classically » is a statement by FOCK [1] but many others, including probably the Copenhagen school, hold this view. This view does not conflict, however, with the notion of the two types of measurements discussed under 5˙3, nor does it essentially conflict with the two equations (2) and (3), describing the final states resulting from these two types of measuremehts. The only difference is that the first factors, A_0 and A_μ, are not vectors in Hilbert space but points, or regions, in the space describing the macroscopic state of the apparatus.

The point of view here attributed to FOCK, but unquestionably held by many others, will be discussed further when the problems of all epistemological ideas will be reviewed. There is, of course, no clear dividing line between macroscopic and microscopic theory and the detail with which the macroscopic variables describe the apparatus remains open. In fact, microscopic phenomena play an important role in many macroscopic apparata. There are the phenomena of superconductivity, of laser action, and the fact that the well rested eye responds to as few as three light quanta. Finally, the theory of the interaction of a quantum system with a classical (macroscopic) system has not been formulated so that the mathematical meaning of the arrows in (2) and (3) is not clear. However, the other epistemological interpretations of the measurement process also have problems, or appear to do so, and it would be a mistake to reject out of hand the duality of the macroscopic nature of the measuring apparatus *vs.* the quantum-mechanical description of the system undergoing measurement. The purpose of the present paragraph is only to justify the use of the quantum-mechanical language for both object and apparatus in most of the rest of this discussion. It is more concrete and leads to an interesting and, apparently, little noted conclusion which will be the subject of the next point.

5˙5. *The quantum-mechanical observables are self-adjoint.* – The arrows in (2) and (3) represent unitary operators. (No equally clear statement is possible in the dualistic interpretation just discussed). Hence, the scalar products of two expressions of the form of the left sides must be equal to the corresponding scalar products of the right sides. Let us write down, therefore, (2) and (3) with μ replaced by ν and form the scalar products of the resulting expressions with (2) and (3). We obtain, in case of (3)

$$(4) \qquad (A_0 \times \sigma_\mu, A_0 \times \sigma_\nu) = (A_\mu \times \sigma_0, A_\nu \times \sigma_0) \;.$$

However

$$(A \times \sigma, A' \times \sigma') = (A, A')(\sigma, \sigma')$$

so that (4) gives

$$(4a) \qquad (A_0, A_0)(\sigma_\mu, \sigma_\nu) = (A_\mu, A_\nu)(\sigma_0, \sigma_0) \;.$$

If $\mu \neq \nu$, the right side vanishes because A_μ and A_ν represent two states of the apparatus which are even macroscopically distinguishable. Hence, for $\mu \neq \nu$, the left side must vanish also. Since (A_0, A_0) cannot vanish, we have

$$(5) \qquad (\sigma_\mu, \sigma_\nu) = 0 \;, \qquad\qquad \text{for } \mu \neq \nu.$$

The characteristic vectors of the operator Q of (1), which is being measured, are orthogonal. If we assume further that the labels of the different outcomes of the measurement are real, the characteristic values of the operator Q become real, its characteristic vectors orthogonal. This is a criterion for the self adjoint nature of Q. Hence, this self-adjoint nature is a consequence of the possibility to describe the measurement process of the second kind by means of quantum mechanics.

The same conclusion is arrived at on the basis of (2), *i.e.* for the measurement process of the first kind. The only difference is that $(A_\mu, A_\nu)(\sigma_\mu, \sigma_\nu)$ appears on the right side of (4a), instead of $(A_\mu, A_\nu)(\sigma_0, \sigma_0)$ but since the first factor of this expression vanishes, the second one's value is irrelevant.

5`6. *Measurement on the general state of the object.* – So far, only the case was considered in which the object, that is the system on which the measurement is being undertaken, is in one of the states which are characterized by a characteristic vector of the operator which is being measured. In the general case, the state vector of the object will be a linear combination of these characteristic vectors

$$(6) \qquad \sigma = \sum_\mu c_\mu \sigma_\mu \;.$$

What will be the final state of object plus apparatus in this case? We shall restrict our attention to the first type of measurement in this case, *i.e.* use eq. (2). Since the operator represented by the arrow in (2) is linear, the initial state

(7)
$$A_0 \times \sigma = A_0 \times \sum_\mu c_\mu \sigma_\mu = \sum_\mu c_\mu (A_0 \times \sigma_\mu)$$

will go over into

(8)
$$A_0 \times \sigma = \sum_\mu c_\mu (A_0 \times \sigma_\mu) \to \sum_\mu c_\mu (A_\mu \times \sigma_\mu) .$$

According to this point of view, the final state of the measurement is not a simple state. It is a linear combination of the states which can be obtained by the measurement process by starting from one of the states σ_μ in which the observed quantity Q of (1) has a definite value. In the final state, neither the apparatus, nor the system on which the measurement is undertaken, are in a pure state. The probability for the quantity Q to assume a definite value such as q_μ is the same as it was originally, namely $|c_\mu|^2$. This is also the probability that the apparatus be in the state A_μ in which it indicates the outcome q_μ for the measurement. There is, furthermore, a statistical correlation between the state of apparatus and object. If we could ascertain, by a measurement thereon, that the apparatus is in the state $A_\varkappa$, we could be certain that the object is the state $\sigma_\varkappa$. One can see this using the standard prescription for calculating outcomes of observations: the probability that the apparatus be in the state $A_\varkappa$ the object in the state σ_ν is the absolute square of the scalar product

(9)
$$\left(A_\varkappa \times \sigma_\nu, \sum_\mu c_\mu (A_\mu \times \sigma_\mu)\right) = \sum_\mu c_\mu (A_\varkappa \times \sigma_\nu, A_\mu \times \sigma_\mu) =$$
$$= \sum_\mu c_\mu (A_\varkappa, A_\mu)(\sigma_\nu \times \sigma_\mu) = \sum_\mu c_\mu \delta_{\varkappa\mu}(\sigma_\nu, \sigma_\mu) = c_\varkappa \delta_{\varkappa\nu} .$$

This vanishes unless $\varkappa = \nu$ which means that, after the measurement, object and apparatus cannot be found in states which do not correspond to each other. The first member of the second line follows from the rule, given before (4a), for calculating scalar products, the vanishing of the scalar product $(A_\varkappa, A_\mu)$ for $\varkappa \neq \mu$ is, as before, a consequence of the distinguishability of the states $A_\varkappa$ and A_μ; the last member is the consequence of (5), established before.

The preceding calculation shows that it is possible to imagine an interaction between object and apparatus resulting in a correlation of the states of these, as demanded by formal measurement theory. In particular, as (9) shows, the probability of the pointer position which corresponds to the eigenvalue q_μ of (1) is $|c_\mu|^2$. Also, in the sense of the first type of measurement, this pointer position is coupled with the state σ_μ, the characteristic vector of Q with the characteristic value q_μ of the system on which the measurement was undertaken. Except for this last point, the second type of measurement, based on (3), would have given the same result. However, a number of problems remains, and we come to these next.

5˙7. Some technical problems of the theory of measurement.

a) We shall be concerned, first, with the technical, that is nonphilosophical, and hence perhaps less serious problems. The first question which naturally arises concerns the arrows in (2) and (3), whether one can find or produce something, with the state vector A_0, which interacts with the system in the way indicated by these arrows. YANASE, and perhaps also FRENKEL, will discuss this question. The answer which they will give refers to the negative side of the picture: YANASE, in particular, will show that the interaction indicated by the arrow can be possible only if the state vector A_0 represents a « large » object, in a sense to be defined by him. This is, perhaps, not unexpected if we recall that most apparata are macroscopic. However, as he will explain, they need to be macroscopic in the sense of being large.

b) As was mentioned before, the discussion of YANASE will give only necessary conditions for the initial state of the apparatus. We cannot ask him, of course, to give a prescription for the construction of an apparatus to measure an arbitrary quantity—such questions are left, as a rule, to the skill of the experimenter and they often have a hard time with them. It is only fair to mention, however, that doubts have been expressed concerning the measurability of most self adjoint operators—in fact paractically all of them, excepting those referring to the linear momenta and the positions of particles [2].

c) The orthodox picture of the measurement process, epitomized in (2) and (3), is highly idealized. Actually, the arrows in (2) and (3) replace a collision matrix—the matrix for the temporary coalescence and subsequent separation of apparatus and object. This is a process that takes time, whereas the idealization of the orthodox measurement theory implies an instantaneous process. This is manifest also in the fact that no time is specified in (2) and (3) whereas—since at least the σ_μ need not be stationary states—it is necessary, in reality, to specify the time after which the state vector of the left side of these equations has transformed into the state vector of the right side. In order to make our equations (2) and (3) meaningful,˙this time must be large as compared with the « time of collision », in this case the duration of the measurement. However, the right side of the equation will depend in general on the exact value of the difference of times to which the two sides of (2) refer. It will depend on this time difference even if we restrict our attention to time differences which are large as compared with the time necessary to carry out the measurement.

It may be worth-while, though, to remark at this point that since no assumptions are made concerning the σ_0 of (3), this is unaffected by the preceding remark. Even as far as (2) is concerned, all the conclusions which we have, or will draw from this equation remain valid if one replaces the σ_μ on the right side by σ'_μ as long as $(\sigma'_\mu, \sigma'_\nu) = \delta_{\mu\nu}$ holds for these. This equation will be valid, however, at all times after the separation of object and apparatus if it is valid at any time thereafter. One is, indeed, tempted to define a measurement of the 1.5 type in which the σ_μ on the right side of (2) is replaced by σ'_μ but $(\sigma'_\mu, \sigma'_\nu) = \delta_{\mu\nu}$ is postulated for these.

The preceding remark about the finite duration of the measurement process assumes added importance if one wishes to adapt the picture of measurement, condensed into (2) and (3), to the theory of relativity. If the measurement were an instantaneous process, which can be described by (2) or (3), it would be a measurement process only in co-ordinate systems which are at rest with respect to each other. The measurement would then have no relativistic meaning. Actually, of course, the measurement occupies a space-time volume and can be viewed also from moving co-ordinate systems. This fact is, however, not incorporated into (2) or (3). It must be admitted that it would be desirable to modify the orthodox picture, epitomized by these equations, so as to respond to the concepts of the theory of relativity and also to a more realistic picture of the measurement process.

5'8. *Observation of the measuring apparatus as a measurement thereon.* – The preceding problems of the measurement theory were of a technical nature and one has the impression that they relate to the actual difficulties of measurements which simple expressions, such as (2) and (3), do not reflect. These idealize the situation but the idealization overlooks only practical difficulties of actually carrying out measurements. The conceptual difficulties of the picture employed will be discussed in some detail in the next Subsection. However, before proceeding to that discussion, it is good to establish another point.

The effect of the measurement, as represented by (8), establishes a statistical correlation between the states of apparatus and object. It does not lead to a state in which the states of either apparatus or object would be definite ones. When the situation given by the right side of (8) is reached, the measurement is not yet completed because no definite « pointer position » A_μ of the apparatus has been reached. This fact is the basis of the most difficult conceptual problem of the quantum-mechanical description of measurements and will be discussed from the conceptual point of view in Subsect. 5'9. It is natural, however, to try to resolve the problem by trying to describe the reading of the pointer, *i.e.* by ascertaining the actual state A_μ of the apparatus.

The actual state A_μ of the apparatus could be ascertained by measuring thereon a quantity P (for pointer position) of which the A_μ are characteristic vectors. If one wishes to describe such a measurement, one will consider the system which has been called, so far, the apparatus, to be the object of the measurement. In other words, one will bring this apparatus into interaction with a new measuring object, the initial state of which may be denoted by B_0. The Hilbert space of the total system, consisting of the original object of the measurement, the apparatus discussed in the last Subsection, and the apparatus introduced now, will be the direct product of the Hilbert spaces of all three. The interaction which leads to the measurement of P, however, will leave the original object of the measurement unaffected—after the measurement leading to the state (8) this will be separated from the apparatus which was used to carry out the measurement thereon. If only the original apparatus and the newly introduced one were considered, their interaction could be represented, in analogy to (2), by

$$(10) \qquad\qquad B_0 \times A_\mu \rightarrow B_\mu \times A_\mu \; .$$

In this case the measurement is surely of the first kind (hence the use of (2) rather than (3)) because the observation of the pointer position, to be carried out by means of the state B_0, surely should not alter that pointer position. Since, furthermore, the interaction between the apparata B and A should leave the original object unaffected, the initial state $B_0 \times A_\mu \times \sigma_\mu$ will go over into

$$(11) \qquad\qquad B_0 \times A_\mu \times \sigma_\mu \rightarrow B_\mu \times A_\mu \times \sigma_\mu \; .$$

If we now consider instead of $A_\mu \times \sigma_\mu$ the actual state of original apparatus plus object, that is the right hand side of (8), as the actual state of these, it follows from the linearity of the interaction between the two apparata that

$$(12) \qquad\qquad B_0 \times \sum_\mu c_\mu (A_\mu \times \sigma_\mu) \rightarrow \sum_\mu c_\mu (B_\mu \times A_\mu \times \sigma_\mu) \; .$$

This is then the state obtained from the measurement of Q of (1) on $\sigma = \sum c_\mu \sigma_\mu$ by the apparatus A if this measurement is succeeded by the measurement of P on the apparatus A by means of the apparatus B.

One sees that all that has been accomplished by the use of the apparatus B is the establishment of a further statistical correlation so that, if the state represented by the right side of (12) is reached, not only the states of the original apparatus and object are correlated, but both are correlated with the state of the apparatus used to ascertain the pointer position of the apparatus used for the original measurement. Clearly, the process could be continued and a third apparatus introduced to measure the state of B, and so on. However, equally clearly, all one obtains by the quantum-mechanical description are correlations between the states of more and more apparata. No final choice of the different μ in (8) or (12) can be obtained this way—as in fact this follows from the causal nature of the quantum-mechanical equations emphasized in Section 1.

As has been pointed out already by VON NEUMANN, whose book contains an excellent discussion of the whole observation process, actual observations in fact usually consist of a succession of several interactions of the type described, with a succession of transfers of information from one apparatus to the next one, of the same kind as is the transfer of information from apparatus A to apparatus B just described. However, if one considers the process of measurement completed only when a definite conclusion, say q_x, and corresponding state σ_x, and corresponding single pointer position is reached, this is not obtainable by means of the quantum-mechanical equations of motion. As was mentioned before, such a conclusion would lead to *one* of several possible alternatives which would be in conflict with the deterministic nature of the equations of motion.

5˙9. *The state of the apparatus may be a mixture.* – The discussion of the last section brings up a point which could have been mentioned among the technical problems. The initial state of the apparatus was assumed to be A_0, that of the object σ. As a result, the final state was determined uniquely by the equations of motion and no room was left for different developments thereof, depending on probabilistic laws. It is quite reasonable to assume, however, and von Neumann's book already mentions this possibility, that the initial state of the apparatus was indeterminate or, as it is called in technical terms, a mixture. One can imagine, therefore, that the indefinite, probabilistic outcome of the measurement is due to this circumstance, that the outcome is different at different occasions because the input was different, different as far as the measuring apparatus was concerned.

The problem is even more serious at the measurement of the pointer position of the first apparatus, leading to the state vector given on the right side of (12). Not only could A_0 have been indeterminate first, the same applies to B_0. Even further: even though the interaction given by (10) should leave the pointer position of the original apparatus A unchanged, it may have changed its quantum-mechanical state because, clearly, many state vectors correspond to any pointer position of a macroscopic system.

All this introduces a great deal of freedom to modify the preceding equations with the hope to account for its probabilistic behaviour. It is, however, not my intention to discuss this in detail; d'Espagnat has consented to do this. He will come to the conclusion that in spite of all this freedom, the probabilistic behaviour of the measurement process, as postulated by the standard theory, cannot be explained. I'll leave this point a bit up in the air but D'ESPAGNAT will clarify it.

14   E. WIGNER

5˙10. *The second part of the measurement process: the collapse of the state vector. Six views.* – How can we explain then the final part of the measurement process, the one following the establishment of a statistical correlation between object and apparatus? This final part consists in the state vector's transition from the sum standing on the right side of (8) (or of (12)) to a single member of this sum. The transition in question is usually referred to as the collapse of the state vector—a very picturesque but not very informative expression. There are, as far as I can see, six different attitudes toward it. Let me enumerate them.

a) It has been claimed, most clearly perhaps by EVERETT [3], that there is no need to assume a reduction of the sum in (8) or (12) to one of its members. The state vector remains this sum after the measurement. We should consider the state vector of the whole universe, including ourselves. This state vector will have many components and we live on one of its components and have the sensations which correspond to that component.

There are doubts in my mind whether the preceding description does justice to the point of view indicated because I am quite unable to endorse it. However, DE WITT will discuss it and will do justice to it.

b) The second point of view to be discussed was mentioned before, it was articulated most concisely by V. FOCK: measuring instruments must be described classically. Again, there is a danger of my not doing justice to this point of view. Some of my objections were mentioned before but they may well be repeated now after we have become a bit more familiar with the problems. First, the dividing line between object and final apparatus can be shifted—we have seen such a shift when we discussed, first, A as the apparatus and obtained (8) as the result of the measurement, then considered A as the object of a measurement which led us to (12). Is A to be considered as a classical or a quantum-mechanical system? Should it be considered classically in one case and quantum-mechanically in the second? What classical properties should we ascribe to the apparatus? The functioning of most apparata, such as counters, cannot be described by classical theory. Do we have a theory of interaction between classical and quantum systems? No such theory is known to me.

As I said, I may not be able to do justice to this point of view and am glad, therefore, that D'ESPAGNAT has agreed to take it up also.

c) The hypothesis that quantum mechanics does not apply to macroscopic systems, but only to microscopic ones, has been advocated by LUDWIG first [4] but has, surely, many other adherents. We must admit that we do not know what equations, and in fact what description, applies in the transition region between the macroscopic and microscopic domains. However, if we admit our present ignorance in this regard—and this point of view admits it—

we have a clearly formulated answer to the question where we should look for the description of the process of transition from the sums in (8) and (12) to a single term of these sums. The answer must be inherent in the unknown laws of the behaviour of macroscopic systems and their interaction with microscopic ones.

It may be better to formulate this point of view in a slightly different fashion. Present quantum mechanics is based on the recognition that the laws of macroscopic theories are approximate laws, valid if the number of particles is very large. Similarly, the point of view now discussed claims that the laws of quantum mechanics are also approximate laws, valid for systems with very few particles. The laws valid in the intermediate region, and more accurately also in limiting cases of very small systems, are yet to be discovered. If they are to explain the transition from the sums in (8) and (12) to a single term in these sums, they must contain a statistical, *i.e.* probabilistic element, foreign to the theories of both limiting cases of very large and very small systems. However, there is surely nothing absurd in this suggestion.

It seems to me that only the future development of physics will answer the question whether this solution of the problem of measurement is fruitful or not.

d) VON NEUMANN, in his well-known book [5], and perhaps with even greater clarity LONDON and BAUER [6], added a postulate to those of the quantum-mechanical equations. If one observes the apparatus A of (8) (or B of (12)), and finds the state of the former to be A_x (or B_x), the state vector of the system composed of apparatus A and the object of measurement go over into

$$(13) \qquad\qquad A_x \times \sigma_x ,$$

or

$$(13a) \qquad\qquad B_x \times A_x \times \sigma_x$$

and the probability of the observation giving these results is $|c_x|^2$. This is what is called the « collapse of the state vector ».

The preceding formulation of the collapse of the state vector considers this frankly, as a postulate of quantum mechanics, to be added to the other postulates. In fact, von Neumann said, explicitly: the state vector changes in two ways: continuously and according to causal laws, in accordance with Schrödinger's time-dependent equation, and discontinuously and according to statistical laws when an observation takes place on the system. Both VON NEUMANN and LONDON and BAUER remarked that it does not matter whether one observes the state of the apparatus A directly, or by means of the apparatus B. As far as the original object of the measurement is concerned, the result is the same state σ_x. It is A_x for the apparatus A, and B_x for that of B, supposing of course that the outcome of the observation corresponded to the value $q_{\varkappa}$ of the operator Q of (1). In no case does a statistical correlation between the three systems remain.

e) The duality in the change of the state vector, as formulated in the preceding paragraph, has been much criticized. One can avoid this duality if one frankly admits that quantum mechanics gives only probability connections between successive observations on a system. This formulation frankly gives primacy to the act of observation; it considers it as the basic quantity between the values of which physics establishes regularities though, according to quantum mechanics, only of a statistical, that is probabilistic, nature.

If one wishes to adopt the last attitude consistently, one will consider the state vector to be only a mathematical tool, useful for calculating the probabilities of the outcomes of observations. One can, naturally, avoid using this tool. The most natural way to do this is to attribute to every outcome of every measurement a projection operator. Let us denote the projection operator which corresponds to outcome $\varkappa$ of the measurement i by $P_{\varkappa}^{i}$. This means, in the language of state vectors, $P_{\varkappa}^{i}\psi = \psi$ if the measurement i on ψ gives surely the result $\varkappa$, if the result is surely different from $\varkappa$ then $P_{\varkappa}^{i}\psi = 0$. The probability for the measurements $2, 3, \dots$ to give the results $\beta, \gamma, \dots$ is then, if the first measurement gave the result α,

$$(14) \qquad w = \frac{\operatorname{Tr} P_{\alpha}^{1} P_{\beta}^{2} P_{\gamma}^{3} \dots P_{\gamma}^{3} P_{\beta}^{2} P_{\alpha}^{1}}{\operatorname{Tr} P_{\alpha}^{1}}.$$

Equation (14) assumes that the measurements are undertaken in the order $1, 2, \dots$. If the times of the measurements are different, that is the measurement 2, for instance, is not carried out immediately after the measurement 1 but some time later, the projection operators have to be properly transformed. Thus P_{β}^{2} at time t_{2} relates to P_{β}^{2} at time t_{1} by

$$(14a) \qquad P_{\beta}^{2}(t_{2}) = \exp\left[iH(t_{1}-t_{2})/\hbar\right] P_{\beta}^{2}(t_{1}) \exp\left[iH(t_{2}-t_{1})/\hbar\right]$$

and similar transformations must be carried out on the other P^{i}. Actually, (14) is more general than the usual formulae expressing probability connections not only because it gives the probabilities for the outcomes of several successive observations, rather than only of two such observations, but also because the possibility of several characteristic vectors, belonging to the same characteristic value is concisely taken into account.

The preceding formulation of the basic law of quantum mechanics does not use state vectors, and hence no « collapse » of this vector occurs therein. As was mentioned before, it is clearly based on the premise that outcomes of observations are the fundamental quantities between which the theory establishes relations. In my opinion, the formulation is entirely satisfactory—either in this or in the preceding form—as long as only a single person's observations have to be considered. If another person undertakes an observation, the outcome of which is later to be communicated to me, one is forced to consider him as an apparatus and ascribe a state vector to him. After his observation, his state will be part of a linear combination of several observational results and this appears to be an unreasonable assumption considering that we ourselvers always felt, after an observation, to have obtained one definite result. There is perhaps no logical or mathematical contradiction here but the conclusion that the state of a « friend » is a linear combination of several states, indicating different contents of his mind, seems very unnatural. This leads at least me to the opinion that quantum mechanics, in its present form, is not applicable to living systems, whose consciousness is a decisive characteristic. This is not surprising, perhaps, if we recall how schematic the picture of quantum mechanics is even when it refers to the measurements of a single observer. It disregards all his mental activities, except his ability to receive outcomes of observations.

Actually, the difference between the point of view discussed now, and that of *c*), is smaller than it may appear. Both points of view come to the conclusion that the validity of quantum mechanics' linear laws is limited. The point of view discussed under *c*) considers all macroscopic objects to lie outside the limits of validity of quantum mechanics. The point of view of the present section places the boundary further out, where life and consciousness begin to play a role.

f) In all the preceding discussion the process of measurement was considered to be similar to a collision process in which the « colliding » systems, the object of measurement and the measuring apparatus, were considered to be isolated systems both before and after the interaction. It was pointed out that the interaction must be of finite duration but except for this the collision picture was assumed to apply so that both object and apparatus were assumed to be free of external effects both before and after the interaction.

It was ZEH who called attention [7] to the circumstance that, as far as macroscopic bodies are concerned, the state of isolation is very difficult to maintain. The energy levels of macroscopic bodies are so closely spaced that even a very small body, at a very large distance, can introduce transitions between them. As a result, if A or B in (8) or (12) are macroscopic apparata, the states on the right side of these equations will not persist for any appreciable time but the terms with different μ will soon be coupled to different states of the matter in the neighborhood. As far as the union of the two systems described by (8) is concerned, this means that their state vector, given by the right side of (8), will soon go over into a mixture of state vectors, a mixture of the terms with different μ on the right side of (8). Naturally, if we considered the state vector of the union of the systems described by (8) and of the surrounding matter, this would remain a state vector—similar to the right side of (12). However, the matter surrounding the object and apparatus of (8) is surrounded with further matter and the extension of our consideration to all that is clearly impossible. Hence, one may as well replace the right side of (8), if A is macroscopic, by the mixture of the states $A_\mu \times \sigma_\mu$ with probabilities $|c_\mu|^2$. If A is not macroscopic but B is, the right side of (12) can be replaced by a mixture of the states $B_\mu \times A_\mu \times \sigma_\mu$ with probabilities $|c_\mu|^2$.

The argument just reproduced leads naturally to the replacement of the state vectors resulting from the measurement by a mixture thus replacing the collapse of the state vector by the process of ascertaining what element of the mixture is actually present, *i.e.* a process increasing our store of information. This renders it similar to the process of observation in ordinary macroscopic theory. One can hope that the process introduced in this way is more acceptable than the process of collapse or the frank admission that the primary concepts of quantum mechanics, between which it establishes correlations, are observation outcomes as implied by (14).

Whether Zeh's observation will be the basis of the final solution of the problem of quantum-mechanical measurements is, perhaps, not yet fully clear. It is predicated on an assumption concerning the initial states actually available in our world, on the impossibility to produce macroscopic systems sufficiently isolated from others so as not to be subject to perturbations by these. It is clear, that the isolation demanded by (8) (or (12)) for a macroscopic measuring apparatus A (or B) is very difficult to achieve. It is not clear that the situation would be different from that discussed in c) and d) if the isolation *could* be achieved.

In my mind the observation which is discussed in this subsection is a very important one and I hope ZEH will further elaborate on it.

5`11. *Return to some general considerations.* – As a last point, it may be well to repeat what the introductory comments emphasized: that our science can not entirely stand on its own feet, that it is deeply anchored in common concepts acquired in our babyhood or born with us, and used in everyday life. As I often say, the law of lepton conservation is surely much deeper than that of the conservation of our keys but, if we look at them frankly, we must admit that science would be more nearly possible without our knowing the conservation law for leptons than the knowledge that our keys are somewhere even if we misplaced them. It often seems to me that the point of view I tried to represent in parts a) and b) of Subsect. 5`10 are responses to this fact.

The fact of our reliance on simple, non quantum-mechanical, everyday observations is perhaps nowhere as evident in the epistemology of quantum mechanics as in our assumption of the knowledge of the apparata used for measurements. We assumed that the initial state vector of our apparatus is A_0 (with properties implied by (2) in particular) but we did not specify how we arrived at this knowledge. If we wished to regard quantum mechanics as a closed, self-contained subject, we would have to say that we made an observation on it and ascertained its state, whence we deduced, by means of the equations of quantum mechanics, its properties. In particular, we deduced that it will react with the system on which we want to undertake a measurement according to (2). This, however raises the question how we knew the state and properties of the apparatus with which we ascertained that the state of the original apparatus was A_0. Clearly, we enter a chain of questions which has no ending and must conclude that the self-contained nature of quantum mechanics is an untenable illusion. However, even if we admit that much or most of the information which we possess did not come to us via our knowledge of quantum theory and on the basis of information acquired by the observations we try to describe, it is well to strive toward a consistent interpretation and an understanding of our and the apparata's role in the process of observations which play such a basic role in our present physical theory. The sessions in which we will participate will be aimed toward an increase of this understanding and toward an improvement of our interpretation of the process of observations.

REFERENCES

[1] V. FOCK: *Classical and quantum physics*, paper SMR 5/8 of the *International Centre for Theoretical Physics* (Trieste, 1968).

[2] The opposite view seems to be articulated by W. E. LAMB: *Phys. Today*, 22, 23 (1969).

[3] H. EVERETT jr.: *Rev. Mod. Phys.*, 29, 454 (1957). Dr. DE WITT will represent this view more in detail. Cf. also his article, *Phys. Today*, 23, No. 9, 30 (1970).

[4] E. LUDWIG: article in *Werner Heisenberg und die Physik unserer Zeit* (Berlin, 1961).

[5] J. VON NEUMANN: *Die Mathematischen Grundlagen der Quantenmechanik* (Berlin, 1932).

[6] F. LONDON and E. BAUER: *La théorie de l'observation en mécanique quantique* (Paris, 1939).

[7] See *e.g.*, *Foundat. Phys.*, 1, 67 (1970).

The Philosophical Problem[*]

E. P. Wigner

B. d'Espagnat (ed.) Foundations of Quantum Mechanics. International School of Physics
"Enrico Fermi" 1970. Academic Press, New York 1971, pp. 122–124

1. *The philosophical problem.* – The principal question which is at the center
of our own as well as Prosperi's discussion concerns the reason for the statistical, that is probabilistic, nature of the laws of quantum-mechanical theory.
The equations of motion of both quantum mechanics and of classical theory
are deterministic; why, then, are the predictions not uniquely given by the
inputs?

As far as I can see, there are three possible reasons for this. The *first,* and
perhaps most natural, one is that the input is always incomplete, that even
if we have as complete a description of the object on which a measurement
is undertaken as quantum mechanics can provide, the same is never true of
the instrument with which we carry out the measurement. As far as I can
see, this is not the reason PROSPERI attributes to the nondeterministic nature
of the theory. It is, indeed, unlikely to be the reason: as D'ESPAGNAT, in
particular, has shown, the probabilities which such an assumption can yield
for the possible outcomes of a measurement cannot be made to coincide with
those postulated by quantum-mechanical theory.

The *second* possible reason for the probabilistic nature of quantum theory's
conclusions concerning the outcomes of measurements is that the theory cannot
completely describe the process of measurement, that some part of the process
is not subject to the equations of quantum mechanics. This is the view which
I believe, most of us, including VON NEUMANN and also myself, accept and
which, I now believe, is also implicit in Prosperi's explanation even though
he does not state it explicitly. If I am right in this, the difference between
our views concerns only the *area* to which quantum mechanics is inapplicable
In von Neumann's view, it is the observer's role which is not subject to the
laws of quantum mechanics, that is, the content of his mind is not obtainable
by means of the laws of the theory. It is either that these laws do not apply
to the functioning of the mind (whatever that word means) or that the con-

scious content of the mind is not uniquely given by its state vector, *i.e.* by the quantity which quantum mechanics uses for the description of all objects. This is what I call the orthodox view. On the contrary, in Prosperi's view, the probabilistic element enters at the interaction of the microscopic object with the macroscopic apparatus used for the measurement. He does not say that quantum mechanics is invalid for macroscopic objects (as does LUDWIG), but says that its description of macroscopic objects is too detailed and, in practice not verifiable by experiments. Hence, it may be replaced, as well, by the cruder description given by classical theory. The translation of the quantum-mechanical description into classical description is, however, not unique: the classical picture, *i.e.* macroscopic positions and velocities, is not uniquely given by the quantum description even though the latter is immensely more detailed. The two are askew to each other: neither determines the other completely. Thus, an extended wave packet, a complete description of the state-in quantum mechanics, if translated into classical theory, gives an incomplete description: at least one variable used by classical theory, the position, is not uniquely given by the state vector of the wave packet. Hence, in Prosperi's view, the probabilistic element enters into the measurement process because we *want* to describe the apparatus, soon after the measurement has taken place, by means of the concepts of classical theory. Those of us who favor the orthodox view agree that it is not unreasonable to assume that the effect of the apparatus on us, that is, on the content of our mind, can be obtained from a classical, that is, quantum-mechanically incomplete, description of the apparatus. Hence, the objections which I wish to voice against Prosperi's picture do not affect the practical consequences thereof. My objection to his treatment is indeed, at least partly pedagogical. It seems to me to obscure the fact that the probabilistic element of the theory is, fundamentally, foreign to quantum-mechanical dynamics which postulates the description of the states of objects by deterministically developing state vectors.

The *third* way to avoid the apparent contradiction between deterministic equations and probabilistic events (outcomes of observations) is not to consider the equations as descriptions of reality (whatever that term means) but only as means for calculating the probabilities of the outcomes of observations. This was discussed in the Section *Basic Quantities of the Various Physical Theories* of my own presentation. The reason for my mentioning this possibility is a sentence in the last formulation of Prosperi's discourse: « Quantum mechanics does not describe a system in itself, but only deals with the results of actual observations on it. » For a positivist (as most of us are) this does not differ basically from the point of view which was called orthodox in the preceding discussion. A positivist, as I understand this term, does not look for the « ultimate reality »; the observations, which I interpret to be the impressions he receives, are his prime concern. Personally, I am increasingly inclined toward favoring this third formulation of quantum mechanics' real role. It admits that the observation is the primary concept of this theory and points to the impressions which the observer receives as the basic entities between which quantum mechanics postulates correlations. These impressions thus assume the role which the positions and the velocities of the particles have played in classical theory. Even though the later parts of Prosperi's aforementioned last account do not seem to agree with the sentence just quoted it seemed desirable to call attention to this mode for the elimination of the observation paradox.

Questions of Physical Theory

E. P. Wigner

B. d'Espagnat (ed.) Foundations of Quantum Mechanics. International School of Physics "Enrico Fermi" 1970. Academic Press, New York 1971, pp. 124–125

2. *Questions of physical theory.* – PROSPERI points out that microscopic experiments on a macroscopic body are very difficult to carry out. He implies that only the values of macroscopic variables are meaningful if one deals with macroscopic objects, such as measuring devices. It is not clear whether he considers the values of microscopic variables to be unobservable in principle. If this were so, quantum mechanics should surely be modified and the quantities which are measurable on given bodies specified. To be sure, the statement that only macroscopic variables are meaningful for macroscopic objects will be a clearly defined statement only after some definition of « macroscopic variables » and « macroscopic objects » has been provided. The phenomena of permanent currents in superconductors, of spontaneous magnetization in different directions, show that the behaviour of large objects can be greatly influenced by quantum effects which are outside the realm of ordinary macroscopic theory. As to the difficulty of observing phases, it might be recalled that the difference between dextro and levorotatory sugar is based on a phase relation—a difference surely observable.

It seems, therefore that Prosperi's postulates cannot be rigorously formulated at present and that their formulation would entail a significant modification of the present theory. This is, of course, in line with our ideas. It is also in line with the fact that Prosperi's postulate cannot be rigorously proved. This is particularly clear from a recent article of LANZ, LUGIATO and RAMELLA: *On the Existence of Independent Subdynamics in Quantum Statistics.*

All these reservations may be, and in my opinion are, entirely valid. In spite of this, I believe, we all feel that there is a great deal of truth in the statement that the behaviour of macroscopic objects follows, in most cases though not always, the causal laws of macroscopic physics. Laser action, the breakdown of supercurrents, are exceptions. *At least* in general, there is no reason to believe that similar effects play a significant role in the functioning of measuring devices and the exceptions are mentioned principally because it may be interesting to investigate situations in which these exceptional effects play a major role.

Thus PROSPERI assumes that for macroscopic bodies the classical description is complete. It follows, as was mentioned before, that the interaction of the microscopic object, on which the measurement is carried out, with the macroscopic apparatus does not obey the law of causality: after the interaction the state of the apparatus is not completely defined and must be described by a distribution function. This is natural since most wave functions, but in particular those resulting from a measurement-type interaction, give finite probabilities to several states which can be distinguished even macroscopically (see for instance, Schrödinger's cat.). This is not in conflict with the fact that the consistent application of quantum mechanics would give from a fully defined original state of object plus apparatus an equally fully defined final state: the statistical nature of the outcome of the measurement derives, according to Prosperi's ideas, from the lack of uniqueness of the macroscopic variables which correspond to a single quantum-mechanical state (a wave function). He blames the statistical nature of the outcome of the measurement on the need to translate the state vector of the apparatus into classical language and the lack of uniqueness of this translation. We blame it on the process of translation of the quantum-mechanically defined state of the apparatus into the sensations of the observer. In either case, the inapplicability of quantum mechanics to some part of the measurement process has to be postulated or admitted—the earlier part of the present discussion also gives the reason for my hesitation to attribute the inapplicability to something as inadequately defined as is the macroscopic nature of something.

That quantum mechanics does not cover in its present form all parts of the measurement process is also apparent from the fact that, in their aforementioned article, LANZ, LUGIATO and RAMELLA propose to « generalize » quantum mechanics so that « an independent dynamics of macroscopic observables holds exactly ».

On Bub's Misunderstanding
of Bell's Locality Argument

S. Freedman and E. P. Wigner

Foundations of Physics *3*, 457–458 (1973)

Received July 7, 1973

Bub's criticism of Bell's locality postulate is discussed. The locality postulate is explained, and it is shown that Bub is in fact arguing against a class of theories which are subject to stronger restrictions than this postulate, and therefore his "refutation" of the latter is misleading.

Since Bub's[1] article seems to be based on a misunderstanding of Bell's locality postulate,[2] it appears to be desirable to spell out this postulate more explicitly than was done before.

The experiment considered[3] in the articles criticized by Bub refers to two spin-$\frac{1}{2}$ particles widely separated in space, but with their spins in opposite directions, forming a singlet state. For example, such a pair of particles would arise from (electric dipole) photodisintegration of a hydrogen atom in its singlet state. The measurements considered concern the positive or negative values of each particle's spin along three possible measurement directions **a**, **b**, or **c**. Altogether, nine Stern–Gerlach measurements on the two-particle system are possible. One of the measurements, for example, refers to the electron spin in the **a** direction and the proton spin in the **b** direction. There are four possible results: The electron spin is in the **a** direction and the proton spin is in the **b** direction; the electron spin is in the −**a** direction and the proton spin is in the **b** direction; the electron spin is in the **a** direction and the proton spin is in the −**b** direction; and the electron spin is in the −**a** direction and the proton spin is in the −**b** direction.

In a hidden-variable theory it is postulated that a set of hidden variables uniquely determines the outcome of all nine measurements. The locality postulate requires that the result of a measurement on one particle be independent of the particular measurement made on the other particle. That is, the result of a measurement of the electron's spin does not depend on the direction chosen for measuring the proton's spin and vice versa. The justification for this postulate stems from the possibility of performing the measurements simultaneously and at different points in space. Relativity theory then seems to demand that the direction of one measurement should not influence the outcome of the other. Bell's argument shows, nevertheless, that there is no single distribution of hidden variables *relating to the two particles* which leads to probabilities for the nine measurements that are in accord with the probabilities given by quantum mechanics. There are now experimental indications[4] of agreement with the quantum mechanical probabilities.

Bub applies a postulate more general than the locality postulate ("the straight phase space reduction rule") to *successive* measurements, in two of the directions **a**, **b**, and **c**, on a single spin-$\frac{1}{2}$ particle. He points out that the quantum mechanical probabilities are not obtained from a single distribution of hidden variables in this case either. This is, however, not relevant: The first measurement could interfere with the values of the hidden variables and affect the outcome of the second measurement without violating the locality postulate. In fact, it is easy to construct a model which does reproduce the quantum mechanical probabilities for successive measurements.[5] Furthermore, such a mechanism could be incorporated in a model in which spatially separated measurements arise from a mechanism in accord with the locality postulate (of course the probabilities for correlated measurements will not agree with those of quantum mechanics).

Bub's article does not question the correctness of the conclusions that follow from the locality postulate, only the reasonableness of such a postulate. This is, of course, a matter of taste and judgement concerning the usefulness of a hidden-variable theory not satisfying such a condition. Bub's "refutation" is misleading, however, because he argues against a class of theories which are subject to stronger restrictions than the locality postulate.

REFERENCES

1. J. Bub, *Found. Phys.* **3**, 29 (1973).
2. J. S. Bell, *Physics* **1**, 195 (1964).
3. E. P. Wigner, *Am. J. Phys.* **38**, 1005 (1970).
4. S. J. Freedman and J. F. Clauser, *Phys. Rev. Lett.* **28**, 938 (1972).
5. J. S. Bell, *Rev. Mod. Phys.* **38**, 447 (1966); J. F. Clauser, *Am. J. Phys.* **39**, 1095 (1971).

Review of the Quantum-Mechanical Measurement Problem

E. P. Wigner

D. M. Kerr et al. (eds.) Science, Computers, and the Information Onslaught.
Academic Press, New York 1984, pp. 63–82

1. Introduction

To review the quantum-mechanical measurement problem, a very old subject, in less than an hour is not an easy task, particularly since I want to add in the last part of my discussion some new ideas. In this first section I shall briefly describe some background from quantum mechanics, although the paper will not be self-contained. But first let me say in a few words why there is a problem, why physicists are all puzzled by some rather obvious facts.

In classical mechanics the "state" of a system was described by the positions and velocities of its parts, in particular of the atoms that constitute it. This has changed drastically with Schrödinger's introduction[1] of the wave function in quantum theory. In classical physics, the state of n atoms was given by $6n$ numbers—the three coordinates of the positions of the atoms and the three components of their velocities. In Schrödinger's quantum mechanics, the "state" is described by a function of the $3n$ position coordinates,† by a wave function, giving a very much greater variety of states than those of classical theory. An important consequence of the much greater complexity of the quantum-mechanical description of the state of the system is that whereas in classical theory the state of the system—the positions and velocities of the objects it contains—can be determined (measured), this is not true in quantum mechanics. Indeed, if a system is sufficiently accurately described by the classical theory—for instance the planetary system—then the positions and velocities of the constituents can be determined rather easily. This is not true for microscopic systems (to which quantum mechanics applies)—for instance the position and ve-

---

†This was modified subsequently to some degree, but these modifications do not really affect the present discussion.

locity of the hydrogen atom's electron cannot be measured. In fact, these concepts are not truly meaningful—as was realized by Heisenberg when he provided the foundation of quantum mechanics.[2]

It follows that a quantum-mechanical measurement, the determination of the "state" of the system, is not a simple process. In fact, there is no way to determine what the wave function (also called the state vector) of a system is. Clearly, such a measurement would have to result in a *function* of several variables, not in a relatively small set of *numbers*. On the other hand, in many cases the system can be brought into a definite state—even if its original state (i.e., its original state vector) cannot be ascertained by any measurement. In this way the quantum-mechanical description is fundamentally different from the classical one: the positions and velocities of macroscopic bodies, which form the classical description of their states, can be obtained by measurements that, in fact, do not affect the state of the system.

The theory of the quantum-theoretical measurement was most precisely formulated by Von Neumann.[3] It was he who postulated most clearly that the state of a system is described by a vector in Hilbert space—the wave functions mentioned before constitute a particular description of such vectors. If ψ and ϕ are two such vectors, it is postulated that a scalar product of them is defined and this is usually denoted by (ψ, ϕ). For the Schrödinger description $\psi(x_1, y_1, z_1, x_2, y_2, z_2, \ldots)$, this scalar product (a number) is

$$(\psi, \phi) = \iint \cdots \int \psi(x_1, y_1, z_1, \cdots)^* \; \phi(x_1, y_1, z_1, \cdots) \; dx_1 \, dy_1 \, dz_1 \cdots \tag{1}$$

but, for the following, the assumption of the existence of the scalar product with the properties (which equations follow from (1)) will suffice. The scalar product obeys the equations

$$(b_1\psi_1 + b_2\psi_2, \phi) = b_1^*(\psi_1, \phi) + b_2^*(\psi_2, \phi) \tag{2}$$

$$(\psi, a_1\phi_1 + a_2\phi_2) = a_1(\psi, \phi_1) + a_2(\psi, \phi_2) \tag{2a}$$

and

$$(\psi, \phi) = (\phi, \psi)^* \tag{2b}$$

the a and b being arbitrary complex numbers. Two vectors are said to be orthogonal if their scalar product is 0. The last relevant characteristic of a Hilbert space is that there is an infinite set of mutually orthogonal vectors, each of unit length, we can call them ψ_1, ψ_2,..., so that

$$(\psi_\kappa, \psi_\lambda) = \delta_{\kappa\lambda} \tag{3}$$

in terms of which every state vector ϕ can be expanded

$$\phi = \Sigma \, a_\kappa \psi_\kappa. \tag{4}$$

Review of the Quantum-Mechanical Measurement Problem   65

It follows then from (2), (2a), and (3) that

$$a_\kappa = (\psi_\kappa, \phi) \tag{4a}$$

and that

$$(\phi, \phi) = \Sigma |a_\kappa|^2. \tag{4b}$$

The a_κ is often called the κ component of ϕ. These components characterize the state as well as the original vector ϕ. In fact, the original definition of a Hilbert space was the vector space of all infinite sequences of "components" a_κ such that the sum in (4b) remained finite. It then further follows from (2), (2a), (3) that the scalar product of

$$\psi = \Sigma b_\lambda \psi_\lambda \tag{5}$$

and of ϕ becomes

$$(\psi, \phi) = \Sigma b_\kappa^* a_\kappa \tag{6}$$

and this was the original definition of the scalar product. It can be shown that the sum in (6) is finite, in fact

$$|(\psi, \phi)|^2 \le \Sigma |b_\lambda|^2 \, \Sigma |a_\kappa|^2. \tag{6a}$$

It can also be shown that there are "unitary operators" that transform the ψ_κ into

$$\psi'_\kappa = \Sigma U_{\kappa\mu} \psi_\mu$$

such that

$$(\psi'_\kappa, \psi'_\lambda) = \delta_{\kappa\lambda} \tag{7}$$

and the expression (6) for the scalar product (ψ, ϕ) is the same in terms of the ψ'_κ components

$$\phi = \Sigma a'_\kappa \psi'_\kappa, \qquad \psi = \Sigma b'_\lambda \psi'_\lambda \tag{8}$$

as in terms of the ψ_κ components

$$(\psi, \phi) = \Sigma b'^*_\kappa a'_\kappa \tag{6b}$$

and that the product of two unitary operators is still unitary. Let us observe also that

$$a'_\kappa = (\psi'_\kappa, \phi) = \left(\sum_\mu U_{\kappa\mu} \psi_\mu, \sum_\lambda a_\lambda \psi_\lambda \right)$$

$$= \sum_{\mu\lambda} U^*_{\kappa\mu} a_\lambda \delta_{\mu\lambda} = \sum_\mu U^*_{\kappa\mu} a_\mu \tag{8a}$$

and that the unitary matrix components obey the equation

$$\sum_\kappa U^*_{\kappa\mu} U_{\kappa\lambda} = \sum_\kappa U^*_{\mu\kappa} U_{\lambda\kappa} = \delta_{\mu\lambda}. \tag{8b}$$

the change of the a_κ resulting from the passage of time is induced by a unitary operator which, naturally, depends on time:

$$a_\kappa(t) = \sum U(t)_{\kappa\lambda} a_\lambda(0). \tag{9}$$

This was a very short and superficial description of the mathematical basis of quantum mechanics. How about the "measurement"? This plays a fundamental role, but it is not as simple a role as in classical physics, because, as was mentioned before, there is no measurement that provides the "state," i.e., the infinity of numbers a_1, a_2, a_3,....

Each "measurement," as was most precisely brought out by Von Neumann,[3] refers to a complete set of orthogonal states, such as the ψ_κ of (3). These sets can be transformed into each other by unitary transformations, but a particular measurement refers to a particular such set. It transforms the original state of the system into one of the elements of this set, in particular if the state vector before the measurement is ϕ, the probability of its being transformed into ψ_κ is

$$P_\kappa = |(\phi, \psi_\kappa)|^2 / (\phi, \phi). \tag{10}$$

It is further postulated that the measuring apparatus shows into which state the system has been transformed—its pointer turning to a point that we may call λ_κ, if the state has been changed to ψ_κ. One says that the apparatus measured an "operator" Λ, the operator being the one which, if applied to a state vector χ, transforms it into

$$\Lambda\chi = \sum (\psi_\kappa, \chi)\lambda_\kappa\psi_\kappa. \tag{11}$$

Clearly, if the original state ϕ was one of the ψ, let us say ψ_μ, this will also be the final state, because the probability of its being transformed into another ψ_κ is 0 by (10) and (7). The pointer of the measuring apparatus will then surely show λ_μ (which is a real number).

This was a very general description of the measurement process; and indeed, as will be discussed later, the measurement of virtually any Λ involves problems. There are a few exceptions, though. In particular, the energy of the system can be measured—the light which an atom emits tells us what its energy was. In fact, quantum mechanics was started by Heisenberg[2] by giving prescriptions for obtaining the possible energy values. The atom's momentum can be measured by a grating, its position can also be measured at least approximately. Hence the properties of the ψ_κ referring to these quantities are defined, and so is the operator Λ relating to them. But the general basis of measurement theory is that all, or at least many, operators of the form (11) can be measured—these being

Review of the Quantum-Mechanical Measurement Problem   67

called Hermitean or self-adjoint operators and having the property that for any two vectors ϕ and χ the equation

$$(\phi, \Lambda\chi) = (\Lambda\phi, \chi) \tag{12}$$

is valid. This can be verified by expanding both ϕ and χ in terms of the ψ_κ and applying (11) and the preceding equations, in particular (3).

The preceding discussion is a very, very brief review of formal measurement theory. It emphasizes the importance of the possible final result of a measurement, namely the ψ_κ in the case of the particular measurement described first. A measurement is always the measurement of an operator—in particular that of the Λ of (11)—but it is clear that the most relevant characteristics of the measurement are the ψ_κ since the λ_κ can be changed by replacing the indicators of the measurement outcome by other numbers, $\lambda'_\kappa = f(\lambda_\kappa)$. But the discussion does show that in order to verify the theory in its full generality, at least a succession of two measurements is needed. There is, in general, no way to determine the original state of a system, but having produced a definite state by an initial measurement, the probabilities of the outcomes of a second measurement are then given by the theory, Hence, it can be said that, fundamentally, quantum mechanics provides "probability connections between the outcomes of subsequent observations" and assumes the possibility of measurements and a knowledge of the quantities we measure as well as the relations between the ψ_κ or ψ'_μ of these. The probability of obtaining the state ψ_κ by the second measurement if the first one gave ψ'_μ is

$$|(\psi_\kappa, \psi'_\mu)|^2, \tag{13}$$

and this is the general result which should be verifiable. Naturally, if the second measurement is delayed by a time interval t, the ψ_κ in (13) should be replaced by $U(t)\psi_\kappa$ where $U(t)$ is the operator that appears in (9).

It may be good to admit, finally, that the preceding discussion is incomplete in at least two regards. First, in some measurements the apparatus shows the same λ for more than one ψ_κ. This is called degeneracy, and its consequences are not difficult to derive. The second point is that in some measurements, it is claimed, the outcome is not a definite number, such as the λ_κ of (11), but an interval of the λ variable. One then says that the operator Λ has a continuous spectrum. Momentum is such a quantity. It is then necessary to replace the ψ_κ by "projection operators," but the result of this is that the probability connections between the outcomes of subsequent observations become much less definite and the whole discussion much more complicated. We shall therefore confine the following discussion to the type of measurement just described, i.e., to measurements that bring the system into a definite state, determined by the result of the measurement.

Eugene P. Wigner

We now proceed to a discussion of the problems of the quantum mechanical description of the measurement process.

2. The Measurement Process—Can Quantum Mechanics Describe It?

It is, of course, possible to claim that the purpose of quantum mechanics is simply to "provide probability connections between subsequent observations," and hence that the concept of observations (measurements) is the primary concept which needs as little explanation as the concepts of classical mechanics, i.e., the positions and velocities. But this is hardly reasonable. First, most "observations" are not quantum-mechanical, and we do not attribute characteristic functions to their outcome. Second, the assumption that observations form the basic concepts between which quantum mechanics establishes probability connections leads to an unreasonably solipsistic philosophy—it is not reasonable to assume that another person's observations are not equally correlated by the laws of quantum mechanics. He appears to be an object whose observations do not appear in this interpretation of the theory, yet they should be describable by it. They are not.

This means, particularly if we want to avoid the solipsistic attitude, that we should not consider an observation to be the basic concept but should try to describe it also, not only ours but also those of others, even including those of animals, using the ideas of quantum mechanics. This is a natural requirement—but it runs into difficulties. These difficulties arise, fundamentally, from the fact that quantum mechanics is a deterministic theory even though the outcome of an observation—the resulting state of observer and observed system—has a probabilistic nature.

Let us first consider the standard quantum-mechanical description of the observation process. The state vector (or wave function) of the initial state of the measuring apparatus will be denoted by $\mathscr{A}_0$—this depends on variables, perhaps $\xi_1, \eta_1, \zeta_1, \xi_2, \eta_2, \zeta_2, \ldots$, different than those of the system to be observed. Let us also first consider the case in which the initial state of the object to be observed is one (ψ_κ) for which the measurement surely gives a definite value (λ_κ). The state vector of the union of apparatus plus system is then denoted by $\mathscr{A}_0 \times \psi_\kappa$—the cross meaning a product of $\mathscr{A}_0$ and ψ_κ, the two depending on different variables (ξ, η, ζ and x, y, z). The object is to remain in the state ψ_κ after the observation but the apparatus' pointer is supposed to show λ_κ, the corresponding state being denoted by $\mathscr{A}_\kappa$. Hence, as the result of the measurement the state $\mathscr{A}_0 \times \psi_\kappa$ goes over into

$$\mathscr{A}_0 \times \psi_\kappa \to \mathscr{A}_\kappa \times \psi_\kappa. \tag{14}$$

Review of the Quantum-Mechanical Measurement Problem   69

This transition is caused by the interaction of the two parts, i.e., is the result of the passage of some time. Hence there is a unitary operator that provides the transition as indicated in (14).

It may be of some interest to observe then that two states, ψ_κ and ψ_μ, which give different results of observation, must be orthogonal. The scalar product of $\mathcal{A}_0 \times \psi_\kappa$ and $\mathcal{A}_0 \times \psi_\mu$ is $(\mathcal{A}_0, \mathcal{A}_0)(\psi_\kappa, \psi_\mu) = (\psi_\kappa, \psi_\mu)$ and this must be equal to $(\mathcal{A}_\kappa, \mathcal{A}_\mu)(\psi_\kappa, \psi_\mu)$. But since $\mathcal{A}_\kappa$ and $\mathcal{A}_\mu$ are even macroscopically distinguishable, their scalar product is 0—the same must hold for the left-hand side, for (ψ_κ, ψ_μ). One can conclude from this that the operator Λ which is measured is self-adjoint, since equation (12) follows from the orthogonality of the characteristic vectors ψ_κ, ψ_μ, . . . and from the real nature of its characteristic values λ_κ, λ_μ, ... This is satisfactory.

But let us consider now the general case in which the original state of the object is a linear combination $\Sigma c_\kappa \psi_\kappa$ of the "pure states" ψ_κ. It then follows from (14) and the linearity of the time development that

$$\mathcal{A}_0 \times \Sigma c_\kappa \psi_\kappa = \Sigma c_\kappa (\mathcal{A}_0 \times \psi_\kappa) \to \Sigma c_\kappa (\mathcal{A}_\kappa \times \psi_\kappa). \qquad (14a)$$

The final state is interesting but complicated. It is a state of the joint object–apparatus-system in which there is a *correlation* between the states of the two. More precisely suppose that a suitable "measurement" of the apparatus is performed and that its state is determined. It will then be certain that the object will be found in the corresponding state if another measurement is carried out thereon. In other words, the measurement of both states, that of the "pointer position of the apparatus" and that of the quantity Λ for the object, will give corresponding results. The probability that the first will give the result λ_μ and the second will be found in the state ψ_ν, is the absolute square of the scalar product of the right-hand side of (14a) with $(\mathcal{A}_\mu \times \psi_\nu)$, this being the absolute square of

$$\sum_\kappa c_\kappa (\mathcal{A}_\kappa \times \psi_\kappa, \mathcal{A}_\mu \times \psi_\nu) = \sum_\kappa c_\kappa (\mathcal{A}_\kappa, \mathcal{A}_\mu)(\psi_\kappa, \psi_\nu)$$

$$= \sum_\kappa c_\kappa \delta_{\kappa\mu} \delta_{\kappa\nu} = c_\mu \delta_{\mu\nu}, \qquad (15)$$

which vanishes if $\mu \neq \nu$. This is very important and satisfactory result. But, of course, this does not mean either that the measurements indicated can be really carried out, nor, more importantly, that the final result thereof can be characterized by the statement that the joint system is, with probability $|c_\kappa|^2$ in the state $\mathcal{A}_\kappa \times \psi_\kappa$. It is in a superposition of such states which has, according to quantum mechanics, important properties different from the "mixture" of these states. This will be shown by an example.

The aforementioned difference can be best illustrated by an example for which the "measurement" is clearly possible and can be easily described. This is the case for the so-called Stern–Gerlach experiment. A particle with spin $\tfrac{1}{2}\hbar$ is sent

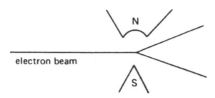

Figure 1. Stern–Gerlach apparatus. The beam enters from the left and splits into two beams.

through an inhomogeneous magnetic field. If the direction of the spin is the same as that of the magnetic field, the particle will be deflected in the direction of increasing magnetic field strength; if the spin has the opposite direction, it will be deflected in the direction of decreasing magnetic field strength.† If the original state is a superposition of the two, the incoming beam will be split into two beams, one deflected in the direction of increasing magnetic field strength, the other in the opposite direction (see Figure 1). If the first beam is then guided into a second inhomogeneous magnetic field, the field and its inhomogeneity being now in the direction of the spin, the beam will not be split but deflected in the direction of increasing field strength—it will behave as it would have behaved in the field it originally passed if its spin had had the direction of the increase of the field (see Figure 2). A similar remark applies to the second beam—the passage through a second inhomogeneous field will deflect it in the direction of decreasing field strength. Thus a correlation is established between the position of the particle and its spin direction—the directions being only two: up or down, not sidewise. This corresponds to our equation (14) for a spin $\frac{1}{2}\hbar$ particle there being only two κ values if the spin component is measured in a definite direction, which in our case is the direction of the magnetic field (of variable strength).

It thus seems that the measurement is completed—one can say that the direction of the deflection of the particle is the "position of the pointer" of the measuring apparatus. If the inhomogeneity of the field causes a deflection by about x_0, the essential part of the final wave function of the particle with initial spin wave function $a\alpha + b\beta$ will be

$$f(x - x_0)a\alpha + f(x + x_0)b\beta, \tag{16}$$

α and β denoting the two spin states, α being the same as direction of the field, and β being the opposite direction. The f is the original position wave-function of the particle, having its maximum at 0. (Naturally $|a|^2 + |b|^2 = 1$.)

Is it really true that the state vector of the system is, with probability $|a|^2$, equal to $f(x - x_0)\alpha$ and with probability $|b|^2$ equal to $f(x + x_0)\beta$? Such a situation would be called a "mixture" of the two states, with probabilities $|a|^2$ and $|b|^2$—

---

†This assumes that the direction of the magnetic field is opposite to that of the spin generated field. If this is not true, directions of deflection are reversed, but the remaining argument applies equally.

Review of the Quantum-Mechanical Measurement Problem   71

and this is what one would expect to result from the measurement. However, this is not true. It is possible to distinguish the state characterized by the wavefunction (16) from the aforementioned mixture. This can be done by putting a second inhomogeneous field in the way of the two beams of (16), with opposite inhomogeneity, which will then transform (16) back into

$$f(x) \, (a\alpha + b\beta) \tag{16a}$$

whereas if the state were $f(x - x_0)\alpha$ with probability $|a|^2$ and $f(x + x_0)\beta$ with probability $|b|^2$, the final state would be $f(x)\alpha$ with probability $|a|^2$ and $f(x)\beta$ with probability $|b|^2$. Are the two situations really distinguishable? Indeed they are: for the state (16a) the measurement of the spin in its original direction would surely give the same outcome as before (call this "positive") whereas for the "mixture" there is no direction in which the result would be positive. In fact if $|a|^2 = |b|^2 = \frac{1}{2}$, the result would be that in every direction, positive and negative occur with equal probabilities $\frac{1}{2}$.

The preceding consideration and calculation are, it must be admitted, quite approximative. Yet the final conclusion is clearly correct: as long as quantum mechanics is valid, a definite state vector is transformed by the passage of time into a definite state vector, whereas the measurement process should transform it, with various probabilities, into several different state vectors. Because the wave function must be (16) as a result of the aforementioned experiment, it must be admitted that the measurement there is not yet fully concluded. On the other hand, if the result of the measurement enters into the consciousness of some person, the measurement is then surely completed[4]—it is not possible to accept the idea that a person is in a superposition of two states: of having seen the particle emit a signal around the position x_0 and of having seen the emission from the position $-x_0$. Hence, if for instance, we accelerate the particle and let it impinge on a plate which then emits a light signal and if this light signal is observed, the measurement is surely completed in the sense that the final state vector is no longer a superposition of the form (16) or even of the form

$$aM_+ \, f(x - x_0)\alpha + bM_- \, f(x + x_0)\beta \tag{17}$$

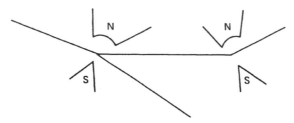

Figure 2. The effect of subsequent deflections using the Stern–Gerlach apparatus.

where M_+ is the state of the observer plus plate if he has seen the flash around x_0 and M_- if he has seen it at around $-x_0$. After the observation, the state vector has certainly "collapsed"—it is $M_+ f(x - x_0)\alpha$ with a probability $|a|^2$ and $M_- f(x + x_0)\beta$ with a probability $|b|^2$. In fact, it is not clear at which point the wave function or state vector collapsed—it may be already when the particle has impinged on the plate and light signals have been emitted. If this is so, quantum mechanics ceases to be valid for the description of the interaction of the particle with the light emitting plate. If the collapse occurs only when the information enters the mind of the observer, the collapse occurred only when this happened. The fact that the wave function (or state vector) "collapsed" when the measurement is completed had been realized already by Von Neumann,[3] and he also pointed out that the point of collapse is uncertain. In our case it may be caused by the interaction of the particle with the plate on which it impinges, or by the entering of the information into the mind of the observer, or it may occur at some intermediate point. (In the last section, the former assumption will be supported.) At any rate, the collapse of the state vector is in contradiction with the usual formulation of quantum mechanics which postulates a linear dependence of the final state vector on the initial one. This problem had also occurred to Von Neumann and probably to several other physicists at about the same time, the early thirties.

The next section will be devoted to a description of the efforts to eliminate the difficulty just described, and the section thereafter to the description of a few other, though less basic, difficulties of the description of the measurement process.

3. Two Proposals to Reconcile Quantum Mechanics with the Measurement Process

The preceding discussion clearly indicates that the deterministic nature and the linearity of the quantum-mechanical equations are in conflict with the probabilistic nature of the measurement process. There are two natural proposals to eliminate this contradiction, and these will be discussed in the present section. It will be shown that neither is adequate.

The first suggestion is that the uncertainty of the measurement's outcome is due to the uncertainty of the initial state of the measuring apparatus. And this really exists if the apparatus is macroscopic. Let us assume that the apparatus is initially with the probabilities $w_1, w_2, \ldots$ in the states $I_1, I_2, \ldots$. It can be shown in fact that it is possible to assume that these are orthogonal—all the properties of a mixture of arbitrary states is the same as that of a mixture of a suitably chosen set of orthogonal states, but this theorem will not be needed to show the inadequacy of the explanation of the probabilistic nature of the measure-

Review of the Quantum-Mechanical Measurement Problem   73

ment. It is sufficient in fact to observe that if the apparatus gives the correct outcome for the states ψ_κ, for which the result of the measurement is definite

$$I_\nu \times \psi_\kappa \rightarrow I_{\nu\kappa} \times \psi_\kappa \qquad (18)$$

where $I_{\nu\kappa}$ is a state of the apparatus in which its pointer has a definite direction (denoted before by λ_κ), then the result of the measurement process if the object is in a superposition $\Sigma c_\kappa \psi_\kappa$ of the ψ_κ states becomes, for every ν,

$$I_\nu \times \Sigma c_\kappa \psi_\kappa = \Sigma c_\kappa (I_\nu \times \psi_\kappa) \rightarrow \Sigma c_\kappa (I_{\nu\kappa} \times \psi_\kappa). \qquad (18a)$$

This means that no matter which I_ν was the original state of the apparatus, or of the observer, its or his final state will not be a definite one, its pointer, or his impression, will be given by a superposition of states. This is pretty evident, but it is good to realise it.

The second attempt at the reconciliation of the measurement phenomenon with the equations of quantum mechanics has attracted much more attention than the one just discussed; it has been favored particularly by D. Bohm.[5] It assumes that there is a more-detailed description of the state of a system than that given by the quantum mechanical wave function or state vector, and if this description, by "hidden variables" were known, the outcome of any measurement could be predicted. In other words, it is assumed that quantum mechanics gives only a superficial description of the system, perhaps similar to that of the classical description of microscopic bodies, and can therefore predict some phenomena, such as measurements, just as little as classical (nonmolecular) theory can explain Brownian motion which is due to the atomic structure of matter.

This is an interesting idea and even though few of us were ready to accept it, it must be admitted that the truly telling argument against it was produced as late as 1965, by J. S. Bell.[6] (See Figure 3.) It must be admitted that Bell's argument uses some non-obvious technical results of quantum mechanics which, however, have been experimentally confirmed. But his basic point is very simple: if there exists a complete description of the state of a system, and a distribution function which gives all these states a non negative value, then no matter how many experiments we are considering, there is a nonnegative probability function that gives the probability of the possible outcomes of each of these experiments.

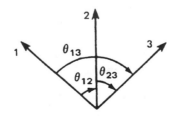

Figure 3. Directions of spin measurement in the Bell experiment.

Thus, for instance, if we have an electron in a definite state, there is a nonnegative function $f(q, p)$ which, if integrated over p gives the probability for the value q of its position and if integrated over q, gives the probability for the value p of its momentum. In the case just mentioned, it is indeed easy to produce such a function: in terms of the wave function $\psi(x)$, the function

$$f(q, p) = \int \psi(q+x)^* \, \psi(q-x)e^{2ip\cdot x/\hbar} \, dx/2\pi\hbar$$

is, for instance, such a function. (And there are many others even though there is none which is first order in both ψ and ψ^* and positive for all p and q.)

The case Bell considers is, however, quite different. He considers two particles with spins $\tfrac{1}{2}\hbar$, the spins forming a singlet (spins in opposite direction, total spin 0), whose wave function is

$$[\alpha(1)\beta(2) - \beta(1)\alpha(2)]\sqrt{2}, \tag{19}$$

where $\alpha(1)$ and $\beta(1)$ are the wave functions of the first particle when it is "up" and "down" respectively, and $\alpha(2)$, $\beta(2)$ have the same meaning with respect to the second particle. The measurements considered involve only the two spins. The measurements are in three different directions, the same for both particles. These directions are assumed, for the sake of simplicity, to be in the same plane and enclose the angles θ_{12}, θ_{23}, and $\theta_{13} = \theta_{12} + \theta_{23}$ with each other. We shall use an abbreviated notation for the outcome of a measurement: if the spin is found in the direction of the field, we call it (as before) α; if it is found to be in the opposite direction, we call it β.

The symbol $(\alpha, \alpha, \beta; \beta, \beta, \alpha)$, for instance, gives the probability that the measurement of the first particle's spin in the directions 1 and 2 gives α, in the third direction it gives β, whereas for the second particle the first two directions give β, the third direction α. We then seem to have $2^6 = 64$ symbols, but most of them vanish since if the measurement of the first particle in a certain direction gives α, the measurement of the second particle's spin in the same direction surely gives β. This follows from the fact that the two spins have opposite directions in *any* coordinate system if the wave function is that of (19)—a fact that will not be proved but is a consequence of the "singlet" character of (19), i.e., of its rotational invariance. Hence there are only $2^3 = 8$ nonvanishing symbols such as $(\beta, \alpha, \beta; \alpha, \beta, \alpha)$.

Let us now proceed to establish J. S. Bell's contradiction between our quantum-mechanical theory and the meaningful nature and nonnegative character of the symbols just defined. First, we write down the probability that particles one and two both give α in directions 1 and 2, respectively. The quantum-mechanical and hidden variables theories equality means

$$\tfrac{1}{2}\sin^2 \tfrac{1}{2}\theta_{12} = (\alpha, \beta, \alpha; \beta, \alpha, \beta) + (\alpha, \beta, \beta; \beta, \alpha, \alpha). \tag{20}$$

The left-hand side is the quantum-mechanical value for the probability that the outcome of the measurement of the two particles' spins in the directions 1 and 2 will be α assuming the wave function is (19). This will not be proved. The right-hand side contains all nonzero symbols which have α for the first particle in direction 1 and α also for the second particle in direction 2. Similarly, the probability that both particles give α in directions 2 and 3 are, according to the two theories,

$$\tfrac{1}{2} \sin^2 \tfrac{1}{2}\theta_{23} = (\alpha, \alpha, \beta; \beta, \beta, \alpha) + (\beta, \alpha, \beta; \alpha, \beta, \alpha). \qquad (20a)$$

The same equation for α in directions 1 and 3 is

$$\tfrac{1}{2} \sin^2 \tfrac{1}{2}\theta_{13} = (\alpha, \alpha, \beta; \beta, \beta, \alpha) + (\alpha, \beta, \beta; \beta, \alpha, \alpha). \qquad (20b)$$

Comparison of the right-sides of (20b) with the sum of (20) and (20a) shows that the latter is larger, so we should have

$$\sin^2 \tfrac{1}{2}\theta_{13} < \sin^2 \tfrac{1}{2}\theta_{12} + \sin^2 \tfrac{1}{2}\theta_{23} \qquad (21)$$

which contradicts the fact that $\theta_{13} = \theta_{12} + \theta_{23}$ by elementary trigonometry as long as all angles are smaller than π. In fact in this case we have

$$\sin^2 \tfrac{1}{2}\theta_{13} > \sin^2 \tfrac{1}{2}\theta_{12} + \sin^2 \tfrac{1}{2}\theta_{23} \qquad (21a)$$

The experimental proof of the validity of the quantum-mechanical inequality (21a) and hence of the violation of the hidden variable inequality (21) took some time and effort, but it now seems unquestionable that the former is valid. This appears to give a convincing argument against the hidden variables theory. Actually, arguments against that theory were given long before the proof just reproduced was given by Bell. The first one is actually due to Von Neumann, but it will not be reproduced here.

The impossibility of giving a causal description of measurement processes already speaks against determinism. We return to this problem later and give reasons for the lack of validity, in fact the lack of meaningfulness, thereof. Before doing that, a few other difficulties will be mentioned.

4. Other Weaknesses of Measurement Theory

Measurement theory as generally accepted, in particular the postulate that every self-adjoint operator can be measured (in our terminology, the transition probability to any member of any complete orthonormal set ψ_κ—see equation (14)), even though founded by Von Neumann, has several weaknesses. These will be mentioned below.

The first and most serious weakness is, in this writer's opinion, that it is not given how the apparatus—the object with the wave function $\mathscr{A}_0$ of (14)—is to be constructed. Its structure would depend, of course, on the set of characteristic vectors ψ_κ into one of which it will transform any state $\Sigma c_\kappa \psi_\kappa$ on which the measurement is carried out. In a few cases, in particular if the momentum is to be measured, one can well think of the apparatus. But in other cases, even if the position is to be measured, there are difficulties. And if we consider an operator such as qpq, the characteristic vectors of which are $x^{-1}\exp(\lambda/ix\hbar)$—it has a continuous spectrum—there is just no suggestion of how it would be measured. It is not surprising, therefore, that the founder of quantum mechanics, W. Heisenberg, came to the conviction that the usual measurability theory is just as nonpositivistic as was N. Bohr's original description of the electronic orbits in atoms. He suggested therefore that collision theory's basic concept, the collision matrix,[7] be the basic concept. Indeed this is measurable—at least if the gravitational effects of the apparata can be neglected.[8]

The second point I wish to mention is that, indeed, it can be proved that with an apparatus of finite "size" only quantities commuting with all additive conserved quantities—energy, momentum, angular momentum, total electric charge, etc.—can be measured.[9] This is, perhaps, most evident for position; in order to measure the position of an object with a certain accuracy, the apparatus' own position must be specified with at least the same accuracy. This means, if the measurement is to be accurate, that the momentum of the apparatus must be spread over a considerable interval which can be called the "size" in this case. But this is only one requirement.[10]

The third point that should be mentioned is that measurement theory is fundamentally nonrelativistic. Thus, it postulates the essentially instantaneous determination of the state vector at a given time. But the wave function can extend to considerable distances—in fact practically all commonly considered wave functions extend to infinity. Since the signal from a distant point takes some time to travel, an essentially instantaneous measurement on a spacelike surface is surely not possible. Of course, at the time the original measurement theory was created, around 1932, there were hardly any efforts to make quantum theory relativistic. It would seem at present that it would be more reasonable to define the state vector or wavefunction on a light-cone, but few attempts have been made in this direction (some by this writer); and the problem does not appear to be easy.

Lastly, it should be admitted that there are difficulties even in defining the most basic relativistic observables. One fundamental such observable would be the position of a space–time point that the world line of a particle goes across. The state vector of a particle that, at a definite time, is at a definite point, has been determined by T. D. Newton and myself,[11] but this state vector is not invariant under Lorentz transformations around that point of space–time. This means that it is not right to claim that we can determine the probability that the particle's path—or any closed and stable system's path—crosses a given space–time point. This point was brought out most clearly by Fleming and by Hegerfeldt.[12] There are other difficulties with localization in space–time, and one can also show that there are limits to the accuracy with which space–time distances can be measured.[13] It is surprising, therefore, that the postulate of the measurability of field strengths at space–time points, which is assumed in quantum field theories, leads to so many interesting and experimentally verified conclusions.

The preceding remarks may well be considered to be crucial for general measurement theory and may even be crucial for relativistic quantum mechanics in general and field theories in particular. But we should not forget the truly spectacular successes of our quantum mechanics, both nonrelativistic and relativistic. It gives amazingly accurate accounts of the magnitudes of the energy levels of atoms and molecules and, though the theory is incomplete, even of nuclei. Actually, these are not energy levels but possible rest masses—quantities that are not subject to the restrictions of the second point of the preceding discussion. Quantum theory also gives very good values for the transition probabilities. It gives good descriptions, based on fundamental data, of the chemical properties of materials and of the physical properties of solids. It gives the collision matrix at least for simple processes. Surely, its successes are marvelous in many areas.

But, equally surely, the problems of measurement theory are serious. Surely, present quantum mechanics does not describe "life," that is, the fact of *consciousness*. After all, it has no operator for pain, pleasure, interest, desire, thoughts—and these are surely basic realities. We must ask ourselves, therefore, where the limits of validity for present quantum mechanics lie. (This will be discussed in the next and last sections.) Let us remember that the validity of all past formulations of the laws of physics had limitations; Newton's theory was essentially restricted to situations in which only gravitational forces played a noticeable role, and similar statements can be made for all past developments of our discipline. The first physics book I read said "Atoms and molecules may exist, but this is irrelevant from the point of view of physics." And this was entirely correct—at that time physics dealt only with macroscopic phenomena and Brownian motion was sort of a miracle. It is good, therefore, to ask what the limits of validity are, even though perhaps these are not the limits of interest for present-day physics.

5. *The Limits of Validity of Our Quantum Mechanics*

This writer thought originally that the phenomenon of life is the only area which is outside of the domain of present-day physics, for it plays no role in the phenomena with which our physics is concerned. It probably plays, I thought, very little role in the behavior of viruses and bacteria, more role in the behavior of mosquitos and butterflies, and a decisive role in that of vertebrates. And this may be true, but the limits of validity of our physics are apparently much further back.

V. Fock already said that "the measuring apparatus must be described by classical theory." He evidently realized that the measurement process cannot be truly described by our present quantum theory, but he gave no reason for this lack of validity which will be attributed here to the measuring apparatus' macroscopic nature. The same applies, in this writer's opinion, to a recent article of Machida and Nanuki,[14] but they go much more into detail in their interesting paper.

This writer's earlier belief that the role of the physical apparatus can always be described by quantum mechanics (as was done before for a particular "measurement," the Stern–Gerlach experiment) implied that the "collapse of the wave function" takes place only when the observation is made by a living being—a being clearly outside of the scope of our quantum mechanics. The argument which convinced me that quantum mechanics' validity has narrower limitations, that it is not applicable to the description of the detailed behavior of macroscopic bodies, is due to D. Zeh.[15] His argument will not be given in detail here—it concerns the structure of the wave functions of the stationary states of macroscopic bodies. Instead we compute an estimate of the length of time that the wave function of a macroscopic body obeys the standard time development equation, i.e., how long the quantum equations accurately reproduce the behavior of such a body.

The point is that a macroscopic body's inner structure, i.e., its wave function, is influenced by its environment in a rather short time even if it is in intergalactic space. Hence it cannot be an isolated system, that is, a system to which our physics equations apply. And the situation is, naturally, even worse in our world, here on the Earth. This means that a radical departure from the established principles and laws of physics is needed. Let us first discuss the existence of the influence of the environment on a macroscopic body in the best possible environment, that is, in intergalactic space.

The macroscopic body chosen will be a cubic centimeter of tungsten which, at the temperature of the 3°K of intergalactic space, does not evaporate. It is, however, subject to several effects, first to the cosmic radiation of 3°K. This contains about 520 quanta per cm^3 as can be easily calculated. It follows that the tungsten cube is hit by $520 \times 3 \times 10^{10} \times \frac{3}{2} = 2.3 \times 10^{13}$ quanta per second. It can be shown then that the cube's vibrational state changes in *very much* less than a second. The other effects include the Van der Waals forces of the hydrogen, the effect considered by Zeh (on a cubic centimeter of a gas). It is not difficult to calculate that the Van der Waals forces will affect the vibrational structure also and change it in about 10 seconds. This assumes a density of about 1 atom of hydrogen per cubic meter. Finally, the atoms impinging on the tungsten cube

Review of the Quantum-Mechanical Measurement Problem   79

will change it also quite rapidly, namely in about 25 seconds. And there may be other effects.

It is natural to propose to put the tungsten into a container and to evacuate that. However, the effect of the container's atoms' Van der Waals forces will play an even greater role—just as the container of Zeh's gas will affect his gas badly. It is clear that no macroscopic body can be "isolated" for any significant time period. Of course, this time period depends on the material and also on the environment; it will be, obviously, very much shorter here in our world than in intergalactic space and also much shorter for most bodies than for the cube of tungsten considered above.

Can an equation for the time change of the state of an apparently nonisolated system be proposed? Quantum evolution equations are apparently deterministic, but they are not if we write one not for the wave function (or state vector) but for the density matrix ρ. As follows from what was said before, this equation cannot be universal because its constants must depend at least on the environment, the effect of which is not, and cannot be, taken into account in a deterministic way.

We shall use a special notation for the specification of the rows and columns of the density matrix. The number of indices will be equal to three times the number of particles, as is the case if the Schrödinger wave function is used to form the density matrix. The first three indices X, Y, Z are the three coordinates of the center of mass. Thus

$$X = \frac{1}{N} \Sigma \, x_i. \tag{22}$$

Y and Z are defined similarly, where N is the number of the particles. The remaining indices are defined in terms of the center of mass coordinates

$$\xi_i = x_i - X, \qquad \eta_i = y_i - Y, \qquad \zeta_i = z_i - Z. \tag{23}$$

The first one of these is the "bulk" of the system

$$\mathcal{L}_0 = \Sigma \, \xi_i^2 + \eta_i^2 + \zeta_i^2. \tag{24a}$$

The next five are the five components of the quadrupole moments:

$$\mathcal{L}_{21} = \Sigma(\xi_i^2 - \tfrac{1}{2}(\eta_i^2 + \zeta_i^2)), \qquad \mathcal{L}_{22} = \tfrac{1}{2}\sqrt{3}\Sigma(\xi_i^2 - \eta_i^2), \tag{24b}$$

$$\mathcal{L}_{23} = \sqrt{3}\Sigma\eta_i\zeta_i, \qquad \mathcal{L}_{24} = \sqrt{3}\Sigma\zeta_i\xi_i, \qquad \mathcal{L}_{25} = \sqrt{3}\Sigma\zeta_i\eta_i.$$

The next seven indices $\mathcal{L}_{31}, \ldots, \mathcal{L}_{37}$ are the components of the octupole moments, and so on. There are no $\mathcal{L}_{1\nu}$ since $\Sigma\xi_i = 0$ and the same holds for $\Sigma\eta_i$ and $\Sigma\zeta_i$. The order of the m components of $\mathcal{L}_{lm}$ is irrelevant, but it should be noted that the $2l + 1$ components of $\mathcal{L}_{lm}$, from $\mathcal{L}_{l1}$ to $\mathcal{L}_{l2l+1}$, are symmetric l degree polynomials of the ξ, η, ζ and transform under rotation of the coordinate system via the $2l + 1$ dimensional representation of the rotation group.

If we were to proceed up to a large definite index l, the number of variables

Eugene P. Wigner

obtained as above would be $(2l+1)^2$ which may exceed the required $3N$ variables. But for the last few l, the m of $\mathscr{L}_{lm}$ can be chosen arbitrarily since they will play no role in the subsequent discussion.

It may be pointed out that the introduction of variables which are symmetric in the coordinates of the particles is not unreasonable—if either Bose or Fermi statistics is valid for them, the elements of the density matrix will be symmetric in the N coordinates x_i, y_i, and z_i if all particles are identical, and for the sake of simplicity we shall restrict this discussion to that—actually unrealistic—case. If several different particles are present, the number of coordinates has to be multiplied by their number and the remaining discussion becomes somewhat more complicated.

Which way are we going to modify the standard quantum-mechanical equation for the time-dependence of ρ?

$$i\hbar \frac{\partial \rho}{\partial t} = H\rho - \rho H. \tag{25}$$

If we insert the variables of the density matrix, we have

$$\rho(X, Y, Z, \mathscr{L}_0, \mathscr{L}_{21}, \mathscr{L}_{22}, \ldots, \mathscr{L}_{25}, \mathscr{L}_{31}, \ldots ; X', Y', Z', \mathscr{L}_0',$$
$$\mathscr{L}_{21}', \mathscr{L}_{22}', \ldots, \mathscr{L}_{25}', \mathscr{L}_{31}', \ldots). \tag{25a}$$

The expression for $\partial\rho/\partial t$ will then contain, in addition to the commutator of H and ρ, as in (25), terms that cause a decay of those constituent state vectors which are superpositions of macroscopically distinguishable states and that cause the conversion of such states into mixtures of states. This is accomplished if the off-diagonal matrix elements at the intersection of those states are made to decay. To recall this, let us consider a two-dimensional density matrix (which could be the density matrix of a spin)

$$\begin{bmatrix} |\alpha|^2 & \alpha\beta^* \\ \alpha^*\beta & |\beta|^2 \end{bmatrix} \tag{26}$$

where $|\beta|^2 + |\alpha|^2 = 1$. This represents a state vector with components α, β. If we multiply the off-diagonal elements with a real $\theta < 1$, we have

$$\begin{bmatrix} |\alpha|^2 & \alpha\beta^*\vartheta \\ \alpha^*\beta\vartheta & |\beta|^2 \end{bmatrix} \tag{26a}$$

which splits up into a mixture of two states when $\theta = 0$, these being: $(1, 0)$ with probability $|\alpha|^2$ and $(0, 1)$ with probability $|\beta|^2 = 1 - |\alpha|^2$. The transition probability of the original state into the two states with components $(1, 0)$ and $(0, 1)$ is equal to these quantities $|\alpha|^2$ and $|\beta|^2$, but the state of (26) is a *superposition* of the states $(1, 0)$ and $(0,1)$, with components α and β, whereas the final state, (26a) with $\theta = 0$, is a *mixture* of those states—the original superposition was changed into a mixture by the decrease of the off-diagonal elements.

Review of the Quantum-Mechanical Measurement Problem   81

It is reasonable therefore to add to the expression of (25), for $\partial\rho/\partial t$, other terms which decrease the off-diagonal elements of ρ and to write

$$i\hbar \ \partial\rho/\partial t = H\rho - \rho H - i\hbar \sum_{\ell m} \epsilon_\ell \ (\mathscr{L}_{\ell m} - \mathscr{L}'_{\ell m})^2 \rho. \tag{27}$$

In (27), the ρ is assumed to have the form (25a) except, for the sake of brevity, that X, Y, Z and X', Y', Z' are replaced by $\mathscr{L}_{1m}$ and $\mathscr{L}'_{1m}$ with m going from 1 to 3. The ϵ_l in (27) have (except for $l = 0$) the dimension sec cm^{-1}. As was mentioned before, their values, which are necessarily positive, depend on the environment; but their effectiveness should decrease with increasing l, in this writer's opinion, except for $l = 0$ and 1, as l^{-2}/a^l where a is a characteristic length.

It may be good to point out that (27) is rotationally and displacement invariant-Galilei invariant as it is often called. It is not relativistically invariant, which is natural as the new term of (27) is caused by the environment, and it is assumed that the coordinates are chosen to be position coordinates in the rest system of the environment. It should also be pointed out that the factor of $i\hbar$ in the new term is self-adjoint—as it must be because ρ must remain self-adjoint. It is less evident that this term preserves the positive definite nature of ρ; the proof for this will not be given here. The proof consists of three steps, the last of which was supplied by Dr. Ron L. Graham who became interested in this question when it was discussed by me at this Laboratory, in Los Alamos.

It should be admitted that there is no clear evidence for the form (27) of the effect of the environment. Even though it seems a reasonable equation, it would not be difficult to find situations in which it is far from valid. For instance, the environment may not show Galilei invariance. But, perhaps, in reasonably empty space it would be a good description of the environment's effect—it does have at least two adjustable constants, ϵ_0 and a. And its validity, even if only approximate, does answer the question of the validity of quantum mechanics for macroscopic objects—the presence of the last term in (27) denying the full validity of (25).

Let us admit, finally, that even if (27) turned out to be a good way to extend the applicability of quantum mechanics to macroscopic objects—and if so it would have significant epistemological consequences, in particular with respect to the meaningful nature of determinism—it would not extend the theory to the phenomenon of life. It offers no description for mental phenomena, for pain or pleasure, for interest and thought, for desire or repulsion. At best, it would extend quantum mechanics to those limits which many of us thought it had already reached. But even that would have significant and, in this writer's opinion unexpected, epistemological consequences.

I find it necessary to close this article by expressing my appreciation for the invitation to this meeting and for the many addresses which I enjoyed, the contents of most of which were new at least to me. It was a pleasure to participate in this conference to celebrate the past director of Los Alamos, Harold Agnew. Let me also offer my apology to the many colleagues whose papers on this subject were not mentioned. It is difficult to mention in an oral address all sources of information which are relevant—even all those from which one has learned a great deal. I can, however, point to an excellent book which describes remarkably well the earlier work on my subject, namely Max Jammer's "The Philosophy of Quantum Mechanics."[16]

References

1. E. Schrödinger, Quantisierung als Eigenwertproblem. *Ann. d. Physik* **79** (1926): 361, 489.
2. W. Heisenberg, *Zeits. f. Physik* **33** (1925): 879; M. Born and P. Jordan, *ibid.* **34** (1925): 858; M. Born, W. Heisenberg, and P. Jordan, *ibid.* **35** (1926): 557.
3. J. von Neumann, *Mathematische Grundlagen der Quantenmechanik.* Berlin: J. Springer, 1932; Princeton, New Jersey: Princeton University Press, 1955.
4. This point has been discussed more elaborately by E. P. Wigner, "Remarks on the Mind-Body Question." In *The Scientist Speculates*, edited by I. J. Good. London: W. Heinemann, 1961. And there are many other articles on the question.
5. D. Bohm, *Phys. Rev.* **85** (1952): 166. See also his books: *Quantum Theory.* Englewood Cliffs, New Jersey: Prentice Hall, 1951. *Causality and Chance in Modern Physics* New York: Van Nostrand, 1957. See also D. Bohm and J. Bub, *Rev. Modern Physics* **38** (1966): 453.
6. J. S. Bell, *Physics* **1** (1965): 195.
7. J. A. Wheeler, *Phys. Rev.* **52** (1937): 1107.
8. F. E. Goldrich and E. P. Wigner. In *Magic Without Magic,* edited by J. Klauder, 147. New York: W. H. Freeman, 1972.
9. E. P. Wigner, *Zeits. f. Physik* **133** (1952): 101.
10. G. Araki and M. Janase, *Phys. Rev.* **120** (1960): 662.
11. T. D. Newton and E. P. Wigner, *Rev. Modern Physics* **21** (1949): 400. See also M. H. L. Price, *Proc. Roy. Soc.* **195A** (1948): 62.
12. G. C. Hegerfeldt, *Phys. Rev.* **D10,** (1974): 3320; G. N. Fleming, *ibid.* **139B** (1975): 963.
13. H. Salecker and E. P. Wigner, *Phys. Rev.* **109** (1958): 571; also E. P. Wigner, *ibid.* **120** (1960): 643.
14. S. Machida and M. Nanuki, *Progress of Theoretical Physics* **63** (1980): 1833.
15. H. D. Zeh, *Found. of Physic* **1,**69 (1970).
16. Max Jammer, *The Philosophy of Quantum Mechanics.* New York: Wiley, 1974.

PART III

Consciousness

Remarks on the Mind-Body Question

E. P. Wigner

Symmetries and Reflections.
Indiana University Press, Bloomington, Indiana, 1967, pp. 171–184

Introductory Comments

F. Dyson, in a very thoughtful article,[1] points to the everbroadening scope of scientific inquiry. Whether or not the relation of mind to body will enter the realm of scientific inquiry in the near future—and the present writer is prepared to admit that this is an open question—it seems worthwhile to summarize the views to which a dispassionate contemplation of the most obvious facts leads. The present writer has no other qualification to offer his views than has any other physicist and he believes that most of his colleagues would present similar opinions on the subject, if pressed.

Until not many years ago, the "existence" of a mind or soul would have been passionately denied by most physical scientists. The brilliant successes of mechanistic and, more generally, macroscopic physics and of chemistry overshadowed the obvious fact that thoughts, desires, and emotions are not made of matter, and it was nearly universally accepted among physical scientists that there is nothing besides matter. The epitome of this belief was the conviction that, if we knew the positions and velocities of all atoms at one instant of time, we could compute the fate of the universe for all future. Even today, there are adherents to this

Reprinted by permission from *The Scientist Speculates*, I. J. Good, ed. (London: William Heinemann, Ltd., 1961; New York: Basic Books, Inc., 1962).

[1] F. J. Dyson, *Scientific American*, 199, 74 (1958). Several cases are related in this article in which regions of inquiry, which were long considered to be outside the province of science, were drawn into this province and, in fact, became focuses of attention. The best-known example is the interior of the atom, which was considered to be a metaphysical subject before Rutherford's proposal of his nuclear model, in 1911.

view[2] though fewer among the physicists than — ironically enough — among biochemists.

There are several reasons for the return, on the part of most physical scientists, to the spirit of Descartes's *"Cogito ergo sum,"* which recognizes the thought, that is, the mind, as primary. First, the brilliant successes of mechanics not only faded into the past; they were also recognised as partial successes, relating to a narrow range of phenomena, all in the macroscopic domain. When the province of physical theory was extended to encompass microscopic phenomena, through the creation of quantum mechanics, the concept of consciousness came to the fore again: it was not possible to formulate the laws of quantum mechanics in a fully consistent way without reference to the consciousness.[3] All that quantum mechanics purports to provide are probability connections between subsequent impressions (also called "apperceptions") of the consciousness, and even though the dividing line between the observer, whose consciousness is being affected, and the observed physical object can be shifted towards the one or the other to a considerable degree,[4] it cannot be eliminated. It may be premature to believe that the present philosophy of quantum mechanics will remain a permanent feature of future physical theories; it will remain remarkable, in whatever way our future concepts may develop, that the very study of the external world led to the conclusion that the content of the consciousness is an ultimate reality.

It is perhaps important to point out at this juncture that the question concerning the existence of almost anything (even the whole external world) is not a very relevant question. All of us recognize at once how meaningless the query concerning the existence of the electric field in vacuum would be. All that is relevant is that the concept of the electric

[2] The book most commonly blamed for this view is E. F. Haeckel's *Welträtsel* (1899). However, the views propounded in this book are less extreme (though more confused) than those of the usual materialistic philosophy.

[3] W. Heisenberg expressed this most poignantly [*Daedalus*, 87, 99 (1958)]: "The laws of nature which we formulate mathematically in quantum theory deal no longer with the particles themselves but with our knowledge of the elementary particles." And later: "The conception of objective reality . . . evaporated into the . . . mathematics that represents no longer the behavior of elementary particles but rather our knowledge of this behavior." The "our" in this sentence refers to the observer who plays a singular role in the epistemology of quantum mechanics. He will be referred to in the first person and statements made in the first person will always refer to the observer.

[4] J. von Neumann, *Mathematische Grundlagen der Quantenmechanik* (Berlin: Julius Springer, 1932), Chapter VI; English translation (Princeton, N.J.: Princeton University Press, 1955).

Remarks on the Mind-Body Question   *173*

field is useful for communicating our ideas and for our own thinking. The statement that it "exists" means only that: (*a*) it can be measured, hence uniquely defined, and (*b*) that its knowledge is useful for understanding past phenomena and in helping to foresee further events. It can be made part of the *Weltbild*. This observation may well be kept in mind during the ensuing discussion of the quantum mechanical description of the external world.

The Language of Quantum Mechanics

The present and the next sections try to describe the concepts in terms of which quantum mechanics teaches us to store and communicate information, to describe the regularities found in nature. These concepts may be called the language of quantum mechanics. We shall not be interested in the regularities themselves, that is, the contents of the book of quantum mechanics, only in the language. It may be that the following description of the language will prove too brief and too abstract for those who are unfamiliar with the subject, and too tedious for those who are familiar with it.[5] It should, nevertheless, be helpful. However, the knowledge of the present and of the succeeding section is not necessary for following the later ones, except for parts of the section on the Simplest Answer to the Mind-Body Question.

Given any object, all the possible knowledge concerning that object can be given as its wave function. This is a mathematical concept the exact nature of which need not concern us here—it is composed of a (countable) infinity of numbers. If one knows these numbers, one can foresee the behavior of the object as far as it *can* be foreseen. More precisely, the wave function permits one to foretell with what probabilities the object will make one or another impression on us if we let it interact with us either directly, or indirectly. The object may be a radiation field, and its wave function will tell us with what probability we shall see a

[5] The contents of this section should be part of the standard material in courses on quantum mechanics. They are given here because it may be helpful to recall them even on the part of those who were at one time already familiar with them, because it is not expected that every reader of these lines had the benefit of a course in quantum mechanics, and because the writer is well aware of the fact that most courses in quantum mechanics do not take up the subject here discussed. See also, in addition to references 3 and 4, W. Pauli, *Handbuch der Physik*, Section 2.9, particularly page 148 (Berlin: Julius Springer, 1933). Also F. London and E. Bauer, *La Théorie de l'observation en mécanique quantique* (Paris: Hermann and Co., 1939). The last authors observe (page 41), "Remarquons le rôle essentiel que joue la conscience de l'observateur. . . ."

flash if we put our eyes at certain points, with what probability it will leave a dark spot on a photographic plate if this is placed at certain positions. In many cases the probability for one definite sensation will be so high that it amounts to a certainty—this is always so if classical mechanics provides a close enough approximation to the quantum laws.

The information given by the wave function is communicable. If someone else somehow determines the wave function of a system, he can tell me about it and, according to the theory, the probabilities for the possible different impressions (or "sensations") will be equally large, no matter whether he or I interact with the system in a given fashion. In this sense, the wave function "exists."

It has been mentioned before that even the complete knowledge of the wave function does not permit one always to foresee with certainty the sensations one may receive by interacting with a system. In some cases, one event (seeing a flash) is just as likely as another (not seeing a flash). However, in most cases the impression (e.g., the knowledge of having or not having seen a flash) obtained in this way permits one to foresee later impressions with an increased certainty. Thus, one may be sure that, if one does not see a flash if one looks in one direction, one surely does see a flash if one subsequently looks in another direction. The property of observations to increase our ability for foreseeing the future follows from the fact that all knowledge of wave functions is based, in the last analysis, on the "impressions" we receive. In fact, the wave function is only a suitable language for describing the body of knowledge—gained by observations—which is relevant for predicting the future behaviour of the system. For this reason, the interactions which may create one or another sensation in us are also called observations, or measurements. One realises that *all* the information which the laws of physics provide consists of probability connections between subsequent impressions that a system makes on one if one interacts with it repeatedly, i.e., if one makes repeated measurements on it. The wave function is a convenient summary of that part of the past impressions which remains relevant for the probabilities of receiving the different possible impressions when interacting with the system at later times.

An Example

It may be worthwhile to illustrate the point of the preceding section on a schematic example. Suppose that all our interactions with the system consist in looking at a certain point in a certain direction at times

Remarks on the Mind-Body Question   175

$t_0, t_0 + 1, t_0 + 2, \cdots$, and our possible sensations are seeing or not seeing a flash. The relevant law of nature could then be of the form: "If you see a flash at time t, you will see a flash at time $t + 1$ with a probability $\frac{1}{4}$, no flash with a probability $\frac{3}{4}$; if you see no flash, then the next observation will give a flash with the probability $\frac{3}{4}$, no flash with a probability $\frac{1}{4}$; there are no further probability connections." Clearly, this law can be verified or refuted with arbitrary accuracy by a sufficiently long series of observations. The wave function in such a case depends only on the last observation and may be ψ_1 if a flash has been seen at the last interaction, ψ_2 if no flash was noted. In the former case, that is for ψ_1, a calculation of the probabilities of flash and no flash after unit time interval gives the values $\frac{1}{4}$ and $\frac{3}{4}$; for ψ_2 these probabilities must turn out to be $\frac{3}{4}$ and $\frac{1}{4}$. This agreement of the predictions of the law in quotation marks with the law obtained through the use of the wave function is not surprising. One can either say that the wave function was invented to yield the proper probabilities, or that the law given in quotation marks has been obtained by having carried out a calculation with the wave functions, the use of which we have learned from Schrödinger.

The communicability of the information means, in the present example, that if someone else looks at time t, and tells us whether he saw a flash, we can look at time $t + 1$ and observe a flash with the same probabilities as if we had seen or not seen the flash at time t ourselves. In other words, he can tell us what the wave function is: ψ_1 if he did, ψ_2 if he did not see a flash.

The preceding example is a very simple one. In general, there are many types of interactions into which one can enter with the system, leading to different types of observations or measurements. Also, the probabilities of the various possible impressions gained at the next interaction may depend not only on the last, but on the results of many prior observations. The important point is that the impression which one gains at an interaction may, and in general does, modify the probabilities with which one gains the various possible impressions at later interactions. In other words, the impression which one gains at an interaction, called also *the result of an observation*, modifies the wave function of the system. The modified wave function is, furthermore, in general unpredictable before the impression gained at the interaction has entered our consciousness: it is the entering of an impression into our consciousness which alters the wave function because it modifies our appraisal of the probabilities for different impressions which we expect to receive in the future. It is at this point that the consciousness enters the theory unavoidably and unalterably. If one speaks in terms of the wave function, its changes are coupled with the entering of impressions into our consciousness. If one formulates the laws of quantum mechanics in terms of probabilities of impressions, these are *ipso facto* the primary concepts with which one deals.

It is natural to inquire about the situation if one does not make the observation oneself but lets someone else carry it out. What is the wave function if my friend looked at the place where the flash might show at time *t*? The answer is that the information available about the *object* cannot be described by a wave function. One could attribute a wave function to the joint system: friend plus object, and this joint system would have a wave function also after the interaction, that is, after my friend has looked. I can then enter into interaction with this joint system by asking my friend whether he saw a flash. If his answer gives me the impression that he did, the joint wave function of friend + object will change into one in which they even have separate wave functions (the total wave function is a product) and the wave function of the object is ψ_1. If he says no, the wave function of the object is ψ_2, i.e., the object behaves from then on as if I had observed it and had seen no flash. However, even in this case, in which the observation was carried out by someone else, the typical change in the wave function occurred only when some information (the *yes* or *no* of my friend) entered *my* consciousness. It follows that the quantum description of objects is influenced by impressions entering my consciousness.[6] Solipsism may be logically consistent with present quantum mechanics, monism in the sense of materialism is not. The case against solipsism was given at the end of the first section.

The Reasons for Materialism

The principal argument against materialism is not that illustrated in the last two sections: that it is incompatible with quantum theory. The

[6] The essential point is not that the states of objects cannot be described by means of position and momentum co-ordinates (because of the uncertainty principle). The point is, rather, that the valid description, by means of the wave function, is influenced by impressions entering our consciousness. See in this connection the remark of London and Bauer, quoted above, and S. Watanabe's article in *Louis de Broglie, Physicien et Penseur* (Paris: Albin Michel, 1952), p. 385.

Remarks on the Mind-Body Question   *177*

principal argument is that thought processes and consciousness are the primary concepts, that our knowledge of the external world is the content of our consciousness and that the consciousness, therefore, cannot be denied. On the contrary, logically, the external world could be denied—though it is not very practical to do so. In the words of Niels Bohr,[7] "The word consciousness, applied to ourselves as well as to others, is indispensable when dealing with the human situation." In view of all this, one may well wonder how materialism, the doctrine[8] that "life could be explained by sophisticated combinations of physical and chemical laws," could so long be accepted by the majority of scientists.

The reason is probably that it is an emotional necessity to exalt the problem to which one wants to devote a lifetime. If one admitted anything like the statement that the laws we study in physics and chemistry are limiting laws, similar to the laws of mechanics which exclude the consideration of electric phenomena, or the laws of macroscopic physics which exclude the consideration of "atoms," we could not devote ourselves to our study as wholeheartedly as we have to in order to recognise any new regularity in nature. The regularity which we are trying to track down must appear as the all-important regularity—if we are to pursue it with sufficient devotion to be successful. Atoms were also considered to be an unnecessary figment before macroscopic physics was essentially complete—and one can well imagine a master, even a great master, of mechanics to say: "Light may exist but I do not need it in order to explain the phenomena in which I am interested." The present biologist uses the same words about mind and consciousness; he uses them as an expression of his disbelief in these concepts. Philosophers do not need these illusions and show much more clarity on the subject. The same is true of most truly great natural scientists, at least in their years of maturity. It is now true of almost all physicists—possibly, but not surely, because of the lesson we learned from quantum mechanics. It is also possible that we learned that the principal problem is no longer the fight with the adversities of nature but the difficulty of understanding ourselves if we want to survive.

[7] N. Bohr, *Atomic Physics and Human Knowledge*, section on "Atoms and Human Knowledge," in particular p. 92 (New York: John Wiley & Sons, 1960).
[8] The quotation is from William S. Beck, *The Riddle of Life, Essay in Adventures of the Mind* (New York: Alfred A. Knopf, 1960), p. 35. This article is an eloquent statement of the attitude of the open-minded biologists toward the questions discussed in the present note.

Simplest Answer to the Mind-Body Question

Let us first specify the question which is outside the province of physics and chemistry but is an obviously meaningful (because operationally defined) question: Given the most complete description of my body (admitting that the concepts used in this description change as physics develops), what are my sensations? Or, perhaps, with what probability will I have one of the several possible sensations? This is clearly a valid and important question which refers to a concept—sensations—which does not exist in present-day physics or chemistry. Whether the question will eventually become a problem of physics or psychology, or another science, will depend on the development of these disciplines.

Naturally, I have direct knowledge only of my own sensations and there is no strict logical reason to believe that others have similar experiences. However, everybody believes that the phenomenon of sensations is widely shared by organisms which we consider to be living. It is very likely that, if certain physico-chemical conditions are satisfied, a consciousness, that is, the property of having sensations, arises. This statement will be referred to as our first thesis. The sensations will be simple and undifferentiated if the physico-chemical substrate is simple; it will have the miraculous variety and colour which the poets try to describe if the substrate is as complex and well organized as a human body.

The physico-chemical conditions and properties of the substrate not only create the consciousness, they also influence its sensations most profoundly. Does, conversely, the consciousness influence the physico-chemical conditions? In other words, does the human body deviate from the laws of physics, as gleaned from the study of inanimate nature? The traditional answer to this question is, "No": the body influences the mind but the mind does not influence the body.[9] Yet at least two reasons can be given to support the opposite thesis, which will be referred to as the second thesis.

The first and, to this writer, less cogent reason is founded on the

[9] This writer does not profess to a knowledge of all, or even of the majority of all, metaphysical theories. It may be significant, nevertheless, that he never found an affirmative answer to the query of the text—not even after having perused the relevant articles in the earlier (more thorough) editions of the *Encyclopaedia Britannica.*

Remarks on the Mind-Body Question   *179*

quantum theory of measurements, described earlier in sections 2 and 3. In order to present this argument, it is necessary to follow my description of the observation of a "friend" in somewhat more detail than was done in the example discussed before. Let us assume again that the object has only two states, ψ_1 and ψ_2. If the state is, originally, ψ_1, the state of object plus observer will be, after the interaction, $\psi_1 \times \chi_1$; if the state of the object is ψ_2, the state of object plus observer will be $\psi_2 \times \chi_2$ after the interaction. The wave functions χ_1 and χ_2 give the state of the observer; in the first case he is in a state which responds to the question "Have you seen a flash?" with "Yes"; in the second state, with "No." There is nothing absurd in this so far.

Let us consider now an initial state of the object which is a linear combination $\alpha\,\psi_1 + \beta\,\psi_2$ of the two states ψ_1 and ψ_2. It then *follows* from the linear nature of the quantum mechanical equations of motion that the state of object plus observer is, after the interaction, $\alpha\,(\psi_1 \times \chi_1) + \beta\,(\psi_2 \times \chi_2)$. If I now ask the observer whether he saw a flash, he will with a probability $|\alpha|^2$ say that he did, and in this case the object will also give to me the responses as if it were in the state ψ_1. If the observer answers "No"—the probability for this is $|\beta|^2$—the object's responses from then on will correspond to a wave function ψ_2. The probability is zero that the observer will say "Yes," but the object gives the response which ψ_2 would give because the wave function $\alpha\,(\psi_1 \times \chi_1) + \beta\,(\psi_2 \times \chi_2)$ of the joint system has no $(\psi_2 \times \chi_1)$ component. Similarly, if the observer denies having seen a flash, the behavior of the object cannot correspond to χ_1 because the joint wave function has no $(\psi_1 \times \chi_2)$ component. All this is quite satisfactory: the theory of measurement, direct or indirect, is logically consistent so long as I maintain my privileged position as ultimate observer.

However, if after having completed the whole experiment I ask my friend, "What did you feel about the flash before I asked you?" he will answer, "I told you already, I did [did not] see a flash," as the case may be. In other words, the question whether he did or did not see the flash was already decided in his mind, before I asked him.[10] If we accept this, we are driven to the conclusion that the proper wave func-

---

[10] F. London and E. Bauer (*op. cit.*, reference 5) on page 42 say, "Il [l'observateur] dispose d'une faculté caractéristique et bien familière, que nous pouvons appeler la 'faculté d'introspection': il peut se rendre compte de manière immédiate de son propre état."

tion immediately after the interaction of friend and object was already either $\psi_1 \times \chi_1$ or $\psi_1 \times \chi_2$ and not the linear combination $\alpha\,(\psi_1 \times \chi_1) + \beta\,(\psi_2 \times \chi_2)$. This is a contradiction, because the state described by the wave function $\alpha\,(\psi_1 \times \chi_1) + \beta\,(\psi_2 \times \chi_2)$ describes a state that has properties which neither $\psi_1 \times \chi_1$ nor $\psi_2 \times \chi_2$ has. If we substitute for "friend" some simple physical apparatus, such as an atom which may or may not be excited by the light-flash, this difference has observable effects and *there is no doubt that* $\alpha\,(\psi_1 \times \chi_1) + \beta\,(\psi_2 \times \chi_2)$ *describes the properties of the joint system correctly, the assumption that the wave function is either* $\psi_1 \times \chi_1$ *or* $\psi_2 \times \chi_2$ *does not.* If the atom is replaced by a conscious being, the wave function $\alpha\,(\psi_1 \times \chi_1) + \beta\,(\psi_2 \times \chi_2)$ (which also follows from the linearity of the equations) appears absurd because it implies that my friend was in a state of suspended animation before he answered my question.[11]

It follows that the being with a consciousness must have a different role in quantum mechanics than the inanimate measuring device: the atom considered above. In particular, the quantum mechanical equations of motion cannot be linear if the preceding argument is accepted. This argument implies that "my friend" has the same types of impressions and sensations as I—in particular, that, after interacting with the object, he is not in that state of suspended animation which corresponds to the wave function $\alpha\,(\psi_1 \times \chi_1) + \beta\,(\psi_2 \times \chi_2)$. It is not necessary to see a contradiction here from the point of view of orthodox quantum mechanics, and there is none if we believe that the alternative is meaningless, whether my friend's consciousness contains either the impression of having seen a flash or of not having seen a flash. However, to deny the existence of the consciousness of a friend to this extent is surely an

----

[11] In an article which will appear soon [*Werner Heisenberg und die Physik unserer Zeit* (Braunschweig: Friedr. Vieweg, 1961)] G. Ludwig discusses the theory of measurements and arrives at the conclusion that quantum mechanical theory cannot have unlimited validity (see, in particular, Section IIIa, also Ve). This conclusion is in agreement with the point of view here represented. However, Ludwig believes that quantum mechanics is valid only in the limiting case of microscopic systems, whereas the view here represented assumes it to be valid for all inanimate objects. At present, there is no clear evidence that quantum mechanics becomes increasingly inaccurate as the size of the system increases, and the dividing line between microscopic and macroscopic systems is surely not very sharp. Thus, the human eye can perceive as few as three quanta, and the properties of macroscopic crystals are grossly affected by a single dislocation. For these reasons, the present writer prefers the point of view represented in the text even though he does not wish to deny the possibility that Ludwig's more narrow limitation of quantum mechanics may be justified ultimately.

unnatural attitude, approaching solipsism, and few people, in their hearts, will go along with it.

The preceding argument for the difference in the roles of inanimate observation tools and observers with a consciousness—hence for a violation of physical laws where consciousness plays a role—is entirely cogent so long as one accepts the tenets of orthodox quantum mechanics in all their consequences. Its weakness for providing a specific effect of the consciousness on matter lies in its total reliance on these tenets—a reliance which would be, on the basis of our experiences with the ephemeral nature of physical theories, difficult to justify fully.

The second argument to support the existence of an influence of the consciousness on the physical world is based on the observation that we do not know of any phenomenon in which one subject is influenced by another without exerting an influence thereupon. This appears convincing to this writer. It is true that under the usual conditions of experimental physics or biology, the influence of any consciousness is certainly very small. "We do not need the assumption that there is such an effect." It is good to recall, however, that the same may be said of the relation of light to mechanical objects. Mechanical objects influence light—otherwise we could not see them—but experiments to demonstrate the effect of light on the motion of mechanical bodies are difficult. It is unlikely that the effect would have been detected had theoretical considerations not suggested its existence, and its manifestation in the phenomenon of light pressure.

More Difficult Questions

Even if the two theses of the preceding section are accepted, very little is gained for science as we understand science: as a correlation of a body of phenomena. Actually, the two theses in question are more similar to existence theorems of mathematics than to methods of construction of solutions and we cannot help but feel somewhat helpless as we ask the much more difficult question: how could the two theses be verified experimentally? i.e., how could a body of phenomena be built around them. It seems that there is no solid guide to help in answering this question and one either has to admit to full ignorance or to engage in speculations.

Before turning to the question of the preceding paragraph, let us note in which way the consciousnesses are related to each other and to the physical world. The relations in question again show a remarkable similarity to the relation of light quanta to each other and to the material bodies with which mechanics deals. Light quanta do not influence each other directly[12] but only by influencing material bodies which then influence other light quanta. Even in this indirect way, their interaction is appreciable only under exceptional circumstances. Similarly, consciousnesses never seem to interact with each other directly but only via the physical world. Hence, any knowledge about the consciousness of another being must be mediated by the physical world.

At this point, however, the analogy stops. Light quanta can interact directly with virtually any material object but each consciousness is uniquely related to some physico-chemical structure through which alone it receives impressions. There is, apparently, a correlation between each consciousness and the physico-chemical structure of which it is a captive, which has no analogue in the inanimate world. Evidently, there are enormous gradations between consciousnesses, depending on the elaborate or primitive nature of the structure on which they can lean: the sets of impressions which an ant or a microscopic animal or a plant receives surely show much less variety than the sets of impressions which man can receive. However, we can, at present, at best, guess at these impressions. Even our knowledge of the consciousness of other men is derived only through analogy and some innate knowledge which is hardly extended to other species.

It follows that there are only two avenues through which experimentation can proceed to obtain information about our first thesis: observation of infants where we may be able to sense the progress of the awakening of consciousness, and by discovering phenomena postulated by the second thesis, in which the consciousness modifies the usual laws of physics. The first type of observation is constantly carried out by millions of families, but perhaps with too little purposefulness. Only very crude observations of the second type have been undertaken in the past, and all these antedate modern experimental methods. So far as it is known, all of them have been unsuccessful. However, every phenomenon is unexpected and most unlikely until it has been discovered—and some of them remain unreasonable for a long time after they have been discovered. Hence, lack of success in the past need not discourage.

[12] This statement is certainly true in an approximation which is much better than is necessary for our purposes.

Remarks on the Mind-Body Question   183

Non-linearity of Equations as Indication of Life

The preceding section gave two proofs—they might better be called indications—for the second thesis, the effect of consciousness on physical phenomena. The first of these was directly connected with an actual process, the quantum mechanical observation, and indicated that the usual description of an indirect observation is probably incorrect if the primary observation is made by a being with consciousness. It may be worthwhile to show a way out of the difficulty which we encountered.

The simplest way out of the difficulty is to accept the conclusion which forced itself on us: to assume that the joint system of friend plus object cannot be described by a wave function after the interaction—the proper description of their state is a mixture.[13] The wave function is $(\psi_1 \times \chi_1)$ with a probability $|\alpha|^2$; it is $(\psi_2 \times \chi_2)$ with a probability $|\beta|^2$. It was pointed out already by Bohm[14] that, if the system is sufficiently complicated, it may be in practice impossible to ascertain a difference between certain mixtures, and some pure states (states which *can* be described by a wave function). In order to exhibit the difference, one would have to subject the system (friend plus object) to very complicated observations which cannot be carried out in practice. This is in contrast to the case in which the flash or the absence of a flash is registered by an atom, the state of which I can obtain precisely by much simpler observations. This way out of the difficulty amounts to the postulate that the equations of motion of quantum mechanics cease to be linear, in fact that they are grossly non-linear if conscious beings enter the picture.[15] We saw that the linearity condition led uniquely to the

[13] The concept of the mixture was put forward first by L. Landau, Z. *Physik*, 45, 430 (1927). A more elaborate discussion is found in J. von Neumann's book (footnote 4), Chapter IV. A more concise and elementary discussion of the concept of mixture and its characterisation by a statistical (density) matrix is given in L. Landau and E. Lifshitz, *Quantum Mechanics* (London: Pergamon Press, 1958), pp. 35-38.

[14] The circumstance that the mixture of the states $(\psi_1 \times \chi_1)$ and $(\psi_2 \times \chi_2)$, with weights $|\alpha|^2$ and $|\beta|^2$, respectively, cannot be distinguished in practice from the state $\alpha(\psi_1 \times \chi_1) + \beta(\psi_2 \times \chi_2)$, if the states χ are of great complexity, has been pointed out already in Section 22.11 of D. Bohm's *Quantum Theory* (New York: Prentice Hall, 1951). The reader will also be interested in Sections 8.27, 8.28 of this treatise.

[15] The non-linearity is of a different nature from that postulated by W. Heisenberg in his theory of elementary particles [cf., e.g., H. P. Dürr, W. Heisenberg, H. Mitter, S. Schlieder, K. Yamazaki, Z. *Naturforsch.*, 14, 441 (1954)]. In our case the equations giving the time variation of the state vector (wave function) are postulated to be non-linear.

unacceptable wave function $\alpha\,(\psi_1 \times \chi_1) + \beta\,(\psi_2 \times \chi_2)$ for the joint state. Actually, in the present case, the final state is uncertain even in the sense that it cannot be described by a wave function. The statistical element which, according to the orthodox theory, enters only if I make an observation enters equally if my friend does.

It remains remarkable that there is a continuous transition from the state $\alpha(\psi_1 \times \chi_1) + \beta(\psi_2 \times \chi_2)$ to the mixture of $\psi_1 \times \chi_1$ and $\psi_2 \times \chi_2$, with probabilities $|\alpha|^2$ and $|\beta|^2$, so that every member of the continuous transition has all the statistical properties demanded by the theory of measurements. Each member of the transition, except that which corresponds to orthodox quantum mechanics, is a mixture, and must be described by a statistical matrix. The statistical matrix of the system friend-plus-object is, after their having interacted ($|\alpha|^2 + |\beta|^2 = 1$),

$$\left|\left|\begin{array}{cc} |\alpha|^2 & \alpha\beta^* \cos\delta \\ \alpha^*\beta \cos\delta & |\beta|^2 \end{array}\right|\right|$$

in which the first row and column corresponds to the wave function $\psi_1 \times \chi_1$, the second to $\psi_2 \times \chi_2$. The $\delta = 0$ case corresponds to orthodox quantum mechanics; in this case the statistical matrix is singular and the state of friend-plus-object can be described by a wave function, namely, $\alpha(\psi_1 \times \chi_1) + \beta(\psi_2 \times \chi_2)$. For $\delta = \frac{1}{2}\pi$, we have the simple mixture of $\psi_1 \times \chi_1$ and $\psi_2 \times \chi_2$, with probabilities $|\alpha|^2$ and $|\beta|^2$, respectively. At intermediate δ, we also have mixtures of two states, with probabilities $\frac{1}{2} + (\frac{1}{4} - |\alpha\beta|^2 \sin\delta)^{\frac{1}{2}}$ and $\frac{1}{2} - (\frac{1}{4} - |\alpha\beta|^2 \sin^2\delta)^{\frac{1}{2}}$. The two states are $\alpha(\psi_1 \times \chi_1) + \beta(\psi_2 \times \chi_2)$ and $-\beta^*(\psi_1 \times \chi_1) + \alpha^*(\psi^2 \times \chi^2)$ for $\delta = 0$ and go over continuously into $\psi_1 \times \chi_1$ and $\psi_2 \times \chi_2$ as δ increases to $\frac{1}{2}\pi$.

The present writer is well aware of the fact that he is not the first one to discuss the questions which form the subject of this article and that the surmises of his predecessors were either found to be wrong or unprovable, hence, in the long run, uninteresting. He would not be greatly surprised if the present article shared the fate of those of his predecessors. He feels, however, that many of the earlier speculations on the subject, even if they could not be justified, have stimulated and helped our thinking and emotions and have contributed to re-emphasize the ultimate scientific interest in the question, which is, perhaps, the most fundamental question of all.

The Place of Consciousness in Modern Physics

Eugene P. Wigner

Consciousness and Reality, C. Muses and A. M. Young (eds.).
Outerbridge and Lazard, New York 1972, chap. 9, pp. 132–141
(Reset by Springer-Verlag for this volume)

The author (Ph. D.) is a Nobel prize winner and one of the leading physicists of our times. The Wigner expansion in quantum perturbation theory was a fundamental contribution, as was Dr. Wigner's work in the relation of group theory to quantum physics. In this connection his studies in fundamental symmetries in nature bring the pioneering work of Herman Weyl to new frontiers. This chapter, compiled from his papers, collects in one place his important thinking on consciousness.

> Modern physics is now in the process of attenuating [in favor of the observer] the separation which it itself once made between object and observer ... in this effort it teaches us very important things.
>
> Charles de Montet, M. D. *L'Evolution vers l'Essentiel*
> Lausanne, Switzerland, 1950.

Until not many years ago, the "existence" of a mind or soul would have been passionately denied by most physical scientists. The brilliant successes of mechanistic and, more generally, macroscopic physics and of chemistry overshadowed the obvious fact that thoughts, desires, and emotions are not made of matter, and it was nearly universally accepted among physical scientists that there is nothing besides matter. The epitome of this belief was the conviction that, if we knew the positions and velocities of all atoms at one instant of time, we could compute the fate of the universe for all future [time]. Even today, there are adherents to this view though fewer among the physicists than – ironically enough – among biochemists ...

It is, at first, also surprising that biologists are more prone to succumb to the error of disregarding the obvious than are physicists. The explanation for this may be [that] ... as a result of the less advanced stage of their discipline, they are so concerned with establishing *some* regularities in their own field that the temptation is great to turn their minds away from the more difficult and profound problems which need, for their solution, techniques not yet available. Yet, it is not difficult to provoke an admission of the reality of the "I" from even a convinced materialist if he is willing to answer a few questions, [e.g.] "If all that exists are some complicated chemical processes in your brain, why do you *care* what those processes are?" ...

There are several reasons for the return, on the part of most physical scientists, to the spirit of Descartes's *"Cogito ergo sum"*, which recognizes the thought, that is, the mind, as primary. First, the brilliant successes of mechanics not only faded into the past; they were also recognized as partial successes, relating to a narrow range of phenomena, all in the macroscopic domain. When the province of physical theory was extended to encompass microscopic phenomena, through the creation of quantum mechanics, the concept of consciousness came to the fore again: it was not possible to formulate the laws of quantum mechanics in a fully consistent way without reference to the consciousness. All that quantum mechanics purports to provide are probability connections between subsequent impressions (also called "apperceptions") of the consciousness, and even though the dividing line between the observer, whose consciousness is being affected, and the observed physical object can be shifted towards the one or the other to a considerable degree, it cannot be eliminated.[1] It may be premature to believe that the present philosophy of quantum mechanics will remain a permanent feature of future physical theories; it will remain a remarkable, in whatever way our future concepts may develop, that the very study of the external world led to the conclusion that the content of the consciousness is an ultimate reality . . .

The property of observations to increase our ability for foreseeing the future follows from the fact that all knowledge of wave functions [the basic tool of quantum physics] is based, in the last analysis, on the "impressions" we receive. In fact, the wave function is only a suitable language for describing the body of knowledge – gained by observations – which is relevant for predicting the future behavior of the system. For this reason, the interactions which may create one or another sensation in us are also called observations, or measurements. One realizes that *all* the information which the laws of physics provide consists of probability connections between subsequent impressions that a system makes on one if one interacts with it repeatedly, i.e., if one makes repeated measurements on it . . .

The important point is that the impression which one gains at an interaction may, and in general does, modify the probabilities with which one gains the various possible impressions at later interactions. In other words, the impression which one gains at an interaction, called also *the result of an observation*, modifies the wave function [that is, the wave forms describing the fluctuating probability of its being observed] of the system. The modified wave function is, furthermore, in general unpredictable before the impression gained at the interaction has entered our consciousness: it is the entering of an impression into our consciousness which alters the wave function because it modifies our appraisal of the probabilities for different impressions which we expect to receive in the

---

[1] J. von Neumann, Mathematische Grundlagen der Quantenmechanik (Berlin: Julius Springer, 1932), Chapter VI; English translation (Princeton, N.J.: Princeton University Press, 1955). Also F. London and E. Bauer, La Théorie de l'observation en méchanique quantique (Paris: Hermann and Co., 1939). The last authors observe (page 41), "Remarquons le rôle essential que joue la conscience de l'observateur . . . " ("Let us note the essential role that the consciousness of the observer plays . . . ")

future. It is at this point that the consciousness enters the theory unavoidably and unalterably. If one speaks in terms of the wave function, its changes are coupled with the entering of impressions into our consciousness.[2] If one formulates the laws of quantum mechanics in terms of probabilities of impressions, these are *ipso facto* the primary concepts with which one deals ... the quantum description of objects is influenced by impressions entering my consciousness ...

The fact that the first kind of reality [i.e. consciousness] is absolute[3] and ... that we discuss the realities of the second kind [perceived objects] much more, may lead to the impression that the first kind of reality is something very simple. We all know that this is not the case. On the contrary, the content of the consciousness is something very complicated and it is my impression that not even the psychologists can give a truly adequate picture of it ... The nature of the first kind of reality is already quite complex and the inadequacy of our appreciation of its properties may be one of the most potent barriers against establishing the nature of universal realities at the present time.

[The author then explains in more technical language that the interaction between *two* (or more) conscious entities is not expressible in terms of present physical theory, however advanced; for the equations then attain a new depth of complexity not explainable on the basis of known concepts, i.e. they become in mathematical parlance "nonlinear", whereas our most advanced physical theories can assume only linear equations.]

It follows that the being with a consciousness must have a different role in quantum mechanics than the inanimate measuring device: the atom considered above. In particular, the quantum mechanical equations of motion cannot be linear [i.e. must be more complicated] ... in fact they are grossly nonlinear if conscious beings enter the picture ...

[The only alternative, Dr. Wigner points out, is for each observer to deny meaning to the consciousness of all others, which, as he also points out, is "unnatural" – and, we may add, totally unwarranted in view of the valid pooling of results of different observers: something we do and rely on successfully every day.]

Measurement is not completed until its result enters our consciousness [after going through some physical stages or instruments]. This last step occurs when a correlation is established between the state of the last measuring apparatus and something which directly affects our consciousness. This last step is, at the present state of our knowledge, shrouded in mystery and no explanation has been given for it so far in terms of quantum mechanics, or in terms of any other theory ...[4]

---

[2] And the consequent expectations (i.e. of future impressions) that they generate in interaction with consciousness. *Eds.*

[3] In that it is presupposed in any assertion about anything else. *Eds.*

[4] It should be noted that some definite basis for such forthcoming explanation appears to lie in hypernumber theory, which, in terms of the two simplest hypernumbers, $\sqrt{-1}$ and $\sqrt{+1}$, has already proved indispensable to quantum theory. Wigner rightly wrote (p. 224 of his *Symmetries and Reflections*) that the "mathematician

The second argument to support the existence of an influence of the consciousness on the physical world is based on the observation that we do not know of any phenomenon in which one subject is influenced by another without exerting an influence thereupon. This appears convincing to this writer. It is true that under the usual conditions of experimental physics or biology, the influence of any consciousness is certainly very small. "We do not need the assumption that there is such an effect." It is good to recall, however, that the same may be said of the relation of light to mechanical objects. Mechanical objects influence light – otherwise we could not see them – but experiments to demonstrate the effect of light on the motion of mechanical bodies are difficult. It is unlikely that the effect would have been detected had theoretical considerations not suggested its existence and its manifestation in the phenomenon of light pressure...

Let us [now] specify the question which is outside the province of physics and chemistry but is an obviously meaningful (because operationally defined) question: Given the most complete description of my body (admitting that the concepts used in this description change as physics develops), what are my sensations?... This is clearly a valid and important question which refers to a concept – sensations – which does not exist in present-day physics or chemistry. Whether the question will eventually become a problem of physics or psychology, or another science, will depend on the development of these disciplines... [5]

It will not be possible to use [the concept of universal reality] meaningfully without being able to give an account of the phenomena of the mind, which is much deeper than our present notions admit. This is a consequence of the fact that, clearly, from a non-personal point of view, other people's sensations are just as real as my own. In all our present scientific thinking, either sensations play no role at all – this is the extreme materialistic point of view which is clearly absurd *and*, as mentioned before, is also in conflict with the tenets of quantum mechanics – or my own sensations play an entirely different role from those of others. It follows that before we can usefully speak of universal reality, a much closer integration of our understanding of physical and mental phenomena will be necessary than we can even dream of at present. This writer sees no cogent reason to doubt the possibility of such an integration...

What I am saying is that... from the point of view of quantum mechanics, the faculty [of self-awareness] is completely unexplained.

It may be useful to give the reason for the increased interest of the contemporary physicist in problems of epistemology and ontology. The reason is, in a nutshell, that physicists have found it impossible to give a satisfactory description of atomic phenomena without reference to the consciousness. This had little to do with the oft rehashed problem of wave and particle duality and

---

fully, almost ruthlessly, exploits the domain of permissible reasoning and skirts the impermissible." The nature of our mysterious cosmos demands nothing less. *C.M.*

[5] The Journal for the Study of Consciousness (vol. 1, no. 2; 1968 and in a note, p. 92, of vol. 2, no. 2; 1969) presented some of the theorems that begin to answer Professor Wigner's profound question. *Eds.*

refers, rather, to the process called the "reduction of the wave packet." This takes place whenever the result of an observation enters the consciousness of the observer – or, to be even more painfully precise, my own consciousness, since I am the only observer, all other people being only subjects of my observations. Alternatively, one could say that quantum mechanics provides only probability connections between the results of my observations as I perceive them. Whichever formulation one adopts, the consciousness evidently plays an indispensable role.[6]

There is, apparently, a correlation between each consciousness and the physico-chemical structure of which it is a captive, which has no [known] analogue in the inanimate world. Evidently, there are enormous gradations between consciousnesses, depending on the elaborate or primitive nature of the structure on which they can lean: the sets of impressions which an ant or a microscopic animal or a plant receives surely show much less variety than the sets of impressions which man can receive. However, we can, at present, at best, guess at these impressions. Even our knowledge of the consciousness of other men is derived only through analogy and some innate knowledge...

It follows that there are only two avenues through which experimentation can proceed to obtain information about [the nature of mind]: observation of infants where we may be able to sense the progress of the awakening of consciousness, and by discovering phenomena... in which the consciousness modifies the usual laws of physics. The first type of observation is constantly carried out by millions of families, but perhaps with too little purposefulness. Only very crude observations of the second type have been undertaken in the past, and all these antedate modern experimental methods. So far as it is known, all of them have been unsuccessful. However, every phenomenon is unexpected and most unlikely until it has been discovered – and some of them remain unreasonable for a long time after they have been discovered... It seems more likely... that living matter is actually influenced by what it clearly influences: consciousness. The description of this phenomenon clearly needs incorporation of concepts into our laws of nature which are foreign to the present laws of physics. Perhaps the relation of consciousness to matter is not too dissimilar to the relation of light to matter, as it was known in the last century: matter clearly influenced the motion of light but no phenomenon such as the Compton effect was known at that time which would have shown that light can directly influence the motion of matter. Nevertheless, the "reality" of light was never doubted...

Our penetration into new fields of knowledge will unquestionably give us new powers, powers which affect the mind more directly than the physical conditions which we now can alter. Poincaré and Hadamard have recognized that, unlike most thinking which goes on in the upper consciousness, the really relevant mathematical thinking is not done in words. In fact, it happens somewhere

---

[6] The fact was pointed out with full clarity first by von Neumann. [See first footnote of this chapter.]

so deep in the subconscious that the thinker is usually not even aware of what is going on inside of him.

It is my opinion that the role of subconscious thinking is equally important in other sciences, that it is decisive even in the solution of apparently trivial technical details. An experimentalist friend once told me (this was some twenty years ago) that if he could not find the leak in his vacuum system he usually felt like going for a walk, and very often, when he returned from the walk, he knew exactly where the leak was.

There are two basic concepts in quantum mechanics: states and observables. The states are vectors in Hilbert space; the observables, self-adjoint operators on these vectors. The possible values of the operators – but we had better stop here lest we engage in a listing of the mathematical concepts developed in the theory of linear operators...

The enormous usefulness of mathematics in the natural sciences is something bordering on the mysterious and there is no rational explanation for it... this uncanny usefulness of mathematical concepts. The principal emphasis [in mathematics] is on the invention of concepts.[7] Mathematics would soon be running out of interesting theorems if these had to be formulated in terms of the concepts which already appear in the axioms. Furthermore, whereas it is unquestionably true that the concepts of elementary mathematics and particularly elementary geometry were formulated to describe entities which are directly suggested by the actual world, the same does not seem to be true of the more advanced concepts, in particular, the concepts which play such an important role in physics.

The complex numbers provide a particularly striking example of the foregoing. Certainly, nothing in our experience suggests the introducing of these quantities... Let us not forget that the Hilbert space of quantum mechanics is the complex Hilbert space [i.e. including $i \equiv \sqrt{-1}$], with a Hermitian scalar product. Surely to the unpreoccupied mind, complex numbers... cannot be suggested by physical observations. Furthermore, the use of complex numbers in this case is not a calculational trick of applied mathematics, but comes close to being a necessity in the formulation of the laws of quantum mechanics. Finally, it now begins to appear that not only complex numbers but analytic functions[8] are destined to play a decisive role in the formulation of quantum theory. I am referring to the rapidly developing theory of dispersion relations. It is difficult to avoid the impression that a miracle confronts us here [i.e. in the agreement between the properties of the hypernumber $\sqrt{-1}$ and those of the natural world].

---

[7] In cordial agreement with the author's brilliant insights, we would prefer *"discovery of concepts"* here. Techniques can be invented – but not the laws of thought, as partially reflected in concepts or axioms, any more than laws of physics. The nature of things, both within and without, is discovered, but not invented by man. That this is so is the chief and ultimate guarantee of the universality of any scientific law, whether that science or knowing be physical or psychological. *C.M.*

[8] The two being closely linked in the theory of the functions of a complex variable. *C.M.*

A much more difficult and confusing situation [than modern physics presents] would arise if we could, some day, establish a theory of the phenomena of consciousness, or of biology, which would be as coherent and convincing as our present theories of the inanimate world. Mendel's laws of inheritance and the subsequent work on genes may well form the beginning of such a theory as far as biology is concerned.

In fact, many feel nowadays that the life sciences and the science of the minds of both animals and men have already been neglected too long. Our picture of the world would surely be more rounded if we knew more about the minds of men and animals, their customs and habits. The second type of shift may mean, however, the acknowledgment that we are unable to arrive at the full understanding of even the inanimate world, just as, a few centuries ago, man came to the conclusion that he has no very good chance to foresee what will happen to his soul after the death of his body. We all continue to feel a frustration because of our inability to foresee our soul's ultimate fate. Although we do not speak about it, we all know that the objectives of our science are, from a general human point of view, much more modest than the objectives of, say, the Greek science were; that our science is more successful in giving us power than in giving us knowledge of truly human interest.

Both physics and psychology claim to be all-embracing disciplines: the first because it endeavors to describe all nature; the second because it deals with all mental phenomena, and nature exists for us only because we have cognizance of it... Both disciplines may yet be united into a common discipline without overtaxing our mind's capacity for abstraction.

New Dimensions of Consciousness

E. P. Wigner

Mimeographed notes, ca. 1978
(Reset by Springer-Verlag for this volume)

Nov. 18, 1978

As our program implies, I am a physicist and, naturally, at a meeting on consciousness I will be interested, first, in the influence of the ideas concerning consciousness on physics, and in the influence of our theories of physics on the understanding of the phenomenon of consciousness. But, finally and most importantly, I will try to explore the question of whether the two disciplines, physics and the study of consciousness, could be united and given a common basis – as was given, in our century, to chemistry and physics.

I will not demonstrate to this gathering that present day physics does not describe the phenomena of consciousness. Even if the physical theories could completely describe the motion of the atoms in our bodies, they would not give a valid picture of the contents of our consciousnesses, they would not tell us whether we experience pain or pleasure, whether we are thinking of prime numbers or of our granddaughters. Rather, I will try to discuss, first, the question which of the fundamental principles of physics would have to be modified if physics were to be extended to encompass consciousness and second, I will say a few words to explain the reasons for the increasing interest of the physicists in the phenomena of consciousness. As to the first question, my answer will be that the most fundamental principles of present day physics would have to be modified. Surely, as I will relate briefly, those principles have undergone at least two basic, though rarely advertised, modifications since the inception of "our science", about 300 years ago. Yet the further modifications needed are, in my opinion, even more fundamental than those which have taken place in the past. But, and this concerns my second subject, even the considerations which are stimulated by present day physics suggest the need for such modifications. They also prompted us to become concerned with the phenomenon of consciousness – removed as this is from the subjects which we discuss in the course of our everyday work.

Let me say first a few words about science in general. Science has, as far as we now know, two functions. To establish correlations between events and to enable us, on the basis of the knowledge of these correlations, to produce machines, equipment, or more generally situations which would be astounding to a man not used to them, as we are, for instance, to the telephone.

Let me expound on this a little more. What I call "our science" started about 300 years ago when Newton discovered the relation between the law of freely falling bodies here on our Earth and the motion of the moon around the Earth. He was concerned with situations in which electric or magnetic forces play no role, only gravitational forces do, and discovered a simple rule to describe the effects of these forces. This surely was a great accomplishment and is generally appreciated as such. Yet I consider another contribution of his, the clear separation of the information which physics does *not* provide, from the information which it does furnish, even more important. The former we call initial conditions, the latter the laws of nature. That the moon is now in that direction, that there is a moon, that I let loose this pencil now from this point, are initial conditions entirely outside the scope of physics. But if you give me the positions of the moon at two different instants of time, physics, the laws of nature, permit me to calculate its position at all other times. The laws of nature enable me to calculate its position at all times if I know them at two distinct instances of time. The same applies to a falling body. If I have two pieces of information about its position, such as initial position and, let us say, its initial velocity, its later positions and velocities are given by the laws of nature, in fact the same laws of nature which permit us to obtain the aforementioned information about the moon.

This is the first point I wanted to make. The point is that science does not even try to give us complete information about the events around us – it gives information about the *correlation* between these events. Thus, it gives a correlation between the positions of the moon at any three instants of time, but, though we may know it, its position at one instant of time is not the consequence of the laws of nature.

The second point I wanted to make is implicit already in what I said before. While the preceding point applies to all science, the one I will articulate now is characteristic of the status of science at a definite period. As I mentioned before, Newton's theory is valid only if only gravitational forces play a role, if the electric, magnetic, and many other types of forces do not affect the system appreciably. This is true for the planetary system of the sun, for the moon, for satellites and comets but is, for instance, not true for the insides of stars. And Newton surely did not believe that his laws are valid under all circumstances. The next big step in physics was initiated by Faraday and Maxwell, describing the time variation of the electromagnetic field. Maxwell's theory also gave a common basis for two types of phenomena which appear distinct: the electric and magnetic effects commonly observed and the phenomenon of light. But as we know today, his theory also represented a limiting case, that of macroscopic bodies, i.e., it was restricted to situations in which all objects contained billions of atoms. The transition to the theory which describes also the behavior of systems containing only few atoms, or of single atoms, took place in the present century with the advent of quantum theory. The question then arises whether this theory, that is the present physics, still represents a limiting case, whether it applies under all conditions or only when some effect is so small that it can be neglected or disregarded. I will try to give evidence that this is so and at least life and consciousness are not described by any present theory of physics, in particular not by quantum mechanics.

I wish to add three observations to the preceding discussion. First, that it is wonderful that nature provided us with situations in which grossly simplified laws of nature have an almost perfect validity. The system of the sun's planets, their satellites, etc., are in such a situation – a theory which neglects all non-gravitational forces describes with an almost incredible accuracy their motion. The macroscopic bodies similarly obey laws, those of classical physics, with a similar accuracy. The inanimate objects seem to obey the laws of quantum theory also very closely. The fact that nature provided us with such situations is a wonderful gift – if these situations did not exist, it might have surpassed human intelligence to discover laws of nature because the consequences of these are very complicated in all situations except in "limiting cases" when greatly simplified laws have high validity. I will try to show that present day physics also deals with such a "limiting case", with situations in which life and consciousness do not influence the events.

My second observation points to the enormous increase of the area in which the known laws of nature have validity. Newton's equations described the motion of planets, satellites, of falling bodies here on Earth – nothing else. Maxwell's equations described all electromagnetic phenomena and, together with the equations of Newton and his successors, all macroscopic objects' behavior. But the physical properties of these objects, the hardness, specific weight, electrical conductivity, etc., etc., of these materials was not given by these theories, these properties played a role similar to the initial conditions – they had to be determined experimentally. Quantum theory, in the course of its development changed this – not only does it gives equations valid also, and particularly so, for microscopic systems, atoms, molecules, etc., – it also permits the derivation of the *properties* of macroscopic bodies, at least as long as no living system is involved. Will this enormous and, on a historical scale, immensely rapid expansion of the area of validity of physical laws continue? Can we hope that it will eventually encompass the phenomena of life? We do not know, but we hope.

This was my second remark. Lastly, I wish to call attention to the way the second revolution in physics, that initiated by Faraday and Maxwell, was related to Newton's physics. An was mentioned before this latter one was based on the knowledge of the motion of planets and satellites – the knowledge being obtained by observing the sunlight scattered by these planets and reaching us. Yet the description of light was not part of the theory – this, the communicator of the information was largely outside physics' range almost until the Maxwellian revolution. It is remarkable that the present situation is similar and this is what I wish to call attention to. Quantum mechanics' function is, from a basic point of view, to give the probabilities of the outcomes of observations. Yet the nature of the observation, the description of the observer, is just as foreign to present day quantum mechanics as was the nature of light to Newton's theory. Are we going to have a revolution, similar to that initiated by Faraday and Maxwell, to incorporate the observer into physics, as Maxwell's theory incorporated light into that discipline? We can hope for it and surely such an extension of the area of "natural science" will modify the fundamental structure of this science at least as much as Maxwellian theory changed the basic concepts of the earlier physics. This is the third remark I wanted to make. Let me now go over to the other subject I wanted to discuss – which of the fundamental principles of present day physics would have to be modified in order to incorporate life and consciousness into the realm of the "natural sciences". Are there signs, within the present scope of these sciences, to suggest, at least vaguely, such modifications? My answer to the last question will be in two cases "yes", in one case "no".

The first point that should be considered in this connection is that, at least physics, deals with isolated systems, that is, systems so far removed from other bodies that these do not exert noticeable influence on the system considered. This, in fact, is a necessary condition for the validity of what is called "causality" that is, for the postulate that the initial conditions of the system determine its later behavior. Clearly, if the system is under the influence of other systems, the behavior of the latter will affect it and its future is no longer determined solely by its own state.

Equally clearly, any entity with a highly developed mind and consciousness cannot be fully isolated from its environment – if we put him out into dark interstellar space, this fact will influence his thoughts. Anyway, we want to know his behavior under less abstruse conditions and these will influence him and he will not be an isolated system. It follows from this that his behavior will not be foreseeable deterministically – if there is a way to foresee his mind's behavior, this must include some probabilistic elements. I believe one is naturally inclined to this conclusion even without the argument just given. It already implies one fundamental change in the kind of description of nature which physics propose for inanimate objects. It assumes that these can be isolated from the influence of their environment.

Yes, present day physics describes the behavior of isolated systems. Yet, there are clear signs that not even physical systems, of an inanimate nature, can all be isolated. The German physicist Zeh has called attention to the fact that the quantum theoretical energy levels of a reasonably macroscopic body are so close to each other that even a single atom or electron, though spatially well separated from it, will influence it in the quantum theoretical sense. Thus the impossibility of true isolation begins much before the interference of life and consciousness with such isolation. Actually your present speaker is proposing an equation describing the probabilistic behavior of a not isolated system but he must admit that it is very unlikely that his equation will remain valid if life and consciousness have an essential role.

The next question which naturally comes up concerns the sharp division between initial conditions and laws of nature. Naturally, the idea that the initial conditions fully determine the later behavior of the system can be meaningful only if the same initial state of the system can be reproduced and it can be ascertained that identical initial conditions have indentical consequences – this is the postulate if the theory is deterministic – or that they give the different possible consequences with the same probabilities. Admittedly, this latter relation may be quite difficult to realise – it requires the establishment of the same initial state a very great number of times. But, as far as living organism of any complexity are concerned, the same initial state hardly can be realised several times. There are no two identical people and if we repeat the same experiment on the same individual the initial conditions are no longer the same – the individual will remember at the second experiment the event of the first one – his mental outlook will have changed thereby. This means that the relevant statements of the theory encompassing life will be terribly different from those of the present natural sciences. There are, of course signs for some degree of causality also in the description of living beings – the last Frontiers of Science statement (by Marvin Harris) reads:

Taboos on the consumption of the flesh of certain edible animals are often used to uphold the view that cultural evolution is not a causally determined process. Examination of such taboos in an evolutionary – ecological context suggests, however, that *deterministic* processes shape many of these apparently random processes and aversions.

Surely an interesting finding, one of many, but very different from those of present-day physics or chemistry, showing the magnitude of the basic changes that the latter disciplines have to undergo if they are to adsorb, or be united with, those referring to life.

The preceding discussion contains the assertion that definite experiments on complex organisms cannot be truly repeated. This is rather obvious but it is surprising that, according to a new theory of Dirac, this is true also in the domain of inanimate nature. The reason is that the fundamental constants of nature, such as the ratio of the gravitational and electric forces between two particles, change in time. The change is very small – about 10^{-10} per year – but even a small change shows that, even as far as inanimate nature is concerned, we are far from a complete understanding and a logically clear formulation of the laws. There are many indications for this but it is not necessary to discuss them at this occasion.

Let me finish by saying more concisely what I believe the relation of our present science, and in particular physics, is to an encompassing theory which describes also life. I do not believe there are two entities: body and soul. I believe that life and consciousness are phenomena which have a varying effect on the events around us – just as light pressure does. Under many circumstances, those with which present-day physics is concerned, the phenomenon of life has an entirely negligible influence. There is then a continuous transition to phenomena, such as our own activities, in which this phenomenon has a decisive influence. Probably, the behavior of viruses and bacteria could be described wiht a high accuracy with present theories. Those of insects could be described with a moderate approximation, those of mammals and men are decisively influenced by their minds. For these, present physical theory would give a false picture even as far as their physical behavior is concerned.

Will man ever have as good a picture of life and consciousness as we now have of the situations in which the role of these is negligible? Of course, I do not know but I am a bit pessimistic. We are, after all, animals and there is probably a continuous transition from the capabilities of other mammals to our capabilities. And the other animals were not able even to develop classical physics – our abilities probably also have boundaries. And this is good – what would man do if his scientific effort were completed? He wants to strive for something and an increase of this understanding is a worthy goal and I hope he will be successful in continuing to increase his knowledge further and further but never making it complete.

One may also ask whether there are other phenomena, even less obvious than consciousness and life? I do not know but I do know that it would be wrong to deny its possibility. I will not.

The Existence of Consciousness

E. P. Wigner

R. L. Rubinstein (ed.) Modernization. Paragon House, New York 1982, pp. 279–285

Thank you very much both for your introduction and also for your participation in the discussion yesterday. It pointed to a number of questions which I should clarify and it also gave me an opportunity to voice my conviction on the questions discussed: my conviction that life and consciousness cannot be described in terms of the concepts of *present-day* physics. I hope that science will eventually develop concepts which give interesting and highly significant information not only about inanimate objects—as does present-day physics—but also about life and consciousness. Whether the so encompassing science will be called physics or will have another name, I do not know. Perhaps it will be called scisif. But that is not important. What is important is whether the three basic principles which physics obeys will be maintained. For this reason, I would like to discuss these principles, the discussion of the second one of which will contain again the demonstration that we are as yet far from a unification of physics and the life sciences.

Before starting on that discussion, I wish to apologize to the physicists among you. You know the principles I will present at least instinctively, even though you may not have heard them presented before. Basically, what I will say will not be really new to you. However, I cannot help because right now I do not talk principally to physicists. And I apologize also to some of the others who attend this meeting because what I'll say has very little to do directly with religion. I also realize that it is very difficult to accept principles on first hearing. If I were giving a class, I would talk about one and a half of the principles one hour, again so much at the next class, and summarize all of them at the third meeting. Because we are not inclined to accept, nor even to ingest, general principles just because someone pronounces them. —Let me now go over to the discussion of the first characteristic of present-day physics which I wish to present.

The Limitation of Physics—Separation of
Initial Conditions and Laws of Nature

The distinction between initial conditions and laws of nature is implicit in Newton's *Philosophiae Naturalis Principia Mathematica* of 1687. Among other accomplishments, this contains the law of gravitation. This permits the calculation of the position and velocity of the planets at all times if the position and velocity at one single time is given. These latter are called initial conditions and the laws of physics give no information about them. In fact, in many parts of physics, in particular in the theory of heat, it is assumed that these are entirely irregular—the positions of the atoms in a gas are as irregular as possible and the same holds of their velocities. The laws of nature, which determine the changes of these quantities as functions of time are, on the contrary, in the words of Einstein, simple and mathematically beautiful.

Let me give a couple of examples. Newton's theory of gravitation says that any two bodies are attracted to each other, the force of attraction being proportional to the product of the masses of the two objects and inversely proportional to the square of the distance between them. And the direction of the force is toward the attracting body. Thus, if one lets loose a body such as this key here, it will fall in the direction of the heaviest body nearby, which is the Earth. If the initial position of the key and its initial velocity—which was zero —are given, the time at which it reaches the table underneath can be calculated. It can also be demonstrated that two objects, even entirely different ones, will fall equally fast—as long as we can neglect other forces, different from the gravitational one, such as air resistance. More than that: the motion of the moon, the existence of its circular orbit, can be demonstrated, and fully calculated, given, of course, its position and velocity at one time. Similarly, Kepler's three regularities of the motion of the planets follow from Newton's laws. It is truly wonderful that regularities of such a great diversity of phenomena can be summarized in a law as simple as Newton's gravitational theory.

I do not know whether I should mention that when Newton first thought of his principle, it did not check. The moon moved too fast. But when the radius of the Earth was remeasured, it suddenly checked. And when the new measurement of the Earth's size was presented to the Royal Society, he at once knew—now my theory is confirmed.

It may be good to reemphasize the point that Newton's theory describes the motion of the moon and of the sun's planets, but does *not explain* their initial states and velocities, not even the existence of the moon or of the planets. These are outside the area of the laws of physics, they are initial conditions. It is also good to realize that Newton's theory has only limited validity—there are forces different from the gravitational one—the air resistance was mentioned before. It is, however, wonderful that we are given situations in which only the gravitational forces play a significant role—such as the motion of the sun's planets. I call these situations "limiting cases." However, in the course of time, physics has expanded greatly, and we can now deal with many situations in which, for instance, other forces, not gravitational ones, also play a significant role. This will be my next subject which will also show that the expansion is, as yet, far from complete and that, in particular, it does not extend to the phenomenon of life.

The Expansion of the Area of Physics

The first enormous extension of the area of physics occurred more than a hundred years ago; it was due principally to Faraday and Maxwell. Their theory describes the regularities of electromagnetic phenomena, the behavior of electric charges and of magnets, their interactions, and even more importantly, the phenomena of radiation, that is in particular of light. It is also based on a simple set of equations called Maxwell's equations, but the distinction of initial conditions and laws of nature remains.

Yet, the whole description of the physical situation is very different from that used by Newton. Newton's description of the "state" of the system consisted in giving the position and velocity of each object; Maxwell's equations describe the electromagnetic field by giving the values of its components at every point of space. The former consisted of a set of numbers, the latter of functions of the three space-coordinates, usually denoted by x, y, and z. I will admit that there was an almost continuous transition between Newton's "classical dynamics" and the electromangnetic field theory, yet the difference remains very great. Even the reformulations of classical dynamics referred to position densities and velocities—very different concepts from field strengths. As all of us know, Einstein fundamentally reformulated Newton's gravitational theory and he made a great effort to give this theory, and that of electromagnetism, a common basis. There is quite general agreement among physicists, including Einstein, that he was not fully successful. It is true, nevertheless, that the perhaps artificial union of the two theories extended the area of physics tremendously. The theory of light describes the way we obtain information about the positions of the planets and the union gives a description of the insides of stars where both gravitational forces and electromagnetic ones, in particular light pressure, play a decisive role.

Neither of the two theories mentioned—that of the essentially gravitational of Newton or the electromagnetic of Maxwell—considered the atomic structure of matter seriously. The extension of the area of physics to this domain is due to a third "revolution," that of quantum theory and quantum mechanics. When I started to be interested in physics, if you wanted to know the density of aluminum, there was a thick book in which you could look it up—it was 2.7. You do the same now but, just the same, the properties of materials are part of present-day physics. They *can* be derived from the atomic structure of the materials, from the interaction of the atoms contained in those which can be derived with the aid of the quantum mechanical theory. All chemistry also, we believe, could be derived on the basis of our understanding of the molecules' atomic structure. This is again a tremendous change—an extension of the area of physics which was just impossible to think of when I first learned about this subject. But, of course, it is still far from describing correlations between all kinds of events—it says nothing about my thought, emotions, desires, etc.

As matter of fact, there is a degree of similarity between the relation of the theory of light to the study of planetary motion and the relation of the hoped-for theory of life and consciousness to quantum mechanics. The theory of light describes our obtaining information about the motions of the planets—their consistency with the theory of gravitation. Similarly, a theory of our consciousness would provide the description of our observation process and the now mysterious "collapse of the state vector," i.e., it would tell us about the "measurement process" which permits us to verify the probabilistic statements of quantum mechanics on the outcomes of observations or statements of quantum mechanics on the outcomes of observations or statements of quantum mechanics on the outcomes of observations or measurements.

Before going to the third characteristic of physics, let me apologize for the gross incompleteness of the present one. In particular, it gave credit for the development of physics only to a very few people when, actually, many dozens of physicists contributed to it decisively. In particular, Einstein's contributions' mention was badly neglected. I did not find it possible to give an even superficially valid description of the many contributions to the expansion of the area of physics.

Physics Deals Best with Limiting Cases

This is a very obvious remark. It is that practically all laws of physics were discovered, and also verified, under conditions when these laws manifest themselves in particularly simple ways. Thus, the law of gravitation was discovered and verified under conditions in which no other but gravitational forces play a significant role. This applies, for instance, to the motion of the planets—there is, of course, a light pressure from the sun's radiation but this has practically no effect. If we verify the law by dropping an object here on the Earth, we choose one of high weight and small size so that the air resistance remains negligible. Where we verify the conclusions of quantum mechanics, we use microscopic objects, that is atoms, because the quantum effects manifest themselves most clearly on these. The skill of the experimental physicist consists in large part in creating circumstances in which the law of nature to be tested manifests itself in a simple way, that is, the effects of which he has no full mastery are very small.

The question then arises: is present-day physics always dealing with a limiting case? Do we carry out all our experiments, and do we do all our theoretical thinking, using circumstances in which some phenomenon is absent or at least plays no significant role? The answer is "yes," we experiment almost always under circumstances in which life and consciousness play no role. As was pointed out before, the mere idea of these is foreign to present-day physics, it does not even ask the question whether I think of my grand-daughter or of prime numbers, whether I am experiencing pain or pleasure. There is, for practically all effects a continuous transition from the situation in which it has virtually no effect, to the situation in which it has a decisive one. Is there such a continuous transition between the situation in which the role of life on consciousness is negligible to one in which it plays a decisive role? The answer is, I believe again "yes": the properties and behavior of viruses and microbes could probably be well described by our physics—and our biologists had considerable success in this regard—the behavior of higher organisms, with clear consciousness I do not really believe follow all laws of physics. Is there some evidence for this? I believe there is. First of all, in every case, except that of heat, when the area of physics was extended, the basic laws, in fact their subjects, have changed. But there is some more mathematical proof for this to which I have alluded before: according to quantum mechanics, I could be put into a state in which I have neither seen, nor not seen, a flash—in which I am in a "superposition" of these states. Yet, the fact is that my mind is never in such a superposition. I realize that this last statement is difficult to swallow for a non-physicist but I wanted to mention it just the same.

Will all this change, are we going to have a science which encom-passes life, so that its validity is not restricted to inanimate nature? We have seen that the area of science has greatly extended in the past and we can hope, and I do, that it will be extended further, to cover the phenomena of life and consciousness. But we can not be sure of this—after all, according to Darwin, we are animals and our ability to "understand," that is to discover important and in-teresting correlations between events, may be limited. I often tell the story that I wanted to explain the associative law of multiplica-tion to a very nice dog but got nowhere—similarly our understanding may also be limited. But I think we should strive to acquire such an understanding and I hope that some day somebody will have a very bright idea, similar to those of Maxwell, of Planck, or Heisenberg, and get us closer to such an explanation or at least to the description of the regularities of events involving life.

With this I should stop and let the commentator tell us where I am wrong.

PART IV

Symmetries

Invariance in Physical Theory

E. P. Wigner

Symmetries and Reflections.
Indiana University Press, Bloomington, Indiana, 1967, pp. 3–13

Initial Conditions, Laws of Nature, Invariance

The world is very complicated and it is clearly impossible for the human mind to understand it completely. Man has therefore devised an artifice which permits the complicated nature of the world to be blamed on something which is called accidental and thus permits him to abstract a domain in which simple laws can be found. The complications are called initial conditions; the domain of regularities, laws of nature. Unnatural as such a division of the world's structure may appear from a very detached point of view, and probable though it is that the possibility of such a division has its own limits,[1] the underlying abstraction is probably one of the most fruitful ones the human mind has made. It has made the natural sciences possible.

The possibility of abstracting laws of motion from the chaotic set of events that surround us is based on two circumstances. First, in many cases a set of initial conditions can be isolated which is not too large a

Address presented at the celebration honoring Professor Albert Einstein on March 19, 1949, in Princeton. Reprinted by permission from the *Proceedings of the American Philosophical Society*, Vol. 93, No. 7 (December, 1949).
[1] The artificial nature of the division of information into "initial conditions" and "laws of nature" is perhaps most evident in the realm of cosmology. Equations of motion which purport to be able to predict the future of a universe from an arbitrary present state clearly cannot have an empirical basis. It is, in fact, impossible to adduce reasons against the assumption that the laws of nature would be different even in small domains if the universe had a radically different structure. One cannot help agreeing to a certain degree with E. A. Milne, who reminds us (*Kinematic Relativity*, Oxford Univ. Press, 1948, page 4) that, according to Mach, the laws of nature are a consequence of the contents of the universe. The remarkable fact is that this point of view could be so successfully disregarded and that the distinction between initial conditions and laws of nature has proved so fruitful.

4   *Symmetries and Reflections*

set and, in spite of this, contains all the relevant conditions for the events on which one focuses one's attention. In the classic example of the falling body, one can disregard almost everything except the initial position and velocity of the falling body; its behavior will be the same and independent of the degree of illumination, the neighborhood of other objects, their temperature, etc. The isolation of the set of conditions which do influence the experiment is by no means a trivial problem. On the contrary, it is a large fraction of the art of the experimenter and on our occasional trips through laboratories all of us theoreticians have been periodically impressed by the difficulties of this art.

However, the possibility of isolating the relevant initial conditions would not in itself make possible the discovery of laws of nature. It is, rather, also essential that, given the same essential initial conditions, the result will be the same no matter where and when we realize these. This principle can be formulated, in the language of initial conditions, as the statement that the absolute position and the absolute time are never essential initial conditions. The statement that absolute time and position are never essential initial conditions is the first and perhaps the most important theorem of invariance in physics. If it were not for it, it might have been impossible for us to discover laws of nature.

The above invariance is called in modern mathematical parlance invariance with respect to displacement in time and space. Again, it may be well to remember that this invariance may have limitations. If the universe should turn out to be grossly inhomogeneous, the laws of nature on the fringes of the universe may be quite different from those which we are studying; and it is not impossible that an experimenter inside a closed room is in principle able to ascertain whether he is in the midst, or near the fringes, of the universe, whether he lives in an early epoch of the expansion of the universe, or at an advanced stage of this process. The postulate of the invariance with respect to displacement in space and time disregards this possibility, and its application on the cosmological scale virtually presupposes a homogeneous and stationary universe. Present evidence clearly points to the approximate nature of the latter assumption.

Invariance

What are the other laws of invariance? One can distinguish between two types of laws of invariance: the older ones which found their per-

Invariance in Physical Theory   5

fect, and perhaps final, formulation in the special theory of relativity, and the new one, yet incompletely understood, which the general theory of relativity brought us.

The older theories of invariance postulate, in addition to the irrelevance of the absolute position and time of an event, the irrelevance of its orientation and finally, the irrelevance of its state of motion, as long as this remains uniform, free of rotation, and on a straight line. The former theorems are geometrical in nature and appear to be so self-evident that they were not formulated clearly and directly until about the turn of the last century. The last one, the irrelevance of the state of motion, is far from self-evident, as all of us know who have tried to explain it to a layman. There would be no such principle of invariance if Newton's second law of motion read "All bodies persist in their state of rest unless acted upon by an external force"; on the contrary, the scope of this invariance could be extended considerably if the bodies maintained their state of acceleration rather than their velocity in the absence of an external force. It is fitting that this principle was first enunciated, in full clarity, by Newton in his *Principia*.

The fact that the older principles of invariance are the products of experience rather than *a priori* truths can also be illustrated by our gradual abandonment of a very plausible principle, the principle of similitude. This principle, formulated perhaps most clearly by Fourier, demands that physical experiments can be scaled; that the absolute magnitude of objects be irrelevant from the point of view of their behavior on the proper scale. The existence of atoms, of an elementary charge, and of a limiting velocity spelled the doom of this principle.

The formulae describing what I am calling the older principles of invariance were first given completely by Poincaré, who derived them from the equations of electrodynamics. He also recognized the group property of the older principles of invariance and named the underlying group after Lorentz. The significance and general validity of these principles were recognized, however, only by Einstein. His papers on special relativity also mark the reversal of a trend: until then, the principles of invariance were derived from the laws of motion. Einstein's work established the older principles of invariance so firmly that we have to be reminded that they are based only on experience. It is now natural for us to try to derive the laws of nature and to test their validity by means of the laws of invariance, rather than to derive the laws of invariance from what we believe to be the laws of nature.

6   *Symmetries and Reflections*

The general theory of relativity is the next milestone in the history of invariance. The fact that it is the first attempt to derive a law of nature by selecting the simplest invariant equation would in itself justify the epithet. More important, in my opinion, is that the general theory of relativity attempts to give the range of the validity of the older theorems of invariance and to replace them with a single, more general theorem. The limitation of the older theorems of invariance is given by the structure of space which manifests itself in a variable curvature. Since the curvature is, in principle, observable, a displacement from a region of low curvature to one with a high curvature does not leave the laws of nature invariant. It is true that the old fashioned physicist can always blame the differences in the laws of nature, as they are valid for different points of the universe, on the absence or proximity of masses. This, however, restores the general validity of the older invariances only by making them meaningless. Clearly, if two points in space-time are equivalent only if they are surrounded by the same distribution of masses, their equivalence will be the exception rather than the rule.

The new principle of invariance which the general theory of relativity substitutes for the older ones is that all actions are transmitted by fields which transmit the perturbations from point to point. Expressed more phenomenologically: the events in one part of space depend only on the fields, i.e., on the measurable quantities, in the neighborhood of that part of space—the effect of events outside moves in only with a finite velocity.* This postulate of invariance is much bolder, and has much less artificiality than the older postulate of invariance with respect to the inhomogeneous Lorentz group. The above formulation is a little more phenomenological than the customary one. The customary requirement of invariance with respect to all differentiable coordinate transformations is, however, included in it. Both postulates express the fact that the laws of physics and of geometry involve only local measurements such as can be expressed by differential equations. In particular, the definition of a preferred Galilean coordinate system, by reference to other, distant Galilean coordinate systems, is barred by the postulate that all the information which is necessary to describe the immediate

---

* It will be noted that the principle postulated is not invariance with respect to general coordinate transformations but the less abstract principle of the absence of action at a distance. This principle, as here formulated, shares most properties of invariance postulates. (Note added with the proofs of this book.)

future of the region in question can be obtained by local measurements. Hence information relating to distant points cannot add anything relevant to the knowledge of local conditions, as would be the case if they would enable one to define preferred coordinate systems.

Invariance in Quantum Mechanics

When the great paradoxes of atomic physics first became apparent about thirty years ago, it was easy to despair to such a degree of our ability to understand the laws of physics as to propose throwing into the winds all laws of physics, excepting the conservation laws for energy and momenta. It was, in fact, Einstein who recommended such a procedure.[2]

The efforts of the past thirty years culminated in having accomplished just that: we now believe that we have a consistent theory of atomic processes, consistent with the older concepts of space and time, and of invariance. This theory is based on an analysis of the measuring process, carried out principally by Heisenberg and Bohr, which emphasizes the effect of the measurement on the measured object. It is thus contradictory to the simple concept of mapping out the field, the concept which underlies the customary formulation of general relativity. In particular, the measurement of the curvature of space caused by a single particle could hardly be carried out without creating new fields which are many billion times greater than the field under investigation.[3]

Very little effort has been made so far to modify the concepts of the general theory of relativity with an appreciation of the effect of the act of measurement on the object of the measurement. However, the older principles of invariance are in harmony with quantum mechanics and this harmony is more complete, the interdependence of quantum equations and the theory of their invariance is more intimate, than it was in pre-quantum theory.

Let me first stress the points of similarity between the role of invariance in classical and quantum theories. The principles of invariance have a dual function in both theories. On the one hand, they give a

[2] H. Poincaré, "Dynamics of Electrons," *Compt. Rend.*, 140, 1504 (1905); "Sur la dynamique de l'électron," *Circolo Mat. Palermo Rend.*, 21, 129 (1906).

[3] An interesting problem in this connection was broached recently by M. F. M. Osborne, "Quantum Theory Restrictions on the General Theory of Relativity," *Bull. Am. Phys. Soc.*, 24, 2 (Berkeley Meeting), Paper A-3 (1949).

8   *Symmetries and Reflections*

necessary condition which all fundamental equations must satisfy: the irrelevant initial conditions must not enter in a relevant fashion into the results of the theory. Second, once the fundamental equations are given, the principles of invariance furnish, in the form of conservation laws and otherwise, powerful assistance toward their solution. The conservation laws for linear momentum and energy, for angular momentum and the motion of the center of mass, can be derived both in classical theory and in quantum mechanics from the invariance of the equations with respect to infinitesimal displacements and rotations in space-time.[4]

However, with these points of analogy, the similarity between the roles of invariance in classical and in quantum physics is pretty much at an end. The reason is, fundamentally, that the variety of states is much greater in quantum theory than in classical physics and that there is, on the other hand, the principle of superposition to provide a structure for the greatly increased manifold of quantum mechanical states. The principle of superposition renders possible the definition of states the transformation properties of which are particularly simple. It can in fact be shown that every state of any quantum mechanical system, no matter what type of interactions are present, can be considered as a superposition of states of elementary systems. The elementary systems correspond mathematically to irreducible representations of the Lorentz group and as such can be enumerated. Since the equations of motion of the states of elementary systems are completely determined by their invariance properties, every state is a linear combination of states the history of which is completely known. However, in the description by irreducible states, the form of almost all physically important operators remains unknown and, in fact, depends on the system, the types of interactions, etc. This leads to a rather strange dilemma: in the customary description the form of the physically important operators is known but the time dependence of the states is unpredictable or difficult to calculate. In the description just mentioned, the situation is opposite: the time dependence of the states follows from the invariance properties, but the form of the physically

---

[4] In classical theory, this observation is due to F. Klein's school. Cf. also F. Engel, "Uber die zehn allgemeinen Integrale der klassischen Mechanik," *Nachr. Kgl. Ges. Wiss. Göttingen,* p. 270 (1916); also G. Hamel, "Die Lagrange-Eulerschen Gleichungen der Mechanik," *Z. Math. Phys.,* 50, 1 (1904), and E. Bessel-Hagen, "Uber die Erhaltungssätze der Elektrodynamik," *Math. Ann.,* 84, 258 (1921).

Invariance in Physical Theory   9

important operators is hard to establish. There is one exception to this; the states of elementary particles are formed by the superposition of the states of a single invariant set. As a result, the possible equations of elementary particles can easily be enumerated and some progress has been made recently also toward the invariant theoretic determination of the operators for the most important physical quantities. The property which makes a particle elementary in the sense of the above statement is that it shall have no internal coordinate which would permit an invariant division of its states into two or more groups. It is certainly no accident that all elementary particles, including the light quantum, obey irreducible equations and hence form elementary systems in the above sense. Since the rigid body is what may be considered classical mechanics' closest analogue to an elementary particle, the group theoretical description of the motion of a rigid body must be considered the closest analogue to the above result.

The second point to which I wish to draw attention in the comparison of quantum and pre-quantum theories concerns the significance of transformations of invariance, such as reflections, which cannot be generated by infinitesimal elements. These had very little role in the classical theory but prove their value both in the discussion of fundamental equations, and also in the attempts to solve these. Into the former category belongs for instance the observation that the theory[5] which identifies the neutrino with the antineutrino, by attributing to the inversion of space coordinates a non-linear operation involving transition to the conjugate complex wave function, cannot be welded into a theory which describes also particles of the conventional type.[6] The applications of the reflection invariance for facilitating the solution of the fundamental equations are even more obvious. They lead for instance to the concept of Laporte's parity quantum number—one of the most important concepts of spectroscopy.

Less specifically, but perhaps not less accurately, one can speak of the general impression of quantum mechanics, and the theory of the invariance of its equations, forming an inseparable entity, almost to the degree to which this is true in the general theory of relativity. Schwinger's quantum electrodynamics gives the latest and starkest

[5] Cf. H. Weyl, "Elektron und Gravitation I," Z. *Physik*, 56, 330 (1929).

[6] Another very interesting set of examples has been given recently by T. Okayama, "On the Mesic Charge," *Phys. Rev.*, 75, 308 (1949).

manifestation of this situation: his theory cannot be formulated at all without developing, unified with it, its theory of invariance. Furthermore, one is inclined to believe that this union is the most important success of the theory; that even the explanation of definite and previously unexpected experimental phenomena is less important to us than the knowledge that we can, in general, carry out our calculations of physical phenomena in an invariant fashion, obtaining the same results if we start with only irrelevantly different initial conditions.

Conservation of Electrical Charge

My account of the role of invariance in quantum mechanics would remain grossly incomplete if I did not mention a dissonant sound in the harmony of quantum mechanics and the older theorems of invariance. This is the conservation law for the electrical charge. While the conservation laws for all other quantities, such as energy or angular momentum, follow in a natural way from the principles of invariance, the conservation law for electric charge so far has defied all attempts to place it on an equally general basis. The situation was, of course, the same in classical mechanics but the simplicity of the connection between invariance and the ordinary conservation laws makes the situation even more conspicuous in quantum mechanics.

A short description of the derivation of the usual conservation laws will make this perhaps more evident than an abstract discussion. In order to derive the conservation law for linear momentum, one first constructs a state in which one component, say the x component of the linear momentum, has a definite value p. For this purpose, one chooses an arbitrary state φ_0 of the system for which one wishes to show the conservation theorem and constructs all states φ_a obtained by displacing the system in the state φ_0 by a in the x direction. One then considers the superposition of the states φ_a with the coefficients e^{-ipa}:

$$\Phi_p = \int_{-\infty}^{\infty} \varphi_a e^{-ipa} da.$$

This state has the property that a further displacement by b,

$$\int_{-\infty}^{\infty} \varphi_{a+b} e^{-ipa} da = \int_{-\infty}^{\infty} \varphi_c e^{-ip(c-b)} dc = e^{ipb}\Phi_p,$$

just multiplies it with e^{ipb}. It is called a pure state with momentum component p in the x direction. The property of Φ_p, of being multiplied

Invariance in Physical Theory   *11*

by e^{ipb} upon displacement by b, will not be lost in time: if φ_0 goes over, after some time, into ψ_0, the state φ_a will go over into the ψ_a which results from ψ_0 by displacement by a. This follows from the invariance of the equations of motion with respect to displacements. As a result of this and the linearity of the equations of motion, Φ_p will go over at the time in question into

$$\Psi_p = \int_{-\infty}^{\infty} \psi_a e^{-ipa} da,$$

which also is multiplied by e^{ipb} upon displacement by b. This property, which characterizes the state with momentum p, is not lost in the course of time, and this constitutes the principle of conservation of linear momentum.

Similar considerations involving the other principles of invariance lead to the other conservation laws. Furthermore, the quantization and the possible values of the quantized quantities also emerge naturally from the above consideration. Thus the quantization of the angular momentum is the result of the condition that rotation by 2π always restores the system to its original state.

No consideration similar in generality and simplicity to the above one is known which would explain the conservation law for electric charges. One can borrow the following argument from classical theory[7]: Suppose we could create charges by some process in a closed system. Let us put then this closed system into a Faraday cage, charge the cage, and create the charge in the closed system. A certain energy E will be necessary for this process. However, inasmuch as no physical phenomenon depends on the absolute value of the potential, the amount of energy E cannot depend on the potential of the Faraday cage inside of which the charge is created. Let us then take our closed system out of the Faraday cage and move it away from it, thereby obtaining a certain amount of work W. Let us then reverse the process which led to the creation of the charge and gain an amount E of energy which is equal to the amount of energy expended in the first place, since the process in a closed system must not depend on the absolute value of the electric potential at which that system is. We now can replace the discharged system into the Faraday cage without the expenditure of any work and have carried out a cycle which resulted in a net gain W of work. This is impossible according to the first law and shows that one of our assumptions must

---

[7] This point was emphasized by J. R. Oppenheimer during the discussion which followed the presentation of this paper.

have been faulty. It is the assumption that electric charges can be created in a closed system.

The above argument shows the connection between the conservation law for electric charges and the assumption of the irrelevance of the absolute magnitude of the electric potential. It has been translated into quantum mechanics and has been given a much more elegant and general form.[8] Nevertheless, it remains less convincing than the consideration leading to the other conservation laws and, certainly, in it fails to account for the quantization of the electric charge.

The lack of full clarity concerning the foundation of the conservation law for charges raises several important points. Is our present scheme of quantum mechanics incomplete in some fundamental respect? In particular, is the Hilbert space with complex coordinates the proper framework for describing state vectors? Would the use of more general hypercomplex wave functions give essentially different results? But the most important question is, undoubtedly: is the existence of a conservation law a particular feature of the electromagnetic type of interaction or are we going to encounter, or perhaps have we already encountered,[9] similar conservation laws for other types of interactions?

[8] F. London, "Quantenmechanische Deutung der Theorie von Weyl," Z. *Physik*, 42, 375 (1927); Cf. also H. Weyl, *loc. cit.*

[9] It is conceivable, for instance, that a conservation law for the number of heavy particles (protons and neutrons) is responsible for the stability of the protons in the same way as the conservation law for charges is responsible for the stability of the electron. Without the conservation law in question, the proton could disintegrate, under emission of a light quantum, into a positron, just as the electron could disintegrate, were it not for the conservation law for the electric charge, into a light quantum and a neutrino. The Gedanken experiment which led to the conservation law for charges would assume the following, admittedly somewhat vague form, if one wanted to prove a conservation law for the number of nucleons: Assuming that there is no such conservation law, two nucleons could first be created at a distance from each other which is large compared with the range of nuclear forces. An amount E of energy would be needed for this. The nucleons then could be permitted to approach each other, furnishing the amount W of work. Finally, they would be permitted to annihilate, which would again release the energy E first expended. A net gain W in energy would result. The impossibility to perform the above operations may, of course, be connected with many physical phenomena, such as the impossibility of localizing sufficiently accurately the systems in which the nucleons are to be created (i.e., the existence of a fundamental length). The impossibility may also be the consequence of the dependence of the energy E, which is necessary to create the nucleons, on the absolute value of the nuclear potential. The point of view which we wish to represent is, however, that the impossibility of the creation of nucleons (without creating antinucleons) is the real resolution of the paradox. It may be mentioned, as a third point of similarity between the two conservation laws, that there is evidence, although contested by some recent experiments, that the "mesonic

Invariance in Physical Theory   13

Relativity theory, to the celebration of which the present paper is intended to contribute, has enriched physics in two ways. It has resolved acute difficulties, presented by the Michelson-Morley, the Fizeau, the Trouton-Noble, and other experiments. It has done this by a profound analysis of the space-time concept, and its results in this connection are part of the store of knowledge of all physicists. Even more lasting and more subtle is probably the contribution which relativity theory has made indirectly. Most important among the indirect contributions of the theory of relativity was its demonstration for the need and of the fruitfulness of the analysis of apparently well established concepts, concepts which have formed a habit of thought for many generations. Its fostering the emergence of the importance of the concept of invariance, its enlarging the scope of this concept, can, I believe, justly claim second position.

It is a pleasure to acknowledge Dr. V. Bargmann's critical comments and remarks on the present paper.

charge" of all nucleons is the same. If this should prove to be true, it would be evidence for the quantization of the mesonic charge. This quantization would be analogous to the well known quantization of the electric charge.

On the Law of Conservation of Heavy Particles

E. P. Wigner

Proceedings of the National Academy of Sciences of the USA, vol. 38.
Washington, D.C. 1952, pp. 449–451

Communicated March 2, 1952

The purpose of this note is to trace in more detail the consequences of treating the conservation law for heavy particles[1] on a par with the conservation law for electric charges. It is thus an attempt to guess at the properties of particles as yet unknown and their interactions, and therefore speculative. However, it does not attempt a classification of elementary particles similar to that proposed recently by several writers, in particular, A. Pais.[2] In this regard it is somewhat more conservative.

We do not know the deeper cause of the conservation law of charges in the same sense as we know, for instance, the cause for the conservation of angular momentum.[3] Perhaps the clearest sign hereof is that the

---

* The contents of this note have been presented November 10, 1951, to the New Jersey Science Teachers Association.

[1] It is difficult to trace the first statement of this principle. It is clearly contained in the writer's article in *Proc. Am. Philos. Soc.*, **93**, 521 (1949), but may have been recognized about that time also by others, cf. T. Okayama, *Phys. Rev.*, **75**, 308 (1949). C. N. Yang informs me that the purpose of introducing an imaginary character to the reflection properties of certain fermions in the paper of J. Tiomno and C. N. Yang (*Phys. Rev.*, **79**, 495 (1950)) was to explain this principle. Cf. also L. I. Schiff, *Phys. Rev.*, **85**, 374 (1952) and, in particular, P. Jordan, *Z. f. Naturf.*, **7a**, 78 (1952).

[2] Pais, A., "On the *V*-Particle." To appear shortly. Also literature quoted there.

[3] Cf., e.g., the discussion in the article quoted in reference 2.

quantization of the angular momentum follows from the same consideration which leads to its constancy in time while no known derivation of the
charge conservation law explains the existence of an elementary charge.
The deeper cause for the conservation law for heavy particles is equally
unknown and no attempt will be made here to find it. It will be assumed
instead that the two conservation laws have similar causes and that these
have similar consequences.

The fundamental reason for believing in the conservation law for heavy
particles stems from the fact that no phenomenon has yet been observed
in which a heavy particle has disappeared without the formation of another
heavy particle. In particular, the proton appears to be entirely stable
even though its disintegration into other particles would not violate any
of the other conservation laws.

There are two ways to determine the electric charge of a particle. One
of these is given by the charge conservation law itself, according to which
the sum of the charges of the products of any reaction must be the same
as the sum of the charges of the particles entering the reaction. So far
this has been the only way to ascertain the "heavy" character of any
particle. Since the neutron transforms by β decay into a proton it must
be a heavy particle and the same is true of the V_0. If we wish to explain
the stability of the proton by the heavy particle conservation law, it must
be the lightest heavy particle. However, it does not follow that all particles which are heavier than the proton are heavy in the sense here used.
It does follow, however, that all heavy particles, in particular also the
V_0, eventually transform into a proton and that they cannot disintegrate
without a heavy particle, and thus eventually a proton, resulting from
the disintegration. Particles which are heavy in the sense here used may
be said to have unit neutronic charge. The neutronic charge should play
the same role for the conservation law of heavy particles as the electric
charge plays in the conservation law of electric charges. It also has opposite sign for particle and antiparticle.

The second way to obtain the electric charge of a particle is to bring it
into interaction with the electro-magnetic field. It is under these conditions that the quantization of the charge manifests itself. The interaction
with the mesonic field seems to be the analog of this for heavy particles.
If this analogy is to hold, the mesonic interaction of all heavy particles
must be the same, apart from the sign. As far as the proton and neutron
are concerned this is known to be correct approximately and leads in
nuclear spectroscopy to the concept of the so-called T multiplet.[4] How-

[4] Cf., e.g., E. P. Wigner and E. Feenberg, *Reports on Progress in Physics*, Vol. VIII,
1942, p. 274, or W. Heitler, *Proc. Roy. Irish Acad.*, **51A**, 33 (1946). Recently K. A.
Brueckner gave a very interesting illustration of the principle given in the last reference (*Bull. Am. Phys. Soc.*, **27**, 1, paper Y9 (1952)).

ever, if the analogy which we pursue is correct, the equality should be as accurate as that between the electric charges of the proton and the electron.[5] Furthermore, the mesonic charge of the V_0 particles should also be equal to that of the proton. From this it follows that the potential between the V_0 and the proton should be the same as between the neutron and the proton, at least as far as both potentials are due to the interaction with π mesons. Under this assumption, the scattering cross-section for the collision of V_0 particles with protons or neutrons is very nearly equal to the proton-neutron scattering cross-section at the same energy. If it should be possible to measure the former cross-sections it would give a clue concerning the validity of the analogy between heavy particle conservation and charge conservation laws.

[5] Schwinger, J., *Phys. Rev.*, **78**, 135 (1950) demonstrated possible causes which give apparent, but only apparent, deviations from this equality.

Symmetry and Conservation Laws

E. P. Wigner

Symmetries and Reflections.
Indiana University Press, Bloomington, Indiana, 1967, pp. 14–27

Introduction

Symmetry and invariance considerations, and even conservation laws, undoubtedly played an important role in the thinking of the early physicists, such as Galileo and Newton, and probably even before them. However, these considerations were not thought to be particularly important and were articulated only rarely. Newton's equations were not formulated in any special coordinate system and thus left all directions and all points in space equivalent. They were invariant under rotations and displacements, as we now say. The same applies to his gravitational law. There was little point in emphasizing this fact, and in conjuring up the possibility of laws of nature which show a lower symmetry. As to the conservation laws, the energy law was useful and was instinctively recognized in mechanics even before Galileo.[1] The momentum and angular momentum conservation theorems in their full generality were not very useful even though in the special case of central motion they give, of course, one of Kepler's laws. Most books on mechanics, written around the turn of the century and even later, do not mention the general theorem of the conservation of angular momentum.[2] It must have

Reprinted by permission from the *Proceedings of the National Academy of Sciences*, Vol. 51, No. 5 (May, 1964).

[1] G. Hamel, in his *Theoretische Mechanik* (Stuttgart: B. G. Teubner, 1912) mentions (p. 130) Jordanus de Nemore (~1300) as having recognized essential features of what we now call mechanical energy and Leonardo da Vinci as having postulated the impossibility of the Perpetuum Mobile.

[2] F. Cajori's *History of Physics* (New York: Macmillan Company, 1929) gives exactly half a line to it (p. 108).

Symmetry and Conservation Laws   15

been known quite generally because those dealing with the three-body problem, where it is useful, write it down as a matter of course. However, people did not pay very much attention to it.

This situation changed radically, as far as the invariance of the equations is concerned, principally as a result of Einstein's theories. Einstein articulated the postulates about the symmetry of space, that is, the equivalence of directions and of different points of space, eloquently.[3] He also re-established, in a modified form, the equivalence of coordinate systems in motion and at rest. As far as the conservation laws are concerned, their significance became evident when, as a result of the interest in Bohr's atomic model, the angular momentum conservation theorem became all-important. Having lived in those days, I know that there was universal confidence in that law as well as in the other conservation laws. There was much reason for this confidence because Hamel, as early as 1904, established the connection between the conservation laws and the fundamental symmetries of space and time.[4] Although his pioneering work remained practically unknown, at least among physicists, the confidence in the conservation laws was as strong as if it had been known as a matter of course to all. This is yet another example of the greater strength of the physicist's intuition than of his knowledge.

Since the turn of the century, our attitude toward symmetries and conservation laws has turned nearly full circle. Few articles are written nowadays on basic questions of physics which do not refer to invariance postulates, and the connection between conservation laws and invariance principles has been accepted, perhaps too generally.[5] In addition, the concept of symmetry and invariance has been extended into a new area—an area where its roots are much less close to direct experience and observation than in the classical area of space-time symmetry. It may be useful, therefore, to discuss first the relations of phenomena, laws of nature, and invariance principles to each other. This relation is not quite the same for the classical invariance principles, which will be called geometrical, and the new ones, which will be called dynamical.

[3] See, for instance, his semipopular booklet *Relativitätstheorie* (Braunschweig: Friedr. Vieweg und Sohn, various editions, 1916-1956).

[4] G. Hamel, *Z. Math. Phys.*, 50, 1 (1904); F. Engel, *Ges. d. Wiss. Göttingen*, 270 (1916).

[5] See the present writer's article, *Progr. Theoret. Phys.*, 11, 437 (1954); also Y. Murai, *Progr. Theoret. Phys.*, 11, 441 (1954); and more recently D. M. Greenberg, *Ann. Phys.* (N.Y.), 25, 290 (1963).

16   *Symmetries and Reflections*

Finally, I would like to review, from a more elementary point of view than customary, the relation between conservation laws and invariance principles.

Events, Laws of Nature, Invariance Principles

The problem of the relation of these concepts is not new; it has occupied people for a long time, first almost subconsciously. It may be of interest to review it in the light of our greater experience and, we hope, more mature understanding.

From a very abstract point of view, there is a great similarity between the relation of the laws of nature to the events on one hand, and the relation of symmetry principles to the laws of nature on the other. Let me begin with the former relation, that of the laws of nature to the events.

If we knew what the position of a planet will be at any given time, there would remain nothing for the laws of physics to tell us about the motion of that planet. This is true also more generally: if we had a complete knowledge of all events in the world, everywhere and at all times, there would be no use for the laws of physics, or, in fact, of any other science. I am making the rather obvious statement that the laws of the natural sciences are useful because without them we would know even less about the world. If we already knew the position of the planet at all times, the mathematical relations between these positions which the planetary laws furnish would not be useful but might still be interesting. They might give us a certain pleasure and perhaps amazement to contemplate, even if they would not furnish us new information. Perhaps also, if someone came who had some different information about the positions of that planet, we would more effectively contradict him if his statements about the positions did not conform with the planetary laws— assuming that we have confidence in the laws of nature which are embodied in the planetary law.

Let us turn now to the relation of symmetry or invariance principles to the laws of nature. If we know a law of nature, such as the equations of electrodynamics, the knowledge of the subtle properties of these equations does not add anything to the content of these equations. It may be interesting to note that the correlations between events which the equations predict are the same no matter whether the events are viewed by an observer at rest, or an observer in uniform motion. How-

ever, all the correlations between events are already given by the equations themselves, and the aforementioned observation of the invariance
of the equations does not augment the number or change the character
of the correlations.

More generally, if we knew all the laws of nature, or the ultimate law
of nature, the invariance properties of these laws would not furnish us
new information. They might give us a certain pleasure and perhaps
amazement to contemplate, even though they would not furnish new
information. Perhaps also, if someone came around to propose a different law of nature, we could more effectively contradict him if his
law of nature did not conform with our invariance principle—assuming
that we have confidence in the invariance principle.

Evidently, the preceding discussion of the relation of the laws of
nature to the events, and of the symmetry or invariance principles to
the laws of nature is a very sketchy one. Many, many pages could be
written about both. As far as I can see, the new aspects which would
be dealt with in these pages would not destroy the similarity of the
two relations—that is, the similarity between the relation of the laws of
nature to the events, and the relation of the invariance principles to
the laws of nature. They would, rather, support it and confirm the function of the invariance principles to provide a structure or coherence
to the laws of nature just as the laws of nature provide a structure and
coherence to the set of events.

Geometrical and Dynamical Principles of Invariance

What is the difference between the old and well-established geometrical principles of invariance, and the novel, dynamical ones? The geometrical principles of invariance, though they give a structure to the
laws of nature, are formulated in terms of the events themselves. Thus,
the time-displacement invariance, properly formulated, is: the correlations between events depend only on the time intervals between the
events, not on the time at which the first event takes place. If P_1, P_2, P_3
are positions which the aforementioned planet can assume at times
t_1, t_2, t_3, it could assume these positions also at times $t_1 + t$, $t_2 + t$, $t_3 + t$,
where t is quite arbitrary. On the other hand, the new, dynamical prin-
ciples of invariance are formulated in terms of the laws of nature. They
apply to specific types of interaction, rather than to any correlation between events. Thus, we say that the electromagnetic interaction is

gauge invariant, referring to a specific law of nature which regulates the generation of the electromagnetic field by charges, and the influence of the electromagnetic field on the motion of the charges.

It follows that the dynamical types of invariance are based on the existence of specific types of interactions. We all remember having read that, a long time ago, it was hoped that all interactions could be derived from mechanical interactions. Some of us still remember that, early in this century, the electromagnetic interactions were considered to be the source of all others. It was necessary, then, to explain away the gravitational interaction, and in fact this could be done quite successfully. We now recognize four or five distinct types of interactions: the gravitational, the electromagnetic, one or two types of strong (that is, nuclear) interactions, and the weak interaction responsible for beta decay, the decay of the μ meson, and some similar phenomena. Thus, we have given up, at least temporarily, the hope of one single basic interaction. Furthermore, every interaction has a dynamical invariance group, such as the gauge group for the electromagnetic interaction.

This is, however, the extent of our knowledge. Otherwise, let us not forget, the problem of interactions is still a mystery. Utiyama[6] has stimulated a fruitful line of thinking about how the interaction itself may be guessed once its group is known. However, we have no way of telling the group ahead of time; we have no way of telling how many groups and hence how many interactions there are. The groups seem to be quite disjointed, and there seems to be no connection between the various groups which characterize the various interactions or between these groups and the geometrical symmetry group, which is a single, well-defined group with which we have been familiar for many, many years.

Geometrical Principles of Invariance and Conservation Laws

Since it is good to stay on *terra cognita* as long as possible, let us first review the geometrical principles of invariance. These were recognized by Poincaré first, and I like to call the group formed by these invariables the Poincaré group.[7] The true meaning and importance of these prin-

[6] R. Utiyama, *Phys. Rev.*, 101, 1597 (1956); also C. N. Yang and R. L. Mills, *Phys. Rev.*, 96, 191 (1954).

[7] H. Poincaré, *Compt. Rend.*, 140, 1504 (1905); *Rend. Circ. Mat. Palermo*, 21, 129 (1906).

ciples were brought out only by Einstein, in his special theory of relativity. The group contains, first, displacements in space and time. This
means that the correlations between events are the same everywhere
and at all times, that the laws of nature—the compendium of the correlations—are the same no matter when and where they are established. If
this were not so, it might have been impossible for the human mind to
find laws of nature.

It is good to emphasize at this point the fact that the laws of nature,
that is, the correlations between events, are the entities to which the
symmetry laws apply, not the events themselves. Naturally, the events
vary from place to place. However, if one observes the positions of a
thrown rock at three different times, one will find a relation between
those positions, and this relation will be the same at all points of the
Earth.

The second symmetry is not at all as obvious as the first one: it postulates the equivalence of all directions. This principle could be recognized only when the influence of the Earth's attraction was understood
to be responsible for the difference between up and down. In other
words, contrary to what was just said, the events between which the
laws of nature establish correlations are not the three positions of
the thrown rock, but the three positions of the rock with respect to the
Earth.

The last symmetry—the independence of the laws of nature from the
state of motion in which it is observed as long as this is uniform—is not
at all obvious to the unpreoccupied mind.[8] One of its consequences is
that the laws of nature determine not the velocity but the acceleration
of a body: the velocity is different in coordinate systems moving with
different speeds; the acceleration is the same as long as the motion of
the coordinate systems is uniform with respect to each other. Hence,
the principle of the equivalence of uniformly moving coordinate systems, and their equivalence with coordinate systems at rest, could not
be established before Newton's second law was understood; it was at
once recognized then, by Newton himself. It fell temporarily into dis-

---

[8] Thus, Aristotle's physics postulated that motion necessarily required the continued operation of a cause. Hence, all bodies would come to an absolute rest if
they were removed from the cause which imparts them a velocity. [Cf., e.g., A. C.
Crombie's *Augustine to Galileo* (London: Falcon Press, 1952), p. 82 or 244.] This
cannot be true for coordinate systems moving with respect to each other. The coordinate systems with respect to which it is true then have a preferred state of
motion.

20   *Symmetries and Reflections*

repute as a result of certain electromagnetic phenomena until Einstein re-established it in a somewhat modified form.

It was mentioned already that the conservation laws for energy and for linear and angular momentum are direct consequences of the symmetries just enumerated. This is most evident in quantum-mechanical theory, where they follow directly from the kinematics of the theory, without making use of any dynamical law, such as the Schrödinger equation. This will be demonstrated at once. The situation is much more complex in classical theory, and, in fact, the simplest proof of the conservation laws in classical theory is based on the remark that classical theory is a limiting case of quantum theory. Hence, any equation valid in quantum theory, for any value of Planck's constant h, is valid also in the limit $h = 0$. Traces of this reasoning can be recognized also in the general considerations showing the connection between conservation laws and space-time symmetry in classical theory. The conservation laws can be derived also by elementary means, using the dynamical equation, that is, Newton's second law, and the assumption that the forces can be derived from a potential which depends only on the distances between the particles. Since the notion of a potential is not a very natural one, this is not the usual procedure. Mach, for instance, assumes that the force on any particle is a sum of forces, each due to another particle.[9] Such an assumption is implicit also in Newton's third law, otherwise the notion of counterforce would have no meaning. In addition, Mach assumes that the force depends only on the positions of the interacting pair, not on their velocities. Some such assumption is indeed necessary in classical theory.[10] Under the assumptions just mentioned, the conservation law for linear momentum follows at once from Newton's third law, and, conversely, this third law is also necessary for the conservation of linear momentum. All this was recognized already by Newton. For the conservation law of angular momentum, which was, in its general form, discovered almost 60 years after the *Principia* by Euler, Bernouilli, and d'Arcy, the significance of the isotropy of space is evident. If the direction of the force between a pair of particles were not directed along the line from one particle to the other, it would not be invariant under rotations about that line. Hence, under the assump-

[9] E. Mach, *The Science of Mechanics* (Chicago: Open Court Publ. Co., various editions), Chap. 3, Sec. 3.
[10] See footnote 5.

tions made, only central forces are possible. Since the torque of such forces vanishes if they are oppositely equal, the angular momentum law follows. It would not follow if the forces depended on the positions of three particles or more.

In quantum mechanics, as was mentioned before, the conservation laws follow already from the basic kinematical concepts. The point is simply that the states in quantum mechanics are vectors in an abstract space, and the physical quantities, such as position, momentum, etc., are operators on these vectors. It then follows, for instance, from the rotational invariance that, given any state ϕ, there is another state ϕ_α which looks just like ϕ in the coordinate system that is obtained by a rotation α about the Z axis. Let us denote the operator which changes ϕ into ϕ_α by Z_α. Let us further denote the state into which ϕ goes over in the time interval τ by $H_\tau\phi$ (for a schematic picture, cf. Fig. 1). Then,

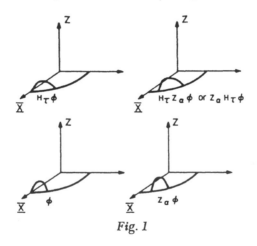

Fig. 1

because of the rotational invariance, ϕ_α will go over, in the same time interval, into the state $H_\tau\phi_\alpha$, which looks, in the second coordinate system, just like $H_\tau\phi$. Hence, it can be obtained from $H_\tau\phi$ by the operation Z_α. It follows that

$$H_\tau Z_\alpha\phi = Z_\alpha H_\tau\phi, \tag{1}$$

and since this is valid for any ϕ,

$$H_\tau Z_\alpha = Z_\alpha H_\tau. \tag{2}$$

Thus the operator Z_α commutes with H_τ, and this is the condition for its being conserved. Actually, the angular momentum about the Z axis

22   *Symmetries and Reflections*

is the limit of $(1/\alpha)(Z_\alpha - 1)$ for infinitely small α. The other conservation laws are derived in the same way. The point is that *the transformation operators, or at least the infinitesimal ones among them, play a double role and are themselves the conserved quantities.*

This will conclude the discussion of the geometrical principles of invariance. You will note that reflections which give rise *inter alia* to the concept of parity were not mentioned, nor did I speak about the apparently much more general geometric principle of invariance which forms the foundation of the general theory of relativity. The reason for the former omission is that I will have to consider the reflection operators at the end of this discussion. The reason that I did not speak about the invariance with respect to the general coordinate transformations of the general theory of relativity is that I believe that the underlying invariance is not geometric but dynamic. Let us consider, hence, the dynamic principles of invariance.

Dynamic Principles of Invariance

When we deal with the dynamic principles of invariance, we are largely on *terra incognita*. Nevertheless, since some of the attempts to develop these principles are both ingenious and successful, and since the subject is at the center of interest, I would like to make a few comments. Let us begin with the case that is best understood, the electromagnetic interaction.

In order to describe the interaction of charges with the electromagnetic field, one first introduces new quantities to describe the electromagnetic field, the so-called electromagnetic potentials. From these, the components of the electromagnetic field can be easily calculated, but not conversely. Furthermore, the potentials are not uniquely determined by the field; several potentials (those differing by a gradient) give the same field. It follows that the potentials cannot be measurable, and, in fact, only such quantities can be measurable which are invariant under the transformations which are arbitrary in the potential. This invariance is, of course, an artificial one, similar to that which we could obtain by introducing into our equations the location of a ghost. The equations then must be invariant with respect to changes of the coordinate of that ghost. One does not see, in fact, what good the introduction of the coordinate of the ghost does.

Symmetry and Conservation Laws   23

So it is with the replacement of the fields by the potentials, as long as one leaves everything else unchanged. One postulates, however, and this is the decisive step, that in order to maintain the same situation, one has to couple a transformation of the matter field with every transition from a set of potentials to another one which gives the same electromagnetic field. The combination of these two transformations, one on the electromagnetic potentials, the other on the matter field, is called a gauge transformation. Since it leaves the physical situation unchanged, every equation must be invariant thereunder. This is not true, for instance, of the unchanged equations of motion, and they would have, if left unchanged, the absurd property that two situations which are completely equivalent at one time would develop, in the course of time, into two distinguishable situations. Hence, the equations of motion have to be modified, and this can be done most easily by a mathematical device called the modification of the Lagrangian. The simplest modification that restores the invariance gives the accepted equations of electrodynamics which are well in accord with all experience.

Let me state next, without giving all the details, that a similar procedure is possible with respect to the gravitational interaction. Actually, this has been hinted at already by Utiyama.[11] The unnecessary complication that one has to introduce in this case is, instead of potentials, generalized coordinates. The equations then have to be invariant with respect to all the coordinate transformations of the general theory of relativity. This would not change the content of the theory but would only amount to the introduction of a more flexible language in which there are several equivalent descriptions of the same physical situation. Next, however, one postulates that the matter field also transforms as the metric field so that one has to modify the equations in order to preserve their invariance. The simplest modification, or one of the simplest ones, leads to Einstein's equations.

The preceding interpretation of the invariance of the general theory of relativity does not interpret it as a geometrical invariance. That this should not be done had already been pointed out by the Russian physicist Fock.[12] With a slight oversimplification, one can say that a geometrical invariance postulates that two physically different situations, such as

[11] See footnote 6.
[12] V. Fock, *The Theory of Space, Time and Gravitation* (New York: Pergamon Press, 1959). See also A. Kretschman, *Ann. Phys.*, 53, 575 (1917).

24   *Symmetries and Reflections*

those in Figure 1, should develop, in the course of time, into situations which differ in the same way. This is not the case here: the postulate is merely that two different descriptions of the same situation should develop, in the course of time, into two descriptions which also describe the same physical situation. The similarity with the case of the electromagnetic potentials is obvious.

Unfortunately, the situation is by no means the same in the case of the other interactions. One knows very little about the weaker one of the strong interactions. The strong one, as well as the weak interaction, has a group which is, first of all, very much smaller than the gauge group or the group of general coordinate transformations.[13] Instead of the infinity of generators of the gauge and general transformation groups, they have only a finite number, that is, eight, generators. They do suffice, nevertheless, to a large extent to determine the form of the interaction, as well as to derive some theorems, similar to those of spectroscopy, which give approximate relations between reaction rates and between energies, that is, masses. Figure 2 shows the octuplet of heavy masses—its members are joined to each other by the simplest nontrivial representation of the underlying group which is equivalent to its conjugate complex.

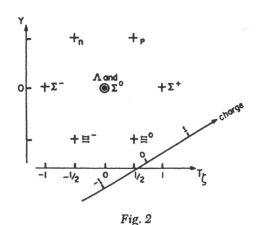

Fig. 2

[13] For the strong interaction, cf. Y. Ne'eman, *Nucl. Phys.*, 26, 222 (1961), and M. Gell-Mann, *Phys. Rev.*, 125, 1067 (1962). For the weak interaction, R. P. Feynman and M. Gell-Mann, *Phys. Rev.*, 109, 193 (1958), and E. C. G. Sudarshan and R. E. Marshak, *Phys. Rev.*, 109, 1960 (1958); also J. J. Sakurai, *Nuovo Cimento*, 7, 649 (1958), and G. S. Gershtin and A. B. Zeldovitch, *J. Exptl. Theoret. Phys. USSR*, 29, 698 (1955).

Another difference between the invariance groups of electromagnetism and gravitation on one hand, and at least the invariance group of the strong interaction on the other hand, is that the operations of the former remain valid symmetry operations even if the existence of the other types of interactions is taken into account. The symmetry of the strong interaction, on the other hand, is "broken" by the other interactions, i.e., the operations of the group of the strong interaction are valid symmetry operations only if the other types of interactions can be disregarded. The symmetry group helps to determine the interaction operator in every case. However, whereas all interactions are invariant under the groups of the electromagnetic and gravitational interactions, only the strong interaction is invariant under the group of that interaction.

We have seen before that the operations of the geometric symmetry group entail conservation laws. The question naturally arises whether this is true also for the operations of the dynamic symmetry groups. Again, there seems to be a difference between the different dynamic invariance groups. It is common opinion that the conservation law for electric charge can be regarded as a consequence of gauge invariance, i.e., of the group of the electromagnetic interaction. On the other hand, one can only speculate about conservation laws which could be attributed to the dynamic group of general relativity. Again, it appears reasonable to assume that the conservation laws for baryons and leptons can be deduced by means of the groups of the strong and of the weak interaction.[14] If true, this would imply that the proper groups of these interactions have not yet been recognized. One can adduce two pieces of evidence for the last statement. First, so far, the conservation laws in question[15] could not be deduced from the symmetry properties of these interactions, and it is unlikely that they can be deduced from

[14] For the baryon conservation law and the strong interaction, this was suggested by the present writer, *Proc. Am. Phil. Soc.*, 93, 521 (1949), and these *Proceedings*, 38, 449 (1952). The baryon conservation law was first postulated by E. C. G. Stueckelberg, *Helv. Phys. Acta*, 11, 299 (1938).

[15] For the experimental verification of these and the other conservation laws, see G. Feinberg and M. Goldhaber, these *Proceedings*, 45, 1301 (1959). The conservation law for leptons was proposed by G. Marx in *Acta Phys. Hung.*, 3, 55 (1953); also A. B. Zeldovitch, *Dokl. Akad. Nauk USSR*, 91, 1317 (1953), and E. J. Konopinski and H. M. Mahmoud, *Phys. Rev.*, 92, 1045 (1953). It seemed to be definitely established by T. D. Lee and C. N. Yang, *Phys. Rev.*, 105, 1671 (1957). See also Fermi's observation mentioned by C. N. Yang and J. Tiomno, *Phys. Rev.*, 79, 497 (1950).

them.[16] Second, the symmetry properties in question are not rigorous but are broken by the other interactions. It is not clear how rigorous conservation laws could follow from approximate symmetries—and all evidence indicates that the baryon and lepton conservation laws are rigorous.[17] Again, we are reminded that our ideas on the dynamical principles of invariance are not nearly as firmly established as those on the geometrical ones.

Let me make a last remark on a principle which I would not hesitate to call a symmetry principle and which forms a transition between the geometrical and dynamical principles. This is given by the crossing relations.[18] Let us consider the amplitude for the probability of some collision, such as

$$A + B + \ldots \rightarrow X + Y + \ldots \ . \tag{3}$$

This will be a function of the invariants which can be formed from the momenta four-vectors of the incident and emitted particles. It then follows from one of the reflection principles which I did not discuss, the "time reversal invariance," that the amplitude of (3) determines also the amplitude of the inverse reaction

$$X + Y + \ldots \rightarrow A + B + \ldots \tag{4}$$

in a very simple fashion. If one reverses all the velocities and also interchanges past and future (which is the definition of "time reversal"), (4) goes over into (3) so that the amplitudes for both are essentially equal. Similarly, if we denote the antiparticle of A by $\bar{A}$, that of B by $\bar{B}$, and so on, and consider the reaction

$$\bar{A} + \bar{B} + \ldots \rightarrow \bar{X} + \bar{Y} + \ldots , \tag{5}$$

its amplitude is immediately given by that of (3) because (according to the interpretation of Lee and Yang), the reaction (5) is obtained from (3) by space inversion. The amplitudes for

$$\bar{X} + \bar{Y} + \ldots \rightarrow \bar{A} + \bar{B} + \ldots \tag{6}$$

[16] For the baryon conservation and strong interaction, this was emphatically pointed out in a very interesting article by J. J. Sakurai, *Ann. Phys.* (N.Y.), 11, 1 (1960). Concerning the conservation of lepton members, see G. Marx, Z. *Naturforsch.*, 9a, 1051 (1954).

[17] See footnote 15.

[18] M. L. Goldberger, *Phys. Rev.*, 99, 979 (1955); M. Gell-Mann, and M. L. Goldberger, *Phys. Rev.*, 96, 1433 (1954). See M. L. Goldberger and K. M. Watson, *Collision Theory* (New York: John Wiley and Sons, 1964), chap. 10.

Symmetry and Conservation Laws   27

can be obtained in a similar way. The relations between the amplitudes of reactions (3), (4), (5), and (6) are consequences of geometrical principles of invariance.

However, one can go further. The crossing relations tell us how to calculate, for instance, the amplitude of

$$\overline{X} + B + \ldots \to \overline{A} + Y + \ldots \tag{7}$$

from the amplitude system of (3). To be sure, the calculation, or its result, is not simple any more. One has to consider the dependence of the reaction amplitude for (3) as an analytic function of the invariants formed from the momenta of the particles in (3), and extend this analytic function to such values of the variables which have no physical significance for the reaction (3) but which give the amplitude for (7). Evidently there are several other reactions the amplitudes of which can be obtained in a similar way; they are all obtained by the analytic continuation of the amplitude for (3), or any of the other reactions. Thus, rather than exchanging A and X to obtain (7), A and Y could be exchanged, and so on.

The crossing relations share two properties of the geometrical principles of invariance: they do not refer to any particular type of interaction and most of us believe that they have unlimited validity. On the other hand, though they can be formulated in terms of events, their formulation presupposes the establishment of a law of nature, namely, the mathematical, in fact analytic, expression for the collision amplitude for one of the aforementioned reactions. One may hope that they will help to establish a link between the now disjoint geometrical and dynamical principles of invariance.

The Role of Invariance Principles
in Natural Philosophy

E. P. Wigner

Symmetries and Reflections.
Indiana University Press, Bloomington, Indiana, 1967, pp. 28–37

What Is the Role and Proper Place of Invariance Principles
in the Framework of Physical Sciences?

A large part of my scientific work has been devoted to the study of
symmetry principles in physics and I was therefore not only flattered but
also greatly pleased at Professor Bernardini's invitation to speak at this
anniversary on the philosophical and epistemological role of these principles. I would like to review the role of symmetry and invariance principles from a somewhat more general point of view than that of the
physicist and appreciate the opportunity also to relate to you some of
the conclusions at which Drs. Houtappel, van Dam, and I arrived in a
series of long discussions.

There is a strange hierarchy in our knowledge of the world around us.
Every moment brings surprises and unforseeable events—truly the future is uncertain. There is, nevertheless, a structure in the events around
us, that is, correlations between the events of which we take cognizance.
It is this structure, these correlations, which science wishes to discover,
or at least the precise and sharply defined correlations. They are refinements and extensions of our everyday knowledge, in some cases so far-
reaching that we do not recognize their origin. In some cases they permit

Address at the 10th anniversary of the Scuola Internazionale di Fisica "Enrico
Fermi," July 14, 1963, Varenna. Reprinted by permission from the *Proceedings of
the International School of Physics* "Enrico Fermi," vol. 29, p. 40 (Academic Press,
1964). Copyright by the Italian Physical Society, Bologna, Italy.

us to foresee some events with certainty. Part of the art and skill of the engineer and of the experimental physicist is to create conditions in which certain events are sure to occur. Nevertheless, there are always events which are unforseeable.

If we look a little deeper into this situation we realize that we would not live in the same sense we do if the events around us had no structure. Even if our bodily functions remained unaltered, our consciousness could hardly differ from that of plants if we were unable to influence the events, and if these had no structure or if we were not familiar with some of this structure, we could not influence them. There would be no way our volition could manifest itself and there would be no such thing as that which we call life. This does not mean, of course, that life depends on the existence of the unbelievable precision and accuracy of the correlations between events which our laws of nature express, and indeed the precision of these laws has all the elements of a miracle that one can think of.

We know many laws of nature and we hope and expect to discover more. Nobody can foresee the next such law that will be discovered. Nevertheless, there is a structure in the laws of nature which we call the laws of invariance. This structure is so far-reaching in some cases that laws of nature were guessed on the basis of the postulate that they fit into the invariance structure.

It is not necessary to look deeper into the situation to realize that laws of nature could not exist without principles of invariance. This is explained in many texts of elementary physics even though only few of the readers of these texts can be expected to have the maturity necessary to appreciate these explanations. If the correlations between events changed from day to day, and would be different for different points of space, it would be impossible to discover them. Thus the invariances of the laws of nature with respect to displacements in space and time are almost necessary prerequisites that it be possible to discover, or even catalogue, the correlations between events which are the laws of nature. This does not mean, of course, that either the precision or the scope of the principles of invariance which we accept at present are necessary prerequisites for the existence of laws of nature. Both are very surprising indeed, even if not quite as amazing as the precision of some laws of nature which I am always tempted to add to Kant's starred sky above us, and the categorieal imperative within us.

This then, the progression from events to laws of nature, and from laws of nature to symmetry or invariance principles, is what I meant by the hierarchy of our knowledge of the world around us.

There are only two points which I feel should be added to the preceding discussion. The first of these amounts to admitting a certain amount of oversimplification. It concerns the way we take cognizance of the events around us—the events between which we wish to establish correlations. Some of these events, such as the rising and the setting of the sun, we perceive directly—though of course it would lead far and would also be difficult to explain what we mean by direct perception. Other events, such as the motion of an α-particle, we perceive only by means of rather complicated machinery, such as a cloud or a spark chamber. We believe, in such cases, that the established laws of nature tell us how the machinery which we employed functions and that we can consider the information obtained by *interpreting* the sense-data furnished by the machinery on a par with the sense-data the interpretation of which we learned during our childhood.

What I am admitting is that the perception of events, as we practice it in physics, is not independent of our knowledge of the laws of nature. This knowledge interprets for us, for instance, what we directly see in the cloud chamber. Hence the separation of our perceptions from the laws of nature is an oversimplification, though we believe a harmless oversimplification. Nevertheless, I felt that this point should not be suppressed.

The second point which I wish to add to the preceding discussion pertains more directly to our subject. It concerns my reference to *laws* of nature rather than one universal *law* of nature. In fact, if the universal law of nature should be discovered, invariance principles would become merely mathematical transformations which leave that law invariant. They would remain, perhaps, useful tools for deriving consequences of the universal law, much as they were used to derive the qualitative rules of spectroscopy from the laws of quantum mechanics. However, if the universal law of nature should be discovered, the principles of invariance would lose their place in the hierarchy described before. Of course, we all try to discover the universal law and some of us believe that it will be discovered one day. Many others believe that our knowledge of the laws of nature will never be complete. Naturally, even the latter alternative only makes the continued significance of invariance principles possible, it does not guarantee it. Let me remark, though, right here that a similar situation exists with respect to the laws of nature. If we had a complete description of all the events with which we shall ever come into contact, the correlations between these events, that is the laws of nature, would have a reduced significance, similar to that of the invariance principles when the universal law of nature is known. However, few if any of us believe in the possibility of the Laplaceian spirit which knows all the events, so that the significance of the laws of nature is more assured than that of the invariance principles.

Let us turn now to two other subjects: first, the nature and development of invariance principles and, second, their continued significance in a possible, and I hope foreseeable, union of the physical sciences with the other areas of human knowledge.

The Nature and Development of Invariance Principles

The classic invariance or symmetry principles are one rung of the ladder removed from direct observations. Nevertheless, they are, and should be, formulated in terms of direct observations. Thus the time-displacement invariance, properly formulated, reads: the correlations between events depend only on the time intervals between those events; they do not depend on the time when the first of them takes place. Thus, if the same relevant conditions are realized at different times, the expectations of further events will be the same, no matter when these relevant conditions were realized. I realize and wish to admit that the qualification "relevant" of the conditions is ill-defined and unprecise. This is unavoidable as long as we expect discoveries of new agents or of new effects of agents with which we are already familiar. In spite of this lack of precision, it remains true that invariance principles are formulated, jumping down over one rung of the ladder, directly in terms of observations. Only in this way can they be general enough to serve as guides to the formulation and testing of new laws of nature. It is in conformity with the formulation of invariance principles directly in terms of observations that invariance principles are also best refuted directly, in terms of observations. Thus the parity principle was refuted by creating a system with reflection symmetry which showed, subsequently, a departure from that symmetry. Only very elementary theory was necessary to see that Wu's experiment was in conflict with the parity principle.

However, even though the classic invariance principles are formulated directly in terms of observations, they are rarely used directly to forecast the future. Rather, they are used to test a theory or law of nature—that is, the nearest rung of the ladder—in order to ascertain whether its consequences will be necessarily in conformity with the invariance principle. Such a test often involves mathematical and conceptual operations of some complexity, using the often elaborate machinery in terms of which the law of nature is formulated. This explains the often erudite nature of some of the considerations in the theory of invariance.

It was mentioned before that it would have been difficult to establish any laws of nature if these were not invariant with respect to displacements in space and time. The same holds, pretty nearly, of rotational invariance, and these invariances were taken for granted even before the concept of the laws of nature had clearly emerged. It is nevertheless very difficult to verify the invariance with respect to rotations—the isotropy of physical space—by direct experiment: there seems to be an obvious difference between up and down and sidewise. However, Newton's theory of gravitation and his equations of motion are consistent with this invariance and they adequately explain the abovementioned differences as due to the attraction by the Earth. Here is then a case in which laws of nature and invariance principles, the two adjoining rungs of the ladder, mutually support each other. The concluding invariance with respect to Galilei transformations, or their modification, that is Lorentz transformations, was not at all anticipated before it was recognized by Galilei and Newton. The symmetry of the laws of nature with respect to reflections in space and time was taken for granted from about the same time on but the effectiveness and value of these symmetries in quantum mechanics was a surprise to everyone. The same applies though, albeit perhaps to a lesser degree, to the other classical symmetries also.

I have spoken so far about the classical symmetries, which are symmetries of the physical space-time continuum, similar to Klein's symmetries of geometrical space. These are, however, not the only invariances of physics any more—the laws of nature may be time-displacement invariant, the concept of invariance does not seem to be. It may not have been desirable to use the same word for the classical and the new invariances—the invariances which will be considered next. However, this

The Role of Invariance Priniciples in Natural Philosophy   33

is clearly a question of semantics. What was called here "new invariances" would better be called, perhaps, nongeometrical invariances, and they include the electromagnetic gauge-invariance, the Ne'eman-Gell-Mann eightfold way, and several others. We all feel that there is a qualitative difference between, let us say, rotational invariance on the one hand, and gauge invariance on the other. Even though we all sense the character of the difference involved, it may be worth while to articulate it and to formulate it as generally in terms of the underlying concepts as I can.

The principal difference is, to put it briefly, that the new invariances are invariances of expressions for specific interactions, not for all laws of nature.

This implies, first, the existence of specific types of interactions, such as gravitational, weak, electromagnetic, and, possibly, two kinds of strong interactions. It implies, second, that the new or nongeometrical types of invariances cannot be formulated directly in terms of correlations between observations, as I emphasized the classical or geometrical invariances are. If they could, they each would have the same relevance for all interactions, as do the classical ones. This is not the case.

The emergence of specific types of interactions as separate and well distinguishable entities is one of the most striking results of the last decade. If a malicious remark be permitted at this point, their number shows an alarming tendency to increase. It is equally striking that each of them is invariant under a specific group. This is the SU_3 group for the strong interaction, the gauge group for the electromagnetic interaction, the somewhat more complex group of the $V - A$ expression for the weak interaction, and, I believe, the general group of co-ordinate transformations for the gravitational interaction. In each case, the invariance group permits the determination of the expression for the interaction by means of a few additional assumptions implying the simplicity of the final equation. If my remark concerning the gravitational interaction is correct, Fock's observation must be interpreted as classifying the invariance with respect to the general co-ordinate transformation of relativity theory as a nongeometrical type of invariance, not a classical one.

Nevertheless, there are very great differences between the relations of the five types of interactions to the corresponding groups—so large that one may be quite uncertain whether we already know the proper group in every case. There must be also a deeper principle which explains the existence of separate types of interactions and the distinct groups which correspond to them.

Let us consider the oldest nonclassical invariance, that of the gauge invariance of electromagnetic interaction. It appears that in order to describe the interaction between electric charges and the electromagnetic field, the introduction of a new concept, that of the electromagnetic potentials, is pretty nearly unavoidable. It is not entirely unavoidable—the use of every concrete concept can be circumvented—and Mandelstam in particular has shown how this can be done in the present case. It is unquestionable, however, that the electromagnetic interaction can be expressed much more simply using the potentials. The potentials are redundant for describing the field, that is, an infinite set of potentials corresponds to the same physical situation. In other words, the physical situation, that is, all observable properties, are invariant under certain transformations of the potentials and, conversely, only those expressions in terms of the potentials are observable which are invariant under these so-called gauge transformations. Up to this point, the electromagnetic potential only appears as an awkward but harmlessly awkward concept for describing the electromagnetic field. Next, one postulates, however, that in order to maintain the same physical situation, every transformation of the electromagnetic potentials to equivalent potentials must be coupled with a certain transformation of the field with which they interact, that is, of the matter field. This is the decisive step. As a result, many expressions in terms of the matter field alone become unobservable because they are no longer invariant under gauge transformations. In order to make them invariant one has to modify them. Furthermore, two different matter-field potential combinations, which can be obtained from each other by a gauge transformation and which are therefore physically equivalent, would change in time into matter-field potential combinations which are not physically equivalent if the unchanged field equations are used to calculate their time-dependence. This would be absurd and it follows that the field equations, determining the time-dependence of the matter-field and of the potentials, must also be modified. Such a modification entails an interaction between the two fields and it appears that the simplest modification which guarantees that equivalent fields remain equivalent throughout the passage of time leads to an accurate description of the electromagnetic interaction.

Hence, the expression for the electromagnetic interaction could be guessed by a rather artificial device. First, redundant quantities for the field can be introduced, namely, the potentials. Second, it can be postulated that the transformations which change a potential into an equivalent one should affect the other physical quantities, namely, the matter‑field, also; and an assumption made what their effect should be. Lastly, the field equations are so modified that they become consistent with the equivalence of the several possible descriptions of the same physical situation. One will recognize a strong similarity between this procedure and that which leads to Einstein's equations of gravitation. There also, one can start from a co-ordinate system which satisfies Fock's requirements, so that the divergence of the metric field is zero. One can then introduce an awkward or redundant field for which the divergence is not zero and postulate that only those quantities are meaningful, that is, observable, which are the same in all these co-ordinate systems. These are the invariants. The decisive step is, in this case, the postulate that all physical quantities, in particular the stress-energy tensor, transform in the same way as the metric field, that is, as tensors. One then has all the postulates of the Riemannian geometry which Einstein used for deriving his equations of gravitation.

One recognizes, on the other hand, also the great difference between the relation of electromagnetic and gravitational interactions with their groups on one hand, and the relation of the weak and the strong interactions with their groups on the other. The latter relations and groups are much more simple and direct. It is possible, of course, that we shall one day understand the reasons for this difference. It is, perhaps, also possible that gradual modifications of the theory will eliminate these differences and also provide us with a more coherent view of the different types of interactions. This is the objective of Utiyama's proposal.

Whereas one may hope that the nonclassical or nongeometric invariances will acquire a common structure and they may even coalesce into a single deeper entity, the difference between classic and nonclassic invariances appears to be much deeper. If one wishes to bring the fact of Lorentz invariance into the form just discussed for gauge invariance, one must look, first, for redundant quantities in the theory. One can say that the usual absolute co-ordinates are redundant in this way, that their use instead of the distances between particles, or the co-ordinates with respect to the center of mass, is in itself "awkward but harmless."

However, the usual absolute co-ordinates can be observed by reference to other physical systems which are too distant to influence the system in question but provide a meaning to the absolute co-ordinates of its constituents. No similar meaning can be attached to the absolute potentials and they are truly redundant.

Outlook to the Future

Our discussion started with pointing to the uncertainty of the future, to the surprises that every moment brings us. I am on uncertain grounds indeed when speaking about the future role of the invariance principles. However, I can perhaps excuse my concluding sentences by pointing out that what one means when speaking about the outlook to the future is not so much a forecast what the future will bring as a visualization of a possible future.

As far as the physical sciences are concerned, the role of invariance principles does not seem to be near exhaustion. We still seem to be far from the "universal law of nature." We seem to be far from it, if indeed it exists, and, to paraphrase Poincaré, the present picture of four or five different types of interactions, with widely divergent properties, is not such as to permit the human mind to rest contented. Hence, invariance principles, giving a structure to the laws of nature, can be expected to act as guides also in the future and to help us to refine and unify our knowledge of the inanimate world.

One is less inclined to optimism if one considers the question whether the physical sciences will remain separate and distinct from the biological sciences and, in particular, the sciences of the mind. There are many signs which portend that a more profound understanding of the phenomena of observation and cognition, together with an appreciation of the limits of our ability to understand, is a not too distant future step. At any rate, it should be the next decisive step toward a more integrated understanding of the world. On the path toward such understanding, we shall not have to treat physical phenomena and phenomena of the mind in such a way that we forget about the tools used for the consideration of one when thinking about the problems of the other. I confess that I have no conception what the structure of this more integrated science may be and it would be surprising if it continued to contain a hierarchy similar to the one described before, in which invariance and symmetry principles have definite places. That a higher integration of science is needed is perhaps best demonstrated by the observation that the basic entities of intuitionistic mathematics are the physical objects, that the basic concept in the epistemological structure of physics is the concept of observation, and that psychology is not yet ready for providing concepts and idealizations of such precision as are expected in mathematics or even physics. Thus this passing of responsibility from mathematics to physics, and hence to the science of cognition ends nowhere. This state of affairs should be remedied by a closer integration of the now separate disciplines.

Events, Laws of Nature, and Invariance Principles

E. P. Wigner

Symmetries and Reflections.
Indiana University Press, Bloomington, Indiana, 1967, pp. 38-50

It is a great and unexpected honor to have the opportunity to speak here today. Six years ago, Yang and Lee spoke here, reviewing symmetry principles in general and their discovery of the violation of the parity principle in particular.[1] There is little point in repeating what they said on the history of the invariance principles, or on my own contribution to these, which they, naturally, exaggerated. What I would like to discuss instead is the general role of symmetry and invariance principles in physics, both modern and classical. More precisely, I would like to discuss the relation between three categories which play a fundamental role in all natural sciences: events, which are the raw materials for the second category, the laws of nature, and symmetry principles, for which I would like to support the thesis that the laws of nature form the raw material.

Events and Laws of Nature

It is often said that the objective of physics is the explanation of nature, or at least of inanimate nature. What do we mean by explanation? It is the establishment of a few simple principles which describe

Reprinted by permission of the Elsevier Publishing Company from *The Nobel Prize Lectures*. Copyright 1964 by the Nobel Foundation.
[1] See the articles of C. N. Yang and of T. D. Lee in *Les Prix Nobel en 1957* (Stockholm: Nobel Foundation, 1958). [Reprinted in *Science*, 127, 565, 569 (1958).]

the properties of what is to be explained. If we understand something, its behavior—that is, the events which it presents—should not produce any surprises for us. We should always have the impression that it could not be otherwise.

It is clear that, in this sense, physics does not endeavor to explain nature. In fact, the great success of physics is due to a restriction of its objectives: it only endeavors to explain the regularities in the behavior of objects. This renunciation of the broader aim, and the specification of the domain for which an explanation can be sought, now appears to us an obvious necessity. In fact, the specification of the explainable may have been the greatest discovery of physics so far. It does not seem easy to find its inventor, or to give the exact date of its origin. Kepler still tried to find exact rules for the magnitude of the planetary orbits, similar to his laws of planetary motion. Newton already realized that physics would deal, for a long time, only with the explanation of those of the regularities discovered by Kepler which we now call Kepler's laws.[2]

The regularities in the phenomena which physical science endeavors to uncover are called the laws of nature. The name is actually very appropriate. Just as legal laws regulate actions and behavior under certain conditions but do not try to regulate all actions and behavior, the laws of physics also determine the behavior of its objects of interest only under certain well-defined conditions but leave much freedom otherwise. The elements of the behavior which are not specified by the laws of nature are called initial conditions. These, then, together with the laws of nature, specify the behavior as far as it can be specified at all: if a further specification were possible, this specification would be considered as an added initial condition. As is well known, before the advent of quantum theory it was believed that a complete description of the behavior of an object is possible so that, if classical theory were valid, the initial conditions and the laws of nature together would completely determine the behavior of an object.

The preceding statement is a definition of the term "initial condition." Because of its somewhat unusual nature, it may be worthwhile to illustrate this on an example. Suppose we did not know Newton's equation for the motion of stars and planets,

---

[2] See, for instance, A. C. Crombie, *Augustine to Galileo* (London: Falcon, 1952), pp. 316 ff. The growth of the understanding of the realm of the explainable, from the end of the 13th century on, can be traced through almost every chapter of this book.

$$\ddot{\mathbf{r}}_i = G \, \Sigma' \, M_j \frac{\mathbf{r}_{ij}}{r^3_{ij}} \qquad \mathbf{r}_{ij} = \mathbf{r}_j - \mathbf{r}_i, \tag{1}$$

but had found only the equation determining the third derivative of the position

$$\dddot{\mathbf{r}}_i = G \, \Sigma' \, M_j \times \frac{\dot{\mathbf{r}}_{ij}(\mathbf{r}_{ij} \cdot \mathbf{r}_{ij}) - 3\mathbf{r}_{ij}(\dot{\mathbf{r}}_{ij} \cdot \mathbf{r}_{ij})}{r^5_{ij}}. \tag{2}$$

More generally, if the forces F_i are nongravitational, one would have written

$$M_i \dddot{\mathbf{r}}_i = (\dot{\mathbf{r}}_i \, \mathrm{grad}) \, \mathbf{F}_i + \dot{\mathbf{F}}_i. \tag{2a}$$

The initial conditions then would contain not only all the $\mathbf{r}_i$ and $\dot{\mathbf{r}}_i$, but also the $\ddot{\mathbf{r}}_i$. These data, together with the "equation of motion" (Eq. 2), would then determine the future behavior of the system just as $\mathbf{r}_i$, $\dot{\mathbf{r}}_i$, and Eq. 1 determine it. The fact that initial conditions and laws of nature completely determine the behavior is similarly true in any causal theory.

The surprising discovery of Newton's age is just the clear separation of laws of nature on the one hand and initial conditions on the other. The former are precise beyond anything reasonable; we know virtually nothing about the latter. Let us pause for a minute at this last statement. Are there really no regularities concerning what we just called initial conditions?

The last statement would certainly not be true if as laws of nature Eqs. 2 and 2a were adopted, that is, if we considered the $\ddot{\mathbf{r}}_i$ as part of the initial conditions. In this case, there would be a relation, in fact the precise relation of Eq. 1, between the elements of the initial conditions. The question, therefore, can be only: are there any relations between what we really do consider as initial conditions? Formulated in a more constructive way: how can we ascertain that we know all the laws of nature relevant to a set of phenomena? If we do not, we would determine unnecessarily many initial conditions in order to specify the behavior of the object. One way to ascertain this would be to prove that all the initial conditions can be chosen arbitrarily—a procedure which is, however, impossible in the domain of the very large (we cannot change the orbits of the planets) or the very small (we cannot precisely control atomic particles). No other equally unambiguous criterion is known to me, but there is a distinguishing property of the correctly chosen—that is, minimal—set of initial conditions which is worth mentioning.

The minimal set of initial conditions not only does not permit any exact relation between its elements; on the contrary, there is reason to contend that these are, or at some time have been, as random as the externally imposed, gross constraints allow. I wish to illustrate this point, first, on an example which, at first, seems to contradict the thesis because this example shows the power, and also the weakness of the assertion, best.

Let us consider for this purpose again our planetary system. It was mentioned before that the approximate regularities in the initial conditions, that is, the determinants of the orbits, led Kepler to the considerations which were then left by the wayside by Newton. These regularities form the apparent counterexample to the aforementioned thesis. However, the existence of the regularities in the initial conditions is considered so unsatisfactory that it is felt necessary to show that the regularities are but a consequence of a situation in which there were *no* regularities. Perhaps von Weizsäcker's attempt in this direction[3] is most interesting: he assumes that originally the solar system consisted of a central star, with a gas in rotation, but otherwise in random motion, around it. He then deduces the aforementioned regularities of the planetary system, now called Bode's law, from his assumption. More generally, one tries to deduce almost all "organized motion," even the existence of life, in a similar fashion. It must be admitted that few of these explanations have been carried out in detail,[4] but the fact that such explanations are attempted remains significant.

The preceding paragraph dealt with cases in which there is at least an apparent evidence against the random nature of the uncontrolled initial conditions. It attempted to show that the apparently organized nature of these initial conditions was preceded by a state in which the uncontrolled initial conditions were random. These are, on the whole, exceptional situations. In most cases, there is no reason to question the random nature of the noncontrolled, or nonspecified, initial conditions,

[3] C. F. von Weizsäcker, Z. *Astrophys.*, 22, 319 (1944); S. Chandrasekhar, *Rev. Mod. Phys.*, 18, 94 (1946).

[4] An interesting and well-understood case is that of "focusing collisions" in which neutrons, having velocities which are rather high but with random orientation, are converted into lower-velocity neutrons but with preferential directions of motion. See R. H. Silsbee, *J. Appl. Phys.*, 28, 1246 (1957); C. Lehmann and G. Leibfried, Z. *Physik*, 172, 465 (1963).

and the random nature of these initial conditions is supported by the validity of the conclusions arrived at on the basis of the assumption of randomness. One encounters such situations in the kinetic theory of gases and, more generally, whenever one describes processes in which the entropy increases. Altogether, then, one obtains the impression that, whereas the laws of nature codify beautifully simple regularities, the initial conditions exhibit, as far as they are not controlled, equally simple and beautiful irregularity. Hence there is perhaps little chance that some of the former remain overlooked.

The preceding discussion characterized the laws of nature as regularities in the behavior of an object. In quantum theory, this is natural: the laws of quantum mechanics can be suitably formulated as correlations between subsequent observations on an object. These correlations are the regularities given by the laws of quantum mechanics.[5] The statements of classical theory, its equations of motion, are not customarily viewed as correlations between observations. It is true, however, that their purpose and function is to furnish such correlations and that they are, in essence, nothing but a shorthand expression for such correlations.

Laws of Nature and Invariance

We have ceased to expect from physics an explanation of all events, even in the gross structure of the universe, and we aim only at the discovery of the laws of nature, that is, the regularities of the events. The preceding section gives reason for the hope that the regularities form a sharply defined set and are clearly separable from what we call initial conditions, in which there is a strong element of randomness. However, we are far from having found that set. In fact, if it is true that there are precise regularities, we have reason to believe that we know only an infinitesimal fraction of these. The best evidence for this statement derives perhaps from a fact which was mentioned here by Yang 6 years ago: the multiplicity of the types of interactions. Yang mentioned four of them—gravitational, weak, electromagnetic, and strong, and it now seems that there are two types of strong interactions. All these play a role in every process, but it is hard, if not impossible, to believe that the laws of nature should have such complexity as implied

---

[5] See, for instance, the section, "What is the state vector?" in E. Wigner, *Am. J. Phys.*, 31, 6 (1963), reprinted in this volume.

by four or five different types of interactions between which no connection, no analogy, can be discovered.

It is natural, therefore, to ask for a superprinciple which is in a similar relation to the laws of nature as these are to the events. The laws of nature permit us to foresee events on the basis of the knowledge of other events; the principles of invariance should permit us to establish new correlations between events, on the basis of the knowledge of established correlations between events. This is exactly what they do. If it is established that the existence of the events $A, B, C, \ldots$ necessarily entails the occurrence of X, then the occurrence of the events $A', B', C', \ldots$ also necessarily entails X', if $A', B', C', \ldots$ and X' are obtained from $A, B, C, \ldots$ and X by one of the invariance transformations. There are three categories of such invariance transformations:

a) Euclidean transformations: the primed events occur at a different location in space, but in the same relation to each other, as the unprimed events.

b) Time displacements: the primed events occur at a different time, but separated by the same time intervals from each other as the unprimed ones.

c) Uniform motion: the primed events appear to be the same as the unprimed events from the point of view of a uniformly moving coordinate system.

The first two categories of invariance principles were always taken for granted. In fact, it may be argued that laws of nature could not have been recognized if they did not satisfy some elementary invariance principles such as those of categories a and b—if they changed from place to place, or if they were also different at different times. The principle c is not so natural. In fact, it has often been questioned, and it was an accomplishment of extraordinary magnitude on the part of Einstein to have re-established it in his special theory of relativity. However, before discussing this point further, it may be useful to make a few general remarks.

The first remarkable characteristic of the invariance principles which were enumerated is that they are all geometric, at least if four-dimensional space-time is the underlying geometrical space. By this I mean that the invariance transformations do not change the events; they only change their location in space and time and their state of motion. One could easily imagine a principle in which, let us say, protons are replaced by electrons and vice versa, velocities by positions, and so on.[6]

The second remarkable characteristic of the preceding principles is that they are invariance rather than covariance principles. This means that they postulate the same conclusion for the primed premises as for the unprimed premises. It is quite conceivable that, if certain events $A, B, \ldots$ take place, the events $X_1, X_2, X_3 \ldots$ will follow with certain probabilities $p_1, p_2, p_3. \ldots$ From the transformed events A', B', C', the transformed consequences $X_1', X_2', X_3', \ldots$ *could* follow with changed probabilities such as

$$p_1' = p_1(1 - p_1 + \Sigma p_n^2),$$
$$p_2' = p_2(1 - p_2 + \Sigma p_n^2),$$

$$\ldots$$

but this is not the case; we always have $p_i' = p_i$.

These two points are specifically mentioned because there are symmetry principles, the so-called crossing relations,[7] which *may be* precisely valid and which surely do not depend on specific types of interactions. In these regards they are, or may be, similar to the geometric invariance principles. They differ from these because they do change the events and they are covariance rather than invariance principles. Thus, from a full knowledge of the cross section for neutron-proton scattering, they permit one to obtain some of the neutron-antiproton collision cross sections. The former events are surely different from the neutron-antiproton collisions, and the cross sections for the latter are not equal to the neutron-proton cross sections but are obtained from these by a rather complicated mathematical procedure. Hence, the crossing relations, even though they do not depend on a specific type of interaction, are not considered to be geometrical symmetry conditions, and they will not be considered here. Similarly, we shall not be

[6] The possibility of an invariance principle in which velocities are replaced by position, and conversely, was studied by M. Born, *Nature*, 141, 327 (1938); *Proc. Roy. Soc.* (London), A165, 291 (1938); *ibid.*, A166, 552 (1938).

[7] The crossing relations were established by M. L. Goldberger, *Phys. Rev.*, 99, 979 (1955); M. Gell-Mann and M. L. Goldberger, *ibid.*, 96, 1433 (1954). For further literature, see, for instance, M. L. Goldberger and K. M. Watson, *Collision Theory* (New York: John Wiley and Sons, 1964), Chap. 10. The relations of the various types of symmetry principles were considered in two recent articles of the present author: *Proceedings of the International School of Physics "Enrico Fermi"* vol. 29, p. 40 (Academic Press, 1964) (reprinted in this volume), and *Phys. Today*, 17, 34 (1964). See also *Progr. Theoret. Phys.*, 11, 437 (1954).

Events, Laws of Nature, and Invariance Priniciples   *45*

concerned with the dynamic symmetry principles which are symmetries of specific interactions, such as electromagnetic interactions or strong interactions, and are not formulated in terms of events.[8]

As to the geometrical principles, it should be noted that they depend on the dividing line between initial conditions and laws of nature. Thus, the law of nature Eq. 2 or 2a, obtained from Newton's principle by differentiation with respect to time, is invariant also under the transformation to a uniformly accelerated coordinate system

$$\mathbf{r}_i' = \mathbf{r}_i + t^2 \mathbf{a} \quad t' = t, \tag{3}$$

where $\mathbf{a}$ is an arbitrary vector. Naturally, this added principle can have no physical consequence because, if the initial conditions $\mathbf{r}_i, \dot{\mathbf{r}}_i, \ddot{\mathbf{r}}_i$ are realizable (that is, satisfy Eq. 1), the transformed initial conditions $\mathbf{r}_i' = \mathbf{r}_i, \dot{\mathbf{r}}_i' = \dot{\mathbf{r}}_i, \ddot{\mathbf{r}}_i' = \ddot{\mathbf{r}}_i + 2\mathbf{a}$ cannot be realizable.

The symmetry principles of the preceding discussion are those of Newtonian mechanics or the special theory of relativity. One may well wonder why the much more general, and apparently geometrical, principles of invariance of the general theory have not been discussed. The reason is that I believe, in conformity with the views expressed by V. Fock,[9] that the curvilinear coordinate transformations of the general theory of relativity are not invariance transformations in the sense considered here. These were so-called active transformations, replacing events $A, B, C, \ldots$ by events $A', B', C', \ldots$, and unless active transformations are possible, there is no physically meaningful invariance. However, the mere replacement of one curvilinear coordinate system by another is a "redescription" in the sense of Melvin[10]; it does not change the events and does not represent a structure in the laws of nature. This does not mean that the transformations of the general theory of relativity are not useful tools for finding the correct laws of gravitation; they evidently are. However, as I suggested elsewhere,[11] the principle which they serve to formulate is different from the geometrical invariance principles considered here; it is a dynamical invariance principle.

[8] See footnote 7.

[9] V. A. Fock, *The Theory of Space, Time and Gravitation* (New York: Pergamon Press, 1959). The character of the postulate of invariance with respect to general coordinate transformations as a geometrical invariance had already been questioned by E. Kretschman, *Ann. Phys. Leipzig*, 53, 575 (1917).

[10] M. A. Melvin, *Rev. Mod. Phys.*, 32, 477 (1960).

[11] See footnote 7.

The Use of Invariance Principles, Approximate Invariances

The preceding two sections emphasized the inherent nature of the invariance principles as being rigorous correlations between those correlations between events which are postulated by the laws of nature. This at once points to the use of the set of invariance principles which is surely most important at present: to be touchstones for the validity of possible laws of nature. A law of nature can be accepted as valid only if the correlations which it postulates are consistent with the accepted invariance principles.

Incidentally, Einstein's original article which led to his formulation of the special theory of relativity illustrates the preceding point with greatest clarity.[12] He points out in this article that the correlations between events are the same in all coordinate systems in uniform motion with respect to each other, even though the causes attributed to these correlations at that time did depend on the state of motion of the coordinate system. Similarly, Einstein made the most extensive use of invariance principles to guess the correct form of a law of nature, in this case that of the gravitational law, by postulating that this law conforms with the invariance principles which he postulated.[13] Equally remarkable is the present application of invariance principles in quantum electrodynamics. This is not a consistent theory—in fact, not a theory in the proper sense because its equations are in contradiction to each other. However, these contradictions can be resolved with reasonable uniqueness by postulating that the conclusions conform to the theory of relativity.[14] Another approach, even more fundamental, tries to axiomatize quantum field theories, the invariance principles forming the cornerstone of the axioms.[15] I will not further enlarge on this question because it has been

[12] A. Einstein, "Zur Elektrodynamik bewegter Körper," *Ann. Phys. Leipzig,* 17, 891 (1905).

[13] A. Einstein and S. B. Preuss, *Akad. Wiss.,* pp. 778, 799, 844 (1915); *Ann. Phys. Leipzig,* 49, 769 (1916). Similar results were obtained almost simultaneously by D. Hilbert, *Nachr. Kgl. Ges. Wiss. Göttingen,* p. 395 (1915).

[14] J. Schwinger, *Phys. Rev.,* 76, 790 (1949). See also S. S. Schweber, *An Introduction to Relativistic Quantum Field Theory* (New York: Row, Peterson, 1961), Sec. 15, where further references can also be found.

[15] See A. S. Wightman, "Quelques problèmes mathematiques de la théorie quantique relativiste" and numerous other articles in *Les Problèmes Mathematiques de la Théorie Quantique des Champs* (Paris: Centre National de la Recherche Scientifique, 1959).

discussed often and eloquently. In fact, I myself spoke about it but a short time ago.[16]

To be touchstones for the laws of nature is probably the most important function of invariance principles. It is not the only one. In many cases, consequences of the laws of nature can be derived from the character of the mathematical framework of the theory, together with the postulate that the laws—the exact form of which need not be known—conform with invariance principles. The best known example is the derivation of the conservation laws for linear and angular momentum, and for energy, and of the motion of the center of mass, either on the basis of the Lagrangian framework of classical mechanics or the Hilbert space of quantum mechanics, by means of the geometrical invariance principles enumerated before.[17] Incidentally, conservation laws furnish at present the only generally valid correlations between observations with which we are familiar; for those which derive from the geometrical principles of invariance it is clear that their validity transcends that of any special theory—gravitational, electromagnetic, and so forth—which are only loosely connected in present-day physics. Again, the connection between invariance principles and conservation laws—which in this context always include the law of the motion of the center of mass—has been discussed in the literature frequently and adequately.

In quantum theory, invariance principles permit even further-reaching conclusions than in classical mechanics and, as a matter of fact, my original interest in invariance principles was due to this very fact. The reason for the increased effectiveness of invariance principles in quantum theory is due, essentially, to the linear nature of the underlying Hilbert space.[18] As a result, from any two state vectors, ψ_1 and ψ_2, an infinity of new state vectors

$$\psi = a_1 \psi_1 + a_2 \psi_2 \tag{4}$$

[16] See footnote 7.

[17] G. Hamel, *Z. Math. Phys.*, 50, 1 (1904); G. Herglotz, *Ann. Physik*, 36, 493 (1911); F. Engel, *Nachr. Kgl. Ges. Wiss. Göttingen*, p. 207 (1916); E. Noether, *ibid.*, p. 235 (1918); E. Bessel-Hagen, *Math. Ann.*, 84, 258 (1921). The quantum theoretical derivation given by E. Wigner, *Nachr. Kgl. Ges. Wiss. Göttingen*, p. 375 (1927), contains also the parity conservation law which was shown, in reference 1, to be only approximately valid. See also the article of reference 15.

[18] I heard this remark, for the first time, from C. N. Yang, at the centennial celebration of Bryn Mawr College. However, see also my article *Proc. Am. Phil. Soc.*, 93, 521 (1949).

can be formed, a_1 and a_2 being arbitrary numbers. Similarly, several, even infinitely many, states can be superimposed with largely arbitrary coefficients. This possibility of superposing states is by no means natural physically. In particular, even if we know how to bring a system into the states ψ_1 and ψ_2, we cannot give a prescription how to bring it into a superposition of these states. This prescription would have to depend, naturally, on the coefficients with which the two states are superimposed and is simply unknown. Hence, the superposition principle is strictly an existence postulate—but very effective and useful.

To illustrate this point, let us note that in classical theory, if a state, such as a planetary orbit, is given, another state, that is, another orbit, can be produced by rotating the initial orbit around the center of attraction. This is interesting but has no very surprising consequences. In quantum theory the same is true. In addition, however, the states obtained from a given one by rotation can be superimposed as a result of the aforementioned principle. If the rotations to which the original state was subjected are uniformly distributed over all directions, and if the states so resulting are superimposed with equal coefficients, the resulting state has necessarily spherical symmetry. This construction of a spherically symmetric state could fail only if the superposition resulted in the null-vector of Hilbert space, in which case one would not obtain any state. In such a case, however, other coefficients could be chosen for the superposition—in the plane case, the coefficients $e^{im\varphi}$, where φ is the angle of rotation of the original state—and the resulting state, though not spherically symmetric, or in the plane case axially symmetric, would still exhibit simple properties with respect to rotation. This possibility, the construction of states which have either full rotational symmetry or at least some simple behavior with respect to rotations, is the one which is fundamentally new in quantum theory. The stationary states of systems at rest have such high symmetries with respect to rotations. Such states play an important role in the theory of simple states such as atoms and the high symmetry of these is also conceptually satisfying.

The superposition principle also permits the exploitation of reflection symmetry. In classical mechanics as well as in quantum mechanics, if a state is possible, the mirror image of that state is also possible. However, in classical theory no significant conclusion can be drawn from this fact. In quantum theory, original-state and mirror image can be superimposed, with equal or oppositely equal coefficients. In the first case the resulting state is symmetric with respect to reflection, in the second case antisymmetric. The great accomplishment of Lee and Yang, which was mentioned earlier,[19] was just a very surprising reinterpretation of the physical nature of one of the reflection operations, that of space reflection, with the additional proof that the old interpretation cannot be valid. The consideration of "time inversion" requires rather special care because the corresponding operator is antiunitary. Theoretically, it does lead to a new quantum number and a classification of particles[20] which, however, has not been applied in practice.

My discussion would be far from complete without some reference to approximate invariance relations. Like all approximate relations, these may be very accurate under certain conditions but fail significantly in others. The critical conditions may apply to the state of the object, or may specify a type of phenomenon. The most important example for the first case is that of low relative velocities. In this case, the magnetic fields are weak, and the direction of the spins does not influence the behavior of the other coordinates. One is led to the Russell-Saunders coupling of spectroscopy.[21] Even more interesting should be the case of very high velocities in which the magnitude of the rest mass becomes unimportant. Unfortunately, this case has not been discussed in full detail, even though there are promising beginnings.[22]

Perhaps the most important case of special phenomena in which there are more invariance transformations than enumerated before is the rather general one of all phenomena, such as collisions between atoms, molecules, and nuclei, in which the weak interaction, which is respon-

[19] See footnote 1.

[20] See E. P. Wigner, "Unitary representations of the inhomogeneous Lorentz group including reflections," in *Elementary Particle Physics*, F. Gürsey, ed. (New York: Gordon and Breach, 1964), for a systematic discussion of the reflection operations.

[21] See E. P. Wigner, *Gruppentheorie und ihre Anwendung auf die Quantummechanik der Atomspektren* (Braunschweig: Friedr. Vieweg, 1931) or the English translation by J. Griffin (New York: Academic Press, 1959).

[22] H. A. Kastrup, *Phys. Rev. Letters*, 3, 78 (1962). The additional invariance operations probably form the conformal group. This was discovered by E. Cunningham [*Proc. London Math. Soc.*, 8, 77 (1909)] and by H. Bateman [*ibid.*, 8, 223 (1910)] to leave Maxwell's equations for the vacuum invariant, that is, the equations which describe light, always propagating at light velocity. For more recent considerations, see T. Fulton, F. Rohrlich, L. Witten, *Rev. Mod. Phys.*, 34, 442 (1962), and Y. Murai, *Progr. Theoret. Phys.*, 11, 441 (1954); these articles contain also more extensive references to the subject.

sible for beta decay, does not play a role. In all these cases, the parity operation is a valid invariance operation. This applies also in ordinary spectroscopy.

In another interesting special type of phenomenon the electromagnetic interaction also plays a subordinate role only. This renders the electric charge on the particles insignificant, and the interchange of proton and neutron, or more generally of the members of an isotopic spin multiplet, becomes an invariance operation. These, and the other special cases of increased symmetry, lead to highly interesting questions which are, furthermore, at the center of interest at present. However, the subject has too many ramifications to be discussed in detail at this occasion.

Events, Laws of Nature, and Invariance Principles

Eugene P. Wigner

Mimeographed notes, ca. 1980
(Reset by Springer-Verlag for this volume)

Introduce

I have often discussed this an related subjects but I still find it difficult to call attention to all the relevant points – points which we all should remember. Actually, in the first part of the discussion, I would prefer to replace in the title, so kindly suggested by Dr. Zichichi, the "Events" by "Initial Conditions".

The sharp distinction between Initial Conditions and Laws of Nature was initiated by Isaac Newton and I consider this to be one of his most important, if not *the* most important, accomplishment. Refore Newton there was no sharp separation between the two concepts. Kepler, to whom we owe the three precise laws of planetary motion, tried to explain also the size of the planetary orbits, and their periods. After Newton's time the sharp separation of initial conditions and laws of nature was taken for granted and rarely even mentioned. Of course, the first ones are quite arbitrary and their properties are hardly parts of physics while the recognition of the latter ones are the prime purpose of our science. Whether the sharp separation of the two will stay with us permanently is, of course, as uncertain as is all future development but this question will be further discussed later. Perhaps it should be mentioned here that the permanency of the validity of our deterministic laws of nature became questionable as a result of the realization, due initially to D. Zeh, that the states of macroscopic bodies are always under the influence of their environment; in our world they can not be kept separated from it.

As to the invariances, they can be characterized as laws which the laws of nature have to obey. If a certain behavior follows from a set of initial conditions, from similar but displaced initial conditions, the same behavior but equally displaced behavior follows.

The three concepts, initial conditions, laws of nature, and invariances, and their changes in the course of the development of physics, will be discussed next in a bit more detail.

Initial Conditions

The initial conditions of a system, together with the laws of nature, are supposed to determine the behavior of the system as long as this remains isolated, i.e. is not subject to the influence of other systems. They are supposed to be independent from each other. The initial conditions of Newton's mechanics are most easily defined in this connection: they are the positions and velocities of all the bodies of the system. This means, of course, that the "bodies" are all point like, i.e. have no variable inner coordinates. This is essentially true for the planets to which Newton applied this theory principally, it could be, but is not generally, true of atoms. But Newton's mechanics was greatly generalized rather soon after its establishment and the basic idea remained unchanged. It may be of some interest to observe in this connection that Aristoteles' initial conditions involved only the positions – be assumed that the velocities are proportional to the forces which the body experiences.

It may be worth reminding here already that Newton's laws are, and were already known by him, to be valid only in a coordinate system which is "at rest", or is in uniform straight motion. And the "rest" or "uniform motion" of the coordinate system can not be definded abstractly – the only simple and general definition is that the coordinate system is such that Newton's laws are valid in connection therewith. This may appear to make his laws meaningless but they become meaningful by the postulate that there *are* such coordinate systems and the validity of Newton's laws with respect to it extends to the whole universe.

Perhaps I should mention also that even though the definition of the initial conditions for Newton's mechanics is very simple, their measurement was very difficult even for the system for which its application was ground-breaking: the motion of the planets around the sun. The full determination of the positions of the planets was difficult, principally because we see only the projection of their three coordinates on the plane perpendicular to the direction at which we see it. But the full determination of the positions is quite old – it seems to have been first carried out by the Greeks – Ptolemeus' work is most often cited. But, of course, the coordinate system in which he specified the position was not the one at rest with the sun but at rest with respect to the Earth.

The next fundamental change in the definition of the initial conditions, together with a fundamental extension of the area of physics, came with Maxwell's equations. The initial conditions for these are more complex mathematically and also, in principle, more difficult to measure – they are *functions* of the three dimensions of space. Only an infinity of numbers can identify a function, even if this is arbitrarily differentiable and the determination of the initial conditions is, therefore, in this case infinitely more difficult in principle than for the Newtonian system. But states with definite properties can be, and have been, produced and the Maxwell equations are well confirmed experimentally. It is perhaps good to remark here that Newton's mechanics was also extended to continuous materials (liquids, gases, also solid bodies) and the same observation applies to these expensions as was made to the electromagnetic field equations.

One other remark should be made concerning the initial conditions of Maxwell's equations. They can be given in terms of the three components of the electric and of the magnetic field strengths at a given time – but the magnetic field's divergence is zero, so these are not free variables. The other description, by means of the four components of the vector potential and their time derivatives, is even less perfect because an arbitrary four dimensional gradient can be added to the vector potential without changes its physical content, i.e. without changing the electric and magnetic field strengths derivable from them. If it is postulated that the divergence of the vector potential be zero, the situation is not better than with the description be means of the electric and magnetic field strength. These are epistemological difficulties but, in practice, do not impair the usefulness and applicability of the equations.

The last fundamental change in the description of the state of a system, and hence of the initial conditions, came with the establishment of quantum mechanics. As you know, this also increased the area of physics, even more than the establishment of Maxwell's equations. The states of quantum mechanics are described by complex so called state vectors in the infinite dimensional Hilbert space – which does not give a greater complexity than the electromagnetic field specification of Maxwell's theory. In fact the vectors in Hilbert space can be replaced, in most practical applications, by "wave functions" which are complex functions in $3n$ dimensional space where n is the member of particles. The correspondance between the physical state and the state vector describing it is almost one-to-one-two state vectors characterize the same state only if all their components differ by the same factor, i.e. if they have the same direction in Hilbert space.

In this regard, the description of the states, hence also of the initial conditions, of quantum mechanical theory comes closer to the ideal described originally than that of Maxwell's theory. But whereas the determination of the state can be done at least approximately for the electromagnetic field, the microscopic state, that is the state vector, of the quantum state can not be determined from the outside even approximately. Every measurement undertaken on the system is very likely to change its state fundamentally. It is true, on the other hand, that many of the possible states of a microscopic system can be produced, i.e. that, particularly systems consisting only of one, two, or perhaps three or even more particles can be pushed into a variety of states. Not into all states which can be described by a state vector – some of these can be proved to be absolutely unproducable (superselection rules). Many, in fact most, can be produced only approximately by a finite apparatus. But, of course, the apparatus producing it has to be large only from a microscopic point of view – not necessarily large as compared with our own bodies.

In sum, the true initial conditions, describing the full microscopic structure of a system, are very difficult to determine – difficult even for microscopic systems. Some such states can be produced and the theory can be tested on them but we must admit that states of most "state vectors" are impossible to create. Just the same, the confirmation of the theory for the obtainable state vectors convinces us of the validity of the theory and its actual significance is, of course, overwhelming. I trust it is unnecessary for me to give evidence for this.

Let me just summarize by admitting that the nature of the initial conditions, that is also the description of the states of the systems in which physics is interested, has changed fundamentally from the old theory, where it was straightforward and simple, to Maxwell's field theory where it remained straightforward but ceased to be simple, then again to quantum mechanics for which it is neither simple nor straightforward. It is remarkable that relativity theory did not introduce such drastic changes – this is perhaps because it did not fundamentally expand the area of physics which, as you know, is now able to describe the properties of all common bodies and indeed contains, in principle, all of chemistry. But, of course, only in principle.

Let me mention, finally, one effect which the theory of relativity should have introduced into the description of the initial conditions and perhaps also into the description of all states. The state vectors, as we use them, describe the properties of the states which they assume at a definite time. But these are unobservable – we can not get signals instantaneously from a distance. It would be, therefore, more reasonable for the state vector, or the wave function, to describe the state on the negative light cone from the points of which signals may reach the observer. I have proposed this and tried to do it also but with very little success, at least so far. Let me now go over to the discussion of the laws of nature.

Laws of Nature

It is not necessary to explain the nature of the laws of nature to this audience. If the initial conditions are given for a system, they predict its future as long as the system is isolated and does not receive signals from outside its negative light cone. The formulation of the laws of nature depends naturally on the type of initial conditions that they receive – in fact one can say that they give the initial conditions which describe the system at times later than the time of the original initial conditions.

Naturally, all our laws of nature are approximate. But there are conditions under which their accuracy is marvelous. Elementary quantum mechanics gives the energy levels of H or He with an accuracy of about 10^{-6} and the present theory improves this further. Newton's law of gravitation, as applied to the planets, gives their positions with a truly surprising accuracy – in the worst case, that of the Mercury, the error is less than $1/20\,000$ after one circulation around the sun. As you know, even this deviation from Newton's laws was eliminated by the general theory of relativity and we hope that other deviations from the present laws of nature will be also eliminated, or at least decreased, by future laws of nature. And that the validity of the present laws of nature will be further extended, surely to high energy phenomena, but perhaps also to the phenomenon of life. But it is truly wonderful that we are provided with situations in which the present laws of nature have such high accuracy. It would have been much more difficult to create science if these situations were not available.

The laws of nature we know are both very interesting and most useful. But, we must admit, they are not perfect, in many cases not consistent. In particular the macroscopic and microscopic theories are not united – we can measure times and distance almost solely macroscopically and we use macroscopic instruments also for quantum mechanical measurements. In fact, the Russian physicist V. Fock declared that the measuring apparata must be described classically, that is not quantum – mechanically, the systems on which the measurement is undertaken are quantum objects. S. Ludwig's defense of quantum mechanics is based on the same demand: measuring and registering apparata must be treated classically but – I must admit – he does not admit that this is a limitation of the validity of quantum mechanics, the hiding of its inability to describe macroscopic objects. But surely, we must realize that the quantum mechanical measurement process, the outcome of which is probabilistic, is not describable by the present quantum mechanical equations which are deterministic. It is natural to try to attribute the probabilistic outcome of the measurement to the uncertainty of the initial state of the measuring apparatus but it is easy to prove that this is impossible, impossible at least if there is no a priori relation between the pre-measurement states of system and measuring apparatus.

Perhaps I should mention that this suggested to me a modification of the quantum mechanical equations as applied to macroscopic bodies. This was stimulated by the article of D. Zeh (1970) according to which a macroscopic body cannot be isolated from the interaction of the environment – not even in intergalactic space. This means that the present deterministic equations of quantum mechanics do not strictly apply to it and I suggested (in 1979) the addition of a term to these equations which should take care of the interaction with the environment. This interaction renders the microscopic state (i.e. the wave function) of macroscopic bodies to be subject to probability laws and would account also for the probabilistic outcome of the measurement process assuming that the measuring apparatus is macroscopic. Although the proposed equation is surely not the final one, it is surely true that the present deterministic equations of quantum mechanics are not strictly valid for macroscopic bodies.

Another conflict between the present microscopic and macroscopic theories, that is between quantum mechanics and general relativity, stems from general relativity's fundamental concept: the space-time distances between space-time points – the basis of the g_{ik}. A point in space-time can be defined only as the crossing point of two world lines, that is by the position and time of a collision. But quantum mechanic's collision theory does not specify the position of the collision precisely and also attributes an extended time interval thereto. Hence, according to the microscopic point of view space-time points cannot be defined with absolute accuracy. Of course, the actual applications of the general theory of relativity do not require the measurement of truly microscopic distances – the distances of the planets from the sun are not microscopic. But the virtual non-existence of such distances shows that the general theory of relativity, that is our present theory of gravitation, is not in harmony with quantum mechanics.

Actually, the postulate of the definability of infinitely small space-time points underlies also the quantum field theories and, in the opinion of some of us, is responsbile for the infinities and for the need of renormalization. There are, therefore, attempts to abandon the concept of the classical and absolutely precise space-time point concept and Dr. T.D. Lee presented to us such an attempt which I hope will be fully successful. My own attempts in that direction were not and we must admit that the departure from the ideas of standard geometry would be as fundamental as Einstein's departure from the idea of absolute simultaneity.

The last part of my discussion points to the lack of absolute consistency of the present laws of physics and their weaknesses. I could have mentioned in this connection also that they do not yet give a truly satisfactory description of nuclear structure and even less or the very high energy phenomena which we heard about a great deal. What bothers me even more is that they make absolutely no reference to the phenomenon of life except for the vague admission that there is an observer behind each measuring apparatus. But we should not forget how amazingly successful they are in the description of certain phenomena, how much further they went than Heisenberg's original postulate of giving the energy levels of atoms and the transition probabilities between them. Their influence was not only scientific, they also had technical applications and changed our mode of life greatly.

Let me mention finally the wonderful property of the laws of nature, not postulated by their definition which is, in Einstein's words that they are "simple and mathematically beautiful". They are given by equations which use only mathematical symbols (though we must admit that some of these were originated by physicists) and use them in a simple way. This is truly remarkable – it could be very, very different and this would reduce the power of man, and his interest in science. We admit that our present laws of nature are approximate and have validity only under certain conditions but these conditions were furnished to us and we recognized at least approximate laws, and this is wonderful. Whether man will ever recognize the true laws of nature is questionable – I hope he will come closer and closer to them but always remain very far. If he recognized them, science could not be developed further and this would deprive man of an important source of pleasure and a deflection from the quest for power. I hope this will not happen.

I will give now a short description of the invariance principles and this will be followed by some reservations which, I hope, describe the limitations of the discussions preceding it.

Invariance Principles

As was mentioned before, the invariance principles are laws which the laws of nature must obey. Surely, if the laws of nature were different at different locations, or if they depended irregularly on time, it would not be possible to recognize them and, in fact, they would not be laws of nature. That the laws of nature are the same in coordinate systems which are uniformly moving with respect to the one defined at the discussion of Newtonian mechanics' initial conditions – and hence uniformly moving with respect to each other – is much more surprising and even some of the greatest scientists originally did not believe it – they questioned Einstein's theory of special relativity. Aristoteles' theory also contradicts it, as was, implied before.

Of course, when we speak about the invariance of the laws we should not forget that the laws apply only to isolated systems and, here on the Earth we are subject to its gravitational attraction. This removes most invariances from everyday experience. It was, therefore, not so easy to recognize all the invariances. But the magnitude of the outside influences can often be judged and their effect taken into account or, under better conditions, such as the Earth's gravitational attraction on fast moving particles, neglected.

As was said before, the initial and most important function of the principles of invariance is to serve as necessary conditions on the validity of laws of nature and hence in establishing, or at least proposing, such laws. But the invariance principles do have other, less obvious applications. The oldest one of these is the establishment of conservation laws. This is usually credited to E. Noether but was proposed even earlier by Hamel. It is far most easily obtained from the quantum mechanical formalism: the commutability of the spatial displacement operators, and of the infinitesimal operators of rotation with the time displacement operator leads at once to the conservation laws of momentum and angular momentum. The commutability of the infinitesimal time displacement operator with the finite such operators establishes the energy conservation law. In pre-quantum mechanical theories the connection is much less obvious and the works of Hamel and Noether do deserve respect.

But the invariance principles have also other, and even more effective consequences in quantum mechanics. If we consider a system at rest – originally an atom was the subject of this consideration – the discrete energy levels' states' state vectors (or wave functions) are all linear combinations of a finite number of state vectors. Naturally it is best to chose these orthogonal to each other. If we subject these states to a rotation, the resulting state vectors can be expressed linear combinations of the original ones. The matrix which transforms the original state vectors into the rotated ones is, apart from a constant factor, uniquely determined. The matrices so obtained for the various rotations form, apart from a constant, a "representation" of the rotation group, i.e. the product of two such matrices gives a matrix which corresponds to the product, that is successive application, of the two rotations. It follows that a representation of the rotation group corresponds to each discrete energy level. These representations describe then several properties of the underlying states – they give, for instance, their total angular momentum. But they give many other properties of these states, the selection rules for optical transitions, the ratios of these transition probabilities between the various states of one energy level to another one, and so on. This is an application of the symmetry principles which is available only in quantum mechanics and is based on the linear nature of the description of the states in that theory. It was first established by the present writer and given in a book in 1931.

Perhaps even more interesting consequences can be derived if one uses an approximate Hamiltonian. In the case of atomic spectra it is natural to assume that the velocities of the electrons are small, hence the magnetic forces also small and that this applies also to the spin interaction. This leads to the L–S coupling theory of spectra. Another assumption leads to the J–J coupling theory in nuclear physics and we have heard of many other approximate symmetry applications in high energy physics. Perhaps it is good if I mention that the weak interaction is neglected as a rule and this leads to the assumption of reflection symmetry and hence to the parity concept.

In summary, in quantum mechanics the symmetry principles lead to an easy establishment of several consequences of the theory and are very useful in this connection also. But the prime function of the symmetry principles remains in my opinion to help in the establishment of the laws of nature and to control the validity of proposed laws.

I will now try to give a critical, very critical, review of the concepts considered to be fully valid in the preceding discussion and allude to the consequences of their possible violations. According to the great philosopher-physicist E. Mach all laws of nature and all concepts have only limited validity and he may be right in this regard.

Reservations

As to initial conditions, it is not clear that they can be obtained in the microscopic sense for all systems. A macroscopic body's energy levels are so close to each other that appears infinitely difficult to determine its exact state. In addition, as will be mentioned next, the difficulty is increased by its vigorous interaction with the environment. If the initial state can not be determined, not even in principle, the concept of "initial state", as we now use it, loses universal validity.

The same is true of the "law of nature" concept if it is impossible to retain the system in an isolated state. As was mentioned before, the difficulty to maintain the isolated state of a macroscopic body was first pointed out by D. Zeh in 1970 and his argument was further strengthened by the present writer around 1979. It was pointed out, in particular, that even if a macroscopic body, such as a cubic centimeter of tungsten, even if it is taken out into intergalactic space, far from all bodies which could influence it much more, the cosmic radiation would affect it so much that it would remain "isolated" only for about a thousands of a second. Of course, this fact does not reduce the practical usefulness of solid state physics, even though this treats even macroscopic systems as isolated, but it does attack the basic principles of our physics.

Another point which was emphasized by various colleagues is the probability that the coordinate system which is at rest with respect to the center of mass of the universe has simpler laws of nature than moving coordinate systems, that is that the Galileo-Newton theory of the invariance with respect to a uniform motion of the coordinate system is not exactly valid. We have, at present, no clear evidence for this, but the arguments denying the absolute validity of this invariance, which is incorporated into the Lorentz transformation and the theory of relativity, are quite reasonable.

What the "reservations" just made mean is that our physics will be fundamentally modified also in the future, as it was several times in the past. And we may hope that it will be also extended further, perhaps even to the phenomenon of life which is not described by our present physics. To finish with an optimistic remark, let me mention one extension of the area of physics which describes properties of objects such as solids – or liquids and gases. The extension resulted from the acceptance of the atomic and molecular structure of bodies – an acceptance opposed by the first physics book I read. It said "atoms and molecules may exist but this is irrelevant from the point of view of physics". It now seems relevant! Physics developed and extended its area greatly in our century! We can hope that it will extend it further.

Violations of Symmetry in Physics

E. P. Wigner

Scientific American *213*, 28–36 (1965)
(Reset by Springer-Verlag for this volume)

Of seven "mirrors" invented by physicists to describe the symmetry of the laws of nature, three have been shattered. Of those remaining, only one may still be wholly intact. It is called the CPT *mirror*

It was just nine years ago this month that physicists learned to their astonishment that left-handedness and right-handedness are built into the universe at the most fundamental level. Until December, 1956, they had assumed that if an event is possible, its mirror image is also possible, and that if one looks at some real event in a mirror, what one sees could also actually happen. This was known as reflection symmetry, and it forms the basis of the parity principle. In the summer of 1956 certain puzzling phenomena in nuclear physics led T. D. Lee and C. N. Yang to question the principle's general validity. In a few months C. S. Wu, Ernest Ambler, Dale D. Hoppes and R. P. Hudson had demonstrated that the phenomena clearly violated the principle.

The parity principle was one of several symmetry principles that physicists had long accepted as axiomatic in developing their mathematical theories. With the fall of parity they became uneasy about the other principles and sought ways to test each of them in turn. As a result of this endeavor at least two more principles have fallen and a third has been called into serious question. This is time symmetry: the principle that nature is indifferent to the direction in which time flows. Physicists have believed deeply that nature is similar to an electric clock that will run forward or backward, depend-ing on which way the starting knob is turned.

The various symmetries can be compared to mirrors that reflect natural events in carefully specified ways. The parity mirror, which we shall designate the *P* mirror, is simply the ordinary mirror of everyday life. It has one property, however, that may puzzle the layman; accordingly we shall also let "*P* mirror" stand for "physicists mirror" in order to distinguish it from the layman's mirror, which we shall call the *L* mirror.

Everyone is familiar with the fact that when an electric current flows in a coil of wire, it induces a magnetic field. We learned in school that the direction of the field—the direction of its north magnetic pole—can be determined by the "right-hand rule." This rule states that if the forefinger of the right hand has the shape and direction of the current flow, the thumb of the right hand points to the north pole of the induced magnetic field.

The early students of electricity defined the direction of current flow as being from the positive terminal of a battery to the negative. Now that we understand that an electric current depends on the flow of electrons, it seems more reasonable to speak of the direction of electron flow, which is from the negative terminal to the positive. Therefore if we are told the direction in which electrons flow in a coil of wire, we must use a *left-hand* rule to determine the direction of north in the magnetic field. (We realize, of course, that "north" and "south" are themselves conventions based on the fact that a compass needle is said to point toward the earth's North Pole.) In any case, in this article the direction of the magnetic field will be considered in relation to electron flow, which requires the left-hand rule. The reader is no doubt familiar with these elementary principles, but reviewing them may help to prevent confusion when we begin to look into mirrors.

Let us examine the parity mirror first. Imagine that we have before us on a table a coil of wire in which electrons are flowing clockwise as seen from above. The left-hand rule tells us that the induced magnetic field is pointing upward [*see illustration on page 347*]. Now imagine that there is a mirror on the ceiling directly over the table; what will we see in it? If we have placed our left hand next to the coil and shaped it to form the left-hand rule, we shall see in the mirror what appears to be a right hand with the thumb pointing down. The hand in the mirror tells us (correctly) that the electron flow as seen in the mirror is counterclockwise, but it also tells us (incorrectly) that the magnetic field is pointing down. The hand in the mirror misinforms us about the direction of the magnetic field because it is (or appears to be) a right hand, and a real field is related to a real electron flow by a left hand. If we accept the right-hand (incorrect) view of the field direction, we can be said to be interpreting the mirror image as laymen; in this sense the mirror is an *L* mirror. If, however, we insist as physicists that the electron flow is the prime reality and that the magnetic field is secondary, we will insist on using the left-hand rule to determine the direction of the magnetic field in the mirror and conclude that it is actually pointing back into the mirror and not out of the mirror. Thus

the magnetic field on the tabletop and the field in the mirror—the physicist's *P* mirror—are pointing in the same direction [*see illustration on page 349*].

We can now describe the parity experiment performed by Miss Wu and her collaborators. In the center of a ring of electric current they placed some radioactive cobalt, which emits electrons and neutrinos when it decays. The whole experimental arrangement has a plane of symmetry: the plane through the ring current. There is nothing to distinguish the upward from the downward direction. It nevertheless turned out that the electrons from the decaying cobalt atoms emerged asymmetrically: almost exclusively in the upward direction. If one were to place a mirror over the experiment, parallel to the plane of the ring current, the electrons would appear to be coming out of the mirror, or downward, which is not the direction they would travel if the current and radioactive material were real rather than mirror images [*see illustration on pages 350 and 351*].

This unexpected result attracted a great deal of attention—and a Nobel prize. The fact that the electrons emerged with a preferential direction, a direction that can be represented by the thumb of the left hand if the forefinger has the shape and direction of the electron flow, meant that the radioactivity of cobalt is partial toward the left hand.

If I may recall days long past, nobody was very happy with this result. It is a fact, of course, that most of us are as partial toward our right hand as cobalt is toward its left. We feel, however, that radioactive cobalt is not entitled to be partial because it should have forgotten its past and because, at the time it emitted the decay particles, it was under no influence that would have favored one side of the plane of the ring of current over the other. This plane was a symmetry plane at the be-

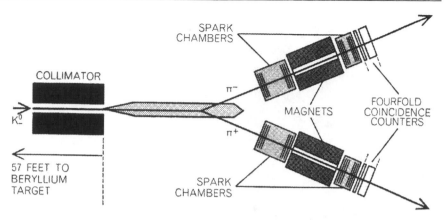

FALL OF TIME REVERSAL seems implied by an experiment showing that the K_-^0 particle (also known as K_2^0) sometimes decays into two pi mesons (π^+ and π^-) instead of into three pi mesons, as required by CP symmetry. If the K_-^0 decays anywhere in the dotted area there is a chance that the decay particles will pass through the two magnets and four spark chambers. The spark chambers are triggered to fire and thereby reveal particle tracks only if all four coincidence counters register the passage of a particle. The magnets produce a deflection (perpendicular to the plane of the page) that indicates the particle's momentum. The experiment was performed at the Brookhaven National Laboratory by James H. Christensen, James W. Cronin, Val L. Fitch and René Turlay of Princeton University.

ginning of the experiment; it should have remained a symmetry plane as long as it was undisturbed by outside influences. This statement is equivalent to the postulate that a possible sequence of real events should remain a possible sequence of events if every event is replaced by its mirror image. Evidently this is not the case for the disintegration process of radioactive cobalt.

Before long a score of physicists had independently proposed a reinterpretation of the Wu experiment that salvaged the principle of reflection symmetry. In essence they proposed that nature does not see itself in the P mirror but in a "magic mirror" where the signs of all electric charges are reversed. In this mirror the mirror image of an electron is a positron (a positive electron) and the mirror image of a radioactive cobalt nucleus is a similar nucleus made of antimatter (antineutrons and antiprotons). If one could view the Wu experiment in this proper mirror, one would see positrons flowing in the

direction that electrons had been assumed to flow. Since a flow of positrons is equivalent to a flow of positive current, one would have to use a *right-hand* rule to see how the magnetic field is pointing. One would then discover that the magnetic field is pointing out of the mirror, or downward, and thus directly opposite to the magnetic field in the real experiment. The decay particles emitted by the antimatter nuclei of radioactive cobalt would also tend to travel out of the mirror, or downward, thereby completing the mirror image of the Wu experiment [*see illustration at right on page 351*].

This reinterpretation of reflection symmetry was originally pure speculation, motivated solely by the desire to maintain the principle of reflection symmetry for the laws of nature. An experimental test was out of the question: even today we are far from being able to produce anticobalt, that is, a cobalt nucleus consisting of antiprotons and antineutrons. It was possible, however,

K-PARTICLE-DECAY EXPERIMENT is being continued at Brookhaven National Laboratory by Cronin, Fitch and their associates, using the apparatus shown here. The motion picture camera atop the scaffold is used to record particle tracks in the spark chamber below it.

to test the new hypothesis in other ways. The reinterpretation turned out to be relevant and was in agreement with all experimental findings until quite recently. These findings include the direction of flight of particles emitted in decay reactions other than that of radioactive cobalt. In particular there is a case in which a radioactive particle as well as

its antiparticle can be produced and observed. These particles are the muon and the antimuon; the decay of the antimuon looks in all details as the image of the decay of the muon would look in the magic mirror just described.

We have not yet mentioned the role of the ring current in Miss Wu's experiment. Its purpose is to create a mag-

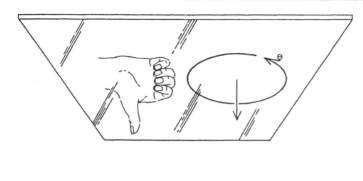

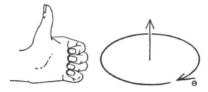

"LAYMAN'S MIRROR" provides incorrect picture of the relation of a magnetic field to the flow of electrons in a coil of wire. The "left-hand rule" applied to an actual electric coil (*bottom*) shows by the thumb direction that magnetic north (*straight arrow*) points up when the forefinger of the left hand is curved in the direction of the flow of electrons (*curved arrow*). When the left hand is reflected in the mirror, however, it becomes a right hand, thereby indicating (incorrectly) that the magnetic field is pointing down. In the "physicist's mirror," shown on the opposite page, the field is shown (correctly) to be pointing up.

netic field perpendicular to the plane of the current. This field in turn orients the spins of the nuclei of the radioactive cobalt atoms. The direction of the decay particles is related directly to the spins of the radioactive nuclei emitting them, and only indirectly by means of the magnetic field to the direction of the flow of electrons in the ring.

The spins of the cobalt nuclei carry an angular momentum, a fundamental property associated with rotating motion. In all studies of rotating motion before Miss Wu's experiment angular momentum was found to have a symmetry plane in the plane of rotation. If this plane were kept horizontal, one would expect the disintegration products (if any) of a rotating object to proceed with equal probability upward and downward. The fact that they do not

when the rotating object happens to be a radioactive cobalt nucleus means that the total symmetry of the laws of nature is smaller than physicists had previously believed. The laws are not invariant if reflected in a P mirror. The magic mirror that gives a true reflection is called a CP mirror; it is a combination of the parity (P) mirror, which reflects the positions of particles, and a "charge conjugation" (C) mirror, which changes the sign of electric charges.

How many mirrors has the theoretician conceived all together? I hope that we will not be suspected of patent-preemption if we claim to have "invented" seven mirrors. They are essentially various composites of the P and C mirrors and a third mirror: the T mirror, which reflects the direction of time. The seven mirrors are P, C, T, CP, CT, PT and CPT.

We have already seen how electric currents and magnetic fields are reflected in the P mirror. Let us now consider how the P mirror reflects the path of a particle as it is scattered, or deflected, by another particle [see illustration on page 353]. We can imagine that the scatterer is a heavy particle, such as an oxygen nucleus, and that the incident particle is a light particle, for example a positron. Thus each particle has a positive electric charge (represented by a plus sign in the illustration). The positron is so light that it will hardly affect the position of the oxygen nucleus; we need be concerned only with the path of the positron as it approaches the oxygen nucleus and is scattered.

We must also take into account the fact that the incident particle has an angular momentum, or spin, and that the axis of spin is parallel to the particle's direction of motion. After the particle has been scattered its direction of motion has changed, and the new direction of motion will be found to correlate with what has happened to the particle's spin angular momentum. If the spin remains pointing in the original direction (remains parallel), the particle's direction of motion will be somewhat above the plane of the original direction. If the spin flips around to point in the opposite direction (becomes antiparallel), the particle's direction of motion will be below the original plane. The particle can traverse either path and will take each path in a certain fraction of all cases observed.

The scattering event in front of the mirror will always be the same, but the image will be different in different mirrors. We shall assume that in each case the mirror is vertical and to the right, that is, between the object and the image. If the mirror is the physicist's P mirror, the reflected paths are just what one would expect. The path seen curving to the right in the actual case is seen curving to the left in the mirror. Moreover, the particle's direction of spin is reversed, so that if the particle seems to be spinning clockwise, as seen from the rear in the actual experiment, it will appear to be spinning counterclockwise as seen from the rear in the mirror image.

If one carried out this experiment, one would unquestionably find that the P mirror is right; with the accuracy of measurement now available one could not detect the difference between a real path and the path as it appears in the mirror. Yet we know from Miss Wu's experiment with radioactive cobalt that the mirror is not really right: in her experiment the P mirror gave an entirely incorrect picture. Hence we know that the P mirror is not quite right in general and that actuality will deviate from what it shows, even though in our scattering experiment its error would be immeasurably small.

Let us consider now how the scattering experiment looks in the C mirror [see illustration on page 354]. The C mirror is not, of course, a material mirror: it does not change the location of points, the direction of motion or the sense of spin direction. All it does is substitute negative electric charges for positive electric charges and vice versa; or, more generally, it substitutes antimatter for matter and vice versa. Thus when we "look" into the C mirror we see that the oxygen nucleus of our scattering experiment is replaced by an antinucleus consisting of antineutrons and antiprotons, and so has an overall negative charge, and that the positron is replaced by an electron, which is also negatively charged.

The situation with the C mirror is quite similar to that with the P mirror. Since no one knows how to make an antimatter nucleus as heavy as the nucleus of oxygen, however, our particular scattering experiment cannot be performed. But it is known from similar scattering experiments with anti-particles that there are no observable differences between actual scattering patterns and their reflections in the C mirror. Nevertheless, it has been established by other experiments that C reflection is no more an exact symmetry than P reflection is. Unfortunately the experimental demonstration of C violation is not as direct as Miss Wu's demonstration of P violation. The argument for the violation of C symmetry is a mathematical one based on the observed spin direction of electrons and positrons that are respectively emitted by negative and positive muons. The experiment was performed in 1957 by G. Culligan, S. G. F. Frank, J. R. Holt, J. C. Kluyver and T. Massam of the University of Liverpool.

If the P and C mirrors are known to be slightly defective when they are tested individually, is it possible that the magic CP mirror mentioned earlier still provides a faithful reflection of reality? The CP image can be obtained in either of two ways: by reflecting the P image in a C mirror or by reflecting the C image in a P mirror. The fact that the image obtained by two such reflections is an excellent picture of reality follows from the fact that the image produced by each mirror is extremely close to reality if reality is reflected in it. Indeed, until recently physicists believed the slight discrepancies in the individual mirrors canceled each other, so that the CP mirror was in exact accord with reality. This certainly seemed to be the case, at least for a number of

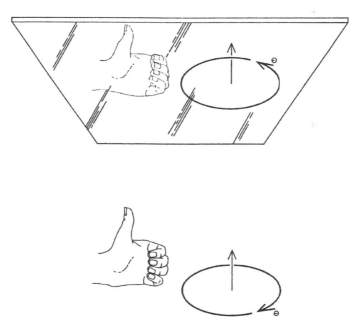

"PHYSICIST'S MIRROR" is also the P, or parity, mirror. It is identical with the layman's mirror, but the physicist is alert to deception. He knows that if he were to substitute a real electric coil for the one he sees in the mirror and send electrons flowing counterclockwise as seen from below, the direction of the magnetic field would still follow the left-hand rule. He recognizes, in other words, that the mirror reflection of a left hand (that is, a right hand) is deceiving him about the direction of the magnetic field. He insists as a physicist that the electron flow is the prime reality and that the magnetic field arises as a consequence.

phenomena that occur in radioactive decays and that violate C and P separately. Before discussing the experiment that has now cast doubt on the CP mirror, I should make a few comments on the T mirror.

Like the C mirror, the T mirror does not change the paths of particles. It merely reverses their direction, thus implying that the time axis is reversed [*see the illustration on page 355*]. In fact the designation T stands for time-reversal symmetry. The concept is hard to accept intuitively because our everyday experience with events that are patently irreversible is so compelling; the pieces of a shattered teacup have never been known to reassemble themselves

spontaneously. Irreversibility of this kind is not at issue. The physicist is concerned rather with the detailed reversibility of events at the atomic and subatomic scale. A model for this kind of irreversibility would be the behavior of a perfectly elastic ball. If such a ball were dropped on a perfectly elastic surface, it would bounce forever. If one were to make a moving picture of this ball as it bounced, there would be no way to tell whether the film were being run forward or backward; the time axis would be completely reversible.

Until recently the T mirror, like the CP mirror, was believed to be exact. And for reasons I shall describe later, physicists are forced to believe that the

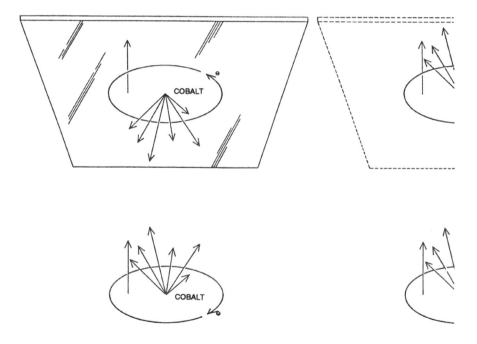

FALL OF PARITY, or reflection symmetry, followed the famous experiment performed in December, 1956, by C. S. Wu, Ernest Ambler and their collaborators. They placed a sample of radioactive cobalt in a magnetic field created by an electric coil and recorded the direction taken by one of the emerging decay products, namely electrons. According to the parity principle, the electrons should have emerged equally up and down; instead they emerged almost exclusively upward, in the direction of the magnetic field, as shown at the bottom in the diagram at left. If the experiment were reflected in the P mirror, the electrons would appear to emerge downward. If the radioactive cobalt and electric current in the mirror were real, the electrons would combination of the *T* mirror with the *CP* mirror—the *CPT* mirror—may still remain exact, even though the *C*, *P* and *T* mirrors appear to fail separately!

Let us turn to the experiment that has cast doubt on the validity of the *CP* mirror and, by implication, on the validity of the *T* mirror. The experiment was carried out a little over a year ago at the Brookhaven National Laboratory by James H. Christenson, James W. Cronin, Val L. Fitch and René Turlay of Princeton University. One of the original purposes of the experiment was the confirmation of *CP* invariance, not a demonstration of its failure. Experiments occasionally give surprising results, however; this one certainly did.

Nonetheless, the evidence for the violation of *CP* invariance is not as direct as the evidence for the violation of *P* invariance furnished by Miss Wu's experiment or even the evidence for the violation of *C* invariance in the experiment of Culligan and his collaborators.

The evidence for the failure of the *CP* mirror stems from one mode of decay exhibited by the *K* meson, or *K* particle. *K* particles are readily produced in a high-energy particle accelerator when a proton beam is directed at a suitable target, such as beryllium. The interaction of a high-energy proton with a neutron (contained in the atomic nuclei of the target) invariably yields two heavy particles, one of which is usually a proton or a neutron. When the bom-

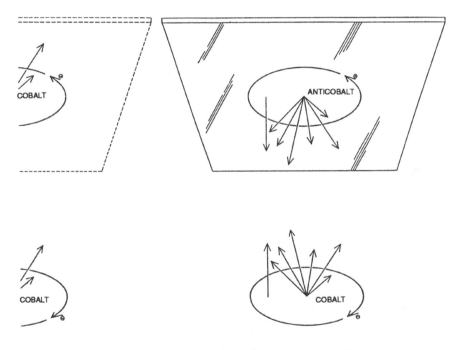

actually go upward (*middle diagram*). Reflection symmetry had been disproved. The principle was salvaged by declaring that nature sees itself in a "magic mirror," or *CP* mirror, in which matter is replaced by antimatter. This symmetric relation is shown in the diagram at right, where radioactive cobalt is replaced by radioactive anticobalt and the electrons flowing in the coil are replaced by antielectrons, or positrons. The decay particles, also positrons, then emerge downward. The magnetic field is likewise reversed because a flow of positive charges gives rise to a magnetic field opposite in direction to that created by a flow of negative charges.

bardment energy is around 30 billion electron volts, as it was in the Brookhaven experiment, the other heavy particle is likely to be one of the so-called strange particles, such as a lambda particle or a sigma particle. Simultaneously the interaction produces a K particle, which can be either positive (K^+), negative (K^-) or neutral (K^0).

The K meson is a very queer particle. It is the same particle whose puzzling decay behavior prompted Lee and Yang to question P invariance. Even before that it was ascertained that the K^0 is not a single particle but two particles that are antiparticles of each other. When a particle has an electric charge, it can easily be separated from its antiparticle because the latter must have an opposite charge. It is therefore impossible to create a charged particle that can be, with some probability, its own antiparticle.

The situation is different if a particle has no electric charge. In this case a state is quite conceivable in which a neutral particle such as a K^0 meson has a 50–50 chance of being either a particle or an antiparticle. An even more surprising result of present quantum-mechanical theory is that there is not one such state but a continuous manifold of "superpositions" of such states. For our purposes, however, we need be concerned only with the two states that are designated $(K^0 + \bar{K}^0)$ and $(K^0 - \bar{K}^0)$. The bar over the second K in each pair signifies antiparticle. It should be emphasized that each state stands for a single particle, but the properties of the two states are different and can be shown to be so by experiment.

The existence of such superposition states is a consequence of the wave nature of matter. Similar superpositions also play an important role in the low-energy region, in particular in the theory of optically active organic compounds, such as optically active amino acids and sugars. For example, one form of sugar can have the property of rotat-ing the plane of polarization of polarized light to the right. Another sugar of identical chemical composition will have the property of rotating the plane of polarized light to the left. The difference in optical activity is accounted for solely by the fact that the two compounds have three-dimensional structures that are mirror images and thus bear to each other the relation of left and right hands [see the illustration on page 356].

The quantum-mechanical interpretation of the position of an atom that determines whether an organic compound is left-handed or right-handed is plotted in the illustration on page 357. The horizontal axis gives the position of the atom, in terms of left or right, in the optically active compound. The vertical axis is the "probability amplitude" for each position of the atom; the probability of finding the atom at any particular position is defined as the square of the probability amplitude. When the curve of the probability amplitude lies entirely to the right of center, the atom is surely to the right, thereby creating an asymmetric situation. This corresponds to a right-handed, or dextro, compound. When the curve of probability amplitude lies entirely to the left, the atom is surely to the left, corresponding to the left-handed, or levo, compound.

The lower pair of curves in the illustration represents probability amplitudes for atoms in which left and right positions are equally probable, with the result that the rotational properties of the compound cancel each other and no optical activity is observed. In the curve at lower left the probability amplitude consists of two symmetrical humps, both positive. This state is optically inactive because it is a mirror image of itself. Such states are stable at low temperature and are described as racemic mixtures. In the curve at lower right the probability amplitude is asymmetric: the hump at the left is positive, whereas

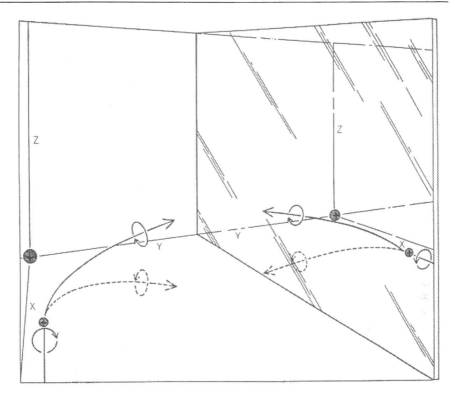

P-MIRROR VIEW OF "SCATTERING" is exactly what one would expect to see in an ordinary mirror. In the actual experiment (*left*) a positron is being scattered, or deflected, to the right as it approaches the positively charged nucleus of a fairly heavy atom, such as oxygen. The positron also possesses spin, or angular momentum, shown by the small curved arrows. If the spin remains unchanged after scattering, the positron will be found above the plane of its original path; if the spin direction reverses, the positron will be found below the plane. In the mirror, as expected, all spins are opposite to those in the real experiment.

the hump at the right is negative. The probability, however, is the square of the vertical displacement of the asymmetric curve and therefore has a positive and equal value on both left and right. This state, although optically inactive also, has different properties from the first optically inactive state; in particular, its energy is very slightly higher.

These four states—two optically active and two inactive—have their counterpart in the neutral *K* mesons. When the neutral *K* meson is first created, it appears as the antiparticle K^0 and corresponds to the optically active left-

handed compound. This state can be considered as the sum of the symmetric and antisymmetric states representing the two optically inactive forms of the compound. In the case of the neutral *K* meson these two states can be designated K_+^0 and K_-^0 (also known as K_1^0 and K_2^0). The only essential difference between the two inactive states of the compound and the two neutral meson states K_+^0 and K_-^0 is that the reflections with respect to which the meson states are symmetric and asymmetric are not ordinary reflections but reflections in a *CP* mirror.

Now, the symmetric state K_+^0 can

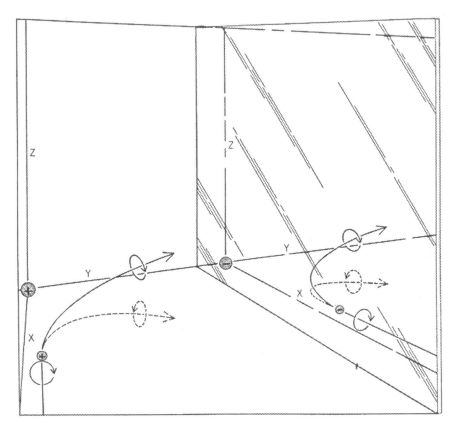

C-MIRROR VIEW OF SCATTERING is unlike that seen in any material mirror. The direction of particle paths and spin remains unchanged, but all charges are reversed.

decay into two pi mesons and will do so in about 10^{-10} second. The asymmetric state K_-^0 can and should decay into three pi mesons, a decay process that takes about 600 times longer than the two-pi decay, or about 6×10^{-8} second. Therefore K_-^0 mesons will still be plentiful long after all K_+^0 mesons have disappeared. The K_+^0 can decay into two pi mesons because all the states of a system composed of two pi mesons are symmetric with respect to CP reflection, as is the K_+^0 state itself. The K_-^0 state, being antisymmetric with respect to CP reflection, should not be able to decay into a symmetric state, but it can decay into three pi mesons. A system of three pi mesons does have states that are antisymmetric with respect to CP. The three-pi decay is a slower process, hence the longer life of the K_-^0.

What Cronin, Fitch and their collaborators observed, however, is that a small fraction of the K_-^0 mesons do decay into two pi mesons, in defiance of CP symmetry. Since only about one in 500 of the K_-^0 mesons decays into two pi mesons, this mode of decay is more than 100,000 times slower for the K_-^0 than for the rapidly decaying K_+^0. Nevertheless, the "forbidden" decay does occur, and this is interpreted as a breakdown of CP symmetry.

One can see that the preceding argument is quite involved and is by no means so simple as that represented in

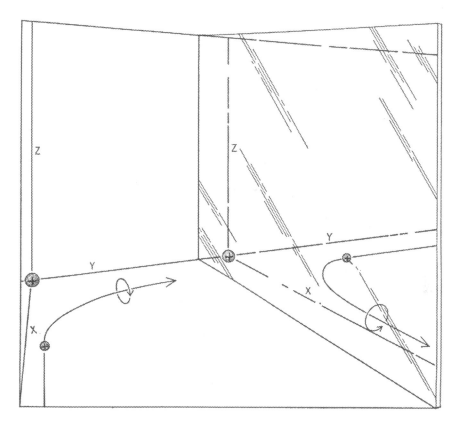

T-MIRROR VIEW OF SCATTERING is also without resemblance to images seen in ordinary mirrors. In the T mirror the scattered particle travels the same path as in actuality, but proceeds in the opposite direction. The imaginary T mirror represents time-reversal.

the breakdown of P symmetry in Miss Wu's experiment. As a result some physicists have been reluctant to accept the Cronin-Fitch experiment as conclusive evidence for the failure of CP symmetry. There may be a way out that preserves CP symmetry—in fact, several ways have been suggested—but the weight of the argument is increasing that CP has failed.

What follows from the violation of CP symmetry? Physicists are left with the belief that the very last mirror, the CPT mirror, is a true mirror. This belief is not based on nature's innate preference for symmetry; it is based on the stubborn fact that we cannot formulate equations of motion in quantum field theory that lack this symmetry and still satisfy the postulates of Einstein's special theory of relativity. If the principle of CPT symmetry is valid, it is evidence for the correctness of the general framework of quantum electrodynamics and of the special theory of relativity, not for nature's preference for any additional symmetry.

It must be noted with some apprehension, however, that in order for the CPT mirror to remain valid the T mirror itself must be invalid. The reasoning, based on the Cronin-Fitch experiment, is this. The $K_-{}^0$ begins in an antisymmetric state and decays into a symmetric state when it is reflected in a *CP* mirror, thereby proving the mirror defective. If the image in the *CP* mirror is now reflected in the *T* mirror, the original asymmetry should be restored—provided that the *CPT* mirror (*C* plus *P* plus *T*) is valid. To turn a symmetric state into an antisymmetric one, however, the *T* mirror by itself must produce an antisymmetric image. This is equivalent to saying that time is not invariant under reflection and that time-reversal symmetry has failed.

Physicists have scarcely begun to examine the implications of this final breakdown. Leaving aside the apparent collapse of *T* symmetry, one can conclude from the failure of *P*, *C* and *CP* symmetry that the laws of nature do show a preference for either the right or the left hand. We are surrounded by many phenomena that appear to show just such a preference, or, more precisely, such a distinction between right and left. Most of us are right-handed and our hearts are on the left side. On a large scale we observe that the earth rotates to the left (counterclockwise) as seen from above the North Pole and proceeds to the left around the sun. The sun, in turn, travels to the right around the galaxy as viewed from above the north galactic pole. Heretofore these asymmetries were attributed to asymmetries in the initial conditions. Now it is possible to attribute the same asymmetries to the laws of motion, that is, to assume that the universe was initially more symmetrical than it is now and that the present state evolved as a result of the asymmetry of the laws of motion. Few people are as yet ready to accept these speculations; I personally do not believe they are valid. Such speculations could nevertheless be test-

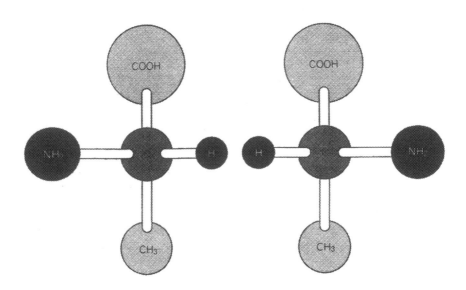

MIRROR-IMAGE MOLECULES, called optical isomers, are well known in organic chemistry. This diagram shows two isomers of alanine, one of the 20 amino acids that living organisms use to build protein molecules. When placed in solution, the isomer at left, known as the levo form, rotates plane-polarized light to the left. Its mirror image, the dextro form, rotates polarized light to the right. Natural proteins are built exclusively from levo amino acids. Many other organic compounds occur in left-handed and right-handed configurations.

ed if we had enough information about the sense of rotation of planets in other solar systems.

The fact that the laws of nature have no pure space-reflection symmetry has one consequence that is unpleasant to admit. It deprives us of the illusion that these laws are—in perhaps a subtle but nonetheless a real sense—the simplest laws that can be conceived and that are compatible with some obvious experience. If one law of nature is possible, an alternative law obtained by reflecting the first law on a plane would be equally possible and equally simple. We had previously thought the law obtained by reflection would be identical with the original law, just as the reflection of a sphere is also a sphere. Now we know that this is not so. The difficulty

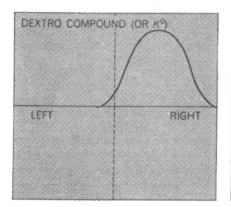

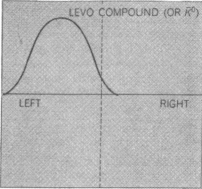

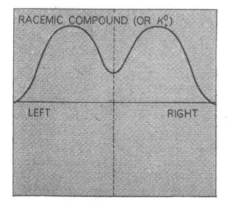

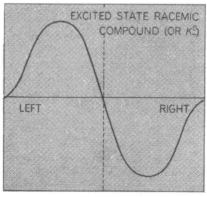

QUANTUM-MECHANICAL INTERPRETATION of dextro and levo organic compounds applies also to states of the neutral K meson. It invokes the concept of "probability amplitude," plotted as the vertical axis in these curves. The horizontal axis gives the position, in terms of left or right, of the atom that determines whether an organic compound is left- or right-handed. The probability of finding the atom at any particular position is the square of the probability amplitude. In the upper pair of curves the atom is surely either to the right or to the left. In the lower pair of curves the atom has an equal probability of being in the left or right position. How these curves apply to the K meson is explained in the text.

began when Miss Wu's experiment showed us that the preferential direction of the decay particles of radioactive cobalt was arbitrarily upward. We got out of this difficulty by postulating another substance: radioactive anticobalt that emits particles downward. This restores the symmetry because we can say that the laws of nature are symmetric but imply two kinds of substance, matter and antimatter. The apparent asymmetry in the laws of nature was thereby reduced to an asymmetry in the initial conditions that allowed matter to predominate over antimatter, at least in the only part of the universe we know at first hand.

The recent experiment of Cronin and Fitch indicates, however, that such an explanation is impossible. The indication, to be sure, is only indirect; we must explore all avenues that may yield another interpretation and preserve the spatial-reflection symmetry of the laws of nature. If these avenues do not lead out of the difficulty, we will have to admit that two absolutely equally simple laws of nature are conceivable, of which nature has chosen, in its grand arbitrariness, only one. The extent to which the laws of nature are the simplest conceivable laws has come to an end—no matter how subtly they may be formulated—as long as they are formulated in terms of concepts that are subject to the symmetry principles we are accustomed to associating with space-time.

The question naturally arises whether or not physics has experienced similar crises before. It has. In classical physics, matter was supposed to be infinitely subdivisible without change in its bulk properties such as specific gravity, viscosity or elasticity. The discovery of the atom put an end to this infinite subdivisibility. The atom therefore increased the complexity of the structure of matter, and this unavoidable new fact was for many people as obnoxious as the lack of reflection symmetry in the laws of nature is for us.

Most of the consequences of atomic structure that first appeared obnoxious were eliminated when physicists and chemists learned to use the atomic scale for their measurements and realized, for example, that atoms provide a natural unit of length. Without such a unit it would be difficult to understand why human beings have an average height somewhere between five and six feet; if all phenomena could be scaled up or down with impunity, men, mice and bacteria could be the same size. Atomic theory also provided explanations for the properties of matter, for its density, viscosity, elasticity and so on. Hence in its end result atomic theory enriched rather than complicated our picture of nature. There is hope, but as yet only a hope, that the present probing into the symmetry of space-time will have a similar result.

Editor's Note: The original of this paper had colored illustrations. They are printed here in black and white and a few small changes in captions and text have been made to accommodate this fact.

Symmetry Principles in Old and New Physics

E. P. Wigner

Bulletin of the American Mathematical Society *74*, 793–815 (1968)

Introduction and summary. Symmetry and invariance considerations have long played important roles in physics. The 32 crystal classes—that is, groups of rotations in three-dimensional space all the elements of which are of the order 2, 3, 4 or 6—were determined 137 years ago, in the same year in which group theory was born. The determination of the 230 space groups, by Schönflies and by Fedorov (these are the discrete subgroups of the Euclidean group which contain three noncoplanar translations) was a masterpiece of analysis and so was the determination by Groth of the possible properties of crystals with the symmetries of these space-groups.

The groups of prime importance in classical physics were subgroups of the Euclidean group and the enumeration of these subgroups and the derivation of the properties which are invariant under them were the principal problems. The invariance groups of the relativity theories were, from the mathematical point of view, much more esoteric but their use by physicists did not contribute greatly to the mathematical theory of groups nor did it point to new interesting mathematical problems. When, however, the invariance arguments were applied to the present century's other great innovation of physical theory, to quantum theory, a score of new problems and several interesting mathematical theorems were uncovered. The basic reason is the difference in the characterization of states in quantum and in pre-quantum theories. In the latter, a state was characterized by the positions and the velocities of particles. These could be specified by points in three-dimensional space. Quantum theory, on the other hand, specifies the states by vectors in an abstract Hilbert space. Symmetry transformations in pre-quantum theories were rather obvious transformations of three-dimensional space; in quantum theory they became unitary transformations of Hilbert space. These form subgroups of all unitary transformations which are essentially homomorphic to the symmetry group in question, *essentially* homomorphic only because a unitary transformation in quantum mechanics is equivalent to any of its multiples by a numerical factor (of modulus 1). However, this essential homomorphy could be reduced, particularly

This is the text of the forty-first Josiah Willard Gibbs lecture delivered before the Annual Meeting of the Society in San Francisco on January 23, 1968; received by the editors April 25, 1968.

as a result of Bargmann's investigations [12], in most cases to a true homomorphy to an extended group which is called, then, the quantum mechanical symmetry group. The quantum mechanical operations of the symmetry group break up the Hilbert space of all states into subspaces each of which is invariant under the operations in question. These operations form, then, within each invariant subspace, a representation of the quantum mechanical symmetry group by means of unitary transformations.

The physicists undertook a much more detailed investigation of the representations of the groups in question than the mathematicians had been interested in previously. In those cases in which the representations were known before in principle, they defined canonical forms of the irreducible representations and determined the invariant subspaces of the direct products (also called tensor products) of invariant subspaces. This led to the theories of the various coupling coefficients, such as three, six, and higher j-coefficients which are of interest from the point of view of pure mathematics also. For several noncompact Lie groups, the unitary representations of which were not known before, these were determined. In many cases, all of the not trivial ones were infinite dimensional. The determination of these for all locally compact groups became a field of mathematical interest.

The theory of group representations was, until a few years ago, at the center of the interest of the physicist investigating the consequences of invariance principles. The invariance group was, in most cases, either the quantum mechanical Poincaré group (LC2I), or a subgroup thereof. In later years, the interest shifted to groups which are only approximate symmetry groups and form extensions of the Poincaré group. Hence, the interest shifted back from the representations of definite groups to the determination of groups, in particular those which contain the Poincaré group as a subgroup. Many interesting results were obtained concerning the existence of such groups and also their representations.

The evolution of the physical sciences. Physics and the natural sciences have changed enormously during the past 100 or 150 years. The spirit has changed, the subject has changed, and the mode of operation has changed.

The change in spirit has been toward increasing sophistication. Whereas a hundred or so years ago, the laws of physics were formulated in terms of directly observable quantities, present day physics uses intricate mathematical constructs—indeed, its analysis of the concept of "directly observable quantities" led to the conclusion that, in the microscopic domain, there is no such thing. It is easy to forget that the first big step in the direction of mathematical sophistication, the introduction of phase space with its billions of dimensions, was due to Willard Gibbs [1]. We all are keenly aware of the other two most important steps: the establishments of the relativity and quantum theories. It is not devoid of the elements of irony that the last two steps were undertaken in order to eliminate not directly observable quantities from the physical theory. What was accomplished is to have eliminated quantities, the impossibility of the direct observation of which was then recognized. What was substituted eventually for these quantities, the state vector and the gravitational metric, are not directly observable either; they are vastly more sophisticated than the earlier concepts both mathematically and conceptually.

The subject of physics, and of the other natural sciences, has changed also. The change can be most easily specified in the case of physics: whereas until the turn of the century physics was concerned only with macroscopic objects, we now consider an article on macroscopic physics, such as the wonderful new theory of friction, that of Bowden and Tabor [2], to be esoteric.

Finally, the mode of operation has changed. We now have Alvin Weinberg's Big Science, with hundred of scientists streaming into the Laboratories each morning to tear away the covers with which nature seeks to hide its secrets. This change has been discussed a greal deal and I do not want to add to the discussion now.

What has not changed is that, as in Galileo's time, and in Galileo's words [3], the laws of nature are spoken in the language of mathematics. What also has not changed—at least not in the last 150 years —is that the concepts of symmetry, of invariance, play a very large role and, it appears, an increasing role in physics.

The parts of mathematics which are of most use to physicists have, on the other hand, changed enormously. Even 50 years ago, the mathematics of the physicist consisted of ordinary differential equations, with a sprinkling of partial ones. The latter, and the theory of Hilbert space, assumed the dominant position about 35 years ago. Nowadays, however, we hear more about the theory of analytic functions, of one or more variables, of the theory of distributions and, last but not least, the theory of groups and their representations, than the theory of Hilbert space and principal axis transformations which, after all, still are believed to provide the language for expressing what we know about the laws of nature.

Crystal symmetry. Let us now turn to the first subject proper:

the role of symmetry in very old physics. This was, essentially, confined to the discipline of crystallography. The reason for my discussing it is not that I have new mathematical results on the subject but that the history of crystallography illustrates the development of the same ideas, side by side, by mathematicians and by natural scientists, first knowing very little about each other, but interacting vigorously later. The natural scientist's role is, to a considerable extent, to furnish the original problems and some of their solutions. The mathematician not only gives a deeper understanding of the solution given by the natural scientist but also greatly generalizes the initial problem. In the early stages, the two disciplines do not know about each other's methods and conclusions; the interaction becomes intimate in the later stages.

My story will be, naturally, mostly confined to the natural scientist whose story may be novel to mathematicians.

It begins in 1830—just 138 years ago—when J. F. C. Hessel determined the 32 crystal classes [4]. These are the finite groups of rotations in three-space, proper and improper, which have only elements of the order 1, 2, 3, 4, and 6. This was just two years before a famous duel took place in Paris and in the same year in which the name group was coined and its concept was precisely formulated.

One may wonder why Hessel confined his attention to groups with elements all of the order 1, 2, 3, 4, 6. The reason was that he knew of a property of crystals, discovered by one of the founders of crystallography, the Abbé Haüy, about 50 years earlier [5]. This law refers to crystallographic planes. It stipulates that, using the three intersections of any three crystal planes as directions of the coordinate axes, the ratio of the intercepts of any crystal plane with respect to these axes, if measured in terms of the intercepts of any other crystal plane, are rational numbers. This so-called law of rational indices is incompatible with any rotational symmetry except rotations by 60 and 90 degrees, and multiples of these angles. This leads to Hessel's condition of all symmetry elements of the rotation group being of order 1, 2, 3, 4, or 6.

Haüy's law of rational indices was an empirical law and, needless to say, it could not have been established were it not true that the rational numbers referred to therein are very simple. If the basic planes are properly chosen, both the numerators and the denominators of the rational numbers are, as a rule, below six, often they are below four. If one were to admit arbitrary rational numbers, the law could not have been verified experimentally. The law could not have been invented at all without the picture of the structure of the crystal, conceived almost 300 years ago (by the bishop, Steno [6]), as a regular, lattice-like arrangement of atoms as we now know it to be. This picture led Haüy to the law of rational indices which was then experimentally verified and this, in its turn, led to the restriction of the rotations, which can be crystal symmetries, to those used by Hessel. When the total symmetries of crystal lattices, that is symmetries including spatial displacements, were investigated just before the turn of the century, by Fedorov and by Schönflies [7], Hessel's groups appeared as the 32 different factor groups of all possible space groups, the displacements forming the normal subgroups. The space groups are discrete subgroups of the Euclidean group which contain three noncoplanar translations. There are 230 of them, as determined by Schönflies and by Fedorov—by means of group theoretical methods which became, by that time, familiar at least to some crystallographers. Hessel's restriction of the groups, to those with elements of orders 1, 2, 3, 4, and 6, was much less arbitrary than might at first appear.

The rest of the story of the role of symmetry in very old physics is interesting only for us physicists. The consequences of the symmetry properties of crystals in terms of the macroscopic properties yielded a fascinating amount of information—information that was derived, principally by Groth [8], with a full understanding of the group property of symmetry operations though, as a rule, not in the language of group theory. It continues to be a pleasure for most of us older physicists to read about the properties of crystals, also to learn about the very few exceptions in which the symmetry, present in the overwhelming number of properties, is violated by a few. It is painful, on the other hand, to admit that the symmetry cannot be formulated in terms of our present, quantum mechanical theory, that it is surely only approximate. Approximate in the sense that it is valid if classical, that is nonquantum, theory is a valid approximation for the motion of the nuclei. It is painful also to admit that none of us has succeeded in finding the limits of the validity of the concept of crystal symmetry and to point to phenomena in which the approximate nature of the symmetry would manifest itself.

Before turning to more modern subjects, and in particular to quantum theory, it may be worth observing that even though the explicit role of symmetry in very old physics was largely confined to crystallography, the intuitive concept of symmetry probably played a great role in the thinking of the early great physicists. Thus, the force between two point-like bodies was assumed to be central, i.e., to have the direction of the line connecting the two bodies. This is the only direction compatible with rotational invariance. The invariance of the laws of physics with respect to translations—of both space and time—was an assumption which pervaded the thinking of natural scientists much before a sophisticated language for its formulation was invented. Some writings of Newton clearly show his awareness of the principle now called Galilean invariance.

At the opposite end of the period under review, Hamel, Klein, and Nöther gave highly elegant and sophisticated derivations of the conservation laws of physics on an invariant theoretic basis. These conservation laws were, of course, well known by that time on the basis of elementary derivations. For this reason, and because the application of the symmetry principle in question is a rather indirect one, and also because the subject has been discussed a great deal before by both mathematicians and physicists, it will not be further elaborated here.

Quantum mechanics. It must have been surprising that symmetry principles played an explicit and direct role only in a very restricted part of very old physics: in crystallography. Surely, the invariance of the equations of motion with respect to the whole Galilei or Poincaré groups [9], including as they do the whole Euclidean group, must have direct consequences for phenomena outside of crystallography. Let us see what those consequences are.

Let us consider a very simple system, a hydrogen atom with the electron moving in a Bohr orbit. What are the obvious and simpleminded consequences of the invariance of the equations of motion with respect to the Galilei or the Poincaré groups? First, that the hydrogen atom could be anywhere else rather than where it actually is and that the electron could be further ahead or farther back on its orbit. Further, that the atom could be in uniform straightforward motion rather than at rest as I imagined it to be. Finally, that the orbit could be tilted in some way in space, rather than being horizontal. All this may be true, but none of this provides us with any of the properties of the orbit.

Let us consider the last conclusion. It is true, perhaps, that the hydrogen atom could be anywhere and could have any velocity. It is surely not true that its orbit could have any orientation in space. If it could, then even a hydrogen atom at rest at a given position could be in infinitely many entirely distinct states which is contrary to all intuition and all experience and also to the vanishing entropy of the internal motion. Everyone felt this, much before the advent of quantum mechanics.

How does quantum mechanics solve this contradiction? Once we know the state of motion of the hydrogen atom, we know its state completely. It then follows that the hydrogen atom (at least if we disregard its spin) is spherically symmetric. Are, however, all the states of the hydrogen atom spherically symmetric? Evidently not; a proton-electron combination must have not-spherically-symmetric states. Since, just as in classical theory, all such states can be subjected to a rotation, there are infinitely many states connected with most excited states of the hydrogen atom. However, and this is the crucial point, all these states can be written as linear superpositions of a finite number of states. The essential point here is that, in quantum mechanics. the physical state, the actual situation of a system, is characterized not by positions and velocities but by a vector in Hilbert space [10]. The equivalence of all directions does entail, in quantum as well as in classical theory, the existence of all states which are obtained from any given state by a rotation. However, unlike in classical theory, this does not conclude the story. In quantum theory, all the states obtained by the rotation of one can be written as linear combinations of certain basic states; in classical theory the linear superposition of states does not exist. The states in quantum theory are vectors in a linear space—which they are not in classical theory. We cannot add two states of a classical system to each other; we can do this for quantum states. Furthermore, the state vector which is the sum of two other state vectors is not really a new state; it is, with a probability $\frac{1}{2}$ the first state, with a probability $\frac{1}{2}$ the second. This is true at least if the state vectors of the latter states are orthogonal; if they are not orthogonal to begin with, they are not entirely different from each other either. It follows from this, in particular, that, when calculating the entropy, only the states with orthogonal state vectors must be counted.

Before formulating the mathematical problems to which this situation, the linear character of the state vectors, leads us, it may be well to complete the preceding picture in some regard. Thus far, we have considered only rotations of a definite state. How about imparting to them a certain velocity? Let us consider the hydrogen atom, first at rest, then moving with various velocities. It follows from the Galilei or the Poincaré invariance that, if a system at rest is conceivable, the same system can be in uniform motion with any velocity, in any direction. We again have to postulate, as a result of the existence of a given state, an infinity of other states, just as in classical theory. Is it true again that, in quantum mechanics, this infinity of states, or rather their state vectors, can be represented as a linear combination of a smaller number of states? No, the answer is, in this case, in the

negative; in fact, the state vectors are orthogonal for all pairs of states
with different velocities. This is natural from the point of view of
physics—one can experimentally distinguish states with different
velocities though not with different orientations of the orbits—but it
is surprising to find such a difference between the consequences of
two types of invariance transformations: rotations in space and con-
ferring a velocity.

Let us look, finally, at the last type of Galilei transformations:
displacements in space and time. The situation with regard to these
is very simple: if the velocity is specified, the state is invariant with
respect to displacements. This is a consequence of Heisenberg's uncertainty relation. If one wishes to obtain a state localized at a given
time near a point, one has to form a superposition of states with different velocities—states which we have just learned to be orthogonal
to each other.

It may be good to repeat in precise mathematical terminology what
was said before in the language of the physicist. The state of a system
is given in classical mechanics by a point in phase space. This has six
dimensions if the system consists of a single particle: the coordinates
are the positional coordinates and the velocity components of the
particle. The phase space has correspondingly more dimensions if the
system consists of more particles [11]. The Galilei and Poincaré
transformations are linear inhomogeneous transformations in phase
space. In quantum mechanics, on the other hand, the state of any
system is characterized by a vector in infinite dimensional Hilbert
space. The invariance transformations are linear, in fact unitary,
transformations in that space. Since the unitary transformation which
corresponds to the product of two invariance transformations is, at
least essentially, the product of the unitary transformations which
correspond to the two factors of the product, these unitary transformations form, at least essentially, a unitary representation of the
symmetry group. The symmetry group is the Galilei group in non-
relativistic theories; it is the Poincaré group in relativistic theories,
but it can be a subgroup of these if some outside influence decreases
the total symmetry of space-time. Thus, for the motion of the electrons in a crystal, the symmetry group is one of the 230 space groups
which were discussed before.

The great difference between classical and quantum transformations is, however, not the difference between a linear inhomogeneous
transformation on the one hand, and a unitary one on the other.
The great difference is, first, that in nonquantum theory the transformation is always the same, or, rather, depends only on the number

of particles the positions and velocities of which we have to transform. In quantum theory, on the other hand, the symmetry transformations, the unitary representation of the symmetry group, are different for different systems; they determine many of the properties of the system. The difference is also that the addition of two states is meaningless in nonquantum theory but meaningful in quantum theory; the state space of quantum theory, its Hilbert space, is a truly linear space.

The first point goes far in explaining the physicist's interest in unitary representations. The unitary representations of the Galilei group are implicit already in Schrödinger's theory; they were shown to be implicit most clearly by Bargmann [12]. The unitary representations of the Poincaré group were determined in the late '30s; except for the trivial one, they were all shown to be infinite dimensional [13]. This is equivalent with the statement that no system can be relativistically invariant unless it can be in an infinity of orthogonal states. By calling attention to the properties of the unitary representations of noncompact Lie groups, the physicists have stimulated the mathematicians' interest in this subject. The mathematicians are now very much ahead of us in this field, and it is not easy to catch up with the results of Gelfand, Neumark, Harish-Chandra, and of many others [14].

The role of the group of rotations in three-space. Let us now return to a question which was alluded to before: the difference between the effect of rotations and that of imparting velocities, to a state. One can say that the total invariance group is composed of three types of elements: displacements, rotations, and the imparting of a velocity. If we consider minimal subspaces which are invariant under displacements, that is, the subspaces of a representation space which form the basis of an irreducible representation of the displacement subgroup, there is one such subspace which remains invariant under rotations. This does not surprise the physicist: the subspace in question is the one the state vectors of which describe the system at rest. In fact, in the relevant irreducible representations of the whole symmetry group, the effect of the rotations on this subspace is that of an irreducible representation of the group of rotations. The transformations which correspond to the imparting of a velocity, on the other hand, transform each minimal subspace which is invariant under displacements into a similar subspace which is, however, orthogonal to the subspace to the states of which a velocity was imparted. Thus, imparting of a velocity has an effect on these subspaces similar to that

which any transformation has in classical theory: it produces an entirely new state. In a sense, therefore, the velocity imparting transformations are trivial, the transformations which correspond to rotations highly nontrivial: they form, for the subspace in question, an
irreducible representation of the rotation group. It follows that the
irreducible representations of the total symmetry group can be
characterized by the behavior of the particular, preferred, minimal
subspace under the influence of displacements and of rotations. This
question was discussed in so much detail in order to give an explanation for the physicist's intense interest in the group of rotations in
three-space, the explanation of this interest starting from the general
Poincaré or Galilei invariance principle.

   If we think of an irreducible representation of the rotation group,
or any other group, and want a firm grasp on it, it is good to define a
coordinate system in the representation space. The way this can be
done most naturally is to specify a sequence of subgroups, G, G_{n-1},
G_{n-2}, $\cdots$, G_1 each being a maximal subgroup of the preceding one
and G_1 being the group consisting of the unit element only. Let us
assume then that the transformations of G which correspond to elements of G_{n-1} which form, therefore, a representation of this subgroup G_{n-1}, contain no irreducible representation of G_{n-1} more than
once. Let us further assume that the same is true of all irreducible
representations of each G_k, if restricted to the subgroup G_{k-1}. Then,
we can specify a direction in the representation space of G uniquely
by enumerating the irreducible representations of G_2, G_3, $\cdots$, G_{n-1}
in the representation spaces of which this direction is contained.
We can then specify a unit vector in this direction; the unit vectors
obtained in this way will be used as basis vectors in the representation
space. They will be orthogonal to each other and will form a complete
set of vectors in the space of the irreducible representation of G from
which we started.

   An example may illustrate the situation. Let us choose for G the
symmetric group of all permutations of n symbols, S_n. If we choose
as the sequence of subgroups the groups S_{n-1}, S_{n-2}, $\cdots$, S_1 where
S_{n-k} is the symmetric group which leaves the last k symbols unchanged, it follows from the classical theories of Young and Frobenius
[15] that indeed no irreducible representation of S_{n-k}, if restricted to
the subgroup S_{n-k-1}, will contain any representation of the latter
more than once. Hence, one can specify, for instance, a vector in the
space of the representation $3+2$ of S_5 by stipulating that it belong to
the representation space of the representation $3+1$ of S_4, to the representation space $2+1$ of S_3, and to the representation 2 of S_2. In the

case of the group of rotations in three-space, the sequence of subgroups contains, in addition to O_3, only O_2 (and O_1), O_2 being the group of rotations which leave the z axis invariant, i.e., the rotations about the z axis. This leads to the usual form of the irreducible representations of O_3.

Before leaving this subject, I would like to mention a new mathematical result related to the preceding consideration [16]. It is a necessary and sufficient condition that a subgroup have the property considered before: that the restriction of the representation to this subgroup contain no representation of the subgroup more than once, and that this be true for all irreducible representations of the original group. In order to formulate the condition, it is useful to introduce the concept of a *subclass*. This is the set of elements which can be transformed into each other by the elements of the subgroup. It is clear that the product of two subclasses consists of complete subclasses, that the ordinary classes of the group consist of one or more subclasses. It is not clear that the subclasses commute as do the ordinary classes. If they do, the subgroup has the property specified above and this is also a necessary condition therefor. It is easy to see that the subclasses will commute if they are self-inverse and it is easy to show that the subclasses of S_n with respect to the subgroup S_{n-1} are self-inverse. Hence, the result of the theories of Young and Frobenius, which was mentioned before, can be obtained in this way also.

The definite form of the irreducible representations of the three dimensional rotation group helps to solve some concrete problems. The squares of the matrix elements have a simple interpretation which is, though, difficult to verify experimentally: they give the probability that a particle with a definite component of its angular momentum in one direction have a given angular momentum component in another direction. Better known, and amply verified, are the Hönl-Kronig rules for the intensity ratios of the transitions between the sublevels into which two levels are split by a magnetic field [17]. The subgroup O_2 must be chosen in this case to be the rotations which leave the direction of the magnetic field unchanged.

Decomposition of the tensor products of representations. The definite form of the irreducible representations also helps in the solution of the problem to which we turn next: the reduction of the Kronecker product (also called inner tensor product) of representations into their irreducible components. It would be difficult to enumerate all the applications of this reduction. The physical basis of most of the applications is the rule for obtaining the state vector of the union of two systems: this is defined in the Hilbert space which is the direct product (also called Kronecker product) of the Hilbert spaces of the separate systems and the state vector is the Kronecker product of the state vectors of the individual systems [18]. Hence, if two systems are united, each being in a state which is one of the basis vectors of an irreducible representation—not necessarily the same irreducible representation—the state vector of the union of the two systems will be in the representation space of the Kronecker product of the two irreducible representations. A similar statement applies if three or more physical systems are united. The first question, from an invariant theoretic point of view, is then: how can the product space be decomposed into subspaces, each of which belongs to an irreducible representation, and what are these irreducible representations?

It may be useful to sketch the way in which an answer to these questions was found. The most important group to which the considerations can be and were applied is again O_3. Somebody said that, unlike the mathematician, the physicist lives in three-dimensional space not only physically but also intellectually. O_3 is an ambivalent group, that is, all its classes are self-inverse. This renders all its characters real and this will be assumed in what follows. The complications which not-real characters entail are minor. The Poincaré group is ambivalent, but some of the groups to be considered later are not.

If the characters of the representations of which we wish to form the direct product are $\chi^{(a)}$, $\chi^{(b)}$, $\cdots$, the character of their direct product is

(1) $\Xi(r) = \chi^{(a)}(r)\,\chi^{(b)}(r)\,\chi^{(c)}(r)\,\cdots$

and the number of times $\chi^{(\nu)}$ is contained in the representation with the character Ξ is

(1a) $N_\nu = h^{-1}\int \Xi(r)\chi^{(\nu)}(r)dr = h^{-1}\int \chi^{(a)}(r)\chi^{(b)}(r)\cdots\chi^{(\nu)}(r)dr.$

h is the volume of the group if the group is compact; it is the order of the group if this is finite. r is a group element and $\int dr$ indicates the invariant group integration for compact groups, summation over all group elements in the case of finite groups. The extension of the last formula to noncompact groups is an interesting problem on which a great deal of progress has been registered but which, to my knowledge, has not been solved completely.

It is noteworthy that the expression for N_ν is symmetric in its factors. We shall denote it by $(a, b, \cdots, \nu)$. This means, in the case of three factors, i.e., the decomposition of the Kronecker product of two representations, that the Kronecker product of a and b contains c just as many times as the Kronecker product of a and c contains b, or that of b and c contains a.

For three factors, the N are, as a rule, not difficult to calculate. In principle, such a calculation permits obtaining the decomposition of the Kronecker product of any number of factors. This follows from the completeness of the character functions which give the relation

$$(2) \qquad \chi^{(a)}(r)\chi^{(b)}(r) = \sum_{\nu} (ab\nu)\chi^{(\nu)}(r).$$

We can introduce this into the expression for the decomposition of the threefold product and obtain

$$(3) \qquad (abcd) = \sum_{\nu} (ab\nu) \int \chi^{(\nu)}(r)\chi^{(c)}(r)\chi^{(d)}(r)dr$$

$$= \sum_{\nu} (ab\nu)(\nu cd).$$

This suggests defining the matrices M^{ϵ}

$$(4) \qquad M^{\epsilon}_{\alpha\beta} = (\alpha\kappa\beta).$$

They are symmetric and their elements are nonnegative integers. Their rows and columns are labeled by the irreducible representations and they all commute, as is evident from the expression for $(abcd)$. In terms of these matrices, one can write

$$(5) \qquad (abc \cdots \nu) = (M^b M^c \cdots)_{a\nu}.$$

The order of the M in this expression can be interchanged and there are several other expressions for the left side. None of these expressions shows the total symmetry of the final expression in any obvious way and, in fact, it is not easy to find a general expression (apart from (1a)) for the left side of (5) which is obviously symmetric.

The possibility of decomposing the Kronecker product of two representations, let us say of $D^{(a)}$ and $D^{(b)}$, into definite components does not answer the principal question of interest for the physicist: how is the decomposition done? This amounts to transforming the product of the representation spaces of $D^{(a)}$ and $D^{(b)}$ into the representation spaces of the irreducible components of their Kronecker product. The transformation in question occurs in concrete problems again and again and has been very closely investigated for the group of rotation in three-space.

The first question which arises is the uniqueness of the transformation. This depends on the uniqueness of the basis vectors in the two spaces to be transformed into each other. Since the basis vectors in both $D^{(a)}$ and $D^{(b)}$ were uniquely specified, this is true also for the direct product of the two spaces. However, if the Kronecker product of $D^{(a)}$ and $D^{(b)}$ contains any representation more than once, not every basis vector in the final space can be specified simply as a definite basis vector of an irreducible $D^{(c)}$. If $D^{(c)}$ occurs in $D^{(a)} \times D^{(b)}$ (a, b, c) times, a unitary transformation of (a, b, c) dimensions will remain free and, if one wants to specify the transformation completely, some added specification of the basis vectors in the final space is necessary. Biedenharn devoted a great deal of thought to this problem, as did also Moshinsky, and the former specified the basis vectors for our problem in the case of the unitary group in n dimensions [19].

In the case of the group of principal interest in physics, the O_3 group, the aforementioned difficulty does not arise: all symbols (a, b, c) are either 0 or 1 in this case. Groups of the nature that (a, b, c) can assume only the values 0 or 1 are called, therefore, simply reducible. Both George Mackey and I have devoted a great deal of attention to them [20].

If more than two systems are to be united, i.e., if one is interested in the Kronecker product of more than two representations, the same irreducible representation may occur more than once in the decomposition even in the case of simply reducible groups. The obvious procedure, in order to specify the transformation uniquely, is to proceed step by step, as we did when calculating the value of the multiple symbols $(a\, b\, c \cdots \nu)$.

Let us look, first, at the reduction of the Kronecker product of two irreducible representations, a and b, of a simply reducible group. The basis vectors of the representation space of a may be denoted by $\alpha_1, \alpha_2, \cdots$; there are as many of them as is the dimension l_a of the representation a. An arbitrary one of the $\alpha_1, \alpha_2, \cdots$ will be denoted by α; summation over α then means that each of the $\alpha_1, \alpha_2, \cdots$ has to be substituted for α and the sum of all the resulting expression taken. A similar remark applies for β, which is one of the l_b basis vectors $\beta_1, \beta_2, \cdots$ of b. In order to express one of the basis vectors γ of the representation c (which is contained in the Kronecker product of a and b) in terms of the basis vectors (α, β) of the product space of the representation spaces of a and b, we have to form a sum

$$(6) \qquad\qquad \gamma = \sum_{\alpha\beta} c_{\gamma\alpha\beta}(\alpha, \beta).$$

Just as into the expression (a, b, c) which we had considered before, the three representations a, b, c enter rather symmetrically into $c_{\gamma\alpha\beta}$. One can express them in the form

$$(6a) \qquad c_{\gamma\alpha\beta} = l_c^{1/2} \begin{pmatrix} a & b & c \\ \alpha & \beta & \gamma \end{pmatrix}$$

where l_c is the dimension of the representation c and the second factor on the right side is called, for reasons which are unimportant now, a three-j-symbol. Except for a possible change in sign, it is invariant with respect to an interchange of the columns. These three-j-symbols, or expressions equivalent to them, were also called Clebsch-Gordan coefficients (though the reason for this is mysterious to me), or vector coupling coefficients, which is the name I prefer. They were calculated for O_3 in great detail and there are tables for them, similar in thickness to logarithmic tables [21].

Now, if we want to couple three systems together, we can first couple two, and then couple the third to the union of these two. In many cases, this is a physically reasonable procedure because the first two systems may be more strongly coupled to each other than either is to the third. Let us denote the three representations the Kronecker product of which we wish to consider by a, b, and c. Then, coupling a and b, we obtain the expansion of the basis vector μ of the representation m in the direct product space of their representation spaces

$$(7) \qquad \mu = \sum_{\alpha,\beta} l_m^{1/2} \begin{pmatrix} a & b & m \\ \alpha & \beta & \mu \end{pmatrix} (\alpha, \beta).$$

We now can take the Kronecker product of the representation m with the remaining representation c. The δ basis vector of the representation d will contain the vector (μ, γ) with the coefficient

$$l_d^{1/2} \begin{pmatrix} m & c & d \\ \mu & \gamma & \delta \end{pmatrix}$$

so that the vector δ becomes

$$(8) \qquad \delta = \sum_{\mu\alpha\beta\gamma} (l_c l_m)^{1/2} \begin{pmatrix} a & b & m \\ \alpha & \beta & \mu \end{pmatrix} \begin{pmatrix} m & c & d \\ \mu & \gamma & \delta \end{pmatrix} (\alpha,\beta,\gamma).$$

One can again proceed further and couple more and more systems together, i.e., obtain the basis vectors of the irreducible parts of the direct product space of several irreducible representations.

It is important to note, however, that the vector δ, obtained above, may not be identical with the vector which would have been obtained, had we reduced out, first, the Kronecker product of the representations a and c, and coupled one of the irreducible spaces in this Kronecker product with the representation space of b. The reason is that a representation d could have been obtained not only by coupling the m part of the direct product spaces of a and b with c, but possibly also by coupling another, m' part thereof with c. Hence, the vector δ above should be denoted by δ^m, its index m specifying the intermediate representation which was then coupled to c. Similarly, the vector

$$(8a) \qquad \delta' = \sum_{\alpha\beta\gamma\mu'} (l_d l_{m'})^{1/2} \begin{pmatrix} a & c & m' \\ \alpha & \gamma & \mu' \end{pmatrix} \begin{pmatrix} m' & b & d \\ \mu' & \beta & \delta' \end{pmatrix}$$

should be denoted by $\delta'^{m'}$, its index specifying the intermediate representation which was then coupled to b. There is no one-to-one correspondence between the δ^m and the $\delta'^{m'}$ but each $\delta'^{m'}$ can be expressed in terms of all the δ^m (and conversely). The coefficients are the same for all basis vectors α, β, γ and δ and are equal, apart from the sign and a factor l_m to the "six-j-symbols", recoupling, or Racah coefficients [22]

$$(9) \qquad \delta'^{m'} = \sum_m l_m \begin{Bmatrix} a & b & m \\ d & c & m' \end{Bmatrix} \delta^m.$$

These Racah coefficients also play important roles in a variety of physical problems, atomic and nuclear spectroscopy being only two of them. They have been extensively tabulated and obey a number of symmetry relations: all columns can be interchanged

$$(10) \qquad \begin{Bmatrix} a & b & e \\ c & d & f \end{Bmatrix} = \begin{Bmatrix} b & a & e \\ d & c & f \end{Bmatrix} = \begin{Bmatrix} a & e & b \\ c & f & d \end{Bmatrix} \text{ etc.}$$

Any pair of columns can be reversed

$$(10a) \qquad \begin{Bmatrix} a & b & e \\ c & d & f \end{Bmatrix} = \begin{Bmatrix} c & d & e \\ a & b & f \end{Bmatrix} = \begin{Bmatrix} a & d & f \\ c & b & e \end{Bmatrix} \text{ etc.}$$

and there are a number of orthogonality relations which, similar to the preceding symmetry relations, are valid for the Racah coefficients of any simply reducible group. For those of O_3, Regge has, in addition, proved a further set of relations the deeper cause of which is still somewhat mystifying

$$(11) \qquad \begin{Bmatrix} a & b & m \\ c & d & m' \end{Bmatrix} = \begin{Bmatrix} \frac{1}{2}(a+c+b-d) & \frac{1}{2}(a-c+b+d) & m \\ \frac{1}{2}(a+c-b+d) & \frac{1}{2}(-a+c+b+d) & m' \end{Bmatrix}$$

if $a, b, c, \cdots$ denote the representations of dimension $2a+1$, $2b+1$, $2c+1$, $\cdots$ of O_3. A similar extension of the relations between the vector coupling coefficients of O_3 was also found by Regge [23].

The considerations of the preceding paragraphs have been extended by various authors to the direct product of more than three irreducible representations. Many authors participated in this work (Biedenharn, Edmonds, Ponzano) the motif for which was no longer the facilitation of the solution of problems in physics but the mathematical exploration of a set of intriguing connections [24]. The most complete review of the subject, together with a significant extension of previous results, was given probably by the Lithuanian group of Jucys, Levinsonas and Vanagas [25]. Chakrabarti, Lévy-Nahas and Lévy-Leblonde and the Australian physicist-mathematician-philosopher Kumar, on the other hand, gave expressions for the δ into which the representations a, b, and c of O_3 entered in a symmetric fashion [26]. Kumar provided such expressions also for irreducible parts of the Kronecker products of more than three irreducible representations of O_3.

When concluding this part of the review, one can safely say that the closer scrutiny and more detailed investigation of the direct products of irreducible representations of at least some groups, in particular of O_3, led to a number of intriguing relations. They are, perhaps, less general and more concrete than would correspond to modern mathematical taste but, undeniably, many of us physicists enjoyed exploring them.

I do not believe that the subject of the reduction of direct products of representations has been exhausted. There are many other contributions, which were not mentioned [27], and there is a number of questions to which I would like to know the answers. It is true, nevertheless, that the problems which are most important for the physicist have apparently been solved—just as the mathematical problems of crystallography were solved. Instead, a new field for the application of symmetry considerations has turned up and it turned up, appropriately enough, in the most modern part of theoretical physics, in particle theory.

Symmetry problems of particle physics. It is not easy to review the symmetry problems of particle physics because the real problems cannot yet be formulated clearly. The observation which we wish to account for is that there are groups of particles—8 heavy particles, for instance—which belong to the same representation of the Poincaré group, except that their masses are somewhat different. The aforementioned 8 particles are the best known example; among the 8

is the proton and the neutron. The other particles are called strange—
they have all been discovered in the reasonably recent past.

It seems reasonable to assume that the particles in question are
related to each other by some approximate symmetry, and the question is only what that symmetry is. We believe that there is a group
which has an 8-dimensional representation and the state vectors of
the 8 particles are basis vectors in that representation space. The
problem is, physically, what the meaning of the operations of the
group is, in the same sense in which the meaning of the operations of
the Poincaré group is displacement, imparting a velocity, etc. No
such meaning is known. It is even doubtful that a real meaning for
the operations of the group can be found because the state vectors
of the particles in this case do not form a linear space—there is no
sense in adding the state vector of a neutron to that of a proton. All
this does not, of course, exclude the possibility that the basic equations are mathematically invariant; at least approximately, under a
group. It only means that the operations of the group have no direct
physical significance. The equations of the oscillator are invariant
with respect to an interchange of the position and velocity coordinates
and such invariances have interesting consequences even though the
underlying mathematical operation is not meaningful physically.

This is the problem of the physical interpretation of the symmetry.
The mathematical problem is to find the group which has an 8-dimen-
sional representation, appropriate for the situation. Gell-Mann and
Ne'eman [28] proposed the unimodular unitary group in three dimensions, SU_3. The first nontrivial real irreducible representation of this
is 8-dimensional. What can the group do for us?

If the operations of the group were true symmetry operations, the
masses of the 8 particles would be equal. This follows from a theorem
of O'Raifertaigh [29] according to which there is no Lie group, containing the Poincaré group as a subgroup, which would have an
irreducible representation such that its restriction to the Poincaré
subgroup would contain a finite number of irreducible representations
of this group, with different masses. It then follows that the mass differences must be due to some inaccuracy of the SU_3 symmetry, some
perturbation thereof. The great success of the SU_3 symmetry is then
to have suggested a simple perturbation operator the matrix elements
of which are in the ratio of the observed mass differences of the 8
particles. I am referring to the Gell-Mann-Okubo mass formula which
is valid within about 6% of the mass differences [30].

The agreement between a mass formula and the observed masses would not be in itself convincing. There are several simple perturbation operators and it is not surprising that one led to a conclusion in good agreement with experiment. However, the 8-fold multiplet is not the only one—at least three others could be identified. It seems most unlikely that all this would be coincidence.

I know that my discussion of this subject is too vague to attract the interest of the mathematician. He likes to start with clearly defined premises and arrive at definite conclusions. We surely understand this predilection. However, the task of the physicist is often just the opposite: he knows the conclusions, the observed phenomena, and wishes to find the premises from which these follow. A great deal of vagueness is unavoidable in the process and it is, I believe, in spite of its somewhat vague nature, or perhaps just because of it, highly interesting. In addition to those who were already mentioned, Michel in France, Roman in Boston, Gürsey in Turkey, Radicati in Italy, and Pais in New York are some of the most effective contributors thereto [31], but I do not wish to go further into detail. The principal point which the preceding discussion was intended to convey is the fact that, after a long period in which the detailed properties of representations were at the center of interest of the physicist and invariance theorist, attention shifted back to the search for the symmetry group, or symmetry groups, most appropriate for the description of the observed phenomena.

It is a pleasure to thank, in conclusion, Professors V. Bargmann, C. C. Gillispie, and T. S. Kuhn for having called to my attention articles with which I was unfamiliar and for useful discussions.

REFERENCES

1. J. Willard Gibbs' *Elementary principles in statistical mechanics*, a finished work 284 pages long, was written in 1901 and published by Yale University in 1902. It was reprinted by Longmans, Green (New York, 1928). An appraisal, by A. Haas (or, rather, two appraisals, a short and a long one) remain worth reading (in Vol. 2 of *A commentary on the writings of J. Willard Gibbs*, Yale University Press, 1936). Gibbs' first writing on the concept to which his name is rightly attached, on phase space, is only one page long. It was published in 1884 (Proc. Amer. Assoc. **33**, 57) and is reprinted, as page 16, in Vol. 2 of *The scientific papers of Willard Gibbs* (Longmans Green, London, 1906).

2. See, e.g., F. P. Bowden and D. Tabor, *Friction and lubrication* (Wiley, New York, 1956).

3. Il Saggiatore (Volume VI of *Le opere di Galileo Galilei*, Firenze, 1896) Section 6: Egei e scritto in lingua matematica. . . .

4. J. F. C. Hessel's paper was published, originally, in Gehler's Physikalische Wörterbücher (Leipzig, 1830); it is reprinted in Ostwald's Klassiker der exakten Naturwissenschaften No. 89 (Leipzig, 1897), see p. 91 ff.

The history of crystallography was described by P. Groth in his *Entwicklungs-geschichte der mineralogischen Wissenschaften* (Berlin, 1926). A modern history, very wide ranging, is J. G. Burke's *Origins of the science of crystals* (University of California Press, 1966).

5. R. J. Haüy's most relevant article is that in the Journ. de Physique **20** (1782), 33. For a complete bibliography, see J. G. Burke, l. c. p. 190. Actually, Burke expresses doubts (see pp. 83–84) concerning Haüy's independence from his predecessors, in particular from T. Bergman.

6. Actually, N. Steno's *De solido intra solidem naturaliter contento dissertationis prodromus* (Florence, 1669) only contains the germs of the ideas of the crystal lattice.

7. A. Schönflies, *Kristallsysteme und Kristallstruktur* (Leipzig, 1891); E. S. Fedorov, Zap. Min. Obsh. (Trans. Min. Soc.) **28** (1891), 1.

8. P. Groth, *Physikalische Krystallographie* (W. Engelman, Leipzig, 1905). A more openly group theoretical attitude is adopted by W. Voigt in his *Lehrbuch der Kristallphysik* (B. G. Teubner, Leipzig, 1910).

9. According to classical, that is nonrelativistic theory, the equations of motion remain invariant if the three positional coordinates x_i ($i = 1, 2, 3$) are subjected to a Galilei transformation

$$x_i' = \sum_{k=1}^{3} O_{ik} x_k + v_i t + a_i \qquad (i = 1, 2, 3)$$

where the O_{ik} form a 3×3 orthogonal matrix and the time t is invariant or subject to a change of origin

$$t' = t + a_0.$$

The vector v gives the velocity of the two coordinate systems with respect to each other; the vector a the displacement of the origin of the second coordinate system with respect to the first one at $t = 0$. The corresponding invariance transformation of relativity theory, the Poincaré transformation, treats the time more on a par with the space coordinates. It introduces, instead of t, the variable $x_0 = ct$ (where c is the velocity of light). The transformation is, in terms of these

$$x_i' = \sum_{k=0}^{3} L_{ik} x_k + a_i \qquad (i = 0, 1, 2, 3)$$

where L is a Lorentz transformation, i.e. an element of $O(1, 3)$. The Poincaré transformation is called also inhomogeneous Lorentz transformation.

10. This is now a commonplace statement. It is due, originally, in the precise form stated, to J. v. Neumann. See his *Mathematische Grundlagen der Quantenmechanik* (J. Springer, Berlin, 1932). English translation by R. T. Beyer, Princeton University Press, 1955.

11. This phase space is the mathematical structure introduced by W. Gibbs and mentioned in Reference [1].

12. V. Bargmann, Ann. of Math. **59** (1954), 1.

13. This was shown, first, by the present writer, Ann. of Math. **40** (1939), 149.

14. Gelfand, Naimark, and their collaborators are extremely prolific writers. The most important books, from our point of view, are: I. M. Gelfand, R. A. Minlos, and Z. Y. Shapiro, *Representations of the rotation and Lorentz groups and their applications*, Macmillan, New York, 1963; M. A. Naimark, *Linear representations of the Lorentz group*, Pergamon Press, London, 1964; I. M. Gelfand and M. A. Neumark, *Unitäre Darstellungen der klassischen Gruppen*, Akademie Verlag, Berlin. See also the articles

of these authors in Vols. 2 and 36 of the Amer. Math. Soc. Translations (Amer. Math. Soc., Providence, R. I., 1956 and 1964).

Harish-Chandra's articles are too numerous for a complete listing and no review of his results is available in book form. His early papers appeared in the Proc. Nat. Acad. Sci. (Vols. 37–40, 1951–54). Among his later writings, we wish to mention his *Invariant eigendistributions on a semisimple Lie algebra* (Inst. Hautes Études Sci. Publ. Math. No. 27, p. 5) and his article in Ann. of Math. **83** (1966), 74.

15. A. Young, Proc. London Math. Soc. **33** (1900), 97; **34** (1902), 361; G. Frobenius, Sitzungsberichte Preuss. Akad. Wiss. 1903, p. 328; I. Schur, ibid. 1908, p. 64. See for a physicist's approach, A. J. Coleman's article in *Advances of quantum chemistry* (Academic Press, New York, 1966).

16. To be published in the Racah Memorial Volume (North-Holland, Amsterdam, 1968), p. 131.

17. R. de L. Kronig, Zeits. f. Physik **31** (1925), 885; **33** (1925), 261; H. N. Russell, Proc. Nat. Acad. Sci. **11** (1925), 314; A. Sommerfeld and H. Hönl, Sitzungsberichte Preuss. Akad. Wiss., 1925, p. 141.

18. The Kronecker product as the representative of the union of two physical systems was implicit already in Schrödinger's papers. See his *Abhandlungen zur Wellenmechanik*, Leipzig, 1927. It was made explicit and precise by von Neumann, [10, Chapter VI, §2].

19. L. C. Biedenharn, J. Math. Phys. **4** (1963), 436 and subsequent papers, ending with L. C. Biedenharn, A. Giovanni, J. D. Louck, ibid. **8** (1967), 691. M. Moshinsky, ibid. **7** (1966), 691, J. G. Nagel and M. Moshinsky, ibid **6** (1965), 682 and M. Moshinsky ibid. **7** (1966), 691, M. Kushner and J. Quintanilla, Rev. Mex. de Fisica **16** (1967), 251. For a summary, see L. C. Biedenharn, Racah Memorial Volume [16, p. 173].

20. E. P. Wigner, Amer. J. Math. **63** (1941), 57. This and a more detailed paper are reprinted in *Quantum theory of angular momentum*, L. C. Biedenharn and H. Van Dam, Editors, Academic Press, New York, 1965. This volume contains reprints of several articles which played an important role in the formation of the ideas of physicists on representation theory, in particular also articles of G. Racah. G. W. Mackey, Amer. J. Math. **75** (1953), 387; Pacific J. Math. **8** (1958), 503. See also his *Mathematical foundations of quantum mechanics* Benjamin, New York, 1963.

21. As far as this writer was able to ascertain, the first calculation of these coefficients is given in his *Gruppentheorie und ihre Anwendungen etc.* (Friedr. Vieweg, Braunschweig, 1931) Chapter XVII. See also the English translation by J. J. Griffin (Academic Press, New York, 1959). The more symmetric expressions for these quantities, the three-j-symbols, were introduced in the article which is reprinted in the *Quantum theory of angular momentum* [20, pp. 89–133]. For an extensive tabulation of the numerical values of these coefficients, see, e.g., Rotenberg, Bivins, Metropolis, Wooten, *The 3-j and 6-j symbols*, Technology Press, MIT, Cambridge, 1959.

22. The first publications containing recoupling (now called Racah) coefficients are due to G. Racah. His work was clearly independent of earlier considerations of the present writer (cf. the second article of [20]). They arose in connection with problems of atomic spectra. Cf. Phys. Rev. **61** (1942), 186; **62** (1942), 438; **63** (1943), 367; **76** (1949), 1352. These articles are all reprinted in the *Quantum theory of angular momentum* [20]. For a numerical table of these coefficients, also called 6-j-symbols, see [21].

23. T. Regge, Nuovo Cimento **11** (1959), 116; **10** (1958), 544. An extension of these relations was given by R. T. Sharp.

24. L. C. Biedenharn, J. M. Blatt, M. E. Rose, Revs. Mod. Phys. **24** (1952), 249;

A. R. Edmonds, *Angular momentum in quantum mechanics*, Princeton University Press, 1957; G. Ponzano, Nuovo Cimento 35 (1965), 1231; 36 (1965), 385; G. Ponzano and T. Regge. Racah Memorial Volume [16].

25. A. Jucys, J. Levinsonas, V. Vanagas, *Mathematical apparatus of the theory of angular momentum*. English translation published by the Israel Program of Scientific Translations, Jerusalem, 1962. The original, in Russian, was published in Vilnius (Vilna) in 1960. See also W. T. Sharp's *Racah algebra and the contraction of groups*, Atomic Energy of Canada Ltd. report 1098 (1960).

26. A. Chakrabarti, Ann. Inst. Henri Poincaré 1 (1964), 301; K. Kumar, Austral. J. Phys. 19 (1966), 719; J. M. Lévy-Leblond and M. Lévy-Nahas, J. Math. Phys. 6 (1965), 1372.

27. A full listing of all the contributions to the subject would require several pages. Furthermore, some noncompact Lie groups, in particular also the Poincaré group, were the subjects of investigations similar to those described for O_4. See, for instance, J. Ginibre, J. Math. Phys. 4 (1963), 720, J. R. Derome and W. T. Sharp, ibid. 6 (1965), 1584; D. R. Tompkins, ibid. 8 (1967), 1502.

Articles dealing with other properties of the representations of the Poincaré group and also extensions thereof include J. L. Lomont and M. E. Moses, J. Math. Phys. 5 (1964), 294; 8 (1966), 837; I. Raszillier, Nuovo Cimento 38 (1965), 1928; J. M. Lévy-Leblond, ibid. 40 (1965), 748; C. George and M. Lévy-Nahas, J. Math. Phys. 7 (1966), 980; J. C. Guillot and J. L. Petit, Helv. Phys. Acta 39 (1966), 281; V. Berzi and V. Goroni, Nuovo Cimento 57 (1967), 207; S. Ström, Ark. Fys. 34 (1967), 215; J. Nilsson and A. Beskow, ibid. 34 (1967), 307; H. Joos and R. Schrader, Comm. Math. Phys. 7 (1968), 21 (the characters of the representations); A. Kihlberg, Nuovo Cimento 53 (1968), 592.

Other articles investigate more complex noncompact Lie groups as they underlie the de Sitter space (the groups $O(4, 1)$ and $O(3, 2)$, or the situation at very high energy $O(4, 2)$ or $U(2, 2)$). See the very early paper of L. H. Thomas, Ann. of Math. 42 (1941), 113. Corrections to this paper and extensions thereof were given by T. D. Newton, Ann. of Math. 51 (1950), 730 and by J. Dixmier, Bull. Soc. Math. France 89 (1960), 9. Other contributions are due to J. B. Ehrman, Proc. Cambridge Philos. Soc. 53 (1957), 290; several articles of A. Kihlberg and S. Ström, including some mentioned before but in particular Ark. Fys. 31 (1966), 491; W. Rühl, Nuovo Cimento 44 (1966), 572; A. Chakrabarti, J. Math. Phys. 7 (1966), 949; A. J. Macfarlane, L. O'Raifeartaigh and P. S. Rao, ibid. 8 (1967), 536; O. Nachtman, Acta Physica Austriaca 25 (1967), 118; H. Bacry, Comm. Math. Phys. 5 (1967), 97. The Trieste Institute of the International Atomic Energy Agency is intensely working on the subject and has issued many reports thereon. The authors include R. Delbourgo, A. Salam and J. Strathdee; R. L. Anderson, R. Raczka, M. A. Rashid and P. Winternitz; D. T. Stoyanov and I. T. Todorov.

A rather complete review of the more important articles from the mathematical point of view, up to 1965, was presented by G. A. Pozzi, Nuovo Cimento Suppl. 4 (1966), 37. See also H. Baumgärtel, Wiss. Z. Humboldt Univ., Berlin, Math.-Naturw. Reihe 13 (1964), 881.

28. Y. Ne'eman, Nuclear Physics 26 (1961), 222; M. Gell-Mann, Phys. Rev. 125 (1962), 1067. See also their reprint collection and review, *The eightfold way*, Benjamin, New York, 1964.

29. L. O'Raifeartaigh, Phys. Rev. 139B (1965), 1952. A mathematically more precise formulation of the O'Raifeartaigh theorem was given by R. Jost, Helv. Phys.

Acta **39** (1966), 369. For an extension of the theorem, see I. Segal, J. Functional Anal. 1 (1967), 1; A. Galindo, J. Math. Phys. **8** (1967), 768.

30. S. Okubo, Progr. Theor. Phys. **27** (1962), 949; M. Gell-Mann, Phys. Rev. **125** (1962), 1067.

31. The interested reader will find an illuminating introduction to, and a collection of, articles dealing with the problems of the extension of the Poincaré group in F. J. Dyson's *Symmetry groups in nuclear and particle physics*, Benjamin, New York, 1966. A more mathematically oriented review of the "higher symmetries" is given in M. Gourdin's book, *Unitary symmetries and their applications to high energy physics*, North-Holland, Amsterdam, 1967. See also A. O. Barut, Proc. Seminar on High Energy Physics and Elementary Particles, Trieste, 1965 (International Atomic Energy Agency, Vienna, 1965).

PRINCETON UNIVERSITY

Symmetry in Nature*

E. P. Wigner

W. O. Milligan (ed.) Proc. R. A. Welsh Foundation on Chemical Research XVI. Theoretical Chemistry, Houston, Texas 1972, 1973, pp. 231–260

Thank you very much for your very kind introduction.

But, even more would I like to express my gratitude to the organizers of this meeting. I don't remember a meeting that I attended and that was more inspiring and more interesting than this one. It is perhaps because it reminds me of so many things that I have halfway forgotten.

It is also because we have heard so many inspiring new ideas that I really feel that we should express our gratitude to the organizers of this symposium and those who made it possible for us to gather here and participate in a meeting. (Applause)

The subject of my own discussion may not be quite in the spirit of the rest of this meeting. It will be partly philosophy. This makes my discussion very different from the preceding ones. If you discuss physics or chemistry, you hope that your audience knows the basic underlying facts, yet does not know what you are going to add to them.

If one talks about philosophy, one must hope that what one says is already implicitly present in the minds of the people to whom one talks because one cannot absorb new basic ideas, new philosophical concepts, unless one has invented them oneself at least three-quarters of the way.

Therefore, I hope that most of what I say is at least latently present in your minds.

1. Conceptual questions and remarks on these.

With respect to the conceptual questions, there is something that we all realize but that is perhaps good to bring back to our minds; namely, that physics deals with three kinds of objects: with events, laws of nature, and symmetry or invariance principles.

The laws of nature are what physics is. Physics doesn't describe nature. Physics describes regularities among events and *only* regularities among events.

---

*An address presented before "The Robert A. Welch Foundation Conferences on Chemical Research. XVI. Theoretical Chemistry", which was held in Houston, Texas, November 20-22, 1972.

232   SYMMETRY IN NATURE

The events are so unbelievably complicated that not only the optimist says that "The future is uncertain." The future is uncertain even if one knows physics, because one does not know the initial conditions which largely determine the events which follow. To recognize regularities, such as that if I have a body now here and if, after half a second, it will be there, then it will reach the ground in one and a half seconds: this is what we mean by physics. It is a set of regularities which in this case involves two previous events from which a third event can be foreseen.

Of course, it is much more common to look at the sky and see the moon at a place, to see it somewhere else in another hour and to foresee where the moon will be in another hour.

These are regularities among the events. What are then the symmetry of invariance principles? The symmetry and invariance principles are regularities not among events but among the laws of nature; that if I know some law of nature, for instance if I can today predict a later position of the moon from two of its earlier positions, then I can make similar inferences tomorrow or the day after. Similarly the inferences about the positions of falling bodies at one place are the same as at other places. The two invariance principles in question are the time displacement and space displacement invariances.

All this really puts us into a rather awkward position because if we knew all the events; past, present and future, the laws of nature would be entirely unnecessary. They would, perhaps, still give interesting connections between the events which we already know about, but they would not provide any new information.

Perhaps it would be refreshing to see these connections prevail, but they would not give new information. Fortunately, it will surely remain impossible to foresee all future events.

The knowledge of the laws of nature may provide us with one other advantage: if somebody came and said, "Oh, you are wrong about that event of tomorrow. It will be entirely different," then by knowing the laws of nature, one could more effectively contradict him.

The relation of symmetry principles to the laws of nature is very similar. If we knew all the laws of nature, the symmetry and invariance principles would not tell us anything new. They would just establish relations between the consequences of the laws of nature.

It would be still interesting to note that the regularities between the events which are valid today will also be valid tomorrow and to make a number of other similar observations. It would remain aesthetically satisfactory to realize the symmetries among the laws of nature. But if we knew all the laws of nature, the symmetry principles would give us no fundamentally new information.

Again, if somebody came with another law of nature and said, "Your law of nature is wrong," perhaps I could more effectively contradict him if I pointed out to him, "Well, your law of nature does not obey a recognized symmetry principle." But, fundamentally, if we really knew the laws of nature — from the point of view of the scientist a not only unlikely but perhaps also undesirable situation — the symmetry principles would be unnecessary.

Well, how about the initial conditions? How do they fit in?

The laws of nature which we can believe — and this was emphasized by Einstein very eloquently — are simple mathematically and have a certain beauty. The regularities are really simple regularities and simple consequences of each other.

The initial conditions are subject to an almost exactly opposite criterion and we will come to that when I come to my second point; the digression which Professor Eyring suggested. This is based essentially on the fact that whereas the laws of nature have a remarkable simplicity and regularity, the initial conditions have an equally fantastic irregularity. And, without that, many of the consequences which we derive everyday from the laws of nature would not be valid.

The irregularity of the initial conditions, which I should not discuss now, but in a minute, is just the opposite of the beautiful regularities of the laws of nature.

I'd like to give next, a microscopic amount of history. The distinctions among events, laws of nature, and invariance principles, were not always present in the scientific thinking. The Greek philosophers expected more regularity among the events than provided by our laws of nature. They were for instance, badly shocked when it was discovered that the planetary orbits are not circular, but elliptic. They felt that this somehow destroyed their picture of the world, that the planetary orbits are not as simple as conceivable.

The first person who realized these differences, and realized it very clearly, was Leibnitz. Following Leibnitz, Newton realized practically all the older kinematic invariance principles. He has an article which I just happened to read in which he describes that if you were on a freely falling body, you would not be able to know it without looking out because the events would have the same form as if you were out far away from the earth's gravitational field. This is a very far-reaching invariance principle, part of the general theory of relativity, but Newton realized it.

234   SYMMETRY IN NATURE

2. The Increase of the Entropy and the Origin of the Universe.

I am afraid that I am so convinced of most that I have said so far that I am probably only repeating myself. Let me, therefore, go over to the next subject, the one suggested by Dr. Eyring in his introductory speech. He points out that the laws of nature — at least those which are relevant for most common events — have time inversion symmetry. That is, if a motion is possible, the opposite motion is possible also, the motion in which the positions of all objects at time $-t$ are the same as the positions are at time t in the original motion, and the velocities opposite. If a position x_3 at time t_3 follows from the positions x_1 and x_2 at times t_1 and t_2, the position x_1 at time $-t_1$ follows from the positions x_3 and x_2 at times $-t_3$ and $-t_2$. How does it come then that there is an important quantity, the entropy, which always increases, that the time symmetry does not apply to it?

This is, of course, a question which intrigues everybody who is interested in semi-philosophical aspects and also knows a little about physical chemistry.

It is probably evident from the foregoing that I'll blame the lack of time inversion symmetry of macroscopic events on the initial conditions which, as I said, are as irregular as possible, considering the macroscopic constraints. In fact, when deriving his h-theorem, Boltzmann assumed that the number of encounters between atoms or molecules in which an atom is aimed at a certain area of its target is proportional to the area of that target, and so on. We have learned to express Boltzmann's assumption in a more sophisticated way, in terms of Gibbs' phase space, but the basic assumption remained the same: There is, except for the constraints imposed macroscopically, no regularity in the initial conditions, as contrasted with the final situation in which there are amazing regularities. These regularities do not manifest themselves as a rule very clearly, but would manifest themselves at once in a fantastic way if we could reverse the direction of motion of every atom, if we could put a mirror in front of every atom so that its velocity would be reversed. Thus if one had a piece of iron which is hot at one end, cold at the other, the temperature of it would begin to equalize. Then, when it is half equalized, if one could put a mirror in front of every atom, it would go back to its original state and the originally hot side would become hot again and the cold side would become cold again. And the entropy principle would be grossly violated.

I once derived an expression for the entropy of a nonequilibrium ensemble. The idea that guided me is hardly new; qualitatively it is contained already in Ehrenfest's famous encyclopedia article. The expression itself may not be new either. I will write it out because even though the formula itself is not terribly unexpected, what it is based upon has some interest. The expression to be given is not the whole entropy, but only the deviation of the entropy

EUGENE P. WIGNER   235

from the equilibrium entropy. The expression for the equilibrium entropy does not hide any mystery. The expression for the difference between the equilibrium entropy and the actual entropy must take cognizance of the fact that one cannot make full use of one's knowledge of a system for extracting work from it at the expense of its energy. Suppose, we mix hot and cold water. In the early stages of mixing, one could quickly subdivide the water and thermally insulate the resulting small amounts from each other. Some of the parts will still be hot, others cold, and one could operate a heat engine on these temperature differences.

But as one mixes the hot and cold water more and more, it will be increasingly difficult to separate it into parts small enough so that the temperature difference between them be still appreciable. Hence, the amount of work that a heat engine will be able to extract will decrease as the mixing progresses. This means that the entropy increases as a result of the mixing process but its magnitude depends not only on the actual state of the system but also on my ability to make use of its properties. In other words, the entropy depends on my knowledge, and ability to make use of the knowledge, concerning the actual state of the system. The expression for the difference between equilibrium entropy and actual entropy depends on the probabilities p_i that the system is in the part i of phase space, the parts being so chosen that one can make use of the knowledge that the system is in a state corresponding to that part of phase space. The actual expression for the entropy difference is

$$S' = k \sum_i \left\{ -p_i \ln p_i + p_i \ln \frac{V_i}{\sum_j V_j} \right\}.$$

This expression is always negative, except in equilibrium when $p_i = V_i/\sum V_j$, in which case it naturally vanishes, V_i being the volume of part i of phase space.

Now, if one claims that the entropy increases, it means that it increases for an ensemble for which the probability was originally uniformly distributed within each region i. In other words, the ensemble was as irregular as is consistent with the knowledge of the p_i, the utilizable knowledge that is available. This theorem is surely implicitly present in earlier treatments of the problem but it is a good illustration of the fact that the irregularity of the initial conditions — the uniformity of the probability in each of the regions i — is something that we always assume and that is necessary for a derivation of the increase of entropy of non-equilibrium systems.

Dr. Eyring mentioned another problem, the origin of our world. I don't know whether this problem is foremost in the minds of other chemists but I do find it extremely puzzling and interesting. It is also a problem to which I can offer no solution.

236   SYMMETRY IN NATURE

The problem is essentially this: Why is it that the entropy is still not the maximum? How old is the world? We can't believe that the world at one time did not exist. How does it come that the entropy has not yet reached its maximum value and that we are not all dissolved in nothing? There are many, many proposals to motivate or explain this, but I must admit that I cannot feel comfortable with any of them.

The most common explanation that cosmologists give is the Big Bang. The question then which naturally comes up, is "What was before the Big Bang?" And there is no answer to that. If you pose it to the cosmologists, some of them say, "It is a fluctuating world; it now expands, but it will contract again; and then, again it will cause a big bang and expand again and so on."

However, at every expansion and contraction, the entropy should increase. And how many contractions and expansions were there? If there was an infinite number, the world should be close to the thermal equilibrium, to the "Wärmetod."

There is another explanation for the present state of the world: That the entropy fluctuates around its maximum and we are now experiencing such a fluctuation, far away from the maximum.

I have two reasons not to believe this. One of the reasons is that if there are fluctuations of the entropy, one is just as likely to be in the part of the oscillation in which the entropy decreases as the one in which it increases. If we were in the decreasing situation then, as explained by Ehrenfest, the chances are that in the next tenth of a second the entropy will reach its minimum and begin to increase again. This would be a catastrophe — all the macroscopic laws of motion would suddenly be reversed and not valid. This doesn't seem to happen, and I hope it won't happen, at least not in my lifetime. (Laughter)

The other reason for my not believing the fluctuation theory is that the universe, as we look at it, has many regularities which are not necessary for a fluctuation. In particular, that it expands practically everywhere.

There is a third explanation which I hesitate to tell you about: Namely, that the entropy increases when the world expands and decreases when it contracts. This would mean that during the contraction people would know the future and not the past, because their memory would be attuned to the future. (Ripple of laughter)

If this were the case, only an expansion of the world would be experienced, because when it really expands, we remember the real past and we see that the world expands. When it contracts, past and future would be interchanged, we would "remember" the future, and therefore, would also believe

EUGENE P. WIGNER   237

that the world expands. However, for this explanation to be valid, we would
have to assume such an intricately arranged set of initial conditions that it is
for me impossible to believe in it. One would have to assume that the presently
experienced irregularity of the initial conditions is only apparent, that they
have some hidden regularity which will manifest itself at the end of the
expansion. Then suddenly regularities would manifest themselves, the entropy
would decrease, which means for me an unbelievable regularity of the initial
conditions.

Do I have any hope? Do I see any distant rays where this difficulty
could find a solution? I don't believe so, but I would like to mention one
point; namely, the point of general relativity about the Black Holes. The
Black Holes are a consequence of the general theory of relativity and their
existence means that some parts of the universe occasionally disappear and
other parts of the universe suddenly appear from nowhere. If we believe
in the existence of such things, there is a distant hope that what we say about
laws of nature and initial conditions and knowledge of the world is all terribly
inaccurate, and there is a meager, but perhaps possible hope that the explanation of our existence is not as incomprehensible as it is to me at present.

I am afraid I dwelt too long on Dr. Eyring's second question concerning
the origin of the universe. Let me conclude, therefore, by repeating that I
continue to be deeply impressed by the enormous irregularity of the initial
conditions and the marvelous simplicity and beauty of the laws of nature,
including their obeying the invariance principles.

3. Different Types of Symmetry Principles. The Kinematic Invari-
ance Principles.

My next subject is the description of the two or three types of symmetry
principles which I like to distinguish: The kinematic, the dynamic, and possibly a third one, which I am not sure about. The kinematic invariance principles are very simple and one could simply note that they are incorporated in
the special theory of relativity, the invariance under the Poincaré group.

It is now a bit surprising how few of the kinematic principles are obvious
and that it was really difficult to realize most of them. This does not apply,
though, to the first ones, the time displacement invariance: that the laws of
nature today are the same as they will be tomorrow; i.e. that into the laws of
nature only time differences enter. If this were not so, it would have been
unbelievably difficult to discover any law of nature, because they would change
and what I find today as the relation between the three positions of the moon
would not be the same tomorrow, so that I could not recognize the relation
between them as a law of nature.

238   SYMMETRY IN NATURE

The other one which is almost equally simple and comes to us equally naturally, but did not come always equally naturally, is the space displacement invariance; that if I drop a body here, it will fall to the ground in just the same way as if I dropped it some other place. Again, if this were not true, it would have been unbelievably difficult to discover laws of nature.

Well, these are the first invariance principles. Mathematically, I should count them as four principles: time displacement and displacements into three different directions.

The next principle is that of the isotropy of space. This was really quite difficult to discover and it was discovered essentially, as I mentioned, by Leibnitz. Before his and Newton's time, the vertical direction appeared to be entirely different from the horizontal directions and the discovery was the recognition of the fact that this difference comes from the accidental fact of an initial condition: that there is an earth underneath us. This reason for the difference was not evident before Newton. The recognition of the fundamental isotropy of space already required a great deal of understanding of the distinction between laws of nature and initial conditions.

The next invariance principle is the invariance principle which Newton discovered, namely, invariance with respect to the transition to a moving coordinate system. This was fantastically surprising. To misquote them, Aristotleists believed that every body remains at rest unless a force is acting on it. They thought that only the state at rest is a natural state. These are not the words which they used, but it is what they believed: that every object remains at rest unless a force is acting on it. If this were true the invariance with respect to Galileo-Lorentz transformations would not be valid.

As you know, when Einstein came on the scene, it was again believed that this principle was not valid. It was believed that there is an ether at rest and that bodies at rest and in motion behaved differently because their relation to the ether at rest is different.

It is interesting the way Einstein discovered the special theory of relativity. It was not postulating a principle, but by investigating what happens relative to a moving coordinate system. And he found that what happens in the moving coordinate system and the coordinate system at rest is the same, but the description in the moving coordinate system was entirely different from that in the coordinate system at rest.

He then said: "This is not reasonable. If it is the same thing that happens, then there must be an invariance principle that connects the two sets of events" and he, essentially, abolished the ether at rest. Maxwell still thought in terms of all sorts of cogwheels and whatnot. But Einstein abolished all that and introduced the invariance principle.

One can say, in fact, that Einstein's greatest discovery was the establishment of the importance of the invariance principles. These were hardly mentioned in earlier physics books and their importance and usefulness were little recognized. The first mechanics book that I read had about half a page devoted to the law of conservation of angular momentum, which is one of the important consequences of the invariance principles. Nevertheless, I believe almost everybody knew that conservation law but few people thought it had much basic significance.

We now have all the kinematic invariance principles. There are ten of them: displacement in time, three displacements in space, rotation about three axes, acceleration for movement in three different directions.

I should explain why I do not mention the principle of general relativity as a kinematic invariance principle. The reason is that, in my opinion, it is not a kinematic invariance principle, but a dynamic invariance principle. It does not refer to the identity of the connection between different events but to the variety of descriptions of the same events by using infinitely many different coordinate systems, with all sorts of curvilinear coordinates. This observation was first made, very soon after Einstein's proposal of his general theory of relativity, by Kretschmann. In the recent literature, it is emphasized most strongly by Fock in his book on the general theory of relativity.

Let me say a few words about the verification or falsification of invariance principles. As a rule, this is not done by setting up laws of nature and ascertaining then whether they obey the invariance principle in question. The laws of nature which Einstein found did not obey the invariance principle he established; the description of the events depended on the motion of the coordinate system. The events did not. And this indicates one method of checking invariance principles. The space displacement invariance, for instance, can be checked by ascertaining whether it's really true that if I drop something at one point it falls in the same way as it falls somewhere else.

An invariance postulate can be checked also by establishing a situation which is symmetric with respect to the postulated invariance and investigating then whether the symmetry remains preserved. The ordinary reflection symmetry was refuted in just this way. A cobalt nucleus placed into a magnetic field has a reflection symmetry with respect to the plane perpendicular to the magnetic field because the field can be generated by a current which lies entirely in the plane perpendicular thereto. In this case, the system did not preserve its symmetry: many more of the emitted electrons travelled toward one side of the initial symmetry plane than toward the other. This showed that the laws of nature cannot preserve the reflection symmetry. However, the lack of the validity of the symmetry tested was not demonstrated by setting

240   SYMMETRY IN NATURE

up a new law devoid of the symmetry but by *the events themselves showing directly that no law of nature can be valid which postulates the symmetry in question.* This is indeed the only truly convincing way to demonstrate the absence of a symmetry.

As all this implies, the invariance principles are empirical, as everything in physics is empirical. It may be good to mention a couple of other invariance principles which were refuted in the course of time. I mentioned already the postulated circular nature of the planetary orbits. Another symmetry principle is due to Fourier and it remains a very useful principle if applied in macroscopic physics. He postulated that no matter how much matter is divided, its properties remain the same. This principle was even present in the books which I first read about physics. Of course, we now know how fantastically it is violated in the microscopic domain.

There is one further problem in connection with the kinematic invariance principles that should be at least touched upon: are these invariance principles, the kinematic invariance principles, unquestionably meaningful? In other words, is it really true that one can check them?

It was mentioned before that one cannot directly verify the full rotational invariance standing here on the Earth. The Earth's gravitational field gives a preferred role to the vertical direction. In order to verify the rotational invariance, one would either have to compare experiments on our Earth with experiments on another planet the axis of which has a different direction, or to move away from the Earth far enough so that its effect become negligible. The second method is, evidently, preferable and it amounts to the postulate of carrying out the experiments on an isolated system.

Can this be done? As far as old-fashioned macroscopic experiments are concerned, the answer is clearly yes. Similarly, the effect of the Earth's gravitational field is negligible on microscopic systems and, as far as these are concerned, the isotropy, though it has not been tested with experiments carried out with this purpose, is surely well established. However, the existence of isolated systems of any reasonably macroscopic size in the true quantum mechanical sense is open to serious doubt. This observation, due to the German physicist Zeh, appears to me particularly relevant for the theory of quantum mechanical measurements, involving the quantum mechanical interaction of a microscopic system with a macroscopic apparatus. The density of the quantum states of the latter is so high that a single electron, a mile away, can cause transitions between its states. Hence, in the quantum mechanical sense, it is in our world truly impossible to produce a macroscopic isolated system.

In my opinion this means that the question of the verifiability of our basic concepts, of the full extent of the quantum mechanical "laws of nature," and hence also of the invariance principles, is not fully resolved. This again is a subject with which I am not in equilibrium. I still believe in the invariance principles. I still believe that they are checked and verified or refuted by connections between events on isolated systems. But we have to admit that the existence of macroscopic isolated systems is, in this particular world, questionable. It is not possible, and perhaps conceptually not possible, to be in such a vacuum that macroscopic bodies are unaffected by their surroundings — in interstellar space, there is about one atom per cubic meter.

4. The Dynamic Invariance Principles.

So far, we have considered only the kinematic invariance principles. They form the essence of the special theory of relativity and they represent, we believe, accurate symmetries. The dynamic invariance principles, Radicati's "internal symmetries" are, in contrast, approximate. The approximation involves either the type of forces which are preponderant — electric, gravitational, etc. — or the state of the system with which one deals — only low or only high velocities, for instance. We shall deal, first, with the former type. The example which comes most easily to mind is the equality of the interaction between protons and protons, protons and neutrons, neutrons and neutrons. In other words, the equality of all kinds of nucleon-nucleon interaction.

This is not accurate. It is valid at best for nuclear forces. It is not valid, clearly, if we take electromagnetic forces into account, it is said to be "broken" by the electromagnetic interaction. Nevertheless, it manifests itself in many surprising regularities of the nuclear energy levels and of nuclear properties.

A symmetry which is even less accurately valid connects not only protons and neutrons, but also lambdas, sigmas and Ξ's. This is a very approximate principle. The neutron-proton mass difference is of the order of one thousandth of the total mass of the proton or neutron. The mass difference between the Λ, Σ, Ξ and the proton is of the order of magnitude of ten percent. Just the same, the recognition of the dynamic symmetry between these particles has been extremely fruitful.

So far, I discussed the dynamic symmetries which are based on the fact that in some physical phenomena only one type of interaction plays a significant role. In nuclei, for instance, the nuclear forces play by far the most important role. There seem to be four or five types of interactions and we do not know why and to what extent this is so, how independent they are from each other and to what extent their effects are additive.

The other type of dynamic symmetries are also approximate, but the approximation is based on the nature of the states to which they apply, not to the preponderance of one of the types of interaction. In other words, these

242   SYMMETRY IN NATURE

symmetries apply if the initial conditions have certain properties. The most common such property is the absence of high velocities, i.e., velocities of the order of magnitude of light velocity, and as a result of this, the possibility to neglect magnetic forces. This is a situation which prevails in almost all systems of interest to chemists. The Russell-Saunders theory of spectra is based on it. The smallness of the multiplet splitting is a consequence of the relatively low velocities of the electrons in light atoms. As chemists, we use this principle every day and when we talk about the collision of two hydrogen atoms, or of a hydrogen atom and a hydrogen molecule, we hardly even mention the fact that we disregard the effect of the orientation of the spins when calculating the forces between these atoms or molecules.

An even more interesting and very amazing symmetry has just the opposite basis. It is a symmetry which prevails when all velocities are very high, close to light velocity. It is the so-called conformal symmetry, discovered originally by the British physicists Cunningham and Bateman but taken up and expanded by the German physicist, Kastrup.

The conformal symmetry implies, first of all, scale invariance. This means that if one expands everything in both space and time, nothing significantly changes. Thus, the angular distribution resulting from a collision will become, at very high energies of collision, independent of that energy. There are many similar consequences. Under ordinary conditions, of course, scale invariance does not apply. The velocities of the electrons in atoms are well below light velocity and this applies particularly for the outer electrons, the ones responsible for the chemical bonds. The same is true of the velocities of nuclei in compounds.

Scale invariance is only one of the elements of conformal symmetry. There are four others which can be described by means of successions of "inversions by reciprocal radii," i.e. substitutions of the form

$$x_l \rightarrow 1/(x_l - a_l)$$

i denoting the four space-time coordinates and the a_l characterising a space-time point. The whole conformal group has 15 parameters, as contrasted with the 10 parameters of the symmetry of the special theory of relativity, or its specialization to low velocities, the symmetry of classical mechanics (Poincaré or Galilei invariances). This is a most remarkable fact: the restriction to low velocities may render the symmetry operations simpler but does not increase their manifold; restriction to velocities in the neighborhood of light velocity increases the ten parametric manifold to a fifteen parametric one.

The extent of the experimental verification of the conformal symmetry is not clear. This symmetry is very rich in consequences and these have not been adequately tested. One of the difficulties is a very fundamental one: even if the relative velocities of the colliding particles are close to light velocity, some of the reaction products may be, and some usually are, emitted at low velocities. This naturally impairs the applicability of the conformal principle. Many interesting ideas were put forward on the application of the principle taking this restriction into account but it would take too long to review these. What I did want to bring out is the enormous increase of the invariance group under the conditions specified, that is if all relative velocities in the system approach light velocity.

5. Other Symmetry and Invariance Principles.

Whether one should call the conformal invariance a dynamic symmetry or an intermediate one is unclear but this is a semantic question. The most important other intermediate symmetries are the reflection symmetries and the crossing rules.

The reflection symmetries have been discussed a great deal the last few years. They can be composed of three primitive ones: reflection of the space coordinates, the operation of "time inversion" which we discussed before when considering its effect on the entropy principle, and the replacement of every particle by its antiparticle, that is the electrons by positrons, the protons by antiprotons, and so on. Present evidence indicates that, at best, the combination of all three primitive symmetries is accurately valid, the other six combinations are valid only for nuclear, electromagnetic, and gravitational interactions, not for the weak interaction. This means, of course, that all reflection invariances are valid in a very good approximation under most conditions and indeed, to date, the asymmetry which manifests itself in the increase of the entropy has not been blamed on the inaccuracy of the time inversion invariance.

I do not believe that the last word has been spoken on the question of the validity of the reflection invariances. The situation as it now appears does not seem to be entirely reasonable. However, there is very little that I could add to the subject.

The crossing relations are regularities of the laws of nature — they represent very general relations between collision matrices referring to different kinds of collisions. However, they can not be tested in the simple and direct ways in which the other symmetry principles are tested and verified: either by creating a symmetric situation and verifying the fact that the symmetry persists, or by ascertaining that a certain relation between two initial states persists in time. In fact, they were discovered — by Gell-Mann and Goldberger — from laws of nature given by the relativistic quantum field theory, as rules which the consequences of this theory satisfy.

If we have a reaction

$$A + B \rightarrow C + D$$

the equality of the rate constant, at different places and for different states of motion of the centers of mass, follows from the general Poincaré invariance. The rate of the inverse reaction

$$C + D \rightarrow A + B$$

can be related to that of the former reaction by time-inversion invariance. If we denote the antiparticle of A, B, etc. by $\overline{A}$, $\overline{B}$, ... the equality of the rate of

$$\overline{A} + \overline{B} \rightarrow \overline{C} + \overline{D}$$

with that of the first reaction follows from particle-antiparticle conjugation invariance. The reflection symmetry also has consequences but we do not need to describe these now. The relations just given are, as was mentioned before, as a rule very well satisfied even though, at least in the case of the relation demanded by reflection symmetry, there are cases when the rule is violated.

The crossing symmetry relates the rate of the processes

$$A + \overline{C} \rightarrow \overline{B} + D$$

and

$$A + \overline{D} \rightarrow \overline{B} + C$$

to that of the process mentioned first. However, the relation is by no means simple. It assumes, first, that the rate is an analytic function of the relevant variables: the energy of collision and the difference between the momenta of incident and outgoing particles. This analytic function is then continued into the region of imaginary values of these quantities and the rate of the last two processes is given by the value of the analytic function for such imaginary values.

The reason I am discussing the crossing symmetry is just because it is so different from the other symmetry principles, both in its origin and also in the nature of the statements it makes. As to the experimental verification of the crossing symmetry, it has not progressed very far. It seems most unlikely that the symmetry is accurate but its verification or refutation is difficult. It is difficult not only because the experiments are difficult to carry out but also because an arbitrarily small change in an analytic function's value in one domain can entail an arbitrarily large change thereof in another domain. It would seem, therefore, that a purely experimental verification or falsification of the symmetry as such is nearly impossible. Nevertheless, there are some encouraging signs and I did want to call attention to this symmetry because of its novel nature and because it differs so fundamentally from the other symmetry principles. In spite of this, the preceding discussion of the dynamic invariance principles is very, very incomplete. It refers too little to the general theory of relativity, and neglects the use of these principles to discover laws of nature, the formulae of interaction.

6. Applications of the Symmetry Principles — and some Desired but not Available Applications.

One of the two most important applications of the invariance principles was mentioned before: to check, and also to discover, the laws of nature. Invariance principles were the main guide to both Einstein and Hilbert when they worked to establish the equations of the general theory of relativity and the present quantum field theories could not exist without the support of the requirement of Poincaré invariance. These determine in which way the infinities, which are inherent in the equations, should be eliminated. They were also the guides for the setting up of the equations.

We must admit that there were cases when the invariance principles misguided us. Cox obtained some data on β decay which were widely questioned because they were in conflict with the reflection invariance principle. It turned out, eventually, that his conclusions were correct and we, his critics, mistaken. The reflection invariance does not hold for β decay.

The second very important application of the symmetry principles is the derivation of some general consequences from the laws of nature. I do not refer to the fact that data obtained in one frame of reference can be duplicated in other frames of reference, though this is of course the most direct, and also the most common, use of the invariance principles. This comes to us so naturally, that we hardly realize that we use an invariance principle. What I am referring to, rather, is that, in spite of the conceptual simplicity of the laws of nature, their consequences are, in most circumstances, very difficult to evaluate completely. *Some* of the consequences of these laws can be obtained, however, and can be obtained most easily and generally, by means of the invariance principles.

The first of these consequences are the conservation laws. Actually, these were known before they were derived from the invariance principles. The demonstration that they are, in classical mechanics, consequences of the kinematic invariance principles, was given, early in this century, first by Hamel, then by Klein and Nöther. The connection between invariance principles and conservation laws is rather obvious in quantum mechanics: the infinitesimal invariance transformations' operators represent the conserved quantities. That there is such a connection also in classical mechanics required a deeper insight and that is what was furnished by Lagrange, Hamel, Klein and Nöther.

246   SYMMETRY IN NATURE

The symmetry principles do have, however, also other consequences. Some of these refer to properties of stationary states, others to transitions between stationary states. The absence of an electric dipole moment of all stationary states is a consequence of the reflection symmetry and there should be small violations of this rule. None were found so far, in spite of serious efforts, in particular of Ramsey and his collaborators. To a chemist, the absence of electric dipole moments of stationary states does sound, at first, surprising: HCl surely has a dipole moment. It does if it has a definite orientation, but a definite orientation implies the superposition of many rotational levels, with different energies, and is, therefore, not a stationary state. This last fact is, perhaps, as remarkable as the former one: the probability of a definite orientation in a stationary state is not an arbitrary function of the orientation but a severely restricted one. Before the relativistic theory of the H atom was established, there appeared to be one exception to the no-dipole-moment rule: the H atom itself, most of the states of which had an "accidental degeneracy." The relativistic effects cause a splitting of this degeneracy and the rule now holds without exception.

Another, and more general, set of conclusions easily arrived at on the basis of invariance principles concerns the effect of small perturbations on the stationary states and the various transition probabilities between these. As to the former, one can, for example, easily show that any energy level's splitting in a weak homogeneous magnetic field (Zeeman effect) results in a number of equidistant levels and the spacing between these levels is proportional to the intensity of the magnetic field. The restriction to a "weak" field need not be taken too seriously — the rule is valid even in the strongest magnetic fields that are available. An example of the latter type of relations is given by the formulae for the ratio of the transition probabilities from the levels resulting from the magnetic (Zeeman) splitting of one level to the levels arising from the same splitting of another level. These are two simple examples for the second type of conclusions from the laws of nature which can be obtained most easily from invariance principles. It would be easy to give a dozen other examples but this is neither desirable nor necessary: there are reasonably thick books devoted to this subject, showing the increased power of the symmetry principles in quantum mechanics as compared with classical theory.

It may be worth while, instead, to mention the area on which the consequences of symmetry principles are much less far reaching in quantum than in classical theory. This area is the area of the first application of symmetry principles: the theory of crystal structures. The flawless derivation of the 32 crystal classes and of the 230 space groups when so much less was known about symmetry, by Fedorov and by Schönflies, continues to impress me as an almost unbelievable accomplishment. The point which I want to bring out now, however, is that the crystal symmetry is, according to classical concepts, a rigorous symmetry, at least if we attribute full spherical symmetry to the constituent atoms. This is not so in quantum mechanics. In fact, the crystal symmetry cannot be rigorously formulated in terms of quantum mechanical concepts, such as a wave function in configuration space. Naturally, if one makes approximations, treats the atoms or the nuclei classically, or at least semiclassically, one obtains the classical results. One clear indication of the approximate nature of crystal theory in quantum mechanics is the fact that one cannot attribute, even approximately, positions to the electrons as one attributes positions to the nuclei. This latter possibility is, evidently, based on the larger masses of the nuclei which renders, in most cases, a classical treatment of their behavior a good approximation. It is called the Born-Oppenheimer approximation. This approximation is not always good: the phenomenon of free rotation, the impossibility in some cases to assign positions to H atoms consistent with crystal symmetry, shows the approximate nature of the theory. The situation with respect to the electrons is, of course, different: here it is the rule rather than the exception that no position can be assigned to them which would be consistent with the symmetry of the crystal and this is not even attempted.

Let me repeat what I find perturbing: it is that crystal symmetry cannot be formulated in quantum mechanical language. Further, we do not know how large the deviations are from the classical symmetry. Clearly, in a crystal, even at the absolute zero, some atoms are not parts of the lattice but wander around. Similarly, some positions are not occupied. The two numbers may not be the same — probably are not the same. We do not know them.

I will now conclude with the enumeration of some general regularities which I feel should, but do not, follow from symmetry principles. There are three such regularities.

The first one is the existence of some conservation laws, apparently rigorously valid, which are not truly consequences of symmetry principles. These are the conservation laws for electric charge, for baryon number, and for two kinds of leptons. It is true that we can derive, for instance, the electric charge conservation law from the so-called gauge invariance. We introduce, instead of the electromagnetic field, electromagnetic potentials. These are not uniquely determined by the physical situation, the addition of a gradient to the potential does not affect the electromagnetic field. We then formulate the interaction of charges with the electromagnetic field in terms of these potentials, and it must be admitted that this simplifies matters. We then introduce the postulate that the interaction shall not be affected by the arbitrariness of the potential and this can be done in a simple and natural way only if there is a conservation law for the charge. However, this hardly appears

248   SYMMETRY IN NATURE

to be a natural derivation. To begin with, it is not clear that the introduction of the potentials to describe the interaction of the field with the charges is necessary. In fact, Mandelstamm has shown that it is not. Second, it is not clear what kind of arbitrariness should be introduced in the field-defining quantities. Someone may propose "potentials" not with four but many more components and not only for the electromagnetic field but also for matter. Where do we stop? The derivation of the other conservation laws — for baryon number and for the two lepton numbers — is even less satisfactory.

The second regularity which I hope will find a more general foundation is the identity of all electrons, and of all elementary particles, as far as their basic characteristics, such as mass and charge, are concerned. The laws of physics could be formulated equally well, and the charge conservation maintained, if the masses and charges were individual properties, not common to all electrons and to all protons. Furthermore, the proton charge is the exact opposite of the electron charge. The experimental verification of this has assumed a fantastic accuracy. One feels that some basic principle is hidden here which we have, as yet, not been able to formulate. Some attempts have been made in this direction but they remained attempts.

The last puzzle which I wish to mention, concerns our starting point, the separateness of the three concepts: events, laws of nature, invariance principles. The separation of the first two, at least, is amazingly sharp and we have contrasted the accuracy of the laws of nature with the randomness appearing in the initial conditions.

Why are there just three? Will their number increase again when some genius discovers a basic modification of the laws of nature? Or, will their number decrease when we bring in other parts into physics, such as the phenomenon of life?

And it is perhaps good that there are phenomena which are far beyond the present range of our understanding and which may remain beyond this range for a long time, because if we knew everything, we would know even less what to begin with our lives.

INTRODUCTION AND DISCUSSION

Dr. G. Herzberg (Discussion Leader), *National Research Council of Canada:* It is my very great pleasure to introduce the last formal speaker in this symposium.

Eugene Wigner is claimed by physicists as well as chemists, as one of their great pioneers. In view of what is written in the program, I will not recount to you his career, but only say that he belongs to that small group of

EUGENE P. WIGNER   249

extraordinarily gifted people who come from a small country, Hungary. He started out as chemist, turned physicist, and proceeded to solve some very basic problems in physics and chemistry.

We are all familiar with Wigner's early recognition of the importance of symmetry in atomic and molecular physics. The wide range of his contributions goes from reaction kinetics to nuclear physics; from molecular structure, for example the correlation rules named after him to solid state physics.

It is particularly appropriate that he should talk to us today on the subject, "Symmetry in Nature."

Professor Wigner.

[The following discussion took place after the presentation of Dr. Wigner's address.]

Dr. Per-Olov Löwdin (Discussion Leader), *University of Uppsala and University of Florida:* We thank Dr. Wigner very much for his wonderful and stimulating lecture.

We all know that the role he has played in the development of the symmetry principles, particularly in quantum mechanics and that no paper dealing in any way with symmetry in Göttingen was left unchecked by Dr. Wigner before it was published.

His paper is now open for discussion. Dr. Guth?

Dr. Eugene Guth, University of Tennessee and Oak Ridge National Laboratory: I hesitate to add to this beautifully simple and critically clear treatment by Professor Wigner of the symmetry principle. However, maybe I can make a few comments. I know they are all known to Professor Wigner, but maybe all of you don't know them.

First, I would like to make a comment on a simple conformal invariance of Newton's equation for the case of gravitational forces.

You can derive Kepler's third law from the equations of motion by integration. This is carried through in any textbook of mechanics. But, you can derive it without any computations.

The equations of motion for the gravitational N-body problem (m_k: masses; G: gravitational constant; r_k: radius vector)

$$m_k \ddot{\vec{r}}_k = G \sum_{i,k=1}^{N} m_i m_k \frac{\vec{r}_i - \vec{r}_k}{r_{i,k}^3}, \qquad (*)$$

are invariant under the conformal transformation

$$x_k' = \gamma^2 x_k$$

$$t' = \gamma^3 t$$

$$(**)$$

Kepler's third law follows immediately from (**) (cf. E. Guth, Contribution to the History of Einstein's Geometry as a Branch of Physics, in *Relativity*, (S. I. Eickler, M. Carmeli, L. Witten, eds.), Plenum Press (1970) l.c. pp. 166-167). As far as I know, this is the simplest way to derive Kepler's third law.

Now, I would like to make just one more comment on the difference between kinematic and dynamic symmetry. This difference is far from being trivial, as Dr. Wigner emphasized. It came up in a discussion between Einstein and a late friend of mine, Friedrich Kottler of Vienna.

Kottler was a man who knew tensor analysis and Riemannian geometry as well as Einstein, but he didn't have Einstein's insight. Einstein thought very highly of him. Einstein started saying that this colleague distinguished himself by his profound understanding of the problems, claiming kinematic symmetry for accelerated systems, instead of dynamical symmetry as Einstein did. Kottler first (before Einstein) formulated the Maxwell-Lorentz equation in generally covariant form. But Kottler was not on the right track. Einstein was, of course, right.

So, you see from this historical fact that the distinction between the dynamic and kinematic invariance is far from being trivial. (cf. A. Einstein, *Ann. d. Physik*, 1916.)

Dr. Wigner: Thank you very much. I did not know about this discussion between Kottler and Einstein and I appreciate your mentioning it.

Dr. Löwdin: Can I ask you a question in connection with the time differences? As you well know, Dirac has now related certain fundamental constants not to the time differences, but to the absolute time since the "Big Bang." Would you care to comment about this?

Dr. Wigner: It is not easy to comment on this problem. Dr. Teller also worked on it, but the latest article which I have read on the question of the variation of the fundamental constants is Dyson's article which will come out very soon in "Perspectives of Physics," a volume dedicated to Paul Dirac.

Dyson, who studied the question more carefully than most anybody else, points out, first of all, that there are several ways it can be looked at. In particular, there is the question of whether the fine structure constant

EUGENE P. WIGNER   251

also changes with time. Dirac at one time thought that it is a logarithmic function of time.

Teller contradicted that in an article based on an analysis of the history of the solar system. I am afraid Dyson's discussion does not arrive at any terribly firm conclusions on the whole problem of the changes of the fundamental constants but on the whole, he is skeptical concerning the change of the fine structure constant.

Dr. Löwdin: Thank you very much.

Dr. E. Bright Wilson, Jr. (Discussion Leader), *Harvard University:* There may be people in the audience more ignorant than I am on this subject, but I doubt it, so let me try this question:

How do you define time? If you define time in terms of some kind of experimental clock, how in the world do you extrapolate time back to those Big Bang days when there wasn't any earth rotating; when there weren't any double stars; when there wasn't any ammonia molecule turning inside out? I suppose all the radioactivity was disturbed and so on?

Dr. Wigner: I am afraid Dr. Wilson has a deeper insight than he is willing to admit. (Laughter)

One can construct two types of clocks. One can construct a clock by having two gravitational masses rotating about each other, or, the way he suggested, construct clocks by quantum systems, by looking at the transition of ammonia he mentioned.

Now, all the discussions are, at present, based on the second type of clock and are intrinsically assuming the validity of equations which apply to ammonia molecules, or other molecules, because, of course, there are other molecules and atoms with which one can measure time. Hence, for atoms and molecules we assume that our usual equations — the equations of quantum mechanics — were always valid and this implicitly defines time. Certainly, there is only one time scale for which these equations are valid and we assume that there is such a time scale and use it to define time.

We then make different assumptions about the gravitational constant, one of them being that it always had its present value. Another assumption is that it was larger in the past and slowly diminished to its present value. We then try to compute the distant past on the basis of both assumptions and choose the assumption which retrodicts a more reasonable past.

Of course, it could have been assumed equally well that the gravitational equations in their present form and with the present gravitational constant were always valid and to define the time that way. However, this is not the usual procedure.

252   SYMMETRY IN NATURE

I do not know whether this fully answers your question. It is an admission of presumptuousness on the part of us physicists when we try to decide which past is more reasonable without being as positivistic about it as we are when we try to describe the present world.

Quantum mechanics says, "Quantum mechanics gives probability connections between observations." Nobody observed the origin of our world; hence quantum mechanics' application thereto is not entirely in the spirit of the theory.

Dr. F. A. Cotton (Discussion Leader), *Texas A&M University:* Well, I certainly wouldn't yield to anybody, certainly not Bright Wilson in my claim to being the most ignorant man in the room on the subject.

However, in these discussions on how you know whether you can extrapolate backwards to the Big Bang and so forth, I am reminded of one of the most elementary things that I, at least, and I think a lot of other people, teach students when we first start telling them about the scientific method and how you use it and don't use it.

One of these prescriptions that we give them is that "Thou shalt not generalize outside the range of thine observations." And it seems to me that when you begin going back beyond the realms of time and space that we have actually observed, you are generalizing, or trying to, beyond the limits of your observations. And this applies, I would think, even to venerable physicists as well as to young students. Isn't this perhaps more a philosophical game than a genuine form of science?

Dr. Wigner: Of course, you are right, and both of you are right. However, if we have laws of physics and equations of motion we want them to be such that they don't lead to a contradiction even in areas where it is very difficult to test them.

It is a different thing to maintain that I know it, and a different thing to see whether they lead to contradictions. The contradictions bother me. I realize, of course, what everybody realizes: that it is not of primary interest to me what happened ten billion years ago. But I would prefer to have equations which don't give an apparent contradiction, even extrapolated to a beginning, and I think you will agree with that.

Dr. Cotton: What was the contradiction ten billion years ago?

Dr. Wigner: What was the entropy then? Why isn't the universe in equilibrium?

Dr. Löwdin: Can I comment on that one?

Dr. Wigner: Please.

Dr. Löwdin: Because when you wrote up this equation here which was based on quantum mechanics, or something similar, there was an interference term of the same type as Dr. Teller had on the blackboard this morning. And this interference term later vanished because you averaged over an assembly where you had enough irregularities.

But the universe, at least to us who live in it, is just one species — not an assembly. There is only one universe, we are living in it, and that is to me the entire explanation.

Dr. Wigner: Is that an explanation?

Dr. Löwdin: Yes. (Laughter) Because you need an assembly of universes to be able to get rid of your interference term.

Dr. Wigner: Well, I don't know. The question is: Why is it still so that the suns are hot and their radiation has not yet filled the world with radiation so that there is equilibrium. Why is that? Somebody knows it? (Laughter)

Dr. Guth: I don't claim to be wiser on this question, but I would like to quote Boltzmann. Boltzmann said that the world as a whole is in equilibrium—and that "our world" corresponds to local fluctuations, and what is beyond that nobody knows.

Dr. Wigner: I realize that it is possible to say our physics is still incomplete. But it is sort of nasty to do that. (Laughter)

Dr. Guth: Absolutely.

Dr. Wigner: It seems to me that there are two very strong reasons against considering this world as a result of a fluctuation, because a fluctuation not only returns to the original — to the so-called equilibrium situation, but also went into the appropriate direction.

Now is it conceivable that people lived on the opposite side of the fluctuation and were afraid every tenth of a second that now it is all over and the entropy will now change the sign of its motion?

That is one argument against the fluctuation theory. The other reason that I don't feel that the fluctuation theory is satisfactory, is that there are many other regularities in the universe, as we now look at it, which would be surprising, if it were a fluctuation. In particular, that all the stars go away from each other, the universe "expands." There is no reason in a fluctuation for having these regularities.

Dr. Guth: I agree with you. I just quoted Boltzmann, and I guess one reason why I quoted him is because there are some, should I say, very distinguished physicists who think that there is a grain of truth in it, *i.e.* George Uhlenbeck.

254   SYMMETRY IN NATURE

When we discuss these questions, one has to be extremely careful. Actually, one should really do it in writing. When I commented to you first, you said we should stick to what is observable.

Well, let's take the question of pulsars. What is the constitution of the pulsars? A neutron-star? Nobody can check it. They can check some of the consequences, but it would be quite possible that there are other consequences observable, perhaps at a later time which might contradict this model. We cannot go there and confirm it.

But the human mind has a sort of compulsion to develop a grand synthesis. We can use as a criterion mathematical beauty. But, actually, we cannot check it. We will never be able really to check it with the accuracy which we can do with the solar system. You see, it is just impossible.

That will be always to some extent, a speculation.

Dr. Wigner: Well, I don't know. We did not go to the sun, which is 150 million kilometers away. Just the same, we are absolutely convinced it is there. (Laughter)

Dr. Guth: We are convinced, but let me mention a serious uncertainty about its structure.

It is not clear whether the sun has a quadrupole moment like Bob Dicke thinks it has and because he found the sun is oblate. Or whether this fact does not imply a quadrupole moment and this would not contribute to the Mercury perihelion advance.

As a matter of fact, at the end of the last paper that I got from Dicke, he says that he doesn't expect that people will believe his interpretation uncritically. As a matter of fact, all he can hope is that with more and more research, one should be able maybe to make his interpretation plausible or contradict it.

Then I come to the question of the dependence of the elementary constants on time. From the papers of Freeman Dyson, which Professor Wigner mentioned, it follows that the only constant, which might have time-dependence, is the gravitational constant G.

It is fairly clear that G cannot depend as strongly on time as Dirac suggested originally. However, the time dependency of G can be decided empirically. I can mention to you two ways. One way is to observe the difference between the astronomical time and the sidereal time. A second way is to measure the gravitational constant with more accuracy. Now, very recently, a new determination of G was made by Jesse Beams and his associates. Jesse told me that he thinks that it is feasible that he can measure it, let's say, with an accuracy one hundred times larger. Then he could possibly find a dependence on time. The scalar-tensor theory of Brans and Dicke assumes a variation of the gravitational constant with time, but weaker than Dirac assumed.

The majority of you are chemists, but maybe some of you are members of the Physical Society. There was an article recently in "Physics Today" by a young physicist by the name Will with the eye-catching title, "Einstein On The Firing Line."

I would like to bring just a very quick comment on that, because I think it is very misleading. I mean, it is good propaganda, but it is very misleading.

I think I can practically prove that Einstein's theory is the only sensible generalization of Newton's theory up to a certain number of decimals. Any improvement over the Einstein theory would be a small correction. I think we know enough by now to state that Einstein's theory is valid within five percent.

Dr. Löwdin: Doctor Buckingham first and then Doctor Eyring.

Dr. A. D. Buckingham (Discussion Leader), *University of Cambridge:* Might I raise a minor and brief point, sir, about the existence of electric dipole moments of systems in stationary states? You pointed out that quantum electrodynamics removed the one and only one, the hydrogen atom. But, chemistry can provide some; namely, the symmetric-rotor molecules where the barrier to inversion is so high that to all intents and purposes, it doesn't occur.

Dr. Wigner: But only to all intents and purposes. (Laughter)

Dr. Buckingham: Even with the smallest imaginable field, there is a linear Stark splitting.

Dr. Wigner: Yes. You are absolutely right. I appreciate the point that was made. Perhaps I should mention that the reason for the phenomenon is that two energy levels of opposite parities coincide. I just amplified what you said.

Dr. Henry Eyring (Speaker), *University of Utah:* If one takes Boltzmann seriously and tries to calculate the probability that one should have fluctuations in a system at equilibrium with the observed differences in temperature and composition we find experimentally it is so fantastically improbable that we must look for some other explanation even if it involves giving up the applicability of the second law of thermodynamics on a cosmic scale.

Dr. Löwdin: Dr. Bright Wilson?

Dr. Wilson: Just a comment on Professor Buckingham's: Of course, chemists are completely uninterested in stationary states, because there aren't any in chemistry, if you get really strict about it. Every molecule can come apart, or change. They are not in exactly stationary states, but of course they may be to a very good approximation. However, the distinction can be important when making general statements.

256   SYMMETRY IN NATURE

Dr. Wigner: I don't believe that the interest of chemists is as restricted as not to extend to stationary states. (Laughter)

Dr. Löwdin: Doctor Davidson? Would you like to make a comment or ask a question?

Dr. Ernest R. Davidson (Discussion Leader), *University of Washington.* I am extremely naive about all this, but let me ask what may be a very stupid question.

I think all of us here, on an operational level, know how to use your formula for the entropy, and we can all calculate it and derive equations with it. That works very nicely.

My question is, in that equation, what do the "p_i's" really mean to you? What do you think of when you see them?   ·

Dr. Wigner: The p_i in many cases is 1 for one domain and 0 for the other domains. But, if I don't have an absolute certainty that the system is in a definite part of phase space, but imagine that it may be in this part or in that part with different probabilities, I still can make use of this knowledge to produce work with a system. And I think this describes the —

Dr. Davidson: Then you interpret the p_i's as a statement of your own knowledge of the system —

Dr. Wigner: Yes.

Dr. Davidson: — and not a property of the system itself, if you were absent?

Dr. Wigner: Yes, or the property of an ensemble which means the same thing.

Dr. Davidson: Wouldn't it in that sense mean then that any time when we actually know the state of any situation, it's going to seem improbable —

Dr. Wigner: Yes.

Dr. Davidson: — and it's our own ignorance which is changing with time rather than the randomness of the system? Is that what that says or not?

Dr. Wigner: No. I may have misstated my point. If one derives it, one obtains that expression.

Dr. Davidson: Yes.

Dr. Wigner: But, if we know that it is in one of them, then we set one $p_i = 1$, the others $= 0$, and that tells us how much work we can extract from it.

Dr. Davidson: Yes.

EUGENE P. WIGNER   257

Dr. *Wigner:* So, perhaps I put down the expression unreasonably generally, and I appreciate your comment.

Dr. *Löwdin:* I noticed that you were rather reluctant about the concept of potentials in connection with the gauge invariance and the constancy of the charge.

I wonder from a practical point of view, perhaps in connection with the Schrödinger equation, what type of potentials one should introduce into the Schrödinger equation, since the Schrödinger equation itself ought to be time reversible.

Then we come to the problem, should we put in retarded potentials, or advanced potentials or symmetric potentials? Could you give us a clue?

Dr. *Wigner:* Well, of course, what we physicists, vaguely familiar with quantum electrodynamics, say is that we should use the field concept and we should attribute a value to the field at every point.

But, when we describe the interaction, then it is necessary, or almost necessary, to use the concept of the electromagnetic potential. It is at least customary. What is the name of the British physicist who wrote a paper on the elimination of the electromagnetic potential?

Dr. *Guth:* Mandelstam?

Dr. *Wigner:* Thank you. Mandelstam. Mandelstam pointed out that it is not absolutely necessary to use it. If we don't use it, the existence of the conservation laws for a charge is even more mysterious. Of course, it follows from the equations, but it is such a general principle, that it should not be only a consequence of the equations which are modified every day, or every other day. (Laughter) But it should be a consequence, I feel, of the symmetry principles. I may be too demanding on the symmetry principles, but I would like to see it.

It is often said that it is a consequence of these principles and that bothers me a little. I could invent 55 other potentials and formulate equations of motion in terms of them and then postulate more conservation laws which are not in existence.

Am I talking sense?

Dr. *Löwdin:* Yes. Thank you.

Now, the question is also whether there is a coupling between the event and the laws in the customary type of discussion of potentials which is rather disturbing from the philosophical point of view.

258   SYMMETRY IN NATURE

Dr. Guth: I think this question which you brought up is very closely related to the question, whether one can formulate a physical theory of quantum mechanics in terms of more or less directly observable quantities like the current and the density; or whether it is more convenient to introduce quantities like the wave function which is not observable.

Actually, for certain purposes, for instance, if you want to discuss the magnetic pole, which Dirac suggested, it is advantageous to get away from potentials and use a sort of generalized Ehrenfest theorem and expectation values. But the trouble is that this formulation is not general enough, it becomes awkward. It is still useful, for example with a Bose type of problem but when we take a Fermi type of problem, then it becomes prohibitively awkward.

Dr. Löwdin: I have one final question to you and it concerns just the ordinary organic chemist.

We are recently learning about the Woodward-Hoffmann rules, and the use of symmetry in connection with chemical reactions, particularly in the area of organic chemistry. Here one has some very simple symmetry problems, and Woodward and Hoffmann have given certain rules which are explicitly conclusive and highly useful from the point of view of practical chemistry.
Have you any comments about those?

Dr. Wigner: I have a question in connection with that. Where could I learn something about it? (Laughter)

Dr. Löwdin: Woodward and Hoffmann have written some excellent, readable papers on it recently in addition to their original work.

Dr. Wigner: I am sorry I know nothing about it, and I can't comment.

Dr. Löwdin: Just a second, if you wanted to comment. Yes, sir?

Dr. Guth: Hoffmann was a week ago at Oak Ridge and gave a talk about these rules. There is an article by Hoffmann and Woodward in "Science," I can send you a Xeroxed copy. Professor Hoffmann is going to send me a longer review from "Angewandte Chemie" and I will be glad to send you a Xerox copy of that.

Dr. Wigner: I'll appreciate it.

Dr. Löwdin: Before turning the microphone over to Dr. Herzberg, I would once more like to express our sincere thanks to Dr. Wigner, both for his illuminating talk and for what he has given us in science, both in physics and in chemistry.

Dr. Herzberg: I have been asked to express the thanks of those of us who have attended these meetings, to those who have organized them.

Professor Wigner has already taken the step of giving thanks for what has been before his talk, but I have now the opportunity to review the whole situation.

I think it is fair to say that this has been an extremely successful conference, successful in the sense that we have, first of all, heard a number of most outstanding presentations, and, at the same time, we have had some extremely interesting discussions.

We have moved from some very specific molecules to some very philosophical problems as in the last lecture of Professor Wigner. I don't think it is possible to review in any way all the lectures that we have had. But, I think all of us feel very much indebted to the speakers who have given us such illuminating and marvelous presentations.

In addition, I think we must express our thanks to The Welch Foundation and the Trustees of The Welch Foundation who have made it possible for us to come together here and to live through these three very stimulating and interesting days.

I would like, on your behalf, to express our thanks to Dr. Milligan and his staff and to the Trustees for all the efforts they have put into the organization of this meeting to let us have this feast, as Dr. Buckingham expressed it this morning.

Dr. Milligan: Thanks very much for your kind remarks Dr. Herzberg. The Foundation's role is like that of a catalyst to get the right people together in one room.

I would like to express the thanks of all of the Trustees and Officers of The Foundation and my own, to the speakers and discussion leaders who were willing to give of their time; and also, to the audience for coming, because, without the audience these Conferences could not exist.

It has been a great pleasure having you here and we hope to see some of you almost a year from now when we will discuss "Inorganic Reagents in Synthetic Organic Chemistry." Thank you very much for coming.

NOTE ADDED IN PROOF

Dr. Wigner: When reading the notes of our discussion I note that my answer to Dr. Davidson's question, concerning the equation on page 235, was grossly incomplete.

If the volumes V_i in phase space are large enough and represent states which can be characterised also macroscopically, so that fluctuations play no role in the course of the events considered, the changes of the system can be described macroscopically. This means that the points of the volume V_i will not spread out over many volumes V_j but will continue to be contained essentially in a single volume. If this volume is, at a later date V_f, so that, at that time, $p_f = 1$, all other $p = 0$, the V_f must be larger than V_i. This follows from Liouville's theorem since V_f must accommodate all points of V_i. On this case, as Dr. Davidson envisaged it, the situation persists that only one p— or perhaps two— are different from 0 and the first term of the equation's right side essentially vanishes but the second one constantly increases.

The several p_i play a role when fluctuations are relevant and the initial volume does not remain concentrated in a single volume but a fraction p_f of it enters the volume V_f, another fraction p_g enters the volume V_g, and so on, so that many parts of phase space are occupied, each with a small probability. In this case, the first term $-p_i \ln p_i$ also becomes relevant and contributes to the increase of the entropy. However, the situation in which several p are finite occurs only when fluctuations are relevant. If the volumes V_i have macroscopic descriptions their points will stay together in one or, at most, a very few volumes and the situation will persist that only one p, or at most a very few, will be different from 0. I believe this is what Dr. Davidson wanted to bring out. However, it is, in my opinion, appropriate to have an expression for the entropy which remains valid even if one allows for fluctuations.

PART V

Relativity

Relativistic Invariance and Quantum Phenomena

E. P. Wigner

Symmetries and Reflections.
Indiana University Press, Bloomington, Indiana 1967, pp. 51–81

Introduction

The principal theme of this discourse is the great difference between the relation of special relativity and quantum theory on the one hand, and general relativity and quantum theory on the other. Most of the conclusions which will be reported on in connection with the general theory have been arrived at in collaboration with Dr. H. Salecker,[1] who has spent a year in Princeton to investigate this question.

The difference between the two relations is, briefly, that while there are no conceptual problems to separate the theory of special relativity from quantum theory, there is hardly any common ground between the general theory of relativity and quantum mechanics. The statement, that there are no conceptual conflicts between quantum mechanics and the special theory, should not mean that the mathematical formulations of the two theories naturally mesh. This is not the case, and it required the very ingenious work of Tomonaga, Schwinger, Feynman, and Dyson[2] to adjust quantum mechanics to the postulates of the special theory and this was so far successful only on the working level. What is meant is, rather, that the concepts which are used in quantum mechanics,

Address of retiring president of the American Physical Society, January 31, 1957. Reprinted by permission from the *Reviews of Modern Physics*, Vol. 29, No. 3 (July, 1957).

[1] This will be reported jointly with H. Salecker in more detail in another journal.

[2] See, e.g., J. M. Jauch and F. Rohrlich, *The Theory of Protons and Electrons* (Cambridge: Addison-Wesley Publishing Co., 1955).

52   *Symmetries and Reflections*

measurements of positions, momenta, and the like, are the same concepts in terms of which the special relativistic postulate is formulated. Hence, it is at least possible to formulate the requirement of special relativistic invariance for quantum theories and to ascertain whether these requirements are met. The fact that the answer is more nearly *no* than *yes*, that quantum mechanics has not yet been fully adjusted to the postulates of the special theory, is perhaps irritating. It does not alter the fact that the question of the consistency of the two theories can at least be formulated, that the question of the special relativistic invariance of quantum mechanics by now has more nearly the aspect of a puzzle than that of a problem.

This is not so with the general theory of relativity. The basic premise of this theory is that coordinates are only auxiliary quantities which can be given arbitrary values for every event. Hence, the measurement of position, that is, of the space coordinates, is certainly not a significant measurement if the postulates of the general theory are adopted: the coordinates can be given any value one wants. The same holds for momenta. Most of us have struggled with the problem of how, under these premises, the general theory of relativity can make meaningful statements and predictions at all. Evidently, the usual statements about future positions of particles, as specified by their coordinates, are not meaningful statements in general relativity. This is a point which cannot be emphasized strongly enough and is the basis of a much deeper dilemma than the more technical question of the Lorentz invariance of the quantum field equations. It pervades all the general theory, and to some degree we mislead both our students and ourselves when we calculate, for instance, the mercury perihelion motion without explaining how our coordinate system is fixed in space, what defines it in such a way that it cannot be rotated, by a few seconds a year, to follow the perihelion's apparent motion. Surely the x axis of our coordinate system could be defined in such a way that it pass through all successive perihelions. There must be some assumption on the nature of the coordinate system which keeps it from following the perihelion. This is not difficult to exhibit in the case of the motion of the perihelion, and it would be useful to exhibit it. Neither is this, in general, an academic point, even though it may be academic in the case of the mercury perihelion. A difference in the tacit assumptions which fix the coordinate system is increasingly recognized to be at the bottom of many conflicting

Relativistic Invariance and Quantum Phenomena   53

results arrived at in calculations based on the general theory of relativity. Expressing our results in terms of the values of coordinates became a habit with us to such a degree that we adhere to this habit also in general relativity, where values of coordinates are not *per se* meaningful. In order to make them meaningful, the mollusk-like coordinate system must be somehow anchored to space-time events and this anchoring is often done with little explicitness. If we want to put general relativity on speaking terms with quantum mechanics, our first task has to be to bring the statements of the general theory of relativity into such form that they conform with the basic principles of the general relativity theory itself. It will be shown below how this may be attempted.

Relativistic Quantum Theory of Elementary Systems

The relation between special theory and quantum mechanics is most simple for single particles. The equations and properties of these, in the absence of interactions, can be deduced already from relativistic invariance. Two cases have to be distinguished: the particle either can, or cannot, be transformed to rest. If it can, it will behave, in that coordinate system, as any other particle, such as an atom. It will have an intrinsic angular momentum called J in the case of atoms and spin S in the case of elementary particles. This leads to the various possibilities with which we are familiar from spectroscopy, that is, spins 0, ½, 1, ¾, 2, . . . , each corresponding to a type of particle. If the particle cannot be transformed to rest, its velocity must always be equal to the velocity of light. Every other velocity can be transformed to rest. The rest-mass of these particles is zero because a nonzero rest-mass would entail an infinite energy if moving with light velocity.

Particles with zero rest-mass have only two directions of polarization, no matter how large their spin is. This contrasts with the $2S + 1$ directions of polarization for particles with nonzero rest-mass and spin S. Electromagnetic radiation, that is, light, is the most familiar example for this phenomenon. The "spin" of light is 1, but it has only two directions of polarization, instead of $2S + 1 = 3$. The number of polarizations seems to jump discontinuously to two when the rest-mass decreases and reaches the value 0. Bass and Schrödinger[3] followed this out in detail

[3] L. Bass and E. Schrödinger, *Proc. Roy. Soc.* (London), A232, 1 (1955).

for electromagnetic radiation, that is, for $S = 1$. It is good to realize, however, that this decrease in the number of possible polarizations is purely a property of the Lorentz transformation and holds for any value of the spin.

There is nothing fundamentally new that can be said about the number of polarizations of a particle, and the principal purpose of the following paragraphs is to illuminate it from a different point of view.[4] Instead of the question: "Why do particles with zero rest-mass have only two directions of polarization?" the slightly different question, "Why do particles with a finite rest-mass have more than two directions of polarization?" is proposed.

The intrinsic angular momentum of a particle with zero rest-mass is parallel to its direction of motion, that is, parallel to its velocity. Thus, if we connect any internal motion with the spin, this is perpendicular to the velocity. In case of light, we speak of transverse polarization. Furthermore, and this is the salient point, the statement that the spin is parallel to the velocity is a relativistically invariant statement: it holds as well if the particle is viewed from a moving coordinate system. If the problem of polarization is regarded from this point of view, it results in the question, "Why can't the angular momentum of a particle with finite rest-mass be parallel to its velocity?" or "Why can't a plane wave represent transverse polarization unless it propagates with light velocity?" The answer is that the angular momentum *can* very well be parallel to the direction of motion and the wave *can* have transverse polarization, but these are not Lorentz invariant statements. In other words, even if velocity and spin are parallel in one coordinate system, they do not appear to be parallel in other coordinate systems. This is most evident if, in this other coordinate system, the particle is at rest: in this coordinate system the angular momentum should be parallel to nothing. However, every particle, unless it moves with light velocity, can be viewed from a coordinate system in which it is at rest. In this coordinate system its angular momentum is surely not parallel to its velocity. Hence, the statement that spin and velocity are parallel cannot be universally valid for the particle with finite rest-mass and such a particle must have other states of polarization also.

---

[4] The essential point of the argument which follows is contained in the present writer's paper, *Ann. Math.*, 40, 149 (1939), and more explicitly in his address at the Jubilee of Relativity Theory, Bern, 1955 (Basel: Birkhauser Verlag, 1956), A. Mercier and M. Kervaire, editors, p. 210.

Relativistic Invariance and Quantum Phenomena   55

It may be worthwhile to illustrate this point somewhat more in detail. Let us consider a particle at rest with a given direction of polarization, say the direction of the z axis. Let us consider this particle now from a coordinate system which is moving in the −z direction. The particle will then appear to have a velocity in the z direction and its polarization will be parallel to its velocity (Fig. 1). It will now be shown that

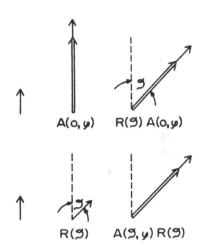

Fig. 1   The short simple arrows illustrate the spin, the double arrows the velocity of the particle. One obtains the same state, no matter whether one first imparts to it a velocity in the direction of the spin, then rotates it $(R(\vartheta)A(0,\varphi))$, or whether one first rotates it, then gives a velocity in the direction of the spin $(A(\vartheta,\varphi)R(\vartheta))$. See Eq. (1.3).

this last statement is nearly invariant if the velocity is high. It is evident that the statement is entirely invariant with respect to rotations and with respect to a further increase of the velocity in the z direction. This is illustrated at the bottom of the figure. The coordinate system is first turned to the left and then given a velocity in the direction opposite to the old z axis. The state of the system appears to be exactly the same as if the coordinate system had been first given a velocity in the −z direction and then turned, which is the operation illustrated at the top of the figure. The state of the system appears to be the same not for any physical reason but because the two coordinate systems are identical and they view the same particle (see Appendix I).

Let us now take our particle with a high velocity in the z direction and view it from a coordinate system which moves in the −y direction. The particle now will appear to have a momentum also in the y direction, its velocity will have a direction between the y and z axes (Fig. 2). Its spin, however, will not be in the direction of its motion any more. In the nonrelativistic case, that is, if all velocities are small as compared

56   *Symmetries and Reflections*

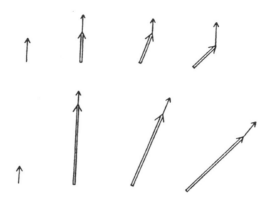

Fig. 2 The particle is first given a small velocity
in the direction of its spin, then increasing velocities
in a perpendicular direction (upper part of the figure). The direction of the spin remains essentially
unchanged; it includes an increasingly large angle
with the velocity as the velocity in· the perpendicular
direction increases. If the velocity imparted to the
particle is large (lower part of the figure), the direction of the spin seems to follow the direction of the
velocity. See Eqs. (1.8) and (1.7).

with the velocity of light, the spin will still be parallel to z and it will,
therefore, enclose an angle with the particle's direction of motion. This
shows that the statement that the spin is parallel to the direction of
motion is not invariant in the nonrelativistic region. However, if the
original velocity of the particle is close to the light velocity, the Lorentz
contraction works out in such a way that the angle between spin and
velocity is given by

$$\tan \text{(angle between spin and velocity)} = (1 - v^2/c^2)^{\frac{1}{2}} \sin\vartheta, \quad (1)$$

where ϑ is the angle between the velocity v in the moving coordinate
system and the velocity in the coordinate system at rest. This last situation is illustrated at the bottom of the figure. If the velocity of the
particle is small as compared with the velocity of light, the direction of
the spin remains fixed and is the same in the moving coordinate system
as in the coordinate system at rest. On the other hand, if the particle's
velocity is close to light velocity, the velocity carries the spin with itself
and the angle between direction of motion and spin direction becomes
very small in the moving coordinate system. Finally, if the particle has
light velocity, the statement "spin and velocity are parallel" remains true

in every coordinate system. Again, this is not a consequence of any physical property of the spin, but is a consequence of the properties of Lorentz transformations: it is a kind of Lorentz contraction. It is the reason for the different behavior of particles with finite, and particles with zero, rest-mass, as far as the number of states of polarization is concerned. (Details of the calculation are in Appendix I.)

The preceding consideration proves more than was intended: it shows that the statement "spin and velocity are parallel for zero mass particles" is invariant and that, for relativistic reasons, one needs only *one* state of polarization, rather than *two*. This is true as far as proper Lorentz transformations are concerned. The second state of polarization, in which spin and velocity are antiparallel, is a result of the reflection symmetry. Again, this can be illustrated on the example of light: right circularly polarized light appears as right circularly polarized light in all Lorentz frames of reference which can be continuously transformed into each other. Only if one looks at the right circularly polarized light in a mirror does it appear as left circularly polarized light. The postulate of reflection symmetry allows us to infer the existence of left circularly polarized light from the existence of right circularly polarized light—if there were no such reflection symmetry in the real world, the existence of *two* modes of polarization of light, with virtually identical properties, would appear to be a miracle. The situation is entirely different for particles with nonzero mass. For these, the $2S + 1$ directions of polarization follow from the invariance of the theory with respect to proper Lorentz transformations. In particular, if the particle is at rest, the spin will have different orientations with respect to coordinate systems which have different orientations in space. Thus, the existence of all the states of polarization follow from the existence of one, if only the theory is invariant with respect to proper Lorentz transformations. For particles with zero rest-mass, there are only two states of polarization, and even the existence of the second one can be inferred only on the basis of reflection symmetry.

Reflection Symmetry

The problem and existence of reflection symmetry have been furthered in a brilliant way by recent theoretical and experimental research. There is nothing essential that can be added at present to the remarks

58   *Symmetries and Reflections*

and conjectures of Lee, Yang, and Oehme, and all that follows has been said, or at least implied, by Salam, Lee, Yang, and Oehme.[5] The sharpness of the break with past concepts is perhaps best illustrated by the cobalt experiment of Wu, Ambler, Hayward, Hoppes, and Hudson.

The ring current—this may be a permanent current in a superconductor—creates a magnetic field. The Co source is in the plane of the current and emits β particles (Fig. 3). The whole experimental arrangement, as shown in Fig. 3, has a symmetry plane and, if the principle of

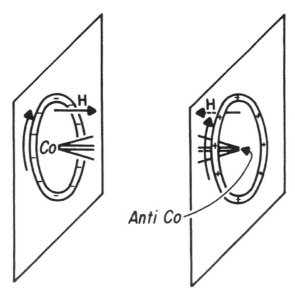

Fig. 3 The right side is the mirror image of the left side, according to the interpretation of the parity experiments,[5a] which maintains the reflection as a symmetry element of all physical laws. It must be assumed that the reflection transforms matter into antimatter: the electronic ring current becomes a positronic ring current, the radioactive cobalt is replaced by radioactive anticobalt.

[5] Lee, Yang, and Oehme, *Phys. Rev.*, 106, 340 (1957).
[5a] The interpretation illustrated has been proposed independently by numerous authors, including A. Salam, *Nuovo Cimento*, 5, 229 (1957); L. Landau, *Nucl. Phys.*, 3, 127 (1957), H. D. Smyth and L. Biedenharn (personal communication). Dr. S. Deser has pointed out that the "perturbing possibility" was raised already by Wick, Wightman, and Wigner [*Phys. Rev.*, 88, 101 (1952)] but was held "remote at that time." Naturally, the apparent unanimity of opinion does not prove its correctness.

Relativistic Invariance and Quantum Phenomena   59

sufficient cause is valid, the symmetry plane should remain valid throughout the further fate of the system. In other words, since the right and left sides of the plane had originally identical properties, there is no sufficient reason for any difference in their properties at a later time. Nevertheless, the intensity of the β radiation is larger on one side of the plane than the other side. The situation is paradoxical no matter what the mechanism of the effect is—in fact, it is most paradoxical if one disregards its mechanism and theory entirely. If the experimental circumstances can be idealized as indicated, even the principle of sufficient cause seems to be violated.

It is natural to look for an interpretation of the experiment which avoids this very far-reaching conclusion and, indeed, there is such an interpretation.[5a] It is good to reiterate, however, that no matter what interpretation is adopted, we have to admit that the symmetry of the real world is smaller than we had thought. However, the symmetry may still include reflections.

If it is true that a symmetry plane always remains a symmetry plane, the initial state of the Co experiment could not have contained a symmetry plane. This would not be the case if the magnetic vector were polar—in which case the electric vector would be axial. The charge density, the divergence of the electric vector, would then become a pseudoscalar rather than a simple scalar as in current theory. The mirror image of a negative charge would be positive, the mirror image of an electron a positron, and conversely. The mirror image of matter would be antimatter. The Co experiment, viewed through a mirror, would not present a picture contrary to established fact: it would present an experiment carried out with antimatter. The right side of Fig. 3 shows the mirror image of the left side. Thus, the principle of sufficient cause, and the validity of symmetry planes, need not be abandoned if one is willing to admit that the mirror image of matter is antimatter.

The possibility just envisaged would be technically described as the elimination of the operations of reflection and charge conjugation, as presently defined, as true symmetry operations. Their product would still be assumed to be a symmetry operation and proposed to be named, simply, reflection. A few further technical remarks are contained in Appendix II. The proposition just made has two aspects: a very appealing one, and a very alarming one.

Let us look first at the appealing aspect. Dirac has said that the number of elementary particles shows an alarming tendency of increasing. One is tempted to add to this that the number of invariance properties also showed a similar tendency. This is not equally alarming because, while the increase in the number of elementary particles complicates our picture of nature, that of the symmetry properties on the whole simplifies it. Nevertheless the clear correspondence between the invariance properties of the laws of nature, and the symmetry properties of space-time, was most clearly breached by the operation of charge conjugation. This postulated that the laws of nature remain the same if all positive charges are replaced by negative charges and vice versa, or more generally, if all particles are replaced by antiparticles. Reasonable as this postulate appears to us, it corresponds to no symmetry of the space-time continuum. If the preceding interpretation of the Co experiments should be sustained, the correspondence between the natural symmetry elements of space-time, and the invariance properties of the laws of nature, would be restored. It is true that the role of the planes of reflection would not be that to which we are accustomed—the mirror image of an electron would become a positron—but the mirror image of a sequence of events would still be a possible sequence of events. This possible sequence of events would be more difficult to realize in the actual physical world than what we had thought, but it would still be possible.

The restoration of the correspondence between the natural symmetry properties of space-time on one hand, and of the laws of nature on the other hand, is the appealing feature of the proposition. It has, actually, two alarming features. The first of these is that a symmetry operation is, physically, so complicated. If it should turn out that the operation of time inversion, as we now conceive it, is not a valid symmetry operation (e.g., if one of the experiments proposed by Treiman and Wyld gave a positive result), we could still maintain the validity of this symmetry operation by reinterpreting it. We could postulate, for instance, that time inversion transforms matter into *meta*-matter which will be discovered later when higher energy accelerators will become available. Thus, maintaining the validity of symmetry planes forces us to a more artificial view of the concept of symmetry and of the invariance of the laws of physics.

The other alarming feature of our new knowledge is that we have been misled for such a long time to believe in more symmetry elements

Relativistic Invariance and Quantum Phenomena   61

than actually exist. There was ample reason for this and there was ample experimental evidence to believe that the mirror image of a possible event is again a possible event with electrons being the mirror images of electrons and not of positrons. Let us recall in this connection first how the concept of parity, resulting from the beautiful though almost forgotten experiments of Laporte,[6] appeared to be a perfectly valid concept in spectroscopy and in nuclear physics. This concept could be explained very naturally as a result of the reflection symmetry of space-time, the mirror image of electrons being electrons and not positrons. We are now forced to believe that this symmetry is only approximate and the concept of parity, as used in spectroscopy and nuclear physics, is also only approximate. Even more fundamentally, there is a vast body of experimental information in the chemistry of optically active substances which are mirror images of each other and which have optical activities of opposite direction but exactly equal strength. There is the fact that molecules which have symmetry planes are optically inactive; there is the fact of symmetry planes in crystals.[7] All these facts relate properties of right-handed matter to left-handed *matter*, not of right-handed matter to left-handed *antimatter*. The new experiments leave no doubt that the symmetry plane in this sense is not valid for all phenomena, in particular not valid for β decay, that if the concept of symmetry plane is at all valid for all phenomena, it can be valid only in the sense of converting matter into antimatter.

Furthermore, the old-fashioned type of symmetry plane is not the only symmetry concept that is only approximately valid. Charge conjugation was mentioned before, and we are reminded also of isotopic spin, of the exchange character, that is, multiplet system, for electrons and also of nuclei, which latter holds so accurately that, in practice, parahydrogen molecules can be converted into orthohydrogen molecules only by first destroying them.[8] This approximate validity of laws of symmetry is, therefore, a very general phenomenon—it may be *the* gen-

---

[6] O. Laporte, Z. *Physik*, 23, 135 (1924). For the interpretation of Laporte's rule in terms of the quantum-mechanical operation of inversion, see the writer's *Gruppentheorie und ihre Anwendungen auf die Quantenmechanik der Atomspektren* (Braunschweig: Friedrich Vieweg und Sohn, 1931), Chap. XVIII.

[7] For the role of the space and time inversion operators in classical theory, see H. Zocher and C. Török, *Proc. Natl. Acad. Sci. U.S.*, 39, 681 (1953), and literature quoted there.

[8] See A. Farkas, *Orthohydrogen, Parahydrogen and Heavy Hydrogen* (New York: Cambridge University Press, 1935).

eral phenomenon. We are reminded of Mach's axiom that the laws of
nature depend on the physical content of the universe, and the physical
content of the universe certainly shows no symmetry. This suggests—and
this may also be the spirit of the ideas of Yang and Lee—that all symmetry properties are only approximate. The weakest interaction, the
gravitational force, is the basis of the distinction between inertial and
accelerated coordinate systems, the second weakest known interaction,
that leading to β decay, leads to the distinction between matter and
antimatter. Let me conclude this subject by expressing the conviction
that the discoveries of Wu, Ambler, Hayward, Hoppes, and Hudson,[9]
and of Garwin, Lederman, and Weinreich,[10] will not remain isolated
discoveries. More likely, they herald a revision of our concept of invariance and possibly of other concepts which are even more taken for
granted.

Quantum Limitations of the Concepts of General Relativity

The last remarks naturally bring us to a discussion of the general
theory of relativity. The main premise of this theory is that coordinates
are only labels to specify space-time points. Their values have no particular significance unless the coordinate system is somehow anchored
to events in space-time.

Let us look at the question of how the equations of the general theory
of relativity could be verified. The purpose of these equations, as of all
equations of physics, is to calculate, from the knowledge of the present,
the state of affairs that will prevail in the future. The quantities describing the present state are called initial conditions; the ways these quantities change are called the equations of motion. In relativity theory,
the state is described by the metric which consists of a network of
points in space-time, that is, a network of events, and the distances between these events. If we wish to translate these general statements
into something concrete, we must decide what events are, and how we
measure distances between *events*. The metric in the general theory
of relativity is a metric in space-time, its elements are distances between
space-time points, not between points in ordinary space.

[9] Wu, Ambler, Hayward, Hoppes, and Hudson, *Phys. Rev.*, 105, 1413(L) (1957).
[10] Garwin, Lederman, and Weinreich, *Phys. Rev.*, 105, 1415(L) (1957); also,
J. L. Friedman and V. L. Telegdi, *ibid.*, 105, 1681(L) (1957).

Relativistic Invariance and Quantum Phenomena   63

The events of the general theory of relativity are coincidences, that is, collisions between particles. The founder of the theory, when he created this concept, evidently had macroscopic bodies in mind. Coincidences, that is, collisions between such bodies, are immediately observable. This is not the case for elementary particles; a collision between these is something much more evanescent. In fact, the point of a collision between two elementary particles can be closely localized in space-time only in case of high-energy collisions. (See Appendix III.) This shows that the establishment of a close network of points in space-time requires a reasonable energy density, a dense forest of world lines wherever the network is to be established. However, it is not necessary to discuss this in detail because the measurement of the distances between the points of the network gives more stringent requirements than the establishment of the network.

It is often said that the distances between events must be measured by yardsticks and rods. We found that measurements with a yardstick are rather difficult to describe and that their use would involve a great deal of unnecessary complications. The yardstick gives the distance between events correctly only if its marks coincide with the two events simultaneously from the point of view of the rest-system of the yardstick. Furthermore, it is hard to imagine yardsticks as anything but macroscopic objects. It is desirable, therefore, to reduce all measurements in space-time to measurements by clocks. Naturally, one can measure by clocks directly only the distances of points which are in time-like relation to each other. The distances of events which are in space-like relation, and which would be measured more naturally by yardsticks, will have to be measured, therefore, indirectly.

It appears, thus, that the simplest framework in space-time, and the one which is most nearly microscopic, is a set of clocks which are only slowly moving with respect to each other, that is, with world lines which are approximately parallel. These clocks tick off periods and these ticks form the network of events which we wanted to establish. This, at the same time, establishes the distance of those adjacent points which are on the same world line.

Figure 4 shows two world lines and also shows an event, that is, a tick of the clock, on each. The figure shows an artifice which enables one to measure the distance of space-like events: a light signal is sent out from the first clock which strikes the second clock at event 2. This clock, in turn, sends out a light signal which strikes the first clock at time t' after the event 1. If the first light signal had to be sent out at time t before the first event, the calculation given in Appendix IV shows that the space-like distance of events 1 and 2 is the geometric average of the two measured time-like distances t and t'. This is then a way to measure distances between space-like events by clocks instead of yardsticks.

Fig. 4   Measurement of space-like distances by means of a clock. It is assumed that the metric tensor is essentially constant within the space-time region contained in the figure. The space-like distance between events 1 and 2 is measured by means of the light signals which pass through event 2 and a geodesic which goes through event 1. Explanation in Appendix IV.

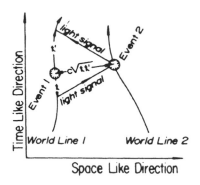

It is interesting to consider the quantum limitations on the accuracy of the conversion of time-like measurements into space-like measurements, which is illustrated in Fig. 4. Naturally, the times t and t' will be well-defined only if the light signal is a short pulse. This implies that it is composed of many frequencies, and, hence, that its energy spectrum has a corresponding width. As a result, it will give an indeterminate recoil to the second clock, thus further increasing the uncertainty of its momentum. All this is closely related to Heisenberg's uncertainty principle. A more detailed calculation shows that the added uncertainty is of the same order of magnitude as the uncertainty inherent in the nature of the best clock that we could think of, so that the conversion of time-like measurements into space-like measurements is essentially free.

We finally come to the discussion of one of the principal problems—the limitations on the accuracy of the clock. It led us to the conclusion that the inherent limitations on the accuracy of a clock of given weight and size, which should run for a period of a certain length, are quite severe. In fact, the result in summary is that a clock is an essentially nonmicroscopic object. In particular, what we vaguely call an atomic clock, a single atom which ticks off its periods, is surely an idealization which is in conflict with fundamental concepts of measurability. This part of our conclusions can be considered to be well established. On the other hand, the actual formula which will be given for the limitation of the accuracy of time measurement, a sort of uncertainty principle, should be considered as the best present estimate.

Let us state the requirements as follows. The watch shall run T seconds, shall measure time with an accuracy of $T/n = t$, its linear extension shall not exceed l, its mass shall be below m. Since the pointer of the watch must be able to assume n different positions, the system will have to run, in the course of the time T, over at least n orthogonal states. Its state must, therefore, be the superposition of at least n stationary states. It is clear, furthermore, that unless its total energy is at least $\hbar/t$, it cannot measure a time interval which is smaller than t. This is equivalent to the usual uncertainty principle. These two requirements follow directly from the basic principles of quantum theory; they are also the requirements which could well have been anticipated. A clock which conforms with these postulates is, for instance, an oscillator, with a period which is equal to the running time of the clock, if it is with equal probability in any of the first n quantum states. Its energy is about n times the energy of the first excited state. This corresponds to the uncertainty principle with the accuracy t as time uncertainty. Broadly speaking, the clock is a very soft oscillator, the oscillating particle moving very slowly and with a rather large amplitude. The pointer of the clock is the position of the oscillating particle.

The clock of the preceding paragraph is still very light. Let us consider, however, the requirement that the linear dimensions of the clock be limited. Since there is little point in dealing with the question in great generality, it may as well be assumed here that the linear dimension shall correspond to the accuracy in time. The requirement $l \approx ct$ increases the mass of the clock by n^3, which may be a very large factor indeed:

$$m > n^3 \hbar t / l^2 \approx n^3 \hbar / c^2 t.$$

For example, a clock, with a running time of a day and an accuracy of 10^{-8} second, must weigh almost a gram—for reasons stemming solely from uncertainty principles and similar considerations.

So far, we have paid attention only to the physical dimension of the clock and the requirement that it be able to distinguish between events which are only a distance t apart on the time scale. In order to make it usable as part of the framework which was described before, it is necessary to *read* the clock and to start it. As part of the framework to map out the metric of space-time, it must either register the readings at which it receives impulses, or transmit these readings to a part of space outside the region to be mapped out. This point was already noted by Schrödinger.[11] However, we found it reassuring that, in the most interesting case in which $l = ct$, that is, if space and time inaccuracies are about equal, the reading requirement introduces only an insignificant numerical factor but does not change the form of the expression for the minimum mass of the clock.

The arrangement to map the metric might consist, therefore, of a lattice of clocks, all more or less at rest with respect to each other. All these clocks can emit light signals and receive them. They can also transmit their reading at the time of the receipt of the light signal to the outside. The clocks may resemble oscillators, well in the nonrelativistic region. In fact, the velocity of the oscillating particle is about n times smaller than the velocity of light, where n is the ratio of the *error* in the time measurement to the *duration* of the whole interval to be measured. This last quantity is the spacing of the events on the time axis; it is also the distance of the clocks from each other, divided by the light velocity. The world lines of the clocks form the dense forest which was mentioned before. Its branches suffuse the region of space-time in which the metric is to be mapped out.

We are not absolutely convinced that our clocks are the best possible. Our principal concern is that we have considered only one space-like dimension. One consequence of this was that the oscillator had to be a one-dimensional oscillator. It is possible that the size limitation does not increase the necessary mass of the clock to the same extent if use is made of all three spatial dimensions.

The curvature tensor can be obtained from the metric in the conventional way, if the metric is measured with sufficient accuracy. It may be of interest, nevertheless, to describe a more direct method for measuring the curvature of space. It involves an arrangement, illustrated in Fig. 5, which is similar to that used for obtaining the metric. There is a clock, and a mirror, at such a distance from each other that the curvature of space can be assumed to be constant in the intervening region. The two clocks need not be at rest with respect to each other; in fact, such

[11] E. Schrödinger, *Ber. Preuss. Akad. Wiss. Phys.-Math. Kl.*, p. 238 (1931).

Relativistic Invariance and Quantum Phenomena   67

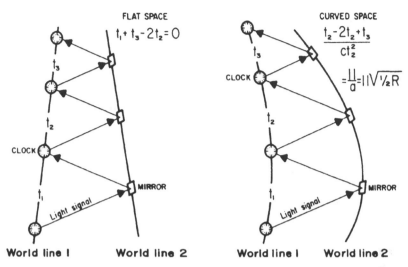

Fig. 5 Direct measurement of the curvature by means of a clock and mirror. Only one space-like dimension is considered and the curvature assumed to be constant within the space-time region contained in the figure. The explanation is given in Appendix V.

a requirement would involve additional measurements to verify it. If the space is flat, the world lines of the clocks can be drawn straight. In order to measure the curvature, a light signal is emitted by the clock, and this is reflected by the mirror. The time of return is read on the clock—it is t_1—and the light signal returned to the mirror. The time which the light signal takes on its second trip to return to the clock is denoted by t_2. The process is repeated a third time, the duration of the last roundtrip denoted by t_3. As shown in Appendix V, the radius of curvature a and the relevant component R_{0101} of the Riemann tensor are given by

$$\frac{t_1 - 2t_2 + t_3}{t_2{}^2} = \frac{11}{a} = 11(\tfrac{1}{2}R_{0101})^{1/2}. \tag{2}$$

If classical theory would be valid also in the microscopic domain, there would be no limit on the accuracy of the measurement indicated in Fig. 5. If $\hbar$ is infinitely small, the time intervals t_1, t_2, t_3 can all be measured with arbitrary accuracy with an infinitely light clock. Similarly, the light signals between clock and mirror, however short, need carry only an infinitesimal amount of momentum and thus deflect clock and mirror arbitrarily little from their geodesic paths. The quantum phenomena considered before force us, however, to use a clock with a minimum mass if the measurement of the time intervals is to have a given accuracy. In the present case, this accuracy must be relatively high unless the time intervals t_1, t_2, t_3 are of the same order of magnitude as the curvature of space. Similarly, the deflection of clock and mirror from their geodesic paths must be very small if the result of the measurement is to be meaningful. This gives an effective limit for the accuracy with which the curvature can be measured. The result is, as could be anticipated, that the curvature at a *point* in space-time cannot be measured at all; only the average curvature over a finite region of space-time can be obtained. The error of the measurement is inversely proportional to the two-thirds power of the area available in space-time, that is, the area around which a vector is carried, always parallel to itself, in the customary definition of the curvature. The error is also proportional to the cube root of the Compton wavelength of the clock. Our principal hesitation in considering this result as definitive is again its being based on the consideration of only one space-like dimension. The possibilities of measuring devices, as well as the problems, may be substantially different in three-dimensional space.

Whether or not this is the case, the essentially nonmicroscopic nature of the general relativistic concepts seems to us inescapable. If we look at this first from a practical point of view, the situation is rather reassuring. We can note first, that the measurement of electric and magnetic fields, as discussed by Bohr and Rosenfeld,[12] also requires macroscopic, in fact *very* macroscopic, equipment and that this does not render the electromagnetic field concepts useless for the purposes of quantum electrodynamics. It is true that the measurement of space-time curvature requires a finite region of space and there is a minimum for the mass, and even the mass uncertainty, of the measuring equipment. However, numerically, the situation is by no means alarming. Even in interstellar space, it should be possible to measure the curvature in a volume of a light second or so. Furthermore, the mass of the clocks which one will wish to employ for such a measurement is of the order of several micrograms, so that the finite mass of elementary particles does not cause any difficulty. The clocks will contain many particles and there is no need, and there is not even an incentive, to employ clocks which

[12] N. Bohr and L. Rosenfeld, *Kgl. Danske Videnskab. Selskab, Mat.-Fys. Medd.*, 12, No. 8 (1933). See also further literature quoted in L. Rosenfeld's article in *Niels Bohr and the Development of Physics* (London: Pergamon Press, 1955).

are lighter than the elementary particles. This is hardly surprising since the mass which can be derived from the gravitational constant, light velocity, and Planck's constant is about 20 micrograms.

It is well to repeat, however, that the situation is less satisfactory from a more fundamental point of view. It remains true that we consider, in ordinary quantum theory, position operators as observables without specifying what the coordinates mean. The concepts of quantum field theories are even more weird from the point of view of the basic observation that only coincidences are meaningful. This again is hardly surprising because even a 20-microgram clock is too large for the measurement of atomic times or distances. If we analyze the way in which we "get away" with the use of an absolute space concept, we simply find that we do not. In our experiments we surround the microscopic objects with a very macroscopic framework and observe *coincidences* between the particles emanating from the microscopic system, and parts of the framework. This gives the collision matrix, which is observable, and observable in terms of macroscopic coincidences. However, the so-called observables of the microscopic system are not only not observed, they do not even appear to be meaningful. There is, therefore, a boundary in our experiments between the region in which we use the quantum concepts without worrying about their meaning in the face of the fundamental observation of the general theory of relativity, and the surrounding region in which we use concepts which are meaningful also in the face of the basic observation of the general theory of relativity but which cannot be described by means of quantum theory. This appears most unsatisfactory from a strictly logical standpoint.

Appendix I

It will be necessary, in this appendix, to compare various states of the same physical system. These states will be generated by looking at the same state—the standard state—from various coordinate systems. Hence every Lorentz frame of reference will define a state of the system—the state which the standard state appears to be from the point of view of this coordinate system. In order to define the standard state, we choose an arbitrary but fixed Lorentz frame of reference and stipulate that, in this frame of reference, the particle in the standard state be at rest and its spin (if any) have the direction of the z axis. Thus, if we

wish to have a particle moving with a velocity v in the z direction and
with a spin also directed along this axis, we look at the particle in the
standard state from a coordinate system moving with the velocity v in
the $-z$ direction. If we wish to have a particle at rest but with its spin
in the yz plane, including an angle α with the z axis, we look at the standard state from a coordinate system the y and z axes of which include an
angle α with the y and z axes of the coordinate system in which the
standard state was defined. In order to obtain a state in which both
velocity and spin have the aforementioned direction (i.e., a direction in
the yz plane, including the angles α and $\frac{1}{2}\pi - \alpha$ with the y and z axes),
we look at the standard state from the point of view of a coordinate
system in which the spin of the standard state is described as this direction and which is moving in the opposite direction.

Two states of the system will be identical only if the Lorentz frames
of reference which define them are identical. Under this definition, the
relations which will be obtained will be valid independently of the
properties of the particle, such as spin or mass (as long as the mass is
nonzero so that the standard state exists). Two states will be approximately the same if the two Lorentz frames of reference which define
them can be obtained from each other by a very small Lorentz transformation, that is, one which is near the identity. Naturally, all states
of a particle which can be compared in this way are related to each other
inasmuch as they represent the same standard state viewed from various
coordinate systems. However, we shall have to compare only these
states.

Let us denote by $A\,(0,\varphi)$ the matrix of the transformation in which the
transformed coordinate system moves with the velocity $-v$ in the z direction, where $v = c \tanh\varphi$:

$$A(0,\varphi) = \left\|\begin{matrix} 0 & 0 & 0 \\ 0 & \cosh\varphi & \sinh\varphi \\ 0 & \sinh\varphi & \cosh\varphi \end{matrix}\right\| \tag{1.1}$$

Since the x axis will play no role in the following consideration, it is suppressed in (1.1) and the three rows and the three columns of this matrix
refer to the y', z', ct' and to the y, z, ct axes, respectively. The matrix
(1.1) characterizes the state in which the particle moves with a velocity
v in the direction of the z axis and its spin is parallel to this axis.

Let us further denote the matrix of the rotation by an angle ϑ in the
yz plane by

Relativistic Invariance and Quantum Phenomena   **71**

$$R(\vartheta) = \begin{Vmatrix} \cos\vartheta & \sin\vartheta & 0 \\ -\sin\vartheta & \cos\vartheta & 0 \\ 0 & 0 & 1 \end{Vmatrix}. \tag{1.2}$$

We refer to the direction in the yz plane which lies between the y and z axes and includes an angle ϑ with the z axis as the direction ϑ. The coordinate system which moves with the velocity $-v$ in the ϑ direction is obtained by the transformation

$$A(\vartheta,\varphi) = R(\vartheta)A(0,\varphi)R(-\vartheta). \tag{1.3}$$

In order to obtain a particle which moves in the direction ϑ and is polarized in this direction, we first rotate the coordinate system counterclockwise by ϑ (to have the particle polarized in the proper direction) and impart it then a velocity $-v$ in the ϑ direction. Hence, it is the transformation

$$T(\vartheta,\varphi) = A(\vartheta,\varphi)R(\vartheta)$$

$$= \begin{Vmatrix} \cos\vartheta & \sin\vartheta\cosh\varphi & \sin\vartheta\sinh\varphi \\ -\sin\vartheta & \cos\vartheta\cosh\varphi & \cos\vartheta\sinh\varphi \\ 0 & \sinh\varphi & \cosh\varphi \end{Vmatrix} \tag{1.4}$$

which characterizes the aforementioned state of the particle. It follows from (1.3) that

$$T(\vartheta,\varphi) = R(\vartheta)A(0,\varphi) = R(\vartheta)T(0,\varphi), \tag{1.5}$$

so that the same state can be obtained also by viewing the state characterized by (1.1) from a coordinate system that is rotated by ϑ. It follows that the statement "velocity and spin are parallel" is invariant under rotations. This had to be expected.

If the state generated by $A(0,\varphi) = T(0,\varphi)$ is viewed from a coordinate system which is moving with the velocity u in the direction of the z axis, the particle will still appear to move in the z direction and its spin will remain parallel to its direction of motion, unless $u > v$, in which case the two directions will become antiparallel, or unless $u = v$, in which case the statement becomes meaningless, the particle appearing to be at rest. Similarly, the other states in which spin and velocity are parallel, i.e., the states generated by the transformations $T(\vartheta,\varphi)$, remain such states if viewed from a coordinate system moving in the direction of the particle's velocity, as long as the coordinate system is not moving faster than the particle. This also had to be expected. However, if the state generated by

72   *Symmetries and Reflections*

$T(0,\varphi)$ is viewed from a coordinate system moving with velocity $v' = c \tanh\varphi'$ in the $-y$ direction, spin and velocity will *not* appear parallel any more, *provided the velocity v of the particle is not close to light velocity*. This last proviso is the essential one; it means that the high velocity states of a particle for which spin and velocity are parallel (i.e., the states generated by (1.4) with a large φ) are states of this same nature if viewed from a coordinate system which is not moving too fast in the direction of motion of the particle itself. In the limiting case of the particle moving with light velocity, the aforementioned states become invariant under *all* Lorentz transformations.

Let us first convince ourselves that if the state (1.1) is viewed from a coordinate system moving in the $-y$ direction, its spin and velocity no longer appear parallel. The state in question is generated from the normal transformation

$A(\tfrac{1}{2}\pi,\varphi')A(0,\varphi)$

$$= \begin{Vmatrix} \cosh\varphi' & \sinh\varphi \sinh\varphi' & \cosh\varphi \sinh\varphi' \\ 0 & \cosh\varphi & \sinh\varphi \\ \sinh\varphi' & \sinh\varphi \cosh\varphi' & \cosh\varphi \cosh\varphi' \end{Vmatrix}. \qquad (1.6)$$

This transformation does not have the form (1.4). In order to bring it into that form, it has to be multiplied on the right by $R(\epsilon)$, i.e., one has to rotate the spin ahead of time. The angle ϵ is given by the equation

$$\tan\epsilon = \frac{\tanh\varphi'}{\sinh\varphi} = \frac{v'}{v}(1-v^2/c^2)^{\frac{1}{2}} \qquad (1.7)$$

and is called the angle between spin and velocity. For $v \ll c$, it becomes equal to the angle which the ordinary resultant of two perpendicular velocities, v and v', includes with the first of these. However, ϵ becomes very small if v is close to c; in this case it is hardly necessary to rotate the spin away from the z axis before giving it a velocity in the z direction. These statements express the identity

$$A(\tfrac{1}{2}\pi,\varphi')A(0,\varphi)R(\epsilon) = T(\vartheta,\varphi''), \qquad (1.8)$$

which can be verified by direct calculation. The right side represents a particle with parallel spin and velocity, the magnitude and direction of the latter being given by the well-known equations

$$v'' = c \tanh\varphi'' = (v^2+v'^2-v^2v'^2/c^2)^{\frac{1}{2}} \qquad (1.8a)$$

Relativistic Invariance and Quantum Phenomena   73

and

$$\tan\vartheta = \frac{\sinh\varphi'}{\tanh\varphi} = \frac{v'}{v(1-v'^2/c^2)^{1/2}}.$$ (1.8b)

Equation (1) given in the text follows from (1.7) and (1.8b) for $v\sim c$.

The fact that the states $T(\vartheta,\varphi)\psi_0$ (where ψ_0 is the standard state and $\varphi \gg 1$) are approximately invariant under all Lorentz transformations is expressed mathematically by the equations

$$R(\vartheta) \cdot T(0,\varphi)\psi_0 = T(\vartheta,\varphi)\psi_0,$$ (1.5a)

$$A(0,\varphi') \cdot T(0,\varphi)\psi_0 = T(0,\varphi' + \varphi)\psi_0,$$ (1.9a)

and

$$A(\tfrac{1}{2}\pi,\varphi') \cdot T(0,\varphi)\psi_0 \to T(\vartheta,\varphi'')\psi_0,$$ (1.9b)

which give the wave function of the state $T(0,\varphi)\psi_0$, as viewed from other Lorentz frames of reference. Naturally, similar equations apply to all $T(\alpha,\varphi)\psi_0$. In particular, (1.5a) shows that the states in question are invariant under rotations of the coordinate system, (1.9a) that they are invariant with respect to Lorentz transformations with a velocity not too high *in* the direction of motion (so that $\varphi' + \varphi \gg 0$, i.e., φ' not too large a negative number). Finally, in order to prove (1.9b), we calculate the transition probability between the state $A(\tfrac{1}{2}\pi,\varphi') \cdot T(0,\varphi)\psi_0$ and $T(\vartheta,\varphi'')\psi_0$, where ϑ and φ'' are given by (1.8a) and (1.8b). For this, (1.8) gives

$$\begin{aligned}
(A(\tfrac{1}{2}\pi,\varphi') &\cdot T(0,\varphi)\psi_0, T(\vartheta,\varphi'')\psi_0) \\
&= (T(\vartheta,\varphi'')R(\epsilon)^{-1}\psi_0, T(\vartheta,\varphi'')\psi_0) \\
&= (R(\epsilon)^{-1}\psi_0,\psi_0) \to (\psi_0,\psi_0).
\end{aligned}$$

The second line follows because $T(\vartheta,\varphi'')$ represents a coordinate transformation and is, therefore, unitary. The last member follows because $\epsilon \to 0$ as $\varphi \to \infty$, as can be seen from (1.7) and $R(0) = 1$.

The preceding consideration is not fundamentally new. It is an elaboration of the facts (a) that the subgroup of the Lorentz group which leaves a null-vector invariant is different from the subgroup which leaves a time-like vector invariant and (b) that the representations of the latter subgroup decompose into one dimensional representations if this subgroup is "contracted" into the subgroup which leaves a null-vector invariant.[13]

[13] E. Inonu and E. P. Wigner, *Proc. Natl. Acad. Sci. U.S.*, 39, 510 (1953).

74   *Symmetries and Reflections*

Appendix II

Before the hypothesis of Lee and Yang[14] was put forward, it was commonly assumed that there were, in addition to the symmetry operations of the proper Poincaré group, three further independent symmetry operations. The proper Poincaré group consists of all Lorentz transformations which can be continuously obtained from unity and all translations in space-like and time-like directions, as well as the products of all these transformations. It is a continuous group; the Lorentz transformations contained in it do not change the direction of the time axis and their determinant is 1. The three independent further operations which were considered to be rigorously valid, were

Space inversion I, that is, the transformation $x, y, z \rightarrow -x, -y, -z$, without changing particles into antiparticles.[*]

Time inversion T, more appropriately described by Lüders[15] as *Umkehr der Bewegungsrichtung*, which replaces every velocity by the opposite velocity, so that the position of the particles at $+t$ becomes the same as it was, without time inversion, at $-t$. The time inversion T (also called time inversion of the first kind by Lüders[16]) does not convert particles into antiparticles either.

Charge conjugation C, that is, the replacement of positive charges by negative charges and more generally of particles by antiparticles, without changing either the position or the velocity of these particles.[17] The quantum-mechanical expressions for the symmetry operations I and C are unitary, that for T is antiunitary.

[14] T. D. Lee and C. N. Yang, *Phys. Rev.*, 104, 254 (1956). See also E. M. Purcell and N. F. Ramsey, *Phys. Rev.*, 78, 807 (1950).

[*] The usual symbol for this is P at present. (Note added with the proofs of this book.)

[15] G. Lüders, Z. *Physik*, 133, 325 (1952).

[16] G. Lüders, *Kgl. Danske Videnskab. Selskab, Mat.-Fys. Medd.*, 28, No. 5 (1954).

[17] All three symmetry operations were first discussed in detail by J. Schwinger, *Phys. Rev.*, 74, 1439 (1948). See also H. A. Kramers, *Proc. Acad. Sci. Amsterdam*, 40, 814 (1937), and W. Pauli's article in *Niels Bohr and the Development of Physics* (London: Pergamon Press, 1955). The significance of the first two symmetry operations (and their connection with the concepts of parity and the Kramers degeneracy, respectively), were first pointed out by the present writer, Z. *Physik*, 43, 624 (1927), and *Nachr. Akad. Wiss. Göttingen, Math.-Physik*, 1932, 546. See also T. D. Newton and E. P. Wigner, *Rev. Mod. Phys.*, 21, 400 (1949); S. Watanabe, *Rev. Mod. Phys.*, 27, 26 (1945). The concept of charge conjugation is based on the observation of W. Furry, *Phys. Rev.*, 51, 125 (1937).

The three operations I, T, C, together with their products TC (Lüders' time inversion of the second kind), IC, IT, ITC, and the unit operation form a group, and the products of the elements of this group with those of the proper Poincaré group were considered to be the symmetry operations of all laws of physics. The suggestion given in the text amounts to eliminating the operations I and C separately while continuing to postulate their product IC as symmetry operation. The discrete symmetry group then reduces to the unit operation plus

$$IC, \ T, \text{ and } ICT, \tag{2.1}$$

and the total symmetry group of the laws of physics becomes the proper Poincaré group plus its products with the elements (2.1). This group is isomorphic (essentially identical) with the unrestricted Poincaré group, i.e., the product of *all* Lorentz transformations with all the displacements in space and time. The quantum mechanical expressions for the operations of the proper Lorentz group and its product with IC are unitary, those for T and ICT (as well as for their products with the elements of the proper Poincaré group) antiunitary. Lüders has pointed out that, under certain very natural conditions, ICT belongs to the symmetry group of every *local* field theory.

Appendix III

Let us consider, first, the collision of two particles of equal mass m in the coordinate system in which the average of the sum of their momenta is zero. Let us assume that, at a given time, the wave function of both particles is confined to a distance l in the direction of their average velocity with respect to each other. If we consider only this space-like direction, and the time axis, the area in space-time in which the two wave functions will substantially overlap is [see Fig. 6(a)]

$$a = l^2/2v_{\min}, \tag{3.1}$$

where $v_{\min}$ is the lowest velocity which occurs with substantial probability in the wave packets of the colliding particles. Denoting the average momentum by $\bar{p}$ (this has the same value for both particles), the half-width of the momentum distribution by δ, then $v_{\min} = (\bar{p} - \delta)\, (m^2 + (\bar{p} - \delta)^2/c^2)^{-1/2}$. Since l cannot be below $\hbar/\delta$, the area (3.1) is at least

76   *Symmetries and Reflections*

$$\frac{\hbar^2}{2\delta^2} \frac{(m^2 + (\bar{p} - \delta)^2/c^2)^{\frac{1}{2}}}{\bar{p} - \delta} \qquad (3.1a)$$

(Note that the area becomes infinite if $\delta > \bar{p}$.) The minimum of (3.1a) is, apart from a numerical factor,

$$a_{\min} \approx \frac{\hbar^2}{\bar{p}^3} (m^2 + \bar{p}^2/c^2)^{\frac{1}{2}} \approx \frac{\hbar^2 c}{E^{\frac{3}{2}}(E + mc^2)^{\frac{1}{2}}}, \qquad (3.2)$$

where E is the kinetic energy (total energy minus rest-energy) of the particles.

The kinetic energy E permits the contraction of the wave functions of the colliding particles also in directions perpendicular to the average relative velocity, to an area $\hbar^2 c^2/E(E + 2mc^2)$. Hence, again apart from a numerical factor, the volume to which the collision can be confined in four dimensional space-time becomes

$$V_{\min} = \frac{\hbar^4 c^3}{E^{\frac{5}{2}}(E + mc^2)^{\frac{3}{2}}}. \qquad (3.3)$$

E is the average kinetic energy of the particles in the coordinate system in which their center of mass is, on the average, at rest. Equation (3.3) is valid apart from a numerical constant of unit order of magnitude but this constant depends on E/mc^2.

Let us consider now the opposite limiting case, the collision of a particle with finite rest-mass m with a particle with zero rest-mass. The collision is viewed again in the coordinate system in which the average linear momentum is zero. In this case, one will wish to confine the wave function of the particle with finite rest-mass to a narrower region l than that of the particle with zero rest-mass. If the latter is confined to a region of thickness λ [see Fig. 6(b)], its momentum and energy uncertainties will be at least $\hbar/\lambda$ and $\hbar c/\lambda$, and these expressions will also give, apart from a numerical factor, the average values of these quantities. Hence $\bar{p} \approx \hbar/\lambda$. The kinetic energy of the particle with finite rest-mass will be of the order of magnitude

$$\tfrac{1}{2}(m^2 c^4 + (\bar{p} + \hbar/l)^2 c^2)^{\frac{1}{2}} + \tfrac{1}{2}(m^2 c^4 + (\bar{p} - \hbar/l)^2 c^2)^{\frac{1}{2}} - mc^2, \quad (3.4)$$

since $\hbar/l$ is the momentum uncertainty. Since $l \lesssim \lambda$, one can neglect $\bar{p}$ in (3.4) if one is interested only in the order of magnitude. This gives for the total kinetic energy,

Relativistic Invariance and Quantum Phenomena   77

$$E \approx \hbar c/\lambda + (m^2 c^4 + \hbar^2 c^2/l^2)^{1/2} - mc^2, \qquad (3.5)$$

while the area in Fig. 6(b) is of the order of magnitude

$$a = (\lambda/c)(l + \Delta v \lambda/c), \qquad (3.6)$$

where Δv is the uncertainty in the velocity of the second particle,

$$\Delta v = \frac{\bar{p} + \hbar/l}{(m^2 + (\bar{p} + \hbar/l)^2/c^2)^{1/2}} - \frac{\bar{p} - \hbar/l}{(m^2 + (\bar{p} - \hbar/l)^2/c^2)^{1/2}}. \qquad (3.6a)$$

This can again be replaced by $(\hbar/l)(m^2 + \hbar^2/l^2 c^2)^{-1/2}$.

For given E, the minimum value of a is assumed if the *kinetic energies* of the two particles are of the same order of magnitude. The two terms of (3.6) then become about equal and $l/\lambda \approx (E/(mc^2 + E))^{1/2}$. The minimum value of a, as far as order of magnitude is concerned, is again given by (3.2). Similarly, (3.3) also remains valid if one of the two particles has zero rest-mass.

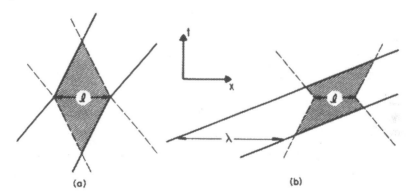

(a)   (b)

Fig. 6a Localization of a collision of two particles of equal mass. The full lines indicate the effective boundaries of the wave packet of the particle traveling to the right, the broken lines the effective boundaries of the wave packet of the particle traveling to the left. The collision can take place in the shaded area of space-time.

Fig. 6b Localization of a collision between a particle with finite mass and a particle with zero rest-mass. The full lines, at a distance λ apart in the x direction, indicate the boundary of the particle with zero rest-mass, the broken lines apply to the wave packet of the particle with nonzero rest-mass. The collision can take place in the shaded area.

78   *Symmetries and Reflections*

The two-dimensional case becomes simplest if both particles have zero rest-mass. In this case the wave packets do not spread at all and (3.2) can be immediately seen to be valid. In the four-dimensional case, (3.3) again holds. However, its proof by means of explicitly constructed wave packets (rather than reference to the uncertainty relations) is by no means simple. It requires wave packets which are confined in every direction, do not spread too fast, and progress essentially only into one half space (one particle going toward the right, the other toward the left). The construction of such wave packets will not be given in detail. They are necessary to prove (3.2) and (3.3) more rigorously also in the case of finite masses; the preceding proofs, based on the uncertainty relations, show only that a and v cannot be *smaller* than the right sides of the corresponding equations. It is clear, in fact, that the limits given by (3.2) and (3.3) would be very difficult to realize, except in the two-dimensional case and for the collision of two particles with zero rest-mass. In all other cases, the relatively low values of a_{min} and V_{min} are predicated on the assumption that the wave packets of the colliding particles are so constituted that they assume a minimum size at the time of the collision. At any rate, (3.2) and (3.3) show that only collisions with a relatively high collision energy, and high energy uncertainty, can be closely localized in space-time.

Appendix IV

Let us denote the components of the vector from event 1 to event 2 by x_i, the components of the unit vector along the world line of the first clock at event 1 by e_i. The components of the first light signal are $x_i + te_i$, that of the second light signal $x_i - t'e_i$. Hence (see Fig. 4)

$$g^{ik}(x_i + te_i)(x_k + te_k) = 0, \qquad (4.1)$$

$$g^{ik}(x_i - t'e_i)(x_k - t'e_k) = 0. \qquad (4.2)$$

Elimination of the linear terms in t and t' by multiplication of (4.1) with t' and (4.2) with t and addition gives

$$2g^{ik}x_ix_k + 2tt'g^{ik}e_ie_k = 0. \qquad (4.3)$$

Since e is a unit vector, $g^{ik}e_ie_k = 1$, and (4.3) shows that the space-like distance between points 1 and 2 is $(tt')^{1/2}$.

Appendix V

Since the measurement of the curvature, described in the text, presupposes *constant curvature* over the space-time domain in which the measurement takes place, we use a space with constant curvature, or, rather, part of a space with constant curvature, to carry out the calculation. We consider only one spatial dimension, i.e., a two-dimensional deSitter space. This will be embedded, in the usual way, in a three-dimensional space[18] with coordinates x, y, τ. The points of the deSitter space then form the hyperboloid

$$x^2 + y^2 - \tau^2 = a^2, \tag{5.1}$$

where a is the "radius of the universe." As coordinates of a point we use x and y, or rather the corresponding polar angles r, ϕ. The metric form in terms of these is

$$(ds)^2 = \frac{a^2}{r^2 - a^2} dr^2 - r^2 d\phi^2. \tag{5.2}$$

Two points of deSitter space correspond to every pair r, ϕ (except $r = a$): those with positive and negative $\tau = (r^2 - a^2)^{1/2}$. This will not lead to any confusion as all events take place at positive τ. The null lines (paths of light signals) are the tangents to the $r = a$ circle.

The experiment described in the text can be analyzed by means of Fig. 7. For the sake of simplicity, the clock and mirror are assumed to be "at rest," i.e., their world lines have constant polar angles which will be assumed as 0 and δ, respectively. The first light signal travels from 1 to 1' and back to 2, the second from 2 to 2' and back to 3, the third from 3 to 3' and back to 4. The polar angle of the radius vector which is perpendicular to the first part 22' of the world line of the second light signal is denoted by ϕ_2. The construction of Fig. 7 shows that angle ϕ_2' which the world line of the mirror includes with the radius vector perpendicular to the second part 2'3 of the second light signal's world line is

$$\phi_2' = \phi_2 + \delta. \tag{5.3}$$

The angles ϕ_1, ϕ_1' ϕ_3, ϕ_3' have similar meanings; they are not indicated

[18] See, e.g., H. P. Robertson, *Rev. Mod. Phys.*, 5, 62 (1933).

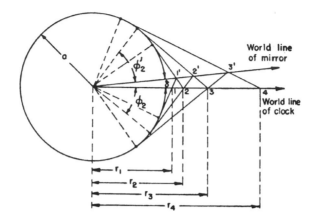

Fig. 7 Analysis of the experiment of Fig. 5. The
figure represents a view of the hyperboloid of deSitter
space, viewed along its axis. Every point of the plane
which is outside the circle corresponds to two points of
the deSitter world with the same spatial coordinate
but with oppositely equal time coordinates. The first
light signal is emitted at 1, reaches the mirror at 1′,
and returns to the clock at 2. The paths of the second
and third light signals are 22′3 and 33′4.

in the figure in order to avoid overcrowding. For reasons similar to
those leading to (5.3), we have

$$\phi_3 = \phi_2' + \delta = \phi_2 + 2\delta, \tag{5.3a}$$
$$\phi_1 = \phi_2 - 2\delta, \tag{5.3b}$$
$$\phi_4 = \phi_3 + 2\delta = \phi_2 + 4\delta. \tag{5.3c}$$

The radial coordinates of the points 1, 2, 3, 4 are denoted by r_1, r_2, r_3, r_4:

$$r_i = a/\cos\phi_i. \tag{5.4}$$

The proper time t, registered by the clock, can be obtained by integrating the metric form (5.2) along the world line $\phi = 0$ of the clock:

$$t = a\ln[r + (r^2 - a^2)^{1/2}]. \tag{5.5}$$

Hence, the traveling time t_2 of the second light signal becomes

$$t_2 = a\ln\frac{r_3 + (r_3^2 - a^2)^{1/2}}{r_2 + (r_2^2 - a^2)^{1/2}} = a\ln\frac{\cos\phi_2(1 + \sin\phi_3)}{\cos\phi_3(1 + \sin\phi_2)}. \tag{5.6}$$

Similar expressions apply for the traveling times of the first and third
light signals; all ϕ can be expressed by means of (5.3a), (5.3b), (5.3c) in

Relativistic Invariance and Quantum Phenomena   *81*

terms of ϕ_2 and δ. This allows the calculation of the expression (3). For small δ, one obtains

$$\frac{t_1 - 2t_2 + t_3}{t_2{}^2} \approx \frac{11}{a},\tag{5.7}$$

and Riemann's invariant $R = 2/a^2$ is proportional to the square of (5.7). In particular, it vanishes if the expression (2) is zero.

Relativistic Equations in Quantum Mechanics

E. P. Wigner

J. Mehra (ed.) The Physicist's Conception of Nature.
D. Reidel, Dordrecht 1973, pp. 320–330

1. RECALLING THE DISCOVERY OF DIRAC'S ELECTRON EQUATION

Let me begin by recalling the way I learned about Dirac's relativistic equation of the electron. It was from a letter to Max Born who asked Dirac to review an article by a third person. The letter started with this review, pointing out that two of the chapters could be united and that this would render the article more concise, that is shorter, and also easier to read. It had a few more remarks on the article in question. Then, in a new paragraph of about ten lines he said that he had been working on the relativistic theory of the electron. He gave his four-component equation, including the interaction with an electromagnetic field.

Both Jordan, who also read it, and I were quite flabbergasted when reading it. We had tried to develop the relativistic equation for a spin 1/2 particle also, but all our attempts were directed toward a two-component equation – after all, the spin 1/2 particle has only two states. We were not satisfied with any of the equations we had found but were not yet ready to give up. The letter to Born changed all that. As Jordan put it 'Well, of course, it would have been better had we found the equation but the derivation is so beautiful, and the equation so concise, that we must be happy to have it.' I still have the letter in question, it is one of my most precious possessions.

2. INTRODUCTORY REMARKS

Actually, Dirac's electron equation was not the first relativistic wave equation: the so-called Klein–Gordon equation, for spin-less particles, was established before. However, the existence of two relativistic wave equations stimulated me to try to find all such equations – at least all which apply to single particles. When Jagdish Mehra suggested that I speak on this subject, I was afraid, first, that all I would be able to say about it would appear as 'old hat'. Worse than that, I was afraid that the subject is closed to such an extent that even Feynman's motto would not apply, the motto according to which the solution of every problem of physics brings, in its wake, a host of new and unsolved problems. However, a closer study of the problem made me realize how much I could learn from the existing literature and how many questions relating to it have remained unanswered. Hence, I am now glad that this is my subject. My discussion will be divided into two parts. The first part will deal with relativistic equations and transformations in the absence of external fields. The effect of fields will be taken up in the second part.

3. EQUATIONS AND RELATIVISTIC TRANSFORMATIONS

As an introduction to this subject, I fear, it will be necessary to repeat very briefly the contents of a very old paper of mine.

The postulate that an equation be relativistically invariant implies that an arbitrary state, represented by a solution of the equations, be transformable to any other coordinate system obtainable by a Poincaré transformation from the coordinate system in which the original specification of the state was given. Since the Poincaré transformations contain the operation of time displacement, if one knows the behaviour of all states under all Poincaré transformations, one already has the relativistic equation of motion – this gives only the change of the state under time displacement. Conversely, however, it is also true that in order to prove that an equation is relativistically invariant, one must exhibit the behaviour of its solutions under all relativistic transformations. Actually, the situation is even worse: the wave equation, i.e. the time displacement operator, does not determine all the other operators of the Poincaré group uniquely. As a rule, though, only one set of such operators will appear natural – few of us would be inclined to interpret the Dirac equation as representing two scalar particles. However, mathematically, this is possible and the totality of the operators of the Poincaré group form a more complete description of the properties of the particle than the time displacement operator, i.e. the wave equation, alone. This is the reason for discussing in this section, at least superficially, the totality of the Poincaré transformations. On the other hand, when we try to describe the behaviour of the particles to the vacuum states of which these transformations apply, when we try to obtain their behaviour in external fields, the equations of motion will prove more informative. It seems that both have their importance, though in different contexts, but the former one will be discussed first.

It is easy to see that the transformations of the state vectors which correspond to the various transformations of the Poincaré group form a unitary representation of that group, but only up to a factor. If we denote the transformations of the Poincaré group by $P_1, P_2, \ldots$, the corresponding operations in Hilbert space by $D(P_1), D(P_2), \ldots$, must satisfy the equations

$$D(P_1)D(P_2) = \omega(P_1, P_2)D(P_1, P_2) \tag{1}$$

the $D(P)$ being unitary operators, the $\omega(P_1, P_2)$ numbers of modulus 1. The only real difficulty in solving these equations was to show that the $D(P)$ can be replaced by $D'(P) = \omega(P)D(P)$, the $\omega(P)$ being again numbers of modules 1, so that

$$D'(P_1)D'(P_2) = \pm D'(P_1 P_2) \tag{1a}$$

holds. Once this is done, the possible sets of operators are easily found – so easily that the argument leading to them can be very quickly reproduced at least in the most important case of a finite mass.

Since the displacements all commute, one can choose basic vectors $|p, \xi\rangle$ in such a way that the effect of a displacement T_a by the four-vector a is given by (we assume (1a) but omit the primes on the D)

$$D(T_a)\,|p, \xi\rangle = e^{ip\cdot a}|p, \xi\rangle \tag{2}$$

p and a are four-vectors in the underlying Minkowski space, $p \cdot a$ is their Minkowski scalar product. The variable ξ can be discrete, it is introduced because there may be several states which transform according to (2) with the same p.

If we apply a Lorentz transformation L to the state $|p, \xi\rangle$, we obtain a state which still transforms by (2) but its momentum p is replaced by Lp.

This follows simply from the formula

$$D(L^{-1})D(T_a)D(L) = D(T_{L^{-1}a}) \tag{3}$$

which is the equation of the Poincaré group connecting Lorentz transformations with displacements. Indeed, it follows from (3) that

$$D(T_a)\,[D(L)\,|p, \xi\rangle] = D(T_a)D(L)\,|p, \xi\rangle = D(L)D(T_{L^{-1}a})\,|p, \xi\rangle =$$
$$= D(L)\,e^{ip\cdot L^{-1}a}|p, \xi\rangle = e^{iLp\cdot a}[D(L)\,|p, \xi\rangle]. \tag{3a}$$

The third member follows from (3), the fourth from (2) as applied to the displacement $L^{-1}a$, and the last one from the linear nature of $D(L)$ and the invariance of the Minkowski scalar product under the Lorentz transformation L. Equation (3a) shows that, indeed, $D(L)|p, \xi\rangle$ is a state with momentum Lp.

Equation (3a) is of basic significance inasmuch as it shows that the magnitude of the momentum is the same for all states which can be obtained from a $|p, \xi\rangle$ state by any Poincaré transformation. Indeed, such a transformation can be written as the product of a Lorentz transformation and a displacement. The latter, as evident from (2), does not change p at all – it only multiplies $|p, \xi\rangle$ by a number. As (3a) further shows, the Lorentz transformation L replaces p by Lp which has the same length as p because of $p \cdot p = Lp \cdot Lp$. One concludes, hence, that if the set of all states is irreducible, the momenta of all will have the same length because states with momenta of different lengths do not mix as a result of the application of Poincaré transformations.

If there are states $|p_0, \xi\rangle$ with a p_0 which is parallel to the time axis – this turns out to be the condition for the finiteness and real nature of the mass – and if R is a rotation, $D(R)$ applied to the states $|p_0, \xi\rangle$ will still give states with momentum $Rp_0 = p_0$. This means that the states $|p_0, \xi\rangle$ transform among themselves under rotations – they must transform by a representation of the group of rotations of space. If we assume that the whole set of operators $D(P)$ form an irreducible manifold, i.e. do not leave any subspace of the Hilbert space invariant, it becomes reasonable to choose the $|p_0, \xi\rangle$ in such a way that they transform by an irreducible representation R^j, with matrix-elements $R^j_{\zeta',\zeta}$, of the rotation group

$$D(R)\,|p_0,\xi\rangle = \sum_{\xi'} R^j_{\xi'\xi}|p_0,\xi'\rangle. \tag{4}$$

One recognizes this j as the spin of the particle, and the length of p_0 will be interpreted, of course, as its mass.

Equation (4) gives only the effect of rotations on $|p_0,\xi\rangle$. In order to obtain the effect of the general Lorentz transformations thereon, one decomposes this into a 'boost' B, that is a symmetric Lorentz transformation, and a rotation, i.e. one sets $L=BR$. The boost changes the momentum p_0 into Bp_0 and one can define $|Bp_0,\xi\rangle$ as

$$|Bp_0,\xi\rangle = D(B)\,|p_0,\xi\rangle. \tag{5}$$

This gives for $L=BR$

$$\begin{aligned}
D(L)\,|p_0,\xi\rangle &= D(B)D(R)\,|p_0,\xi\rangle = D(B)\sum_{\xi'} R^j_{\xi'\xi}|p_0,\xi'\rangle = \\
&= \sum_{\xi'} R^j_{\xi'\xi}|Bp_0,\xi'\rangle.
\end{aligned} \tag{5a}$$

Finally, we have to apply the Lorentz transformation L not only to the states $|p_0,\xi\rangle$ the momentum vectors of which are parallel to the time axis, but to all states $|p,\xi\rangle$ though we can assume p to have the same Minkowski length as p_0. In order to do this, we note, first, that the boost Bp is uniquely defined by p and p_0 as a result of the equation $p=B_p p_0$. For the calculation of $D(L)|p,\xi\rangle$ we decompose, therefore, the Lorentz transformation LB_p into a boost and a rotation

$$LB_p = BR. \tag{6}$$

We can write, therefore,

$$\begin{aligned}
D(L)\,|p,\xi\rangle &= D(B)D(R)D(B_p^{-1})\,|p,\xi\rangle \\
&= D(B)D(R)\,|p_0,\xi\rangle = \sum_{\xi'} R^j_{\xi'\xi}|Bp_0,\xi'\rangle.
\end{aligned} \tag{6a}$$

The second member is a consequence of (6), the third of (5), applied for B_p instead of B and multiplied by $D(B_p)^{-1}=D(B_p^{-1})$. The R and B in (6a) are given by (6).

This completes the determination of the transformations of all the state vectors under all Poincaré transformations. Every such transformation can be written as the product of a Lorentz transformation and a displacement $P=LT_a$ and the effect of $T(a)$, that is the operator $D(T_a)$, was given already by (2).

This completes the determination of the Poincaré transformations in the case that there is at least one time-like momentum vector, i.e. for finite real mass of the particle. Once (1a) is established – this was used when obtaining (4) – the calculation is indeed easy. The calculation is not significantly more difficult if the momenta are null-vectors, i.e. for light-like momenta. The same applies in the case of states – probably not realized in nature – in which the momenta are space-like ('tachyons'). The case that all four components of the momentum vanish is somewhat more difficult to deal with, and this has first been done by the Russian mathematicians, Gelfand, Neumark, and their collaborators. However, surely, there is no real particle all states of which are invariant under all displacements. Such a particle would be present always and everywhere with the same probability.

4. WHAT HAVE WE ACCOMPLISHED?

What we have accomplished by the determination of the Poincaré invariant manifolds in Hilbert space is the character of possible elementary particles. The emphasis is on possible, not actual. Just as the laws of physics do not tell us what phenomena actually take place but only what succession of events is possible, the invariance considerations do not tell us what particles actually exist but only give some of the properties of the particles which may exist. In particular, the invariance considerations do not tell us why the ratio of the masses of the two stable finite-mass particles is 1836.1, why their electric charges are so precisely oppositely equal, and so on. Before expressing reservations even with regard to what we seem to have accomplished, it may be worthwhile to give the reasons for doubting the existence of particles with imaginary mass, and of zero-mass particles with infinite spin. The reasons are semi-experimental.

If particles with imaginary mass, that is space-like momenta, existed, we could take any two of them, give them momenta with oppositely equal spatial components and extract any amount of energy from them. Their energy would become negative but the energy spectra of such particles would extend from $-\infty$ to ∞. If such particles existed, no thermal equilibrium would be possible. Their existence, or the possibility of their being produced, would give rise to phenomena which are fantastic and which would have been observed. This does not mean that equations with imaginary mass should not be used in calculations – Sudarshan, for instance, used them to advantage and so did Hadjioannou even more – but it does mean that one has to consider the existence of particles with space-like momenta most unlikely.

The same objection does not apply to particles with zero mass, that is of light-like momenta, and infinite spin. Their energy spectrum can be very naturally so chosen that it does not contain negative values. In fact, G. Mack here in Trieste, and Melvin at Drexel, have given serious consideration to the existence of such particles. It is true, nevertheless, that their virtual existence would give an infinite heat-capacity to a finite volume of space. This is not impossible – the doubling of the heat capacity due to the existence of neutrinos was hardly noted. Nevertheless, it is good to realize that the existence, or the possibility of production, of such particles would also render any true thermal equilibrium impossible. This is not in conflict with any experimental fact and, if we think of the origin of the world and the fact that it is still far from equilibrium, may even have attractive features. It is contrary to accepted views.

Let us now return to our principal question: are the results which we obtained concerning the limitations on the possible types of particles unquestionable. There are two questions in this connection the elucidation of which would be of interest. The first concerns the validity of the Poincaré group as the basic invariance group. It would seem that even if the space were finite, and the world had perhaps de Sitter character, the Poincaré group would be an adequate approximation, at least in microscopic physics. However, this is not a priori clear because the invariance group of de Sitter space, for instance, is very different from that of Minkowski space. In particular, the representations of the de Sitter group do not permit the definition of positive definite operators which could play the role of energy. The problems which an energy spectrum containing negative values would raise were mentioned before, at the discussion of the existence of tachyons. Nevertheless, Thirring, who has investigated this question most thoroughly, came to the conclusion that the conclusions derived from Poincaré invariance are valid to a very high approximation even in de Sitter space. It may be good to return to this question later, it seems to me that there are problems here which are not definitely and finally settled.

The second assumption which entered the considerations which were sketched in the preceding section is that the states are described by vectors, or rays, in a complex Hilbert space. What I have in mind in this regard when voicing reservations is not the replacement of the complex Hilbert space by a real one – such a space would be even more restrictive – but something like a quaternionic Hilbert space, or, more generally, a Hilbert space based on some other algebra but that of complex numbers. Such Hilbert spaces may permit other types of invariant manifolds, the existence of other types of particles. Jauch and his collaborators have given a good deal of thought to such possibilities and so did Gürsey but again, it seems to me, that the possibilities are not fully exhausted and that there remain problems of unquestionable mathematical interest and possibly of physical significance.

5. SOME FURTHER PROBLEMS

The manifolds of states, their various types of momenta which the transformation properties of their state vectors provide, are still a very meagre information. They determine the probabilities of the outcomes of very few measurements and we know that, at least in principle, the statements of quantum mechanics are formulated in terms of outcomes of measurements. One can say, of course, that the measurements should be described as operators in the Hilbert spaces which we found, but this requirement, though the equivalent of much repeated statements on the foundations of quantum mechanics, means very little in terms of the actual design of the measuring apparata.

We have learned, of course, from dispersion theorists, that the measurement of precisely defined operators is unnecessary, and one could content oneself with rather inaccurate position measurements. Goldrich and I have discussed this point most recently. It seems worthwhile, nevertheless, to discuss the possibility of accurate position measurements – the position is surely the most simple classical variable. In addition, a short discussion of a new 'observable' will be given even though its role is by no means fully clear.

There are two types of position operators which have been discussed. From an invariant-theoretic point of view, it is most natural to consider an operator, or a quartet of operators, which refer to a position in space-time. The meaning of a position in space-time was not always made clear by the proponents of such operators. Clearly, the position in question is not the position of a particle. If it were, and if one had observed the particle at position x, y, z, t, the probability would be zero of observing it at any point x', y', z', t' different from x, y, z, t. This is surely not so: even if we found the particle somewhere at time t, it will be somewhere also at time $t' \neq t$. In quantum mechanics, however, the states which are characteristic vectors of the quartet of position operators with characteristic values x, y, z, t are orthogonal to the characteristic vectors of any other set of characteristic values x', y', z', t' and this is true, for instance, as soon as $t' \neq t$.

The physical meaning of the space-time location was most clearly explained by A. Broyles. The localization refers not to a particle but to an event. The search for them constitutes a departure from the now accepted concepts of quantum mechanics! This does not contain observables referring to events. In fact, the greatest conceptual difficulty in the reconciliation of the quantum mechanics with the general theory of relativity is the basic difference in the observables they consider. This is the coincidence of two particles in general relativity, that is an event, whereas the quantum mechanical scattering and reaction theories do not attribute a space-time point to such an event. Hence, the interpretation of the space-time operators is still foreign to the present structure of quantum mechanics and it is, in fact, never specified what kind of events these space-time operators localize. A conceptual problem remains to be solved here – apparently, the mathematical problem is easier in this case than the interpretation of the formulae obtained. It may be added that, in addition to Broyles, several other colleagues have greatly contributed to this question.

The other concept of localization conforms with standard, old-fashioned ideas. It refers to the position of a particle at a definite time – definite time implying also a coordinate system in a definite state of motion. Newton and I determined these operators first but our ideas were made more precise by Wightman and given more elegant relativistic formulation by Fleming. He characterized these operators as functions of a vector in space-time, the vector being perpendicular to the space-like plane in which the particle's position is to be determined and ending at the position of the particle in that plane.

One may well have reservations concerning the basic nature of these position operators. First, the measurement on a space-like plane implies that one can receive the measurement signal instantaneously, in other words that the signal that the particle is at a given point in space reaches the observer with over-light velocity. This is the first reason for trying to replace these position measurements by others. The second reason is that though our postulates for the position operators could be satisfied for particles with finite mass, it is not possible to satisfy them for particles with zero mass and spin larger than 1/2. In particular, they cannot be satisfied for light quanta. The most decisive efforts in this direction were made by Bertrand, and by Suttorp and de Groot, but their conclusions were very much the same as ours. In particular, it proved to be necessary for the definition of the position operators to extend the original

Hilbert space, referring only to positive energy states, to include also negative energy states. Even this constitutes significant progress – the usual position operators, that is multiplication with x, y, and z, also mixes positive and negative energy states. However, this is just what Newton and I wanted to avoid.

Conceptually, it would appear most natural to define the position of a particle on a light cone rather than on a space-like plane. It is possible to receive signals at the tip of the cone from any point thereof. It is perhaps not obvious, but it is true nevertheless, that the state vectors which correspond to states traversing a light cone at different points are orthogonal to each other. This means that the operators of which they are eigenfunctions, that is the operators of the position on the light cone, are self-adjoint. However, they do not seem to be very simple in terms of the operators commonly used. Nevertheless, it seems to me that it would be worthwhile to determine the characteristics of these operators more closely – as I consider altogether the light cone physics very promising.

The operators in the Hilbert space of single particles – not only the positions operators – are the first subject about which it would be good to know more. The second subject concerns the behaviour of such particles in external fields. This will naturally lead to the more customary forms of the relativistic wave equations, to their forms not as representations of the Poincaré group but as old-fashioned equations. There are even more unsolved problems in this connection than in the areas we have touched upon so far. However, before taking up this subject, it may be useful to take up the question under what conditions an outside effect can be described as an 'external field'.

6. WHEN CAN AN INTERACTION BE DESCRIBED BY AN EXTERNAL FIELD?

The question under what conditions the effect of a system on a particle can be described as that of an external field is a very general one but will be discussed only briefly. The condition includes, evidently, that the interaction leave the particle in a pure state, i.e. that no correlation become established between the state of the particle and the state of the system to the influence of which the particle is subjected. The field interaction is, therefore, the opposite extreme of the measurement interaction since the purpose and characteristic of the latter is just the establishment of a correlation between the states of the object, in our case the particle and the measuring device, i.e. the system with which it is to interact. Further, the interaction which is to be described by a field must leave the scalar product of any two possible initial states of the particle unchanged. Since these states are rather arbitrary, this means that the field-producing system must be left, in spite of its interaction with the particle, in the same state, no matter what the initial state of the particle was. This can be accomplished in a simple manner only if it remains in its initial state in spite of its having influenced the state of the particle by the interaction. This will be true for all states of the particle if it is true for a complete set of states thereof.

A complete set of states of our particle can be characterized by the momentum and helicity of the states. These will be changed by the field and this implies a change in the momentum and angular momentum of the field-producing system. This can leave the state of this system unchanged only if its initial state's momentum spectrum extended over a range considerably larger than the largest momentum change of the particle that can be expected and if the same is true of its angular momentum spectrum. This also must give significant probabilities to a wider range of angular momenta than the maximum angular momentum transfer can be expected to be, and the probabilities of the various angular momenta within that range must not show appreciable variations. These are only necessary conditions but their specification may suffice in the present context. They also suggest that strong interactions cannot be described by fields under any conditions and, indeed, the fields which appear in the Klein–Gordon and Dirac equations are only the electromagnetic fields.

7. EXTENSION OF THE HILBERT SPACE OF A FREE PARTICLE TO DESCRIBE THE EFFECT OF THE FIELD

It is important to remember that the general state vector of the free particle

$$\sum_{\xi} \int \frac{\mathrm{d}^3 p}{p_0} \, \phi(p, \xi) \, |p, \xi\rangle \tag{7}$$

is a state vector in Heisenberg's sense, that is, in Wightman's words, sub specie aeternitatis. When we derive the various operators, such as the position operator, these will depend on the time of the measurement of the quantity in question. If these operators are the same as for the free particle, all the properties obtained will be the same as those of the free particle. In particular, if p is considered to be the momentum operator – this is independent of time – the momentum distribution will remain the same throughout time. Hence, if one wants to introduce the action of external fields on the particle, one has to make one of two possible modifications. One either has to change the operators which correspond to the various physical quantities, such as position, velocity, etc., the change to depend on the external field, or else one has to extend the Hilbert space and introduce state vectors different from (7). In the literature only the second alternative is considered and, also almost invariably, one works with the Fourier transform of (7), the wave function in Minkowski space.

The question naturally arises concerning the extent to which the Hilbert space is to be extended. In the case of the Dirac and Klein–Gordon equations, the restriction on the length of the momentum is abolished – this can range over all possible values. The restriction on the spin is, on the other hand, maintained. Furthermore, if the action of the field is of finite duration, the Fourier transform of the wave function will continue to contain a part which contains a Dirac delta function of the square of the momentum minus the mass square $\delta(p_t^2 - p_x^2 - p_y^2 - p_z^2 - m^2)$ so that both at $t = -\infty$ and at $t = \infty$ only the coefficient of this delta function will be relevant. But whereas one assumes that, before the interaction, the coefficient of the delta function at negative p_t was zero, this is no longer true after the interaction. The interpretation of this, in terms of the creation of antiparticles, was given of course by Dirac and Oppenheimer, and this interpretation is generally accepted now.

The point to which I wanted to call attention is different, though. It concerns particles with spins larger than 1/2. For these, no equations including the effect of a field have been proposed which would leave the spin unchanged. The extension of the Hilbert space to account for the effect of the field includes not only the whole four-dimensional momentum space but also all spin values, in integer steps, up to, and naturally including, the initial spin value. Furthermore, even if the interaction is of only finite duration, the lower spins do not disappear but remain just as the negative p_t part, caused by the electromagnetic interaction, remains.

It is not clear what this means. No two particles with identical masses but different spins are known. Does this mean that the interaction of particles with a field cannot be described for particles with spin 1 or more? Is this perhaps the seed of an explanation that no elementary particles with such spin exist? Or is it simply that we have not yet discovered all methods to describe the effects of fields? This is another puzzle worth adding to those mentioned before.

DISCUSSION

C. Fronsdal: Concerning Professor Wigner's remark about the instability of the spin for the higher wave equation, it is my understanding that the Fierz–Pauli theory was developed exactly to avoid this problem.

E. P. Wigner: The Fierz–Pauli theory, to my knowledge, has never been generalized to higher spins. Perhaps you are right that somebody will come around and find a very beautiful equation for particles with higher spin, and with interaction with the electromagnetic field, and perhaps even with other fields, but it looks very difficult.

C. Fronsdal: It has been done. Difficulties of a deeper nature have been pointed out, but the difficulty that you mentioned, I think, is not present.

E. P. Wigner: I do not know of equations which do what you indicate for particles with higher spin, but I hope I will learn from you about it.

A. Martin: I have a question about stable versus unstable particles. Professor Wigner mentioned that unstable particles are not orthogonal to the Hilbert space built out of stable particles and, of course, from a mathematical point of view it's certainly nice to build the Hilbert space exclusively with stable particles. However, I think, that from a physical point of view, we are really not in a satisfactory situation, because it appears that stability is a kind of dynamical accident, of which the most impressive example is the prediction of the omega-minus particle by Gell-Mann. The omega-minus particle is stable with respect to strong interactions, but it was predicted from the existence of resonances. So, somehow, the omega-minus particle would be more honourable than these resonances which were used to predict it, in spite of the fact that it is perfectly true to say that you can build a Hilbert space out of stable particles. It appears to me to be a very unsatisfactory situation.

E. P. Wigner: Dr. Martin is referring to an impromptu remark I made about unstable particles. The description of unstable particles as resonances is, from the point of view of present basic theory, the logical description. Surely, the neutron's state vector in Hilbert space is not orthogonal to all states of the trio of proton, electron, and antineutrino. It is, in fact, a linear combination of the state vectors of the triple. Nevertheless, since the neutron's properties are so very different from the ordinary states of the proton-electron-antineutrino triple, one is not used to describing the neutron as a reso-nant state of the triple. The situation is very different for the usual nuclear resonances the properties of which are expected to follow from the properties and interactions of the constituents. For unstable 'particles', such as the neutron or the omega-minus, this is not the case and it is natural, therefore, to disregard the basic resonance character of these and to treat them as independent particles. We have no theory which gives their properties in terms of those of the constituents. For this reason, I will at least temporarily and perhaps permanently, throw up my hands.

PART VI

Nuclear Physics

On the Development
of the Compound Nucleus Model

E. P. Wigner

Symmetries and Reflections.
Indiana University Press, Bloomington, Indiana, 1967, pp. 93–109

The Compound Nucleus Model

The compound nucleus model pictures the nuclear reaction as a succession of two events. The first event is the union of the colliding nuclei into a single unit, the so-called compound nucleus. This compound nucleus, although not stable, has many of the properties of stable nuclei. In particular, it has rather well defined energy levels. The second event is the disintegration of the compound nucleus, either into the nuclei from which it was formed, or into another pair of nuclei. In the first case, no reaction, only a scattering process has taken place; the second case corresponds to a real reaction.

The probability of the formation of the compound nucleus is very small unless the energy of the colliding pair coincides very closely with one of the energy levels of the compound nucleus. On the other hand, if the coincidence is perfect, the cross section for the formation of the compound nucleus is very large: its impact parameter corresponds to angular momentum $\hbar$ of the colliding pair about their common center of mass. Hence, the cross section for the formation of the compound nucleus shows sharp maxima but drops to very small values between these. The disintegration of the compound nucleus is subject to probability laws: once the compound nucleus has been formed, the probability

Richtmyer Memorial Lecture, January 28, 1955. Reprinted by permission from the *American Journal of Physics*, Vol. 23, No. 6 (September, 1955).

of a particular mode of disintegration is independent of the mode of
formation. As was mentioned before, the different modes of disintegration lead to different reaction products; if the disintegration leads
to the same nuclei which formed the compound nucleus, no reaction
has taken place.

The model just described was proposed also for chemical reactions,[1]
the result of the first step being called in this case the compound
molecule. However, certainly not all chemical reactions show a cross
section with high maxima and very low values between the maxima.
Hence, the compound molecule picture is a valid picture only for a
limited class of chemical reactions. Other mechanisms, i.e., other models, are more suitable for the description of other reactions.[2] Whether
and in what sense the compound nucleus picture is a general one for
all nuclear reactions is one of the questions which I wish to bring up
in the course of this discussion.

I do not want to give a detailed history of the origin of the compound
nucleus model. Those of us who have read Bohr and Kalckar's paper,[3]
for instance, do not need to be reminded of it. The reminder would
mean very little for those whose acquaintance with the subject is of
later date and who are unfamiliar with the early papers on the subject.
Experimental work constituted, in my opinion, the most important step
in the development. The experiments of Moon and Tillman, of Bjerge
and Westcott, of Szilard, and of several other investigators[4] demonstrated that the slow neutron absorption of many nuclei shows the
characteristics of high maxima and low values between the maxima
which were known consequences of the compound molecule theory of
chemical reactions. It was natural, therefore, to attempt a two stage
theory of transformations to describe at least some nuclear reactions.

The process of the formation and subsequent disintegration of a compound is also the customary picture for the description of the scattering

---

[1] Cf., e.g., M. Polanyi, Z. *Physik,* 2, 90 (1920).

[2] The most useful model for the simplest type of exchange reactions is the adiabatic model. Cf. F. London, *Sommerfeld Festschrift,* p. 104 (Leipzig: S. Hirzel,
1928).

[3] N. Bohr and F. Kalckar, *Kgl. Danske Videnskab. Selskab, Mat.-Fys. Medd.,*
14, 10 (1937).

[4] P. B. Moon and J. R. Tillman, *Nature,* 135, 904 (1935); L. Szilard, *Nature,*
136, 150 (1935); T. Bjerge and C. H. Westcott, *Proc. Roy. Soc.* (London), A150,
709 (1935); E. Amaldi and E. Fermi, *Ricerca Sci.,* 1, 310 (1936); J. R. Dunning,
G. B. Pegram, G. A. Fink, and D. P. Mitchell, *Phys. Rev.,* 48, 265 (1935).

and fluorescence of light. The absorption of light leads to an excited state of the absorber; this excited state corresponds to the compound nucleus. The reemission of light by the excited state corresponds to the disintegration of the compound nucleus. Hence, the compound nucleus model shows great formal similarity with the process of light absorption and reemission. The model which had been introduced to explain the absorption of light and its reemission could have been taken over, in fact, verbatim but for one new element. This new element is the energy dependence of the probabilities of the different modes of disintegration of the compound nucleus, in particular the proportionality between the probability of neutron emission and the square root of the energy with which the neutron is to be emitted. Even though the disintegration probabilities of the compound nucleus, at a given energy, do not depend on its mode of formation, they do depend on its energy, sometimes quite critically.

Extensions of the Applications of the Compound Nucleus Model

Even after the compound nucleus model had proved its worth for the description of neutron absorption phenomena, at least at low energies, there were those of us who, mindful of the limited validity of this model in describing chemical reactions, doubted its usefulness as a general framework into which all nuclear reactions could be fitted. The more courageous camp to which we owe most of the extensions of the compound nucleus model, headed principally by Bethe, Breit,[5] and Weisskopf, derived much encouragement from the experiments of Hafstad, Heydenberg and Tuve, and of Herb and his collaborators,[6] who showed, almost simultaneously with the slow neutron experiments which were mentioned before, that reactions induced by protons of about 1 Mev show the same type of resonance structure as exhibited by reactions induced by neutrons of a few ev. This was indeed impressive demonstration of the wide scope of applicability of the com-

[5] H. A. Bethe, *Rev. Mod. Phys.*, 9, 71 (1937); G. Breit, *Phys. Rev.*, 58, 1068 (1940), 69, 472 (1946).

[6] L. R. Hafstad and M. A. Tuve, *Phys. Rev.*, 48, 306 (1935); L. R. Hafstad, N. P. Heydenberg, and M. A. Tuve, *ibid.*, 50, 504 (1936); R. G. Herb, D. W. Kerst, and J. L. McKibben, *ibid.*, 51, 691 (1937); E J. Bernet, R. G. Herb, and D. B. Parkinson, *ibid.*, 54, 398 (1938).

pound nucleus model. The reactions which attracted Weisskopf's attention principally were, however, not these but those in which the resonances were so numerous and their widths so great that they overlapped to give a continuous energy dependence to the cross section. The continuous energy dependence might have been taken as an indication that the compound nucleus model does not apply—it was considered to be such an indication, and rightly so, in the case of chemical reactions—but Weisskopf and Ewing suggested a way to apply the concepts of the compound nucleus theory in the case of overlapping levels.[7] If the levels of the compound nucleus are very closely spaced, it becomes impossible to investigate their properties individually and, in fact, Weisskopf and Ewing's theory deals with joint properties of a very large number of levels of the compound nucleus and, usually, with joint properties of many levels of the product nucleus.

Weisskopf and Ewing's statistical model is one of the most encompassing models that were ever put forward. It gives an expression for virtually every nuclear cross section and it has stimulated experimental work on almost every type of nuclear reaction. It is natural that it contained several important assumptions and it may be useful to test these assumptions separately in the light of our present knowledge of nuclear transformations.

The statistical model is on safest ground in the energy region where the levels of the compound nucleus are sharp enough to be easily distinguishable. This restricts the energy of the incoming particles to a few Mev in heavy nuclei but leaves much more leeway in light nuclei. In this case the only relevant assumption of the statistical model is that when the compound nucleus disintegrates, all possible states of the product nucleus are formed with essentially the same probability. "Essentially the same" means that it is the same if the energy dependence of the disintegration probability, which was mentioned before, is disregarded. Although there are many indications that this assumption is invalid under certain conditions, there are at least equally many indications that it is a useful guide, if not much more.

The statistical theory is on much less safe ground in the energy region in which the levels of the compound nucleus are broad enough to overlap. It was mentioned before that the compound molecule theory of chemical reactions does not apply in such cases. In such a case, even the

7 V. F. Weisskopf and D. H. Ewing, *Phys. Rev.*, 57, 472, 935 (1940).

On the Development of the Compound Nucleus Model   97

concept of the compound nucleus is questionable in the sense that its energy determines its properties, in particular the probabilities of the various modes of its disintegration. Just as a very high energy electron can traverse an atom without much energy loss, a very high energy proton will be able to traverse a nucleus without being much affected by it. This is contrary to the compound nucleus model as used in the statistical theory, which postulates that the emergence of the proton should be just as probable no matter whether the compound nucleus originates from the collision of a very fast proton or of a very fast neutron with the appropriate target nucleus. It is natural to expect, in the second case, that the neutron would emerge from the target. It is clear, therefore, that the validity of not only the statistical model, but also of the original ideas of the compound nucleus theory, is limited toward the high energy region. The statistical theory demands not only that the reaction products depend only on the energy (and angular momentum) of the compound nucleus, it demands further that all energetically possible reaction products appear with essentially the same probabilities.

Many experiments gave strong support to the statistical theory even under the most adverse conditions. However, more recently, the conflicting experiments began to preponderate. They can be brought to a common denominator: the probability of the formation of the product nucleus in its various states of excitation is not equal but there is a definite preference for the formation of states of low excitation. This is directly demonstrated in the experiments of Gugelot and B. Cohen.[8] The preferential emission of protons and α particles, first discovered by the Swiss school,[9] can also be explained in this way.

The question naturally arises whether there is an energy region in which the naive form of the compound nucleus theory remains valid but the specific hypothesis of the statistical theory is invalid. Personally, I am inclined to be doubtful concerning the basic postulate of the statistical theory with respect to the equality of the probabilities of all modes of disintegration of the compound nucleus. Nevertheless, I wish to emphasize that, in my opinion, there is no unequivocal evidence that the statistical theory breaks down before the compound nucleus theory, as

[8] P. Gugelot, *Phys. Rev.*, 93, 425 (1954); B. L. Cohen, *ibid.*, 92, 1245 (1953).
[9] O. Hirzel and H. Waffler, *Helv. Phys. Acta*, 20, 373 (1947); E. B. Paul and R. L. Clarke, *Can. J. Phys.*, 31, 267 (1953).

discussed above, ceases to be valid. Courant's considerations[10] suggest, on the contrary, that the statistical assumption is correct whenever the compound nucleus picture can be used in the form discussed above.

The principal objection to the assumption of equal transition probabilities to all possible states is, however, the same as that to all theories which we inherited, mostly from ourselves, from the years before the War. All these consider all nuclear properties to be smooth functions of the mass and charge numbers and of the energy. As far as the normal states of nuclei are concerned, this view was refuted by the observations of Mayer[11] and of Haxel, Jensen, and Suess[12] and their theories of nuclear shell structure. However, the situation is not too different even as far as nuclear reactions are concerned. The large scattering cross sections of the iron group, the large absorption cross sections of the rare earths, are too systematic to be accidental and have been reinforced by the total cross section measurements of Fields, Russell, Sachs, and Wattenberg.[13] Bohm and Ford[14] tried to interpret these measurements on the basis of the independent particle model. However, the independent particle model which does reproduce the gross structure cannot account for the fine structure, i.e., the resonance character, of the cross section curve. When the empirical situation was fully clarified as a result of the experiments of Barschall and his collaborators,[15] it was

---

[10] E. D. Courant, *Phys. Rev.*, 82, 703 (1951); H. McManus and W. T. Sharp, *ibid.*, 87, 188 (1952).

[11] M. G. Mayer, *Phys. Rev.*, 74, 235 (1948); also W. Elsasser, *J. Phys. Radium*, 5, 625 (1934).

[12] O. Haxel, J. Jensen, and H. Suess, *Z. Physik*, 128, 295 (1950).

[13] R. Fields, B. Russell, D. Sachs, and A. Wattenberg, *Phys. Rev.*, 71, 508 (1947).

[14] K. W. Ford and D. Bohm, *Phys. Rev.*, 79, 745 (1950).

[15] The information on low energy (50 to 3,000 kev) cross sections was obtained by Barschall and his collaborators as a result of a series of investigations, starting back in 1948. Cf. H. H. Barschall, C. K. Bockelman, and L. W. Seagondollar, *Phys. Rev.*, 73, 659 (1948) (Fe, Ni, Bi); R. K. Adair, H. H. Barschall, C. K. Bockelman, and O. Sala, *ibid.*, 75, 1124 (1949) (Be, O, Na, Ca); C. K. Bockelman, R. E. Peterson, R. K. Adair, and H. H. Barschall, *ibid.*, 76, 277 (1949) (Zr, Ag, In, Sb, I, Ta, Pb); R. E. Peterson, R. K. Adair, and H. H. Barschall, *ibid.*, 79 (1950) (lead isotopes); C. K. Bockelman, D. W. Miller, R. K. Adair, and H. H. Barschall, *ibid.*, 84, 69 (1951) (Li, Be, B, C, O); H. H. Barschall, *ibid.*, 86, 431L (1952) (review); D. W. Miller, R. K. Adair, C. K. Bockelman, and S. E. Darden, *ibid.*, 88, 83 (1952) (review); N. Nereson and S. Darden, *ibid.*, 89, 775 (1953) (higher energies); M. Walt, R. L. Becker, A. Okazaki, R. E. Fields, *ibid.*, 89, 1271 (1953) (Co, Ga, Se, Cd, Te, Pt, Au, Hg, Th); A. Okasaki, S. E. Darden, R. B. Walton, *ibid.*, 93, 461 (1954) (Nd, Sm, Er, Yb, Hf); M. Walt and H. H. Barschall, *ibid.*, 93, 1062 (1954) (angular distributions); cf. also R. K. Adair, *Rev. Mod. Phys.*, 22, 249 (1950); A. Langsdorf, *Phys. Rev.*, 80, 132 (1950).

On the Development of the Compound Nucleus Model   99

again Weisskopf, in collaboration with Feshbach and Porter, who gave the solution: the independent particle model—it would be more accurate to say, *an* independent particle model—gives only the average cross section, that is, the gross structure, by determining the product of the density and strength of the levels of the compound nucleus.[16] The reaction, itself, proceeds via the compound nucleus mechanism. In other words, it is not the compound nucleus model which is at fault; the discrepancy has to be blamed on the statistical assumptions which were so very plausible but which have to be modified so that the average cross section be at least in rough accord with the independent particle picture. While these statements appear to imply some mystical correspondence principle role of the independent particle model, this implication is incorrect and Scott, Thomas, Lane, and others have shown very concretely how the results of Weisskopf's model can be obtained as vestigial traces of the independent particle model.[17] Whether this interpretation of the Barschall maxima, as made more precise by these authors, will stand the test of time is as yet uncertain.

Since this model is not yet very generally known, let me spend a few minutes outlining it. If we assume, in the sense of the independent particle model, that the effect of the target nucleus on the incident particle can be accounted for by a suitable potential, one can describe any state of the compound nucleus by specifying a state of the target nucleus and giving also the state of the incident particle in the potential created by the target nucleus. Figure 1a shows a set of levels of the compound nucleus at about 8 Mev excitation. Most of the levels correspond to an *excited* state of the target nucleus plus a suitable state of the incident particle. It is assumed, however, that the particular state marked by crosses corresponds to the *normal* state of the target nucleus and a state of about 8 Mev energy of the incident particle. This state of the compound nucleus—if it is worthy of that name—will disintegrate very fast into the initial target nucleus and the incident particle with its original energy. In fact, the compound state will last only as long as it takes the incident particle to cross the potential generated by the target nucleus. The disintegration of this particular compound state will always pro-

[16] H. Feshbach, C. E. Porter, and V. F. Weisskopf, *Phys. Rev.*, 90, 166 (1953); 96, 448 (1954).
[17] J. M. C. Scott, *Phil. Mag.*, 45, 1332 (1954); E. P. Wigner, *Science*, 120, 790 (1954); A. M. Lane, R. G. Thomas, and E. P. Wigner, *Phys. Rev.*, 98, 693 (1955).

ceed by reemission of the incident particle with its original energy; it
is a property of the extreme independent particle model that it gives no
reaction, only scattering. The other states of the compound nucleus will
have a short life-time also; each of them will disintegrate only into one
state of the target nucleus, and this will be an excited state for all of the
not-crossed states. Conversely, only the crossed state will form if the
incident particle strikes the target nucleus in its *normal* state. The probability of disintegration of the various compound states, resulting in the
target nucleus in its normal state, is illustrated on the second line

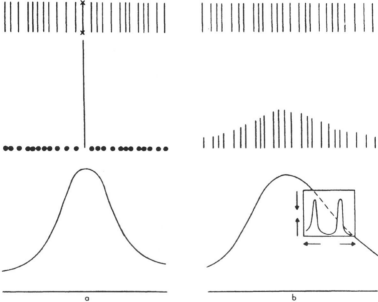

Fig. *1a* refers to the independent particle model in which the effect
of the target nucleus on the incident particle can be described by a
potential. This potential depends, naturally, on the state of excitation
of the target nucleus. The top of the figure gives the energy levels
of the compound nucleus; the level with crosses corresponds to the
normal state of the target nucleus plus a suitable state of the incident
particle. The other levels correspond to excited states of the target
nucleus. The heights of the lines in the middle of the figure give
the transition probabilities from the state of the compound nucleus
above them, to the normal state of the target nucleus (plus the incident particle). This probability vanishes for all states of the compound nucleus except the crossed one. The bottom part of the figure
gives the cross section as function of energy; the width of the line
is due to the large disintegration probability of the crossed state.

On the Development of the Compound Nucleus Model   101

of the graph. These probabilities also give the partial widths for the formation of the compound state from the incident particle and the target nucleus in its normal state. Hence, the bottom graph represents the scattering cross section of the target nucleus according to the independent particle model; it is the cross section curve of Bohm and Ford.

If we admit that the independent particle picture is inaccurate, it will not be possible to label the states of the compound nucleus by the states of the target nucleus from which they can be obtained. Rather, every state in the neighborhood of the crossed state will acquire some of the characteristics of that state. The probability of disintegration into the normal state of the target nucleus will be, therefore, much smaller for the originally crossed state but it will be, on the other hand, finite for many compound states for which it vanished in the independent particle model. The same holds for the partial widths of these states, so that the cross section will have resonance character. The situation in this model, as contrasted with the extreme independent particle model, is given in Figure 1b. The average cross section of this model, given in the bottom diagram, will be the cross section of the Feshbach–Porter–Weisskopf clouded crystal ball model.

At the same time, our model indicates a possible reason for systematic deviations from the statistical theory in the sense indicated by the ex-

---

Fig. 1b represents the conditions in the model considered in the text. Chosen a state of the compound nucleus, the state of excitation of the target nucleus is not uniquely given but can be, with various probabilities, one of several states. As a result, there is no definite crossed state of the compound nucleus. Conversely, many states of the compound nucleus can disintegrate into the normal state of the target nucleus (plus the incident particle), the probabilities for such a disintegration of the compound nucleus states are given by the ordinates in the middle of the figure. They are largest for those levels of the compound nucleus which are close, on the energy scale, to the crossed level of the independent particle picture of Figure 1a. The bottom of the figure gives the average cross section of the model; a portion magnified in energy scale but contracted on the cross section scale is given by the insert. The width of the maximum in the average cross section is due, principally, to the fact that the levels of the compound nucleus which can disintegrate into the normal state of the target nucleus are spread out on the energy scale. Nevertheless, the *average* cross section of Figure 1b shows a similarity to the *actual* cross section of the independent particle model, given in Figure 1a.

perimental data. Under certain conditions, the transition probabilities
into the various states of the product nucleus are not equal. Rather, the
transition probability into all states which lie within unit energy interval
is independent of the position of that interval. Under these conditions,
the high energy region, although it contains a large number of energy
levels, does not any more receive the lion's share of the transitions. The
very few levels in a unit range at low energy receive as many transitions
as the much more numerous energy levels in a high energy unit interval.
This is the preference for the formation of the low energy levels which
was mentioned before as explaining the systematic deviations from the
statistical model.

The above summary, though far from complete, may give some idea
of the variety of the problems to which the compound nucleus model
has been applied. Let me proceed, therefore, to the last subject on which
I wish to speak, the foundation of the compound model and the basic
problems on which it sheds some light.

Basic Aspects of the Compound Nucleus Model

The story of the gradual extension of the compound particle model,
always guided by the experimental results, appears to me a story worth
telling. The story of the theoretical considerations behind it seems to
me even more interesting.

A very restricted problem, the anomalous behavior of the slow neutron
absorption of Cd, was the principal problem that Breit and I wanted
to explain originally.[18] We made two rather far-reaching assumptions.
The first of these is the existence of a somewhat mysterious "compound
state," the second the absence of a direct coupling between the states
of the continuum. Our formula contained three adjustable parameters,
and we now know that the agreement between formula and observation is remarkable even if one considers that three adjustable constants
are at one's disposal.

The rest of the story is similar in its main lines, and on a small scale,
to the story of almost every physical theory. First, the ephemeral and
the general elements of the theory were separated. Second, the general
elements were formulated in such a general way that they could serve

[18] G. Breit and E. P. Wigner, *Phys. Rev.*, 49, 519, 642 (1936).

On the Development of the Compound Nucleus Model   103

as a framework for the description of a wide variety of phenomena. This, unfortunately, also means that the physical content of the framework is so small that almost any experimental result could be fitted into it. The interest, therefore, shifted away from the framework to the body of information which could be described in the language given by the framework. Third, an attempt was made to deduce from very general principles the physical content which remained in the framework. The story of almost every physical theory goes along this way; some run through it fast, the more important ones less rapidly.

It may be worthwhile to illustrate this by means of an example. Galileo's discovery of the laws of free fall led to a concrete and definite equation. What was ephemeral in this equation was the constancy of the acceleration. The general part of it was formulated as Newton's second law, the proportionality between acceleration and force. This is the first phase mentioned above. However, Newton's second law is so general that it permits almost any type of motion; it provides a language to describe motions rather than determining the motion by itself. Hence, the problem after the establishment of the second law shifted from the description of the motion to the determination of the forces between bodies. This shift of the problem is what was called the second phase of the theory. The last phase of the theory is the search for the fundamental reason for those elements of the general physical law which give it whatever physical content it does have. This is, actually, Newton's first law. Its fundamental basis, the equivalence of moving observers, was embodied eventually in the Galilei and Lorentz transformations.

The most important ephemeral part of the compound nucleus theory was soon recognized as the assumption of a single compound state which entailed the existence of a single resonance level. However, when the theory was extended to assume many compound states, to describe many resonance levels, the number of adjustable parameters also became arbitrary and it became apparent that almost any experimental result on nuclear transformation might be consistent with the theory. As the generality of the resonance formula increased, its specificity decreased and so did the specificity of the assumptions entering the theory. The possibility of carrying out this development ad extremum, and of deriving a framework from practically no specific assumption, was first recognized by Kapur and Peierls.[19] Their theory was little understood

[19] P. L. Kapur and R. Peierls, *Proc. Roy. Soc.* (London), A166, 277 (1938).

at first. It contains, however, many ideas which reappear in practically all subsequent work.

It may be of some interest to describe why and how I returned to the subject. About a year before the end of the last war, Fermi commented to me on the anomaly that the compound nucleus theory had become widely accepted without having a good foundation in basic theory. I considered this as a challenge and made an attempt to formulate assumptions from which the one level formula could be derived. The assumption of the somewhat mysterious "compound state" was replaced by the assumption that the wave function is energy independent in the part of the configuration space in which the incident particle is inside the nucleus or very close to it. Instead of the assumption that there is no interaction between the states of the continuum, it was assumed that, outside of the immediate neighborhood of the target nucleus, the incident particle behaves like a free particle. From these assumptions, the one level formula with its three parameters, a rather specific formula, could be rederived. This corresponds to the first stage of the development which I sketched before.

It was tempting, however, to eliminate the first assumption and to replace it with the more general one that the wave function in the interior part of the configuration space, that is, where target nucleus and incident particle are close together, is a linear combination of several wave functions, with energy dependent coefficients, rather than an energy independent function. This virtually abolished the first of the earlier assumptions because any function can be written as a linear combination of a sufficiently large number of given functions. The only relevant assumption which then remained was that of the finite range of interaction. It should not have surprised me that the development could be carried out with the more general assumption concerning the wave function in the interior part of configuration space: the considerations of Kapur and Peierls and other considerations of Breit clearly foreshadowed this. The simplicity of the final formula was surprising, though. It suggested a simplified approach which was given, almost simultaneously by Eisenbud and by Schwinger and Weisskopf.[20] Similar ideas were formulated even before in radioengineering.

Let me remark, only parenthetically, that the assumption of a finite

---

[20] L. Eisenbud and E. P. Wigner, *Phys. Rev.*, 72, 29 (1947); T. Teichmann and E. P. Wigner, *ibid.*, 87, 123 (1952).

On the Development of the Compound Nucleus Model   105

range of interaction was also eliminated subsequently, by Thomas.[21] The model, as described for instance in the book of Blatt and Weisskopf, or in that of Sachs,[22] retains the division of the configuration space into two parts. All of the typically nuclear interaction takes place in the internal region; the waves spread essentially freely in the external domain. The nuclear interaction is replaced by a formal connection between the normal derivative and the value of the wave function on the boundary of these two regions,

$$v_s = \sum_t R_{st}\, d_t. \tag{1}$$

It may be well to repeat that (1) expresses the properties of the internal part of configuration space, depends on the interaction in the internal region, and is in fact a substitute for the usual description of this interaction by means of potentials, etc. It does not depend on any interaction which may take place outside the internal region. The v_s are expansion coefficients of the value of the wave function on the boundary of the internal region, the d_t are the expansion coefficients of the normal derivative. The R_{st} form a matrix, the R matrix, the dependence of which on the energy is given by

$$R_{st} = \sum_\lambda \frac{\gamma_{\lambda s}\,\gamma_{\lambda t}}{E_\lambda - E}. \tag{2}$$

The E_λ are the energy values of the compound states; they are not to be confused with the total energy E of the system. The $\gamma_{\lambda s}, \gamma_{\lambda t}$ determine the probabilities that the compound state λ disintegrate in the mode s or t. The R becomes very large if E approaches one of the E_λ; this corresponds to a large cross section at the coincidence of the energy E of the system with an energy level E_λ of the compound nucleus. The one level formula will be valid if one term in the sum (2) predominates. In this case the fact that the numerator is a product of two factors, one depending only on s, the other only on t, assures that the probabilities of the various modes of disintegration of the compound nucleus are independent of the mode of formation of the compound nucleus. Thus, many of the properties of the simple compound nucleus theory remain

[21] R. G. Thomas, *Phys. Rev.*, 100, 25 (1955).
[22] J. Blatt and V. F. Weisskopf, *Theoretical Nuclear Physics* (New York: John Wiley and Sons, 1952), Chapters VIII and X; R. G. Sachs, *Nuclear Theory* (Cambridge: Addison-Wesley Publishing Co., 1953), pp. 290-304.

apparent in (2). At the same time, (2) contains in general so many parameters—actually an infinite number of them—that it has hardly any direct physical content. In spite of this, (2) is not useless: it gives a framework, that is a language, for the description of nuclear transmutation processes. The reduced widths $\gamma_{\lambda_s}^2$ and the resonance energies E_λ constitute a simpler description of the reaction process than all the cross section versus energy curves. These quantities can be compared, interpreted, and even calculated more easily than the cross sections themselves. In fact, a theory has been put forward, and a wealth of data is being accumulated on the magnitude of the $\gamma_{\lambda_s}^2$ and E_λ. The validity of (2) has ceased to be an interesting question and has been replaced by the problem of the magnitudes of the γ_{λ_s} and E_λ.

This, it seems to me, answers the question which was posed at the beginning of this talk, concerning the generality of the compound nucleus model as a description of nuclear reactions. The generality is almost complete if we consider the model as a framework for describing nuclear reactions, as a language in which the results can be formulated.[23] The physical content which we originally associated with the model is not general but can be expressed in the language of the model in a particularly simple way. Other, equally simple and important physical pictures are much more difficult to express in that language. Though Thomas formulated the assumptions of the statistical model within the framework of equations (1) and (2), the equally simple picture of direct interaction between incident and target particles, proposed by Courant,[24] has not been formulated in that language. It still appears to me that the language is really useful only when the energy levels of the compound nucleus are well separated, when at least some of the features of the original compound nucleus picture apply. If this is not the case, the language becomes clumsy and often gave me the impression that an equation such as $(\sqrt{(1 - x^2)})^2 = 1 - x^2$ is being proved by means of the power series expansion for $\sqrt{(1 - x^2)}$.

Here ends the description of the second phase of the theory. If it is permissible to recognize similarities between an elephant and an ant, I would point to the fact that Newton's second law is similarly a frame-

---

[23] It has been shown, in particular, that the process of spontaneous disintegration can also be described by means of Eqs. (1) and (2). See L. Eisenbud and E. P. Wigner, *Nuclear Structure* (Princeton, N.J.: Princeton University Press, 1958), Sec. 9.5.

[24] See footnote 10.

On the Development of the Compound Nucleus Model   *107*

work. Once it was established, the problem became the determination of the forces, their dependence on the material, and distance separating the bodies between which they act. In the case of the compound nucleus theory, the corresponding problems were discussed in the preceding section.

What remains, then, is the third phase in the development, a deeper understanding of the rather meager information that is contained in (2). Because even though (2) represents a very general function of the energy E, it is not an entirely arbitrary function. One can see, for instance, that R_{ss} always increases with increasing energy wherever it is' finite.

The basic reason for this and other properties of R has puzzled many of us. It was suggested, finally, by Schultzer and Tiomno.[25] Their work, inspired by earlier investigations of Kronig and of Kramers, indicated that unless R has the form (2), one may have the paradoxical situation that the outgoing wave begins to leave the internal region before the incoming wave has reached it. The principle that such a thing cannot happen has come to be called the "causality condition." Its emergence has contributed a great deal to our understanding of equation (2).

The Causality Condition

The surmise of Tiomno and Schutzer was proved for a wide variety of phenomena by Van Kampen[26] and by Gell-Mann, Goldberger, and Thirring,[27] whose work also contributed greatly to our understanding of the connection between the operator R and the collision matrix.

The following lines appear to me a rather general derivation of the result of Tiomno, Schutzer, Van Kampen, Gell-Mann, Goldberger, Thirring, at least for the non-relativistic case. The basic equation which will be used postulates that the time derivative of the probability of finding the system in the internal region of configuration space is equal to the probability current across the surface of the internal region. If one wishes to use no concepts which refer to the internal region, one

[25] W. Schutzer and J. Tiomno, *Phys. Rev.*, 83, 249 (1951). R. de L. Kronig, *J. Opt. Soc. Am.*, 12, 547 (1926); H. A. Kramers, *Atti. congr. intern. fisici Como 2*, p. 545 (1927).

[26] N. G. Van Kampen, *Phys. Rev.*, 89, 1072 (1953); 91, 1267 (1953). Also J. S. Toll, Princeton University Dissertation, 1952.

[27] M. Gell-Mann, M. L. Goldberger, and W. E. Thirring, *Phys. Rev.*, 95, 1612 (1954).

can postulate instead that the negative derivative of finding the system in the external region is equal to the current across the boundary separating internal and external regions. In order to make use of this postulate, one evidently cannot restrict oneself to stationary wave functions but must consider a superposition of states with different energy values. For the sake of simplicity, the normal derivatives of all the states which will be superposed will depend in the same way on the position on the boundary, i.e., the normal derivative of the total wave function on the boundary shall have the form

$$\mathrm{grad}_n \, \varphi = \sum_k a(E_k) f(x) e^{-iE_k t/\hbar} \text{ (on the boundary)}, \tag{3}$$

the summation to be extended over several energy values E_k; the variable x describes the points of configuration space. The value of the wave function on the surface will then be given by the R operator

$$\varphi(x) = \sum_k a(E_k) R(E_k) f(x) e^{-iE_k t/\hbar} \text{ (on the boundary)}. \tag{3a}$$

The current into the internal region is the integral of $-i(\varphi \, \mathrm{grad}_n \, \varphi^{\circ} - \varphi^{\circ} \, \mathrm{grad}_n \, \varphi)$,

$$\text{current} = -i \sum_{lk} a(E_l)^{\circ} a(E_k) \, e^{i(E_l - E_k)t/\hbar}$$

$$\times \left(\int f^{\circ} \, R(E_k) f \, dS - \int (R(E_l)f)^{\circ} \, f dS \right). \tag{4}$$

dS indicates integration over the boundary between internal and external regions.

Although the formulae which were used above do not give an explicit expression therefore, the wave function in the internal region must be completely determined once the normal derivative of a monoenergetic wave is given on the boundary. It then follows from the principle of superposition and from the quadratic nature of all expressions for probabilities in quantum mechanics that the probability of finding the system in the internal region is given by an expression of the form

$$\sum_{kl} P_{lk} \, a(E_l)^{\circ} \, a(E_k) \, e^{i(E_l - E_k)t/\hbar}. \tag{5}$$

The P can depend on the function f, which underlies our considerations, but not on time. Furthermore, since the last expression is a probability, the P matrix must be positive definite. Equating the time derivative of this last expression with the expression for the current gives

$$(E_l - E_k) P_{lk} = (R(E_l)f, f) - (f, R(E_k) f). \tag{6}$$

On the Development of the Compound Nucleus Model   109

The scalar product here indicates integration over the boundary. It can be shown by rather elementary arguments that all R as operators on functions on the boundary S are real and symmetric. Hence, the matrix with the general element

$$P_{lk} = (f, \frac{R(E_l) - R(E_k)}{E_l - E_k} f) \tag{7}$$

is positive definite, no matter how the energy values E_l, E_k are chosen and no matter how many such energy values were chosen, i.e., what the order of the matrix is. This is, however, the condition that

$$(f, R(E) f) = \sum_\lambda \frac{\gamma_\lambda^2}{E_\lambda - E} \tag{8}$$

have an expansion as given above, with positive γ_λ^2 and the fact that such an expansion exists for all f establishes our basic formula.[28]

The only point at which essential use was made of the non-relativistic nature of the underlying theory is where the expression for the current was given. However, a very similar derivation applies for much more general expressions for the current, such as are compatible with the theory of relativity. The only point at which the derivation treads on doubtful ground is where it assumes that the E_k are arbitrary, e.g., that they can be also negative. If negative E_k are excluded, the result becomes less specific and an integral over negative energy values may have to be added to the sum in the expression for R. This again duplicates Van Kampen's result. It follows that if one wishes to consider (2) to be a consequence of the "causality principle," it is necessary to make another physical assumption.

The assumption which appears most natural to me is that it is permissible to consider any constant potential to prevail in the external region and that the R_{st} in (1) remain the same, no matter what the values of these potentials are. Similar assumptions, involving semipermeable walls, etc., are familiar to us from thermodynamics but these assumptions can now be shown, by the general methods of statistical mechanics, never to lead to contradictions. Whether this will be true of the assumption of arbitrary constant potentials for the various possible reaction products remains to be seen.

[28] K. Loewner, *Math. Z.*, 38, 177 (1953). The considerations of this article have been simplified by E. P. Wigner and J. v. Neumann, *Ann. Math.*, 59, 418 (1954).

Summary of the Conference
(Properties of Nuclear States)

Eugene P. Wigner

Proceedings of International Conference on Properties of Nuclear States, Montréal 1969.
Les Presses de L'Universite de Montréal, Montréal 1969, pp. 633–647
(Reset by Springer-Verlag for this volume)

As I said a couple of days ago, as far as my own knowledge of nuclear physics is concerned, this conference was very effective. Often, I think that meetings like the present one will assume the role which publications used to play 30 or 40 years ago. At present, it is impossible to read and digest all publications; it is impossible even if one does not consider oneself to be a physicist but only claims to be a nuclear physicist. Frequently one may be tempted to call oneself a nuclear reaction theorist, or something akin to that, but perhaps we do not have to go quite that far in specialization. Anyway, a meeting such as the present one helps the participants greatly to reacquire a footing in their subject, to become acquainted with more recent developments which they may overlooked and also with the possible weaknesses of these developments as well as of their own work. Not only the official presentations help in this regard; conversations in corridors or on the bus help equally if not more. I myself wish to express my indebtedness to several members of the conference in this regard.

Let me now come to my subject proper: the summary of our conference. There are two ways to summarize the proceedings of a conference. One can try to summarize each paper that has been presented, or one can present a picture of the subject (or subjects) that has emerged in one's mind as a result of what one has heard. I hope I am acting in accordance with the wishes of the participants when I use the second method, and that my summary will also give some indication of the extent to which the questions posed by Vogt in his very stimulating introductory address have been answered.

The papers presented to us can be divided into two categories, even though these are far from being sharply separated. We had experimental and theoretical papers.

Of the two, I found the progress in experimental techniques the more impressive. The most spectacular component of this is the general availability of high energy particles. These include not only protons, deuterons and α particles, the projectiles of old-fashioned experimental physics, but also heavy ions, and the information provided by the use of these projectiles seems decisive. Dr. Gove and Dr. Stephens discussed these, and Dr. Gove went so far as to propose U projectiles. One of the advantages of experiments with heavy ions is, of course, that one can obtain states with very high angular momenta, another that peripheral reactions, illustrated by Dr. Bromley, become more prominent.

Heavy-ion bombardment may also provide the best route toward super-heavy nuclei, in the projected stability region of 114 protons, 164 neutrons.

Even more impressive than the spectacular progress toward higher energies and heavier projectiles are the smaller scale developments of experimental techniques, that is, more accurate and more effective detection and registration devices. These made it possible to implement a suggestion, due originally to Breit, to measure the quadrupole moment of excited states. Dr. Broude reported on the absorption of polarized neutrons by oriented nuclei – not long ago, such an undertaking would have been considered fantastic. He and Stephens also reported on the measurement of life-times in the nano-second and lower region, by what he called the plunger method. The recoiling nucleus in an excited state is permitted to traverse freely a fraction of a millimeter; if it emits its γ ray during this traverse, the γ ray is Doppler-shifted. If it emits it after being stopped, there is no Doppler shift. The ratio of the Doppler-shifted to the unshifted line's intensity gives the fraction of decays during the time of traverse – a known time. This, then, gives the life-time.

Many other examples could be mentioned. Dr. Bromley reported on improvements in his apparatus, and the accuracy which James could achieve in the $(p, 2p)$ experiments also surprised and delighted me. The same applies to the data Barrett reported upon, even though his conclusions on the difference between proton and neutron densities in nuclei do not yet appear to be definitive.

As you know, I am no experimentalist and, I fear, this means that my appreciation of the development of experimental techniques is less than it should be. It is great, nevertheless.

Before going over to the subject about which I should know more, that is the discussion of the theoretical papers, let me make two observations. The first one is that the theories of nuclear structure and of nuclear reactions cannot be quite separated. Our information on nuclear structure derives to a very great extent from nuclear reaction experiments and the interpretation of these requires some reaction theory. This became evident several times in the course of the discussion. Dr. Stein's measurement of the purity of the ^{208}Pb shell-theory wave function depends on the interpretation of stripping reactions; this is only one example of our dependence on the theory of stripping and, more generally, peripheral reactions. Dr. Macfarlane articulated this point very clearly. Dr. Broude complained of the absence of an adequate theory of stopping power as far as heavy ions are concerned and a critical review of most experimental conclusions will support my statement of our dependence on reaction theories. The dependence of the interpretation of experiments on theory is, of course, a general situation; it prevails not only in nuclear physics. What is, perhaps, unique is that we have to depend so strongly on theories, the accuracy of which is still subject to doubt.

Let me go on to my second observation: as long as we do not have a more complete understanding of nuclear forces, we cannot have a theory of nuclear structure in the same sense as we have a theory of atomic structure. Our present ideas on nuclear forces derive, in last analysis, principally from scattering experiments – experiments measuring proton-proton and neutron-proton scattering. These experiments have not been reviewed in detail at our conference, though Brueckner did give an overall picture of the situation and both J.P. Elliott and Breit referred to them. Breit also mentioned some problems of interpretation. Now it is true that there is a theorem, due I believe to Bargmann, that a complete knowledge of the bound states and of the phase shifts, at every energy and at a given angular momentum, can be represented by only one potential, depending, of course, on the angular momentum. However, we do not have a complete knowledge of the phase shifts, and at high energies, where other than scattering processes become possible, the interaction cannot be described by a phase shift. Hence, the theorem I mentioned is not really applicable in our case. Second, we all believe that the forces should not be represented by an ordinary potential; they are velocity dependent or non-local. Hence, the same scattering data can be interpreted in several ways. This was also pointed out by Brueckner, who, in addition, discussed the meson field-theoretical basis of our knowledge of nucleon-nucleon interaction in a very comprehensive way. Uncertainties remain, and they become evident to any reader of Moravcsik's book. There are as many interactions as authors proposing them; nay, there are more. They are also terribly complicated – Brueckner said that they contain 8 parameters – now, as Einstein said, we can truly believe only simple rules... It is not even necessary to mention that there may be three-body forces, further complicating the situation.

I have enlarged so much on this point because I wanted to bring out the fact that our theories of nuclear structure and properties are not theories in the same sense as our atomic theories are. The latter take only a few numbers from the body of experimental material they are intended to interpret: the values of the fundamental constants e, h, and so on. There is no similar uniqueness in nuclear theories. Most of them are relations between different properties rather than derivations or calculations of the properties themselves. There is a great deal of arbitrariness in the choice and number of the properties which are taken from experiment and also in the number and identity of the properties which are "explained" in terms of these. In this regard, there is a similarity between nuclear theories and dispersion theory. The latter relates different cross sections but does not calculate any *ab initio* as atomic theory, for instance, tries to calculate all properties from the Schrödinger equation. In nuclear theories, there is a good deal of arbitrariness in the choice of the data which form the input – none of the Slater integrals is calculated in Talmi's theory but all are obtained from the data to be explained. Dr. Harvey made the difference between nuclear and atomic theories particularly clear. If one does accept a particular set of nuclear forces, the situation becomes similar to that one faces in atomic theory. If one does not, as apparently neither Gillet, nor Harvey, nor Arima did, one establishes only relations within sets of data and verifies or refutes schemes of approximations. Some of these have high accuracy, as do some of Gillet's conclusions, or those of R-matrix nuclear reaction theory. The calculations which are based on fundamental but as yet uncertain theories, and which use essentially only these theories as input, explain more but do so with lower accuracy. This applies also to the reaction theories of Weidenmüller and of Danos and Greiner. The accuracy increases as an increasing number of parameters is taken from observations – such as the positions and all partial widths of the resonance levels in R-matrix theory, or all the Slater integrals in nuclear structure theories. This difference in the character and objectives of the various theories or approximation schemes is my second observation. I will call the former theories, which make significant use of our knowledge of nuclear interaction, basic; the second type, with a dispersion-theoretical character, will be called phenomenological. Naturally, these categories are not sharply separated; there are many transitions between them.

Let me now review the nuclear structure theories individually. My discussion will have a somewhat negative tone. The successes of the theories are well brought out by their protagonists; what they yet have to accomplish should be brought out by their reviewers.

The shell or independent particle model in its original form was little discussed and seems to be somewhat in disrepute. We know from Talmi's work – and this has been confirmed by Harvey also – that, if one considers all the Slater integrals as parameters to be taken from experiment, the positions of all energy levels within a shell can be obtained in many cases very accurately. Harvey reported on a case in which there were 63 such integrals, but the number of levels was much higher. With few exceptions, all the energy levels were reproduced with the 63 constants and some of the constants had little effect on the results.

The principal objection to old-fashioned shell theory is that it does not reproduce the electromagnetic transition probabilities. It gives, as a rule, a fraction of a Weisskopf unit for them, in the best case a very few Weisskopf units. We have heard of transition probabilities well in excess of 20 units. Harvey and Khanna propose, therefore, that modified operators be used for the calculation of the transition probabilities. This is a departure from the classical independent particle model and we hope it will be successful.

Another area in which the shell model is bound to fail is that of high excitation, particularly in heavier elements. This is not a valid objection if one compares the shell model with other nuclear theories: none of them is valid in the region in which there are thousands of excited states. However, if the independent particle model of nuclei is compared with the similar model for the electronic shells, there is a very significant difference. The terms resulting from an atomic configuration are more or less separated in energy from the terms arising from other configurations. This is not true for the energy levels of nuclear configurations. Thus the lowest configurations for $A = 28$, the $2s - 1d$ configurations, give not only the normal state of ^{28}Si, which is a $T = 0$ state, they also give a $T = 6$ state, the "analogue" of the normal state of ^{28}O – if such a nucleus exists. In other words, the terms of the nuclear configurations are spread out over enormous energy ranges, and one does not know for how many terms one should take the shell model seriously.

Hand in hand with this last problem goes the problem of the enormous proliferation of the number of terms of some configurations. Both Vogt and

Harvey mentioned this. Thus, the aforementioned lowest configurations for $A = 28$ give 6706 terms with $J = 3$, $T = 1$. Needless to say, nobody expects to find the location of this many terms either in ^{28}Si or in ^{28}Mg. It is necessary in such cases to truncate the set of states furnished by the model, and ingenious schemes of truncation have been proposed by both Arima and J.P. Elliott. It would be a pleasure to describe these proposals, but my description would add nothing to what they said. It would be good, though, to understand more clearly the relation of the two papers. This is not clear to me now, even though Arima touched upon this subject elsewhere.

The developments which were just reviewed were essentially independent of the detailed nature of the interaction between nuclei. This is most evident where the Talmi rules are concerned, but is essentially true also of the other contributions which were mentioned. The question then arises: how accurate is the independent particle model? What fraction of the total nuclear wave function can be represented by a single Slater determinant of single-particle functions with definite j values, or by a linear combination of such Slater determinants differing from the first one in the m-values (projections of j in a given direction) of the single-particle functions? I shall call linear combinations of the nature just specified standard linear combinations; they are used also in atomic spectroscopy. In fact, Slater's original paper is based on their use. The first question, i.e., the question of how closely a standard linear combination of Slater determinants can approximate the true wave function of the stationary states, remains largely unanswered. I will say a few words, however, on the second question: what is the general nature of the admixtures to the best standard linear combination of Slater determinants. There is a third question, which should have been the first question and which was raised by both Breit and Palevsky: to what extent can the state of nucleus be represented by a wave function which depends only on the coordinates of nucleons. But let us postpone the discussion of this question until the end.

Turning then to the second question: there appear to be two types of admixtures to the standard linear combination of Slater determinants. The first type of admixtures are similar Slater determinants but with single-particle functions which represent nearby excited states, i.e., belong to slightly excited configurations. The two-particle-one-hole, etc., states belong in this category. One can hope to obtain them by ordinary perturbation methods. The second type of admixtures describe principally short-range correlations, for example, those resulting from the short-range repulsion. The so-called Jastrow functions give a visualizable description of the resulting deformation of the original Slater determinants; but the short-range correlations are taken into account, as a rule, by replacing the actual interaction by an effective interaction, preferably using Brueckner's method. If the second kind of modification of the standard independent-particle wave function were represented by additional Slater determinants, there would be a very great number of these representing very highly excited configurations, each appearing with a small coefficient. It would be interesting to know what fraction of the total wave function consists of these functions, and the answer must be contained implicitly in the papers of Brueckner, Bethe, and their collaborators. It is not known generally. It would also be interesting to know what Jastrow (correlation) function describes the modification of the standard Slater determinants introduced by the second type of admixtures. This also is unknown to me.

When one speaks of admixtures to the Slater determinants representing a given basic configuration, one usually disregards this second type of admixtures. As Kerman also observed, this is true of Stein's estimates; these refer only to the first kind of admixtures. Stein's contribution did, however, give some interesting information on the first kind of admixtures, although no entirely definitive one. He quoted earlier estimates ranging from $1/2$ to 9 and even to 44% in the doubly-magic nucleus ^{208}Pb, but his own work seems to support the intermediate figures of the order of 5%. This is much lower than most of us expected. Stein found that one particular 2-particle-2-hole state constituted 0.9% of the total wave function. Naturally, the admixtures of the first type will play a greater role in non-magic nuclei.

The theories the conclusions of which were discussed so far belong rather to the category which I characterized before as phenomenological. The theories which I call basic have, of course, a tougher job and even though progress in the field is great, the results remain somewhat controversial. There was some discussion of the binding energy of a nucleon in nuclear matter and the deficiency of the value calculated for this quantity. What surprised most of us, I believe, is the proximity of this value to the observed one. This should be considered as a real success of the theory. The means toward this and other successes, in the hands of Brown, Kerman, Elliott, and many others not participating in our conference, is the accounting for the second type of deviation from the independent particle model by replacing the actual interaction by an effective interaction. This is a much smoother interaction than the actual one and the solution of the Schrödinger equation with this interaction would contain much smaller departures of the second type from the standard linear combination of Slater determinants. On the other hand, as far as the energy values of the low-lying levels are concerned, those given by the effective interaction and by the true interaction should be the same.

Since the second kind of departure from the independent-particle model is greatly diminished by the use of an effective interaction, it becomes much more reasonable to calculate these characteristic values starting from a standard linear combination of Slater determinants and modifying this by taking into account only the first type of departure from the model. This can be done by more or less standard perturbation methods and leads to the modification of the standard linear combination by the addition of not many and not too heavily excited configurations, some-particle-some-hole states, etc. There are some strong indications that the energy values of these states approximate the observed energy values closely. Nevertheless, I do not think I am unreasonable when saying that the situation is not entirely clear. Neither can one say that the effective interaction has reached its final form. Dr. Elliott obtains the matrix elements of the effective potential directly from the phase shifts. Dr. Brown derived the effective potential from the Hamada-Johnston form by Brueckner's method and considered three-body forces important while Dr. Brueckner characterized them as insignificant. One can well hope that the agreement between effective interaction theory and experiment will improve as the form of the interaction is refined and as experience in its use accumulates.

Let me admit, on the other hand, that the possibility, not entirely excluded, that the effective interaction now in use is not of the saturating type, perturbs me greatly.* Surely, we do not want to use an interaction which, if applied consistently, gives absurd results – we do not want to use it even if, for the present, we do not expect to use it entirely consistently. There are too many not altogether consistent methods of calculation and it is difficult to specify the one which should be used. It seems to me, therefore, that proving that the binding energy accurately calculated with the effective interaction is proportional asymptotically only to the first power of the number of nucleons is an urgent task.

The possibility of replacing the real interaction in the Hamiltonian by an effective interaction is a very important step but it is not the only, nor even the most widely used, departure from the standard Hartree-Fock-Slater scheme of nuclear theory. The most fruitful departure, as far as direct contact with experimental results is concerned, was suggested by the frequent occurrence of rotational spectra and is patterned on Löwdin's modification of that scheme, originally proposed by him for the quantum mechanical treatment of molecules. It is based on the concept of an intrinsic state, i.e., a Slater determinant which is a solution of the Hartree-Fock equations, the single-particle states of which do not, however, necessarily conform with the symmetry of the problem. In our case of spherical symmetry, they do not have definite j-values but may be quite asymmetrical. As a rule, however, they retain axial symmetry. One forms from the Slater determinants obtained in this way wave functions with definite J-values just as one forms them from the usual Slater determinants whose single particle functions have definite j-values: one rotates them in space and superposes the resulting wave functions so that the superpositions have definite J-values (total angular momenta).

Thus, the theory of the intrinsic state differs conceptually only slightly from the standard Hartree-Fock theory. As far as consequences are concerned, the principal difference is in the number of terms resulting from one solution of the Hartree-Fock equations. This is finite in the standard theory, there is an infinity of "rotational" levels in the intrinsic state theory. The multiplicity of the rotational bands, present in almost every nuclear spectrum, is the principal support of this theory.

The difficulties of interpretation and differences from experiment of the conclusions arrived at with the intrinsic state model may well originate from the fact that one does not start out with a solution of the Hartree-Fock equations but assumes, as a rule, arbitrarily that the average potential is that of an anisotropic oscillator and uses the single-particle states appropriate for this

---

* It just occurs to me that a recent article of Richard L. Hall may be useful in this connection.

potential in the original Slater determinant. (A similar remark applies to many calculations based on the original shell model.) At large deformations and low angular momenta, the kinetic energies of the levels are proportional to $J(J+1)$ independently of this assumption, and this is a very important result of the theory. However, at low deformations and high J the result depends very much on the single-particle functions of the original Slater determinant, particularly on their behavior at large distances from the center. If the single-particle wave functions are sharply cut off at large r, the kinetic energy increases, at least as a rule, faster than $J(J+1)$ – in contradiction to the experimental findings. This has been pointed by Y. Sharon in a contribution to our conference. He therefore concludes that the change in the potential energy must also be taken into account and by assuming simple interactions, he does indeed eliminate the discrepancy. It seems to me that the situation would be quite different if the single-particle functions of the intrinsic state had a tail beyond the nuclear radius. The nuclear volume calculated with such an intrinsic state would increase with increasing angular momentum J and the kinetic energy would increase slower than $J(J+1)$. At very high J, the kinetic energy would be linear in J if the single-particle states are generated by an oscillator potential; the coefficient of J in that region is just one quantum of oscillator energy. The magnitude of J at which this occurs depends on the deformation: the more nearly spherical the nucleus is, the lower the J-value at which the $J(J+1)$ formula for the kinetic energy becomes invalid. In the limiting case of very small deformation, the energy is proportional to J even at the lowest J-values.

We were reminded by Dr. Barrett that the conditions on the surface of the nucleus are far from simple and the details of the preceding discussion should not be taken too seriously. It is true, however, that at high J the conditions near the surface of the nucleus become important and that the mass distribution of the nucleus will then depend on J if the intrinsic state theory is literally accepted. We were reminded by Le Tourneux that it need not be accepted literally, that the intrinsic state may well be a function of J. The same observation is implicit in the article of Mariscotti, Sharff-Goldhaber, and Buck as mentioned by Elbeck. Mariscotti, Sharff-Goldhaber and Buck are one step further than Le Tourneux in the phenomenological direction. They attribute to the intrinsic state an elastic resistance against deformation and can account for the deviation from the $J(J+1)$ rule by means of the additional parameter characterizing this elastic resistance – the elastic constant being taken from the spectra themselves.

What are the principal problems of the intrinsic state model? It was mentioned before that the intrinsic state should be a solution of the Hartree-Fock equations but is in practice not obtained that way. This is surely excusable in view of the complexity of the nucleon interaction. The ease with which calculations with the oscillator potential can be carried out is an added incentive to use that potential. It is perturbing, on the other hand, that the calculation of the energy of the "projected" states, i.e., the states with definite J obtained from the intrinsic state, takes only the kinetic energy into account. Such a procedure is justifiable (as a first approximation) in the theory of molecules because, if ψ_0

is the intrinsic state then $O_R\psi_0$, that is ψ_0 rotated by R, is orthogonal to both ψ_0 and to $V\psi_0$ unless R is a rotation about the internuclear axis: here V is the operator of the interaction. It is easy to see then that the expectation value of the interaction energy is the same for ψ_0 as for the various states with definite J. The reason for the orthogonality is that the positions of the nuclei are different in ψ_0 and $O_R\psi_0$ (unless R is a rotation about the internuclear axis) and the molecular V does not change the positions of the nuclei. Neither of these orthogonalities holds in nuclear theory and $O_R\psi_0$ is not even nearly orthogonal to ψ_0 or $V\psi_0$ unless ψ_0 is a strongly deformed state and the rotation R is far from the unit element. It is true that, unless one uses a very much simplified interaction (as Sharon does, for instance,) the potential energy as a function of J is difficult to calculate, and the result will remain somewhat uncertain. Nevertheless, a bit more attention to this point would be desirable.

In an area in which one has rigorous mathematical methods or an understanding of the accuracy of approximations, it is not necessary to approach a problem from several angles. This situation does not yet prevail in nuclear theory and one would like to see the same problem occasionally approached by means of different theories – after all, they are all approximations to the same quantum mechanical equations. It seems to me that the wave functions which the intrinsic state theory yields for low values of J should be obtainable, with a little more trouble, from ordinary shell theory. Even a strongly deformed 1σ oscillator wave function, with its major axis 1.5 times its minor axis, can be obtained as superposition of $1s$ and $1d$ wave functions of an isotropic oscillator with an accuracy higher than 95 %. Obtaining similar answers on the basis of two theories, at least for some relatively simple problems, would be very reassuring for those of us who are not directly involved in one or another type of calculation and, I suspect, would be reassuring also for those who are.

There is a continuous transition from the intrinsic state theory to the collective model, and the former is, in a sense, a child of the latter. One might consider it as a more mathematical elaboration thereof. The collective model is itself a child of the liquid drop model. Similar to the latter, it appeals more to the imagination than the independent particle and the intrinsic state models. It has been very successful in many ways but is, just because of its appeal to the imagination, more difficult to discuss than the other models. If the present discussion is inadequate, there is some comfort in the fact that the first volume of the theory's originators, A. Bohr and B. Mottelson, is available. The basic ideas are discussed in a short article, by Baranger and Sorensen, in the *Scientific American*.

Strutinsky's suggestion that the energies of heavy, fissionable nuclei have two minima if considered as functions of the deformation is the most interesting idea suggested in recent years by the collective model and may be the most interesting idea in theoretical nuclear physics proposed in the last few years. It has stimulated many similar suggestions and we heard some of them in the course of our conference.

The collective model's explanation of the vibrational spectra is different from that discussed before. It suggests that, so to say, the intrinsic state vibrates, that if we take an instantaneous picture of the nucleus in its second vibrational state, we may find the nucleons to form some of the time a prolate, at other times an oblate shaped cloud. This picture seems to be different from that furnished by the intrinsic state model: one would expect that this would project always at least roughly the same instantaneous picture, either prolate, or oblate. Actually, there is no contradiction: if the deformation is small, the anisotropic part of the Gaussians of the oscillator wave functions can be expanded into a power series. In the simplest case of no angular momentum about the symmetry axis, the $J = 2$ projection then has a factor, additional to the $J = 0$ state, of $\sum_i r_i^2 \cdot P_2(\theta_i)$ and the absolute value of this is large both when all θ_i are close to 0 or π, and also when all the θ_i are close to $\pi/2$.

A novel element was introduced into the discussion by the phenomenon of β vibrations. There are two rotational bands, displaced from each other, and optical transitions are allowed, though not extraordinarily strong, between them. The collective model's interpretation is that in the states of the upper band the shape of the nucleon cloud oscillates around the equilibrium shape, whereas no such oscillation takes place in the states of the lower band. If one tries to visualize such explanations in terms of a wave function in configuration space, one cannot help wondering how well the shape of the nucleon cloud is defined. The root mean square quadrupole moment of A particles moving independently and at random in a sphere of radius r is $(12/7)^{1/2}r^2$ and this is only about 5 times smaller than the largest quadrupole moments. It does not seem that the nuclear surface is sufficiently well defined to be regarded quite on a par with the surface of a liquid. It is true that the large quadrupole deformations have always the same sign, i.e., give the same average shape, whereas the fluctuations just mentioned have random signs. It also seems to me that the correlations, i.e., the deviations from the Hartree-Fock picture, decrease the fluctuations and give the nucleus a more rigid shape. This last point was mentioned also by Sorensen – it is about the only point in which I would like to differ from Dr. Vogt's introduction address. In spite of this increased rigidity of the nuclear shape, the fact that, even in heavy nuclei, the nuclear surface cannot be very sharply defined, deserves some attention. Without such a discussion, the limits of the validity of the collective model cannot be very clear.

There is one more nuclear model which was mentioned repeatedly and which surely has a very great "truth content": the α particle model. However, except for this fact, there is nothing that I can say about it. Let me instead, before closing the subject of nuclear models, repeat a point made very strongly by Breit and mentioned also by Grodzins. This point is actually very old and also very negative: we cannot be sure that nuclei are best described by a wave function in the configuration space of the constituent nucleons. It is entirely possible that a description in Fock space, with a sequence of wave functions, each corresponding to a definite number and character of mesons present, would give a more natural description. It would be very awkward to describe a metal in terms of a wave function depending only on the coordinates of the atoms as constituent entities. The fact that the number of glue particles, electrons in one case and mesons in the other, is not constant in the latter case should certainly not matter. The complexity of the interaction between the nucleons may not be very different from that of the similarly calculated interaction between metallic atoms and, as Breit told us, the adiabatic approximation is rather worse in the case of nucleons. However, this is just a dream for the present.

Before concluding, let me respond to a suggestion made by Dr. Ewan. He asked me to propose a few experiments the results of which I would consider interesting. He asked me to do this even though I am sure he was well aware that experiments proposed by theoreticians are, as a rule, impossible to perform. Well, here are three proposals:

(a) It would be useful to measure all reaction cross sections leading to the same compound nucleus, whether or not the reactions proceed via the compound nucleus mechanism. Thus, the complex ^{24}Mg is obtained in the reactions ^{23}Na $+ {}^1$H, ^{20}Ne $+ {}^4$He, ^{21}Ne $+ {}^3$He, ^{12}C $+ {}^{12}$C, and several others. If not only the cross sections were known for processes in which these pairs are products, but also for processes in which they are reacting partners, we would see whether or not our reaction theories induce us to introduce too many parameters, perhaps in the form of simulated resonances. (Naturally, there are reciprocal relations between the cross sections so that only about half of them would need to be measured.)

(b) As we hinted before, and as was suggested also by Dr. Litherland, it would be very informative to follow the rotational spectrum to its very end. According to the standard shell model, there should be a definite break at the highest angular momentum consistent with the configuration originating the rotational band. However, the earlier discussion shows that there is also a good deal of ambiguity in the prediction of the other theories.

(c) When one reads of the miracles of modern experimental methods, one cannot help wondering whether it would not be possible to obtain an instantaneous picture of the nuclear cloud in the normal and also in some excited states. The significance of such pictures was mentioned several times in the course of the discussion of the collective model. Any information, even though incomplete, would be useful and interesting.

In conclusion: you will have noted that this review did not cover the subject of nuclear reactions, even while emphasizing their importance. Nuclear reactions form very important means in the research on the structure of nuclei but this review is long enough even when restricted to problems of nuclar structure. I wish to apologize to those whose results and conclusions I overlooked or neglected. It would have been very difficult to appreciate fully all that we heard during this week. In addition, I must admit that there were a few presentations which were too difficult for me to follow; I am looking forward to reading them in print. We all appreciate the care with which the papers were presented and the readiness of the speakers to answer all our questions.

I am indebted to Dr. Cusson for careful reading of the manuscript and for making a number of valuable suggestions.

Summary of the Conference
(Polarization Phenomena, Madison 1971)

Eugene P. Wigner

Polarization Phenomena in Nuclear Reactions, H. H. Barschall and W. Haeberli (eds.).
University of Wisconsin Press, Madison 1971, pp. 389–395

Before starting on this review, let me express my pleasure at having attended a conference together with so many interested people with such a wide range of points of view. For some of us, the wealth of the material presented, the variety of the ideas and experimental arrangements, was the greatest surprise. Let me also mention the usefulness of private conversations, which are part of the attraction of conferences. I am particularly indebted to Drs. Darden and Weidenmüller for illuminating discussions.

The following review of the conference will consist of two parts. The first part will be concerned with questions of central theoretical interest. Not all of these relate directly and exclusively to polarization phenomena; for many of them, the study of the polarization phenomena is only a tool to obtain answers to questions in which one would be interested anyway, even if polarization phenomena did not exist. The second part of the review will recall some of the experimental procedures and phenomena which have been described and which are so striking and ingenious that even a theoretician cannot forget them.

1. Polarization Phenomena and the Collision Matrix

All of us theoreticians know that the collision is only incompletely described by giving the colliding and the separating particles, together with their momenta, that the complete description of the initial and of the final states includes the dependence on the spins or, more modernly, helicities. However, one is often inclined to disregard this dependence, or at least to postpone its consideration to some indefinite future. It was good, therefore, to learn that a beautiful theory of this dependence has been developed and, even more important, that a wealth of experimental material has been created to be interpreted by that theory.

Since much of the notation used both by theoreticians and by expermentalists is different from that used in other parts of physics, in particular in particle theory, it may be useful if a contact with the other notation is established. This will be restricted to the case in which only two particles collide and the reaction can also yield only two particles, not three or more. In this case, the variable specifying the columns of the collision matrix are, first, the description of the colliding particles and their states of excitation. Next come the momenta of the particles and, finally, their helicities, that is, their angular momenta about their directions of motion. These last variables can assume only discrete values, each $2S + 1$ values for a particle with spin S so that there correspond $(2S_1 + 1)(2S_2 + 1)$ columns to each pair of particles colliding with definite momenta. The rows of the collision matrix are similarly labeled, with the description of the reaction products, their momenta, and their helicities. Hence, to each pair of momenta of each type of reacting particles and reaction products there corresponds a submatrix of the collision matrix with $(2S_1 + 1)(2S_2 + 1)$ columns and $(2S_1' + 1)(2S_2' + 1)$ rows if S_1' and S_2' are the spins of the reaction products. The criterion for time inversion invariance, for instance, implies the symmetry with respect to the interchange of all labels of the rows with all labels of the columns, including the labels of the particles. Naturally, the elements of the collision matrix vanish unless the sum of the momenta of initial and final states are the same, and the collision matrix is also rotationally and Galilei or Lorentz invariant, depending on whether one uses classical or relativistic theories. In the simplest case of zero spins this means that the collision matrix depends only on the scalar product of the momenta of the incoming particles – this specifies the energy of the collision – and the scalar product of one of the initial and one of the final momenta – this specifies the momentum transfer. The elements of the collision matrix with rows and columns referring to different total momenta are zero.

The notation of the collision matrix, usual in other parts of physics, has been discussed in such detail because it would appear to me to be useful if a review article were to be written, comparing the notation just described with that used in the studies which we heard discussed. Such a review might well derive the consequences of the various conservation laws, space and time inversions, as they manifest themselves in the two descriptions. It might also contain the observation of Dr. Simonius, who answered at least partially a question which has puzzled many of us for some time: to what extent is the collision matrix truly observable? It may be recalled that Heisenberg's motivation for dealing with the collision matrix was that he wanted to use quantities which are truly observable. In fact, the absolute values of the off-diagonal elements of the collision matrix are reaction cross sections and hence observable. Similarly, the real parts of the diagonal elements are essentially scattering cross sections and also observable. However, one can have serious doubts on the observability of the complex phases of the off-diagonal elements since they relate to relative phases of different reaction products and such phases, for instance the phase of an $^3H + {}^1H$ pair relative to the phase of an $^2H + {}^2H$ pair, are apparently unobservable. (I have some reason for saying "apparently".) What Dr. Simonius has taught us is that the relative phase of the $^3H + {}^1H$ pair with respect to the $^2H + {}^2H$ pair may be unobservable, but the relative phases of the different polarization states of the $^3H + {}^1H$ pair, and of the $^2H + {}^2H$ pair, are observable. This means, more generally, that the relative phases of the elements of the aforementioned submatrix of the collision matrix, with $(2S_1 + 1)(2S_2 + 1)$ columns and $(2S_1' + 1)(2S_2' + 1)$ rows, are observable. The absolute values are, of course, all observable so that the number of unobservable phases in any column of the collision matrix is equal to the different reaction products which the collision can yield – different states of polarization not counting as different products. Similarly, the only phases in a row of the collision matrix which he did not show to be observable are the phases of the elements referring to different colliding pairs, again a change in the state of polarization not constituting a change in the pair in the sense considered. Naturally, it was assumed by Simonius that all the various measurements of polarizations and asymmetries which were discussed can truly be carried out and also that the particles of the colliding pair can be produced in all states of polarization. However, the measurements which we have heard about from the Los Alamos group and from Dr. Postma, give clear justification for these assumptions. The Los Alamos groups call the reactions in question "polarization transfer reactions" – they are the ones assumed to be feasible by Dr. Simonius' arguments.

2. The Distorted Wave Born Approximation

The distorted wave Born approximation (DWBA) played a very important role in the discussions we have heard. For most of us who are not active in this field, this approximation method is a bit foreign. The authors do not always betray how many parameters of their distorted waves were taken from the experiments they are interpreting, and how well these parameters reproduce the elastic scattering cross sections. This is a criticism. A glance at the angular dependence of some of the cross sections, experimentally determined, does on the other hand clearly suggest the validity of a picture at least akin to that used. The oscillations in the angular dependence of the cross sections strongly indicate a diffraction phenomenon. All this has been well known for some time. What is novel is the new measure of the validity of this DWBA method because a new set of experimental data is to be explained with the same theory which was adjusted, in the past, to a smaller number of data. The new data are the results of polarization measurements and of measurements of cross sections of polarized particles.

The preceding statement does contain some exaggeration. Some new parameters unavoidably had to be introduced to calculate the results obtained by polarization measurements. Also, the theory was modified somewhat, but modified in a very natural way, by taking into account the shape dependence of the spin-orbit force. Dr. Satchler called the modified spin-orbit force the "full Thomas form". There is agreement that the DWBA, after these rather minor modifications, reproduces the results of polarization experiments at least qualitatively. In view of the complexity of the experimental results, the strong fluctuations of the polarizations as functions of the directions of the outgoing particles, this is no small success. Nevertheless, the agreement between polarization experiments and DWBA theory is not quantitative and the deviations, at larger deflection angles, are considerable. Gorlov, Satchler, Blair, Graw, Goldfarb, Santos, Haeberli, Grotowski, and Mayer all agree on this. It is also perturbing that it was necessary, in some cases, to assume a greater deformation of the spin-orbit force than is present in the central force. Blair advocates a deformation 1.5 times too large in ^{28}Si. To quote Dr. Goldfarb,

"polarization phenomena present the DWBA with a very difficult challenge".
I do hope, though, and similar views were voiced by others, that the challenge
will be met successfully.

At this point, it would be difficult for me to resist expressing my pleasure
at the measurements, reported by Haeberli, for at least one set of reactions
of all the cross sections and all the polarizations not involving more than two
particles. The reactions in question lead to ^{54}Fe as compound, the compound
being obtainable by bringing ^{52}Cr into collision with ^{2}H, or ^{53}Cr with a proton.
Altogether, four processes were investigated at Wisconsin and at Saclay – two
scattering processes and two reactions. Such complete investigations are more
likely to provide convincing evidence for or against a reaction theory than the
measurements of only some of the processes, because when more processes —
that is, more cross sections and polarizations – are measured, the number of
adjustable parameters does not increase but the number of results does. I admit
that I have said this before.

3. Nuclear Forces and Spin-Orbit Interaction

Dr. Signell's discussion of this subject caused a good deal of apprehension and
raised some doubts concerning ideas which were often considered established.
He mentioned, in particular, calculations by Landé and Svenne, who, using
accepted and in many applications apparently successful potentials, obtained
for some nuclei a density about threee times too high and a binding energy
about 25 % too large. This seems to present a very serious problem. Too large
a binding energy at too high a density indicates a very serious weakness of the
theory. This may be either in the interaction – in the Hamada-Johnston, or Yale,
or Tabakin potentials – or in the calculation of the properties of nuclei, using
these potentials. This is the Hartree-Fock method, based on Brueckner-Bethe
theory, and, considering the success of the potentials in question in explaining
various properties of spectra, one will suspect this method in the first place.
Needless to say, this view is put forward with much hesitation, and surely
this writer cannot propose an alternate for the Brueckner-Bethe theory. It
would, however, be wrong to leave this difficulty unmentioned. Possibly we must
revert to the point which Breit has reemphasized, that the nucleus contains
not only protons and neutrons but also a cloud of pions and other mesons with
fluctuating densities so that a more natural description of the nucleus would
explicitly refer to these additional particles. This would mean, presumably, the
introduction of a Fock space as far as pions and other mesons are concerned.

Altogether, Signell's report on what we do and what we do not understand
about the origin of the spin-orbit interaction of nucleons was quite pessimistic.
Concerning the description of the nucleon-nucleon scattering experiments, I am
a bit disappointed that these are not described, for each angular momentum
$J \geq 1$ and each parity, by 2 by 2 matrices rather than by phase shifts. If we do
not yet have all the data to do this, it would be better to admit it rather than
to use an incomplete description. Dr. Baker made a start in this direction.

4. Charge Independence

The problem of obtaining the consequences of the charge independence of nuclear forces – one should say, the near charge independence of nuclear forces – has been with us for a long time. Weidenmüller discussed it in passing, and it was taken up again in McKee's report. We learned that many (d, p) cross sections are similar to the (d, n) cross sections on the same element, but this is not very relevant in the present context unless the element on which the cross section is measured contains an equal number of protons and neutrons. It is relevant, on the other hand, that the $^3H(p, p)$ and the $^3H(n, n)$ cross sections are quite different – surely this statement was meant to imply that they are different even after a correction is applied on account of the Coulomb forces. This is definitely in conflict with what one would have expected on the basis of charge independence. Finally, McKee mentioned that the polarizations resulting from the $^2H + {}^2H \rightarrow {}^3H + n$ and the $^2H + {}^2H \rightarrow {}^3H + {}^1H$ reactions are quite different, the former being significantly lower. We also heard, from Drs. Perey and Greenlees, that the effective interaction of at least two nuclei with equal proton and neutron contents, of ^{16}O and of ^{40}Ca, is greater for protons than for neutrons – again, after a correction for the Coulomb effect has been applied. This seems to contradict the conclusion which one might arrive at from the validity of the isotopic spin concept, but Dr. Satchler later called my attention to an additional correction which may restore the agreement between experiment and theory. Nevertheless, these are not good tidings for those of us who would like to see the validity of the isotopic spin quantum number confirmed and who would like to apply this concept in an easy and effortless way. There are so many phenomena, including the mere existence of the analog states, with energies quite accurately given by the charge independence assumption, that the validity of the isotopic spin concept seems to us assured. In a private conversation, Dr. McCullen, of the University of Arizona, told me about another recent confirmation. This refers to the β transition probabilities from the ground state of ^{42}Sc, a nucleus with $J = 0$, $T = 1$ ($T_\zeta = 0$) to two $J = 0$ states of ^{42}Ca. The ground state of this latter nucleus is the $T_\zeta = 1$ member of the isotopic spin triplet of which the normal state of ^{42}Sc is the $T_\zeta = 0$ member. The other $J = 0$ state of ^{42}Ca belongs to another isotopic triplet. Hence, if the isotopic spin is a valid quantum number, the transition from ^{42}Sc to the normal state of ^{42}Ca should be favored, that to the excited state forbidden. In fact, Dr. McCullen told me, it is at least $10\,000$ times weaker. These β transitions are suited for testing the validity of the isotopic spin concept because, between two $J = 0$ states, only Fermi transitions are possible. If the isotopic spin is a valid quantum number, these should take place only between members of the same T-multiplet – at least if the wave lengths of the electron and of the neutrino which are emitted are large as compared with the size of the nuclei involved. Thus, Dr. McCullen's observation confirms that validity.

5. Remarkable and Novel Experimental Methods and Observations

Dr. Weidenmüller mentioned Lobashov's measurement of the circular polarization of the $5/2^+ \to 7/2^+$ transition in ^{181}Ta. It amounted to about 10^{-6}. If parity were a truly rigorous concept, the radiation of such a transition would show no circular polarization at all. The degree of circular polarization, around 10^{-6}, is of the order which the weak interaction might well generate.

Kaminsky's method for producing polarized particles is surely the last method a non-professional would have thought of. It will be recalled that he passes a beam of particles through a crystal, its direction parallel to one of the crystal axes. Under such conditions the energy loss is, as Dr. Kaminsky has shown, a good deal smaller than for a beam of particles of random direction. This is the so-called channeling phenomenon. Furthermore, if the crystal is magnetized – Kaminsky used a nickel crystal – the emerging beam of particles is polarized. In fact, it is polarized more strongly than was expected.

Dr. Ebel gave at least a tentative explanation for this phenomenon. He said that the traveling particle picks up an electron from the nickel, interacts with its spin, then loses it again to the crystal. This process is gone through several times. On each occasion, the electron is more likely to be picked up from the filled electron band and can be discharged most easily into the unfilled band so that an angular momentum is transferred to the crystal. This must come, ultimately, from the particle traveling through the crystal, which as a result acquires a polarization. If this explanation is correct, the polarization is related principally to the magnetization of the crystal, and its direction with respect to the direction of the traveling particle can be varied greatly. This would render the method a very flexible one. What is puzzling about Ebel's explanation is that one does not see why the method works only if the direction of the particle coincides with one of the crystal axes or, more generally, is a channeling direction.

Kaminsky's method of producing a polarized beam of particles is not the only experiment making use of the interaction with a solid; Brooks' anthracene crystal provides another one, and it too is a very clever one.

I now come to the method of producing a polarized beam of particles, protons or deuterons, which we believe we understand fully and which seems to be the most efficient one. I am referring to Donnally's Lambshift method, which he described so clearly that it is unnecessary to repeat it. Further, as we have learned from Brückman, Glavish, and Walter, the polarization of the protons or deuterons can be transferred by collisions also to other particles. This has a varying effectiveness – generally a rather low one – but Dr. Walter, quoting results obtained by Dr. Simmons and his associates at Los Alamos, mentioned that a fully polarized deuteron beam can produce neutrons with a polarization of 0.9.

Dr. Jeffries discussed other methods of producing polarized beams of particles. Some of these were skillful modifications of older methods, but I found his dynamic polarization method particularly impressive. It also seems very effective. Nevertheless, the problem of producing good beams of particles with arbitrary energy, and also in an arbitrary state of polarization, is far from being completely solved. We can except to hear more about this question at future polarization conferencs.

The present review of our conference is an incomplete one in many respects; its writer can only hope that he will be forgiven by those whose work and results were left unmentioned. Most of the speakers excused themselves for the incompleteness of their discussions, caused by the necessary brevity of their presentations. It is only appropriate that the summary shares this weakness of incompleteness, and the apologies therefore. Even more appropriate is it to thank Dr. H. H. Barschall for his indefatigable help when preparing the final version of this summary. There would be many more mistakes in it had he not stood by.

Introductory Talk

E. Wigner

Garg, J.B. (ed.) Statistical Properties of Nuclei. Plenum Press, New York 1972, pp. 7–23
(Reset by Springer-Verlag for this volume)

1. Introduction and Specification of Our Subject

I have worked, in the course of time, on sufficiently many books so that I should
have realized that the Introduction is the part of the book which one writes
last. If there will be any merit in my Introductory Talk today, this will be
due mainly to the help which I received from the committee which organized
this symposium and from the participants who submitted papers – papers
short enough so that I could read them. Incidentally, their rather large number
indicates the magnitude of the interest in our subject, and even a quick perusal
of the papers submitted shows the variety of the problems tackled.

Perhaps the first thing we should admit is that our subject is not precisely
defined. In statistical mechanics, the subject of which *is* clearly defined, we are
interested in the time averaged properties of systems large or small, which have
a definite energy and, parenthetically, which are at rest and not in rotation. The
quasi-ergodic theorem, if valid, assures us that these time averaged properties
depend only on the energy and are independent of the other initial conditions.
Thus the problem is quite clearly defined. On the other hand, if we wish to
specify the subject of Statistical Properties of Nuclei, we must say that we
consider situations in which we are not interested in as detailed a picture of
the nucleus as one is generally interested in physics but tries to find properties
and rules which are reasonably simple, and we believe very interesting, and
which are very general, shared by most all nuclei under appropriate conditions.
The appropriate condition is, in practice, principally a rather high energy even
though, as long as we remain nuclear physicists, we do not go to extremes in
this regard, and stay away from energy regions in which the creation of the
particle physicists's particles become an important process.

To how much detail our interest extends is, of course, a question which
different physicists will answer differently. The analogue states give herefor an
example. These are members of an isotopic spin multiplet the extreme member
of which, that is the member with the largest neutron to proton ratio, is either
the normal state, or a state of low excitation, of the nucleus which has one more
neutron, and one less proton, than the nucleus with which we are concerned.
The member of the isotopic spin multiplet in which we are interested has,
however, a rather high excitation energy – of the order of 9 MeV in the region
of Fe. Since the state in question has a wave function resembling that of a

low energy state, it has high transition probabilities to the states which truly have low energies. The point which I am trying to make is that one may be interested in the explanation of these high transition probabilities, or not be interested in it and consider such a high transition probability as part of the fluctuations of the transition probabilities. In the latter case, the excessive transition probability is a subject of the statistical theory of nuclei, in the former case it is not – for most of us it is not. However, if we go to even heavier nuclei, the analogue state is not a single state but is dissolved into a reasonably large number of states nearby, the isotopic spin of these states having been, before the dissolution of that state, that of the normal state. In such a case, there is a group of states, rather than only one, with rather large transition probabilities to low lying states and one is more inclined to consider these large transition probabilities as fluctuations and hence subjects of the statistical theory. Instead of the analogue states, I could have quoted Feshbach's doorway states which, in heavy nuclei, create a situation qualitatively similar to that of analogue states. Both examples illustrate the fuzziness of the boundary of the statistical theory – this boundary depends on our personal interest for details.

2. The Statistical Theories' Ensembles of Hamiltonians

Some of the very interesting laws of statistical mechanics can be derived from the simple postulate that the equations of motion have a Hamiltonian form. The entropy theorem and the equipartition theorem are in this category. A great deal of other work in statistical mechanics is based on a reasonably detailed knowledge of the Hamiltonian which is, in most practical instances, known. There are, to my knowledge, no theorems in the statistical theory of nuclei which would have as general a basis as the entropy or equipartition theorems and what is worse, we do not know the nuclear Hamiltonian. Further, some relevant properties of the Hamiltonian are quite complicated. The spectrum of the Hamiltonian has a lower bound but extends, on the positive side, to infinity. Near the lower bound, the spectrum is discrete but there is a threshold at which a continuous spectrum sets in. The character, that is multiplicity, of the continuous spectrum changes, however, at every threshold drastically. These are properties of the Hamiltonian which we know; as was mentioned before, we do not know its exact form. It is not surprising, therefore, that one of the chief quandaries of the statistical theory concerns the properties of the Hamiltonian to be made use of.

The Hamiltonian is, of course, a matrix in Hilbert space, and the most natural set of properties to be made use of are those shared by practically all matrices, or rather self-adjoint matrices, in Hilbert space. This then leads to the concept of ensembles of self-adjoint, or of real symmetric, matrices in Hilbert space that is a definition of the measure for such matrices in Hilbert space so that the concept of "vast majority of all self-adjoint matrices" or of "practically all self-adjoint matrices" be mathematically defined.

The definition of such a measure seems to be strongly facilitated by the fact that, once we restrict ourselves to a single angular momentum and zero spatial momentum, there seems to be no coordinate system in Hilbert space which plays a preferred role, except the one in which the Hamiltonian is diagonal. Since we do not wish to define the measure of matrices with respect to that coordinate system, it is natural to demand that the measure be invariant with respect to unitary transformations. This leads to a measure which can be an arbitrary function of the invariants of the matrix, multiplied with the differentials of the independent components of the matrix elements. If one further demands, as has been done in their well-known paper by Porter and Rosenzweig, that the probabilities of the independent components of the matrix elements be independent of each other, one obtains, essentially, the Wishart distribution. More precisely, the ensemble obtained is such that the number of matrices within unit interval of the independent components of the matrix elements M_{ik} is proportional to

$$P = \exp\left(\alpha \sum M_{ii} - \beta \sum |M_{ik}|^2\right) \tag{1}$$

α and β being arbitrary constants. (See Fig. 1.)

The trouble with this distribution is that practically none of its matrices has characteristics similar to those observed for actual Hamiltonians. In particular, and I find this most decisive, the density of the characteristic values of most matrices, as function of energy has, in the neighborhood of the lower bound (which can be arbitrarily adjusted by a proper choice of α and β) a negative second derivative, practically right from the start. This has been most conclusively demonstrated by B. Bronk. The actual density, of course, has a positive second derivative with respect to energy. Since the ensemble with the density (1), which is essentially identical with the Wishart ensemble (long known to the mathematicians) does not give the actual density distribution of the characteristic values, it is natural to question all other consequences derived theoretically by means thereof. This has been done most eloquently by Uhlenbeck. As a result, Dyson who, along with Gaudin-Mehta, developed most consequences of the Wishart model, has recently decisively turned away from it.

The aforementioned facts have been known for a long time but attention has been focussed on them only a recent years, but in recent years attention has been focussed on them increasingly. No true solution has been found to date but we do have at least two proposals to which I wish to add another one. Let me, however, mention first the others.

(a) Return to the Independent Particle Model

A statistical theorem in nuclear physics was, of course, first postulated by Weisskopf. However, just about the same time, Bethe proposed an expression for the density of energy levels in nuclei, and this expression was based on an extreme independent particle model, on the model of the degenerate Fermi gas. As we know, this model gives a fair picture of the density of levels, as function of energy. In particular, except for the very lowest energy region, the second derivative of the density with respect to the energy is positive.

Recently, the independent particle model, or an approximation thereto, was applied to the region in which we usually expect a statistical theory to be valid. The original suggestion to do this is due to French and Wong and a very interesting comparison with experimental data is due to Bohigas and Flores. The principal difference between the model considered by these authors, and the Wishart model considered earlier, is that a great many matrix elements of the new model's Hamiltonian are 0. These include the matrix elements which connect independent particle states which differ in more than two orbits. If I understood them right, French and Wong mention a case in which less than 10 percent of all the matrix elements of a 50 by 50 matrix are different from 0. In the usual ensemble, that is the Wishart ensemble, the probability of the vanishing, or even approximate vanishing, of so many matrix elements is practically zero.

I hope it will not be resented if I say a few words about the conclusions of the new theory which, incidentally, will be reviewed in more detail by its authors. The first conclusion is that the density of the levels, as function of the energy, is a Gaussian rather than the semicircle characteristic of the Wishart and related ensembles. This is, of course, very different from the actual density but it does have the property that, at least at its low energy end, its second derivative is positive. Hence, the actual density distribution can be obtained by an infinite succession of such distributions, corresponding to configurations with increasingly highly excited single particle states. The second conclusion, due to Bohigas and Flores, is that the distance of the second and further neighbors is subject to much larger fluctuations than is the case for the matrices of the Wishart ensemble. Bohigas and Flores find an experimental confirmation of this greater fluctuation in the spectrum of Th.

The number of zeros in the French Hamiltonians is so great that I was worried, at first, that the argument given by von Neumann and myself for the repulsion of the levels, that is for zero probability of coincidence, becomes invalid. This is not the case; I could prove that as long as there is any, no matter how indirect, way to get from one diagonal element to every other, going through non-vanishing off-diagonal elements, the chances for the coincidence of two characteristic values vanish in the same sense as if all non-diagonal elements had a chance for not being zero.

Nevertheless, and in spite of my full realization of the significance of the new picture, I find it difficult to fully reconcile myself thereto. The independent particle picture has surely very little validity when the spacing between the energy levels is many thousands times smaller than the spacing of the single particle levels. Furthermore, even though the probability of zero spacing remains zero also in the new picture, the spacing between two levels which belong to drastically different configurations will have a very good probability of being very small, and in disagreement with the present experimental evidence. The experiment supports in the energy region to which we usually apply the statistical model the distribution between the nearest neighbors which applies for the Wishart ensemble. As to the discrepancy with respect to the spacing distribution of second and more distant neighbors which was found by Bohigas and Flores for Th, a recent paper of Garg, Rainwater and Havens, if I understand it right, finds the opposite kind of disagreement in the spectra of Ti, Fe and Ni. I hope I will be corrected if I misunderstood that paper.

Surely, I do not want to claim unrestricted validity for the Wishart model. As I emphasized before, it gives an absurd level density. However, I believe that the French-Wong model also has restricted validity – perhaps restricted to lower energies than we like to think of when applying purely statistical considerations – and I believe the proponents of the model realize its limitations fully as well as I do.

Let me go over, next, to another model which may replace that of Wishart, that is to another ensemble.

(b) Brownian Motion Toward Prescribed Level Density

Unfortunately, this time I do not have to apologize to the originator of the idea for discussing his work: Dr. Dyson does not attend our meeting. You may recall his 1962 paper in which he pictures the matrix elements as particles in Brownian motion. Needless to say, this is a picture only, but a very attractive one. Dyson's particles, representing the matrix elements, execute Brownian motion on a line and the statistical distribution of their position on that line represents the statistical distribution of the value of the matrix element which corresponds to the Brownian particle in question. The particles are subject to two agents. One of them is an elastic force pulling each toward a zero point. The other agent provides the irregular momenta which correspond, in the case of the Brownian particles, to the irregular momenta imparted by the thermal motion of the atoms of the medium in which they are suspended. These irregular momenta prevent the Brownian particles from settling down at the equilibrium point of the elastic force, that is at zero. The different particles do not interact so that there is no statistical correlation between their positions, i.e., between the different matrix elements. In equilibrium, the Brownian particles will have a Gaussian (normal) distribution about 0 – the ensemble of matrices is, in the stationary state, the Wishart ensemble.

So far this is only a picture to illustrate the Wishart ensemble. Next, however, Dyson derived an equation for the motion of the characteristic value of the matrices of this ensemble, and this was the truly surprising contribution. The characteristic values also obey an equation – the Smoluchowski equation – which allows us to picture them also as Brownian particles. The term representing the irregular momenta provided by the temperature motion of the medium is essentially the same as in the case of particles representing the matrix elements. The restraining force, which is in the case of the matrix elements proportional to their value, is, however, for the particles representing the characteristic values augmented by another force. This has the same form as the electrostatic repulsion between two point charges in two-dimensional space, i.e., is inversely proportional to the distance between them. This term provides the repulsion between the characteristic values so that the distributions of these are not independent of each other – as they cannot be because of the well known effect of level repulsion. For the distribution of the characteristic values Dyson obtains, naturally, Wishart's formula which does express the level repulsion. The 1962 paper of Dyson, though it contains many interesting results of detail, and also a new derivation of Wishart's results, gives the Wishart distributions as the distributions of the matrix elements and of the characteristic values. It does consider the approach to this distribution if the distribution was different to begin with – the approach as it would take place if the Brownian motion picture corresponded to reality.

Dyson's *new* article proposes a significant departure from the Wishart picture. He does not refer any further to the matrix elements as Brownian particles but only to the characteristic values. These are subject, as in the article which I just quoted, to the irregular temperature motion of the medium as well as to the electrostatic-like repulsion by the other characteristic values. However, the elastic force acting on them is replaced by the requirement of a force causing their density to become a definite function of the energy – presumably the experimentally observed function.

Dyson mentions that proposals similar to his were made also by Leff, Fox and Kahn, and Mehta. What has been accomplished? It has been proved that one can find a Brownian motion model for any level density expression. However, this is perhaps of only incidental interest. Much more important is that Dyson has adduced evidence that the local properties of the level distribution, the probability of a definite spacing between neighbors, and between second and further neighbors, is the same as for the Wishart ensemble in the region of the same overall density. Dyson does not claim to have proved this rigorously but he did make it very plausible. In this conclusion, his theory is in less conflict with the theory which we discussed before than with that of French and Wong, and Bohigas and Flores.

Dyson does not define explicitly the matrix ensemble of which his Brownian particles are characteristic values. The fact that he imposes the requirement of a definite density of these characteristic values as function of their magnitudes shows, however, that the ensemble is not invariant under unitary or real orthogonal transformations. This is quite reasonable: surely, the ensemble of Hamiltonians need not be invariant under all transformations. Nevertheless, it may be of some interest to consider ensembles which are so invariant. If they are to give a density of characteristic values different from the semicircle law, the matrix elements cannot be independent from each other; there must be correlations between them.

(c) Invariant Ensembles of Positive Definite Hamiltonians

It has been mentioned before that the actual Hamiltonians all have a lower but no upper bound. This suggests using positive definite matrices for the ensemble of self-adjoint matrices characteristic of Hamiltonians since a lower bound, different from zero, can be obtained from such matrices by adding to them the unit matrix with a suitable coefficient. Such an addition does not change the relevant properties of the matrices.

Up to this point the distinction between ensembles of real symmetric matrices and of hermitean, that is in general complex, matrices was not clearly brought out – the discussion could be held sufficiently general to cover both cases. This will not be true in the present instance and we shall concentrate on the real symmetric case. The simplest form for a positive definite hermitean matrix is $m^\dagger m$ where m can be any matrix with, in general, complex coefficients. There are two similarly simple forms for a real symmetric matrix: $H = r^T r$ is one, r being an arbitrary real matrix, and

$$H = m^\dagger m + m^T m^* \tag{2}$$

the other, where m is again an arbitrary, in general complex, matrix. It is tempting to use for r the usual, real, non-symmetric, Wishart ensemble, or if one adopts (2), for m the complex again non-symmetric similar ensemble.

Since both of the aforementioned Wishart ensembles are invariant under orthogonal transformations, $r \to R^{-1} r R$ and $m \to R^{-1} m R$ where R is a real orthogonal matrix, this will be true also of the two ensembles, $r^T r$ and (2). We, therefore, have to choose between them. Now we know from Ginibre's work that the density of the characteristic values of the Wishart ensemble of real matrices r is finite at 0. One can conclude from this that the density of the characteristic values of the ensemble of matrices $r^T r$ is in fact infinite at 0 (A.T. James has considered such ensembles). On the other hand, if the H of (2) has characteristic value 0, one easily sees that its characteristic vector ψ satisfies both equations

$$m\psi = 0 \quad \text{and} \quad m^*\psi = 0 \tag{3}$$

so that it can be assumed to be real. This imposes a number of conditions on m which increases with its dimensions so that, in the limit of very many dimensional matrices, – and the Hamiltonian is infinite dimensional – the density of the characteristic values as function of the energy will be tangent to the energy axis in a very high order. This means that many of the derivatives of the density of characteristic values will be positive – the essential condition which the original Wishart distribution failed to satisfy.

It seems to me, therefore, that along with the suggestions of French and Wong, and of Dyson, the properties of the ensemble of matrices given by (2) may be worth exploring. This does not have the physical basis which recommends the proposal of French and Wong, it does not have the flexibility of Dyson's proposal even though this could be built in. This would be, though, at the expense of its mathematical simplicity which is the principal element that renders it to me attractive.

Before closing this subject, let me recall what I said about our general subject at the start, that it is not clearly defined. It is quite possible that, from the point of view of one view of our subject the first, from another the second, and possibly from a third point of view the third ensemble will appear most attractive.

3. Comparison of the Statistical Theories with Experiments

My discussion so far was very much the discussion of a theoretical physicist who is fascinated by the theoretical problems, particularly if they have some mathematical attraction. I would like, though, to say a few words about the comparison with experiment which, even we theoreticians must admit, is after all the ultimate touchstone of our ideas.

As the program of our meeting indicates, our subject can be divided, at least roughly, into two categories: the investigation of the average properties, such as level density, average transition probability is the first and, for practical applications, probably the more important subject. The other subject, which was in the foreground of the preceding discussion because it appears more interesting to the theoretician, concerns the statistical distribution of these quantities, of the spacings between levels, of the transition probabilities. It is interesting how different our understanding of these subjects is.

Let me begin with the average properties. As to the level densities, we do have a fair understanding from first principles, if not a complete one. Our Session 6 will be devoted to this subject. The situation is different with respect to transition probabilities. The average transition probability with respect to particle emission can easily be given in terms of what is called the strength function. The strength function, however, is the fraction of the states of the reaction's product nuclei which is present in the states of the compound nucleus within unit energy range. I cannot help but feel that it should be possible to estimate this fraction, at least in crude approximation, from first principles. However, no full scale effort in this direction is known to me. A similar remark could be made with respect to γ-ray emission though the problem appears more difficult in this case. In summary, our knowledge of the average partial width of the levels in the statistical region is less satisfactory from the theoretical than from the experimental point of view.

Let us now look at the distribution of spacings and of transition probabilities about their average values. As for the spacings they seem to agree rather well with the law first deduced by Gaudin and Mehta from the Wishart ensemble, which was, as mentioned before, generalized by Dyson. I mentioned, as exception, the Th spectrum, and there are other surprising situations. However, everyone who occasionally plays with cards knows that the truly surprising hand is the one which does not show *any* surprising feature. There may be one exception from the general agreement which I should mention. According to theory, levels with different total angular momenta J should show no correlations, there should be no level repulsion between them. In disagreement with this, Garg, in 1964, claimed a strong repulsion, and this claim was reaffirmed, even though in a much milder form, in some later publications. It is true that, according to Dyson, it should be very difficult to distinguish between a weak and a strong repulsion. However, the only source of weak repulsion that I could think of originates via the interaction with the electronic shells. This renders the J values of the nucleus itself an unprecise quantum number. If this is the reason for the repulsion, it should depend on the chemical structure and should, in particular, disappear in the gas phase, at least if the molecules of the gas contain only one atom of the element the spectrum of which is investigated and if the electronic state carries no angular momentum. It may be that there are some experimental data on this question with which I am unfamiliar. If not, some experimental work would be useful because the phenomenon, if it is real, is truly puzzling.

The area for the comparison between experiment and theory which remains to be discussed concerns the distribution of the magnitudes of the transition probabilities around their average. For the neutron widths, and presumably for the widths of particle emissions which lead to definite states of the product nuclei, the simplest possible distribution, that proposed by Porter and Thomas, seems to be, on the whole, well confirmed. There are some exceptions which may be handled either by an improvement of the statistical theory, or by searching for their causes as does the doorway theory of Feshbach, or the isotopic spin theory. This ambiguity of the borderlines of the statistical theory was discussed in the Introduction and there is no need to repeat what was said there.

Let me mention, instead, the very great progress that was made in the treatment of reactions in which the state of the final products is not unique. The fission process is a prime example for this. Before the work of Bohr, Mottelson, and perhaps some others, one could have thought, either, that the final state of the fission process is uniquely defined, i. e., that there is only one fission channel in configuration space. One also could have thought that there are as many fission channels as are possible pairs of fission products. What we have learned from Bohr, Mottelson and others is that neither of these two extreme views is correct; the number of fission channels is greater than one but very small. In the case of fission phenomena, statistical considerations permit the interpretation of experimental data, to support and to direct the theory, very effectively. I would not be a true theorist if I did not wish that further experiments be performed in this area. It would be interesting, it seems to me, to ascertain whether the different fission channels lead to states in which the probabilities for the different pairs of fission products are equal, i.e., whether the channel functions differ principally in the phases, or both the phases and amplitudes, of the components representing the different fission products. A careful comparison of the abundances of the different products, as furnished by the different resonances, would give an answer.

Let me stop here with the comparison of experiment and theory. As I said, such comparisons are the lifeblood of all physics but we all hope to learn about the contact between experiment and theory, as far as our subject is concerned, in the course of our conference. Let me, instead, come to my last subject which points, I am afraid, to a fundamental weakness of our present statistical theory of the nucleus.

4. Hamiltonian or Collision Matrix?

It was mentioned before that, at least above the binding energy of any particle, the Hamiltonian's spectrum is continuous and if we admit the interaction with the radiation field, this applies in fact to the whole spectrum. One can ask, therefore, what the energy spectrum of the compound nucleus really represents, what is the Hamiltonian the properties of which we are trying to recapture with our ensembles. As far as I can see, there are only two possible answers to this question.

The first answer is that we should not take the whole question, and in fact the statistical theory, too seriously. It applies in an intermediate energy region in which the level structure is too complex to be interesting but not yet dense enough for the resonances to overlap significantly. In this point of view, the statistical theory is, by its very nature, an approximate theory.

The second answer appears to be that of R matrix theory. This claims that one should use a Hamiltonian which is confined to a finite region of configuration space (if one disregards the center of mass coordinates), with boundary conditions on the wave function which render the Hamiltonian self adjoint even if restricted to that finite region. It will then have a point spectrum and the statistical theory applies to that.

The difficulty with this interpretation is that it depends on the "finite region" to which the R theorem is applied. If we change that region, both the density of the characteristic values, and also the statistics of their spacings, changes. If we increase the region, the density of the characteristic values increases. The distribution of the spacings, in terms of the average spacing, remains approximately invariant, but only approximately. If one goes to the limit of a very large internal region – this is the name of the region to which one applies the R theorem – the spacing becomes very nearly the average (very small) spacing throughout and the fluctuations converge to zero. This is, perhaps, not surprising because, in this case, we look at the characteristic value distribution of the Hamiltonian as applied to some very large, essentially empty, domain in which the nucleus occupies a very small volume. There is a good deal to be added and elaborated on this last statement but, on the whole, it seems to me uncertain that this second point of view is much more attractive than the first which simply admits that there is no precise formulation for the statistical theory.

The difficulties which we are having with the statistical theory based on the Hamiltonian remind us that, when dealing with the continuous spectrum – and we are, except for the normal state, always in the continuous spectrum if we do not disregard the coupling with the electromagnetic field – the collision matrix provides the appropriate characterization of the situation. This suggests that if we wish to extend the theory to regions of higher energy, and if we wish to give it a more rigorous formulation, we should try to do this in terms of the collision matrix rather than the Hamiltonian which we have to restrict, quite artificially, so that it has a discrete spectrum. To formulate the theory in terms of the collision matrix is, naturally, relatively easy in the region of well separated resonances where we can simply translate the present theory into the language of the collision matrix using, perhaps, the Humblet-Rosenfeld theory. The translation though, I fear, will not have the simplicity of the present formulation, nor its natural character. Also, I should admit *some* difficulties if we wish to be truly precise and wish to take reactions such as $(\gamma, 2\gamma)$ and similar processes into account. This may be, though, unnecessary. The serious difficulty, at any rate, is to be expected in the higher energy region where any new theory will have to stand on its own feet. Thus a statistical theory in terms of the collision matrix, if at all feasible, is a task for the future.

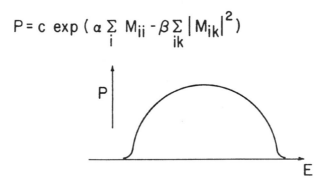

Fig. 1. Wishart ensemble

Discussion

Krishnaiah: I wish to make a comment about the terminology "Wishart Distribution". Nuclear physicists call the distribution $p \times p$ matrix M as Wishart Distribution where the elements of M are independently and normally distributed with zero means and the variances of diagonal elements are equal to 2 while the variances of the off diagonal elements are equal to 1. The distribution of M is *not* Wishart Distribution. The distribution of $S = XX'$ is known as Wishart Distribution where S is a $p \times n$ matrix whose columns are distributed independently as multivariate normal with zero mean vector and covariance matrix Σ; here X' denotes the transpose of X. We note that a random vector y is distributed as a multivariate normal with zero mean vector and covariance matrix Σ if its distribution is given by:

$$f(y) = C \exp[-\frac{1}{2}y^{1}\Sigma^{-1}y]$$

where C is a normalizing constant and Σ^{-1} is the inverse of Σ. A very brief summary of some of the literature on the distribution problems connected with the eigenvalues of the matrix S is given in the following paper: P.R. Krishnaiah and T.C. Chang. "On the Exact Distributions of the Extreme Roots of the Wishart and Manova Matrices." *Journal of Multivariate Analysis, Vol. 1, pp. 108–117, 1971.*

Moldauer (Argonne): I just wanted to make the comment that one respect in which all the matrix models differ from physics is that the matrix models all deal with finite matrices, while physics deals with an infinite matrix, and it seems to me that one must always, in interpreting the results of the matrix model, be careful not to use those results which specifically have to deal with the finiteness of the matrix. I just wanted to make that clear.

Wigner (Princeton): Dr. Moldauer is absolutely right. Except I am firmly convinced that Dyson's Distribution could be made infinite, and so could be this $m^\dagger m$ matrix be made infinite. In that case, you start with an infinite dimension, but nobody has done it right.

Bloch (Saclay): I would like to mention that a way to derive distributions which differ from (how shall we call it – Wigner Distributions if not Wishart) in such a way that one obtains a level density which is not the semicircle law which was presented I think two years ago by Balian. I don't have the exact reference now, but it was presented in the *Kyoto Conference on Statistical Mechanics.* It was based on the use of information theory. One states as a postulate that the distribution should contain the minimum information – information being defined as usual as $\int P \log P$ where, it is consistent with some constraints. It is very easy to show that if as constraints you take the values of the averages of the matrix and its mean square, then you obtain the Wigner Distribution. But, you could take different constraints. In particular, you could take the level density as the constraint. Then it is very easy to obtain the full distribution, and one can see that it will also give rise to level repulsion as usual.

Wigner: Do I understand you right that this anticipates to a considerable extent the work which I attributed to Dyson?

Bloch: I think it has a very different starting point. It is based on a rather arbitrary postulate that Dyson also used but it gives a way to introduce a method for obtaining different level density from just the semicircle law distribution.

Wigner: Is there any information available that the spacing distribution is the same as calculated by Gaudin and Mehta?

Bloch: The spacing distribution itself is rather difficult to evaluate, but the complete probability distribution P of $(E_1, E_2, ... E_n)$ is very simple. It's very similar to the Wishart Distribution. It involves the product of all the differences of eigenvalues two by two multiplied by a function of E_1 times the same function of E_2 times the same function and so on, exactly as in the Wigner Distribution except that here it is not a Gaussian function but an arbitrary function which has to be determined by the level density.

Wigner: That is very much the same as Dyson's work, and I am sorry that I did not mention Balian's work. I was not familiar with it. I am still not familiar with it. Thank you very much.

Feshbach (M.I.T.): Just a comment about your very last paragraph of Professor Wigner's talk. There is another procedure which was carried out by a student of mine, Victor Newton. One takes a model of coupled equations in the continuum – in other words, a scattering problem determined by the solution of many Schrödinger equations coupled together. In this model, one takes the coupling part of the potential matrix as random and solves the resultant equation. In this way, one simultaneously considers both the continuum and the statistical hypothesis.

Wigner: That seems rather opposite to what French and Wong do who put in many zeros. Thank you.

Mehta (Saclay): I just wanted to give the reference of Balian's work. It is *Nuovo Cimento B57, 183 (1968)* which deals with the information theory treatment of the random Hamiltonians.

Wigner: Thank you.

Rosenzweig (Suny/Albany and Argonne): First one comment, I would like to tell you, that during the last few weeks Mehta analyzed more recent data than was available to Bohigas and Flores to test the difference between their model and the conventional theory. We hope that in tomorrow's session, we will be able to show our results maybe after or during Bohigas' talk. We find that the disagreement between the experiment and the conventional theory is not great at all. That was just a comment since you brought it up at this time. The other things I would like to ask you: Are the ensembles considered by French and Wong and by Bohigas and Flores – could they conceivably be generated as a special case of Dyson's ensembles?

Wigner: I believe not, because in Dyson's ensembles, you have a direct repulsion between any two levels, and I shouldn't speak about it, but I think French should be here to answer these questions.

Rosenzweig: But Dyson is not here.

Wigner: But Dyson I can represent. In Dyson's ensemble, essentially all the matrix elements are present and there are no zeros. It is much too general for that.

Rosenzweig: So the specific things that French and Wong find as consequences could not be obtained.

Wigner: Not as consequences – as assumptions different from the assumptions of Dyson. In Dyson's theory there is an electrostatic repulsion between any two levels.

French (Rochester): Perhaps I should make something clear about the zeros in matrices. It is indeed true that if one uses a picture or a representation in space which is simple, a Hartree-Fock kind of thing, of describing it in configurations. Then when you put in two body interactions, you do get a fair number of zeros. I don't think you get as many as has been suggested. But, that we regard as not the real essence of the thing, that is done. The real essence instead, is that constraints on all the matrix elements are imposed by the condition that we have a two-body interaction. And we have, for example, verified that if you take large matrices and put them and think of them as being in block form and impose only the zeros corresponding to the fact that you cannot move two particles from here to there, that is much more likely to give you a semicircular form than to have a Gaussian form. If now, however, you impose a further condition that the matrix elements have to satisfy a large number of constraints then the distribution becomes Gaussian. In other words, I don't think that one should put too much stress on inserting zeros. That doesn't really convert you from semicircular to Gaussian.

Wigner: But isn't it a necessary consequence of your theory that there are those zeros. In fact, the higher up you go in energy the more zeros you have because at the higher energy you have many configurations which differ by more than two particles.

French: Yes, indeed I feel it is a necessary consequence, but it is not however, sufficient, I believe, to produce the Gaussian form – necessary, but not sufficient.

Wigner: Thank you.

Concluding Remarks

E. P. Wigner

Symmetry Properties of Nuclei
(Proceedings of the 15th Solvay Conference on Physics, Brussels 1970).
Gordon and Breach, New York 1974, pp. 351–362

THIS IS the third conference in nuclear physics that I have attended within a year. I have here been learning very, very much again and this is nice, but the fact that one has learned so much on three successive occasions gives one the impression that he still knows an awfully small part of all that he would like to know. This is somewhat discouraging.

Dr. Amaldi was so kind as to ask me to summarize what I have learned and experienced during the conference. I will follow the tradition of these conferences and be very open, voicing my reservations or lack of understanding clearly.

In this spirit, let me begin with a general remark. For much of the time, up to this morning, I had the impression that we did not form a fully homogeneous group, that we were, more or less, divided into three groups, none with a full appreciation for the ideas and conclusions of the members of the other groups. This, I fear, interfered to a certain extent with the ultimate purpose of the conference: to acquaint all of us with all that is worth while to know about the symmetry relations' role in nuclear physics. Today, however, all this has changed. We all participated with keen interest in what we heard presented. Today's sessions were very much in the spirit of the Solvay Congresses, as their purpose was originally visualized by the founders of this institution.

Perhaps I am turning in the wrong direction when I mention, next, two subjects which are very important in nuclear physics and about which we heard very little, at least up to this morning. Perhaps, as a result of the rapidity with which information now accumulates, we had too many and not too few subjects under discussion. Let me mention, nevertheless, the two subjects about which it would have been nice to hear a bit more. The first of these is the $j - j$ coupling shell model. It would have been good to learn a little more about the present status of this model, its recent successes, and its difficulties. It would have been good, also, to hear about the interface of this and the collective model from the viewpoint of the shell model.

The other subject which was, until this morning, sort of pushed under the table is the giant resonance phenomenon. In some cases, single levels exhibit

352   *E. P. Wigner*

some very marked properties. Thus, from any level of an isotopic spin multiplet, the Fermi β transition leads, at low excitation, preponderantly to a single level. In other cases, a conglomerate of energy levels shares a similar property. The original example for this is the γ ray dipole absorption. This is a giant resonance because it does not lead to a single but to a multiplicity of levels all situated, though, within a relatively narrow energy range of a few MeV. Another example is the one which I happened to mention this afternoon: the dissolution of the analog state in the many states, with lower T, in its neighborhood. Actually, this last subject did not go unheeded at our conference—Dr. French discussed it in some detail this morning. In fact, he arrived at conclusions which I found quite surprising.

Let me now review the three subjects which were discussed during the first few days.

THE COLLECTIVE MODEL

Let me first tell you what was difficult for me to accept of the discussion of this subject. There were two such things: the first one is that it uses a classical picture a little too freely. We all admire Niels Bohr's original picture of the hydrogen atom, how he explained the Balmer series with such ingenuity. Nevertheless, today we do not use his picture any more and we talk in terms of the wave functions. It seems to me that the collective model is too often represented in terms of a classical picture which is just as far from or just as close to reality (to the wave functions) as are Bohr's orbits. Thinking in terms of Bohr orbits, I would be led to believe that I can put the H atom into rotation, just as I can any nucleus. This, of course, is impossible. However, thinking in classical terms may lead to similar difficulties in nuclear physics. A case in which it leads to such a difficulty was actually recognized by Dr. Bohr a long time ago. If the deformation of the nucleus is small, then a rotation of the nucleus and a shape vibration are represented by exactly the same wave function. In this case the appellations, rotational state or vibrational state, are therefore misleading; in some ways they merge. One concludes that even at higher deformations the two are not as distinct as the terminology indicates or as the picture which it creates in one's mind. They are not as different as a ball put into a kind of oscillation and a ball which is spun around.

The second difficulty which I have—and it is my fear that this difficulty is shared by many others—is that it is rarely clear to me whether a type of motion that is described illustrates a stationary state or a giant resonance. Surely, those in the field would not confuse the two situations, but the lack of specification is confusing many of us.

Let me say at this point that no one surpasses me in my admiration for the collective model and the ingenuity and intuition that went into its development. But this is not always the case for the language that is used. Before discussing the new achievements of the model, permit me now to compliment Dr. Bohr's analogy between translational and rotational motion. He pointed out that if we have a nucleus, let us say at rest, we can put it into motion. This is a rigorous concept and the stationary nature and the exact state vector of the moving nucleus follow from Galilean or Lorentz invariance. Similarly, he said we can put it into rotation, though this is not a rigorously defined process, because rotating coordinate systems are not equivalent to nonrotating coordinate systems. Nevertheless, the analogy is very appealing.

Let us now turn to the new conclusions about which we learned from Drs. Bohr and Mottelson. One of the very interesting observations is that, if we express in a rotational band the energy as function of ω, it is a simpler function than as function of the angular momentum. Of course, initially the energy is given in terms of the angular momentum I but, as Dr. Bohr explained, if we make a plot of the energy as function of angular momentum or as function of $I(I + 1)$ (which gives a somewhat simpler plot) and if we take the derivative of this and denote it by ω and then plot the energy as function of ω, the functional dependence is much simpler than in terms of I. In particular (if we look at it as a power series) this converges very much faster. I do not know whether we fully know what this means physically, but it is one of the interesting facts.

Let me insert next something which I did not learn here. This is a result of Y. Sharon. He considered the rotational spectra from the point of view which was used also by Drs. Bohr and Mottelson but was first proposed, I believe, by Peierls and Yoccoz. Sharon assumed a rigid intrinsic state. We heard today from Dr. Bouten about similar calculations. The change in kinetic energy is, as we know, in first approximation proportional to $I(I + 1)$. In second approximation there is a term proportional to the square of this. However, the coefficient of this square is positive definite, and that means that if one plots the energy as function of $I(I + 1)$, it is convex from below. The experimental curve is concave from below, the coefficient of the square of $I(I + 1)$ is negative. Sharon then calculated the change of the potential energy as function of I and found that it decreases and can render the coefficient of the square of $I(I + 1)$ negative. I am bringing this out because it may be important to realize that, in calculating the rotational spectra of nuclei, not only the change in the kinetic but also that in the potential energy has to be taken into account. In this regard, there is a great difference between molecular rotation and nuclear rotation. In the case of molecular rotation, the potential energy is, in the approximation of a rigid molecule

E. P. Wigner

(which is, of course, an approximation) independent of the angular momentum I and equal to the potential energy of the state with fixed nuclei. This plays the role of the intrinsic state in the theory of molecular spectra. The reason for the independence of the potential energy on I is that, no matter how little the intrinsic state is rotated, the resulting state is orthogonal to the initial one. Since the operator of the potential energy does not change the position of the nuclei, the potential energy can be calculated for every position of the intrinsic state separately, and since all these potential energies are equal to that of the intrinsic state, the total potential energy becomes equal to that of the intrinsic state, independent of I.

This is not so in the case of nuclei because a small rotation of the intrinsic state does not render its state vector orthogonal to the state vector of the unrotated intrinsic state. The amount of rotation needed to render the rotated intrinsic state essentially orthogonal to the initial intrinsic state depends, of course, on the deformation but is surprisingly large even for a significantly deformed intrinsic state. The lack of orthogonality makes the calculation of the potential energy as function of the angular momentum I more difficult and, at any rate, the potential energy is no longer independent of I.

Next, we learned from both Dr. Bohr and Dr. Mottelson about the possibility of a number of new phenomena. Dr. Bohr, in particular, mentioned the possibility of a phase transformation, at a definite I. This would be a very interesting phenomenon. I presume it would mean that the energy, as function of I, or of ω, would not be smooth but would show a kink or discontinuity of some sort. This, however, was not the only new phenomenon anticipated and Dr. Mottelson, in particular, spoke about vibrations with very high amplitudes and about the interaction of these vibrations with single particle motion. The phenomena foreseen by Dr. Mottelson are numerous and would be most interesting to observe. To be sure, we all heard before about the interaction between collective and single particle motion but never as concisely as this time.

Let me mention next a problem presented by the collective theory which always puzzled me. It is due to the fact that, after all, the nucleus does not consist of a very large number of constituents as does a liquid drop or a solid body. How closely can its shape be defined? I once made a calculation of the definition of the shape presented by a number of particles in a sphere and randomly distributed therein. The average value of the deformation, that is the difference between the maximum and minimum moments of inertia, is much larger than one would expect. This means that the statistical fluctuations of the shape are not negligible as compared with the departure of most intrinsic states from spherical. Now the statistical fluctuations of the shape, calculated this way, may not give an entirely fair picture of the actual fluctuations because the nuclear matter does have properties similar to liquid matter and the fluctuations in density and shape are smaller than if the particles moved independently of each other. There is a repulsive core to begin with and the exclusion principle also has an effect. Nevertheless, the shape fluctuations are surely real and obscure the picture of the nuclear shape. It would be good to clarify it.

Basically, the same question arises in another context and it arose also in today's discussion. If the wave function of the $J = 0$ state is given, to what extent does this determine what we call the nuclear shape? If many photographs of the nucleus are taken, and these superposed on each other, the resulting shape is, of course, spherical. The individual pictures may be so elongated that this is at once evident but they are, on the average, less elongated than the intrinsic state. The question remains: is there a real definition of the deformation of the nucleus? Naturally, many intrinsic shapes give the same $I = 0$ state and may give even the same $I = 2$ state, $I = 4$ state, and so on. Since we talk so much about the shape of the nucleus, it would be good to have a really solid definition of what we do mean by that shape.

The next problem which I wish to raise is: how far do the rotational spectra go? We heard today that in beryllium, according to calculations, levels up to $I = 8$ exist and we heard about experimental data indicating rotational levels up to $I = 10$. The experimental data and the calculations do not quite agree: according to the calculations there is no break in the I dependence of the energy at $I = 6$, which is the last angular momentum permitted by shell theory. On the other hand, if I understood Dr. Wilkinson correctly, the experimental data do show a considerable jump in the energy between the $I = 6$ and the $I = 8$ states. Since I believe in a certain validity of the shell model, I would not be surprised by such a jump. Dr. Wilkinson told me though that the experimental data are not yet definitely established.

GROUP THEORY AS TOOL FOR CALCULATIONS

This subject is more closely related to the subject which was, originally, meant to dominate our conference. We heard about two types of calculations. Both types were concerned with the mathematics of the shell model. The first type was directed toward an improved classification of the states that particles within a shell or within two shells, perhaps, can produce. Dr. Judd's discussion was, I think, exclusively concerned with this and he wanted to develop a scheme in which the vanishing of so many matrix elements appears natural. Dr. Moshinsky's talk on the group theory of the oscillators of which we have the 73-page manuscript was also concerned with this but went further toward actual calculations. The usefulness of his model always surprises me.

356   *E. P. Wigner*

The density distribution of the nucleons, from which his calculations start, is very different from the density distribution of actual nuclei. The density distribution, in the limiting case of very large number of nucleons, is, to begin with, $\varrho(r) = (R^2 - r^2)^{3/2}$, R being the nuclear radius. This is, essentially, the density distribution of the nuclei in an oscillator potential. This distribution is, of course, modified by the admixtures of excited states, caused by the interactions between the nucleons. The admixtures unquestionably modify the $(R^2 - r^2)^{3/2}$ density and the purpose of the calculation, which makes such far-reaching use of group theoretical theorems, is just to calculate the admixtures. It is surprising that this can be done adequately when the final density differs as much from the initial one as it seems to do in this case. I happen to have a paper by Dr. Heisenberg's son here, showing the density of the protons in the Nd^{142} nucleus, as measured by the Stanford group of which he is a member. The density, as measured, assumes its maximum around $\frac{3}{4}R$ where the model's initial density is less than a third of the maximum density assumed at the center of the nucleus.

Nevertheless, as we know, the theory reproduces the experimental data in many cases very accurately—surprisingly so. We must hope that this is not the manifestation of something which, at one time or another, bothered all of us: that occasionally theories which are quite far from reality give good agreement. Lawson pointed this out most eloquently by means of his pseudonium theory. I do not believe, though, that this fear is justified in this case.

The last remark brings us naturally to Dr. Hecht's report. The theory he reported on is truly surprising. It starts by replacing, in a particular instance, the $d_{5/2}$ and $g_{7/2}$ orbits with $f_{5/2}$ and $f_{7/2}$ orbits. For some quantities, this surely gives entirely incorrect results. Thus, the density of a particle in the $d_{5/2}$ orbit goes to zero at the center as r^2, for the g orbits as r^4. If there are particles in both orbits, the density near the center is proportional to r^2. The replacement of these orbits by the f orbits would give an r^3 density dependence. The density at the origin is only one particular quantity for which the replacement leads to incorrect conclusions. However, as far as the energy levels are concerned, the model has amazing usefulness.

Before leaving this subject, let me return for a minute to Dr. Judd's report and be more specific concerning the significance of the groups he used. Dr. Judd was concerned with the filling of 14 single particle states. Since each of these states can be filled or empty, we have 2^{14} orthogonal states, and the transformations of these are given by $U(2^{14})$. This is, of course, an enormous group, in fact clearly unnecessarily large; it can be broken up into 15 subgroups, the kth of which is $U\binom{14}{k}$ and refers to the case of k particles and $14 - k$ holes. Even these are extremely large groups, the largest one being $U(3432)$. What Dr. Judd showed is that it is not necessary to use these groups in order to specify the possible states. Instead of the transformations of the 2^{14} states of the system, he considered the transformations of the bilinear forms of the creation and annihilation operators. The operator $a_\varkappa^+ a_\lambda$ annihilates the orbit (single particle state) λ and fills the state $\varkappa$. There are 14^2 such operators. The operators $a_k^+ a_\lambda^+$ fill both orbits $\varkappa$ and λ but, because of $(a_\varkappa^+)^2 = 0$, $a_\varkappa^+ a_\lambda^+ = -a_\lambda^+ a_\varkappa^+$, there are only $14 \times 13/2$ such operators. Similarly, there are $14 \times 13/2$ operators $a_\varkappa a_\lambda$. We have, altogether, $14 \times (14 + 13) = 28 \times 27/2$ operators. Their transformations form the group $0(28)$—a much smaller group than most of those mentioned before. What Dr. Judd found is that all the 2^{14} states of the 14 orbit problem can be uniquely characterized by means of this group and one can explain by means of it the vanishing of all the matrix elements which do vanish. Why this is true is a mathematical question and a mathematical question of some depth to which I, at least, do not know the answer.

Let me now turn to the last subject—the old-fashioned application of group theory to derive the consequences of symmetries and invariances, accurate and approximate.

CONSEQUENCES OF THE ACTUAL SYMMETRIES —ACCURATE AND APPROXIMATE

The difference between the two types of applications of group theory was brought out in our discussions most clearly by Dr. Heisenberg. What gives our present subject a new life is the much improved knowledge of the forces between nucleons. This was brought out, perhaps most dramatically, by Dr. Wilkinson who compared the three nucleon-nucleon interactions. He said that the $p - p$ and $n - n$ interactions—responsible for charge symmetry—are now believed to be equal within a couple of tenths of a per cent. I believe he will be the first to admit that it is very difficult to be absolutely sure of this because the $n - n$ interaction can be measured, at present, only as final state interaction. Even the measured direct interactions $p - p$ and $p - n$, are difficult to interpret in terms of an interaction Hamiltonian, and the difficulty is much greater if the information comes from final state interaction measurements. However, a good case can be made for the equality of $p - p$ and $n - n$ interactions also from other data.

Similarly, Dr. Wilkinson had a measure for the difference between $p - n$ and $p - p$ interactions. It was long suspected that the former is stronger but the accurate measure of the difference, about 2.8 per cent, is certainly quite new.

Isotopic Spin-SU$_2$ Symmetry was Dr. Wilkinson's next subject. This symmetry is, of course, a consequence of the equality of the three nucleon-nucleon interactions. The first of the consequences considered was the second order dependence of the masses of the members of isotopic spin multiplets on T_z (which is half of the difference between neutron and proton numbers). Dr. Wilkinson said that the coefficient of the third order term, T_z^3, is only about 10 keV—very small indeed.

It is not easy to draw inferences from this on the equality of the three kinds of nucleon-nucleon interaction because any interaction that has vector character in the isotopic spin space gives (in first and second approximation) a similar T_z dependence of the masses and the same is true in first approximation even of a tensor operator in that space. (An operator of scalar nature—we hope that the overwhelming part of the interaction is of this nature—does not give rise to any difference between the masses of the members of the isospin multiplet.) Hence, what the second order dependence of the masses in T_z proves is only that the third approximation of a vector-like operator in the isotopic spin space, and the second approximation of any tensor-like operator, do not play a significant role in the Hamiltonian. Something that does give information and evidence for the equality (and hence scalar nature) of the three nucleon-nucleon interactions is the β decay of spin zero angular momentum $(J = 0)$ nuclei into their isotopic spin partners. To these transitions, the Gamow-Teller matrix element does not contribute because it has vector character in ordinary space. Wilkinson pointed out that the so-called ft values of all these transitions are very nearly the same (about 3050), thus giving evidence for the equality of the aforementioned three interactions. As a slight exception, Dr. Wilkinson mentioned Al^{26}, decaying into Mg^{26}, with an ft value apparently somewhat below the others. However, this decreased ft value is not absolutely sure and, second, it concerns a degree of accuracy which is beyond that of ordinary nuclear physics. We are usually satisfied if two transition probabilities, which should be equal according to theory, agree within about 15 per cent and the apparent difference in this case is much less than that.

I would like to mention something here that I learned from Dr. J. D. McCullen of the University of Arizona. He investigated the β decay of Sc_{21}^{42}, a $J = 0$ nucleus. The decay product is Ca_{20}^{42}, with a $J = 0$ normal state, the isotopic spin partner of the Sc^{42} normal state. The second excited state of Ca^{42} also is a $J = 0$ state but it is not the isotopic spin partner of the Sc^{42} normal state. Hence, the β transition thereto is forbidden by the isotopic spin rules and it is, in fact, 10,000 times weaker than the transition to the partner, that is the normal state. It is, naturally, possible that other circumstances also play a role in this much decreased transition probability but that two such probabilities should differ by a factor 10,000 remains remarkable.

The regularities which one can derive from the validity of the isotopic spin concept for nuclear reaction cross sections and polarizations are not equally well confirmed. Dr. Wilkinson discussed the polarization resulting from two nuclear reactions which should yield equal polarizations. The two polarizations, as he showed them to us, did look alike. However, his data were confined to small angles. At larger angles, the polarizations resulting from $H^2 + H^2 \rightarrow He^3 + n$, and from $H^2 + H^2 \rightarrow H^3 + H^1$ are quite different. So are, I understand, the cross sections of the $He^3(p, p)$ and of the $H^3(n, n)$ reactions. The conflict with the validity of the isotopic spin concept may be more apparent than real because, occasionally, very small deviations give very large effects. Nevertheless, it is my impression that we yet have to learn a great deal about the consequences of the very close equality of the three nucleon-nucleon interactions, when are these consequences valid, and when not.

I should mention, perhaps, the limitation of the validity of the rules derived from the isotopic spin concept: this validity is restricted, except for the lightest nuclei, to low energies. Dr. Wilkinson mentioned this many times. I happened to read, a few days ago, about the $N^{14} + \alpha$ reaction, going through the F^{18} compound nucleus. The yields of the different reaction products obtained make it evident that, for excitation energies of 15 Mev or so, the isotopic spin of the levels is no longer pure. This will always be the case if levels with different isotopic spins are close to each other—in such a case even a relatively small interaction, such as the electrostatic one, causes a significant amount of mixing. We know, in fact, that for heavier nuclei the states with isotopic spin $T = T_{\zeta} + 1$, i.e., exceeding that of the normal state, are dissolved into giant resonances.

Of the not strictly valid symmetries, that of charge independence is most accurate in nuclei. When I say this, I consider the rotational (and of course the translational) invariance to be strictly valid, which is, as far as we know, truly correct. I also count the reflection invariances among the strictly valid ones, as they are in practice valid as far as nuclear physics is concerned. Nevertheless, I should recall some of the beautiful experiments about which we heard and which exhibit tiny violations of the parity concept. May I add, to what we heard in this connection, a reference to Lobashov's beautiful work on the circular polarization of yet another transition, the $5/2^+ \rightarrow 7/2^+$ transition in Ta^{181}. It is of the order of 10^{-6}, in consonance with the usual assumption that the parity-violating forces play a subordinate role in the parts of nuclear physics, other than β decay.

The SU_3 symmetry, my next subject, deals with a symmetry which is less closely valid than the isotopic spin's SU_2 symmetry. Dr. Elliott was the first to recognize the usefulness of the model, based on the assumption of the validity of this symmetry, and it was discussed at our meeting by Dr. Elliott

360   *E. P. Wigner*

himself. Dr. Elliott also referred to the SU_4 model and showed that this has, in the cases considered by him, an accuracy of about 90 per cent, that is, 90 per cent of the total wave function belongs to a single representation of SU_4. The SU_3 symmetry is a great deal better yet and appears to have, in many cases, a truly surprising validity. There would be little point in my repeating what Dr. Elliott said on this subject—his presentation was lucid and complete.

There is one reservation, though, on the confirmation of this and partly also of the SU_4 model which will be discussed later. The reservation is that most of the confirmation is derived by comparison with other calculations, and only to a small degree by comparison with experimental data. This is a little bothersome. The agreement between two calculations is not as significant as the agreement with experiment and we were only too often reminded of this in the course of our discussions. It is, of course, much easier to calculate something on the basis of a more or less confirmed theory than to persuade an experimental physicist to measure something which may be very difficult to measure. Nevertheless, there is no real substitute for experimental confirmation and if the agreement of a model only with theory, to be sure a more elaborate theory, can be established, the model becomes a calculational tool, similar to those discussed before.

The Supermultiplet (or SU_4) Theory. Our next subject concerns the symmetry which is least accurate: the SU_4 symmetry. This would be strictly valid if the forces between nucleons were independent of both isotopic and ordinary spin. The independence of isotopic spin was discussed in detail before; it is a reasonable approximation. The dependence on ordinary spin is, on the other hand, much greater. We know this ever since Rabi's measurement of the quadrupole moment of the deuteron. One can well be surprised, therefore, by the degree of validity of this model, as discussed particularly by Dr. Radicati. Let me discuss only that part of the evidence which is based on comparison with experimental information. This comparison will show a remarkable similarity with the comparison of the SU_2 theory with experiment, inasmuch as it will deal with the same two types of data. It will be less convincing.

The mass formula of the supermultiplet model has a much wider applicability than that of the isotopic spin theory: it extends to the normal states of all nuclei. However, neither its derivation from the basic postulates of the theory, nor its confirmation by Radicati, are equally clear. I am sure Dr. Radicati will agree with this statement.

As to β decay, it does follow from supermultiplet theory that only the transitions between members of the same supermultiplet are allowed. The Fermi, that is $\Delta J = 0$, transitions are, as we saw, allowed only between members of the same T multiplet, the Gamow-Teller ones also allowed between members of those T multiplets which are united into the same supermultiplet. It follows that the ft values of β transitions between members of different supermultiplets should be much larger than those within the same supermultiplet, i.e., should be well above 3000. Actually, they are, as Dr. Radicati told us, as a rule about 100 times larger. Again, this could perhaps be attributed to many other cases and, as I believe I pointed out in the course of the discussion of Dr. Radicati's paper, it can be deduced from postulates somewhat less far-reaching than the total SU_4 symmetry. It remains true that this conclusion of the supermultiplet theory seems to be confirmed.

Incidentally, the large ft values, that is the small values of the corresponding Gamow-Teller β decay matrix elements, are in many cases in sharp conflict with the $j - j$ coupling shell model. This is very confusing because this model gives, *at least* for the energy levels, regularities—what I call Talmi rules—which are, in many cases amazingly accurate. The fact that the β decay probabilities are in many cases so much lower than this model would postulate, is very puzzling. The magnetic moments, as I learned from Dr. Talmi, are not given by the model as well as the energy levels but not as poorly as the β decay matrix elements.

The low decay rate of C^{14} was brought up in the discussion as to be in disagreement with the postulates of supermultiplet theory. It was assumed that the normal state of the daughter nucleus, of N^{14}, is in the same supermultiplet as that of C^{14} so that the transition should be rapid. A possible explanation of this discrepancy is that the normal state of N^{14} is a $^3D_1$ state and that it is, therefore, not part of the supermultiplet which contains the $^1S_1$ normal state of C^{14}. Dr. Bertsch called my attention to a paper by Mangelson, Harvey, and Glendenning which seems to confirm this assumption. Nevertheless, I admit that, even if we accept this reasoning, the very low transition rate remains puzzling. It appears to indicate a validity of the supermultiplet model which is much closer than one would expect, a surprisingly low transition rate between the S and D supermultiplets. One would expect the ^3S and ^3D states to mix to a reasonable extent—in fact, the $j - j$ model postulates such a mixing.

CLUSTER STRUCTURE

The discussion of the cluster model was very interesting and, again, the enormous range of applicability, and the visualizability of its pictures of the nuclei are amazing. Naturally, one worries about the fact that the state vectors given by these models correspond to $L - S$ coupling, just as those of the

362   *E. P. Wigner*

supermultiplet theory. Furthermore, it appears to be more difficult to incorporate the departures from $L - S$ coupling into this model than into supermultiplet theory.

My final observation draws attention to the new tools of nuclear physics about which we heard, from several speakers, usually somewhat incidentally. I am referring to the tools furnished by particle physics, μ decay data, hypernuclei of various kinds, probing of the nucleus not only with electrons but also with muons, pions, and other particles. These new tools promise to furnish new insights into nuclear structure, and perhaps also into the Symmetry Properties of Nuclei which was our original subject.

We had an interesting meeting and we'll all return home enriched in knowledge of nuclear physics. As the last speaker, I wish to express the appreciation of all of us to the organizers of the meeting, to the group here for its constant help and hospitality, and last but not least to those who stand behind them.

PART VII

Broader Philosophical Essays

The Limits of Science

E. P. Wigner

Symmetries and Reflections.
Indiana University Press, Bloomington, Indiana, 1967, pp. 211–221

The present discussion is not put forward with the usual pride of the scientist who feels that he can make an addition, however small, to a problem which has aroused his and his colleagues' interest. Rather, it is a speculation of a kind which all of us feel a great reluctance to undertake: much like the speculation on the ultimate fate of somebody who is very dear to us. It is a speculation on the future of science itself, whether it will share, at some very distant future, the fate of "Alles was entsteht ist wert dass es zu Grunde geht." Naturally, in such a speculation one wishes to assume the best of conditions for one's subject and disregard the danger of an accident that may befall it, however real that danger may be.

The Growth of Science

The most remarkable thing about Science is its youth. The earliest beginning of chemistry, as we now know it, certainly does not antedate Boyle's *Sceptical Chemist*, which appeared in 1661. More probably, one would place the birthyear of chemistry around the years of activity of Lavoisier, between 1770 and 1790, or count its years from Dalton's law in 1808. Physics is somewhat older; Newton's *Principia*, a rather finished work, became available in 1687. Archimedes discovered laws of physics around 250 B.C., but his discoveries hardly can be called the real begin-

Reprinted by permission from the *Proceedings of the American Philosophical Society*, Vol. 94, No. 5 (October, 1950).

ning of physics. On the whole, one is probably safe in saying that Science is less than 300 years old. This number has to be compared with the age of Man, which is certainly greater than 100,000 years.

The number of people who devote years of their life to the acquisition of knowledge had an equally spectacular rise. Thus, about ten per cent of the American youth are graduated from college, a percentage that has lately doubled in every twenty years. Harvard College was founded in 1636 and it was certainly not a scientific college at that time. The American Association for the Advancement of Science is one hundred years old and had originally 461 members. Today, it has more than half a million and its membership increased by almost 10,000 in a single half year. The growth of college attendance was less spectacular in some other countries, probably more spectacular in Russia.

Man's increased mastery of the Earth can be directly traced to his increased knowledge of the laws of nature. The surface of the Earth, as a whole, was not affected by man for 99,700 years but vast areas have been deforested or the surface's store of some minerals depleted since the birth of science. For 99,700 years, a man equipped with a good telescope on the moon might not have discovered man's existence on the Earth. He could not have overlooked it during the last three hundred years. There is no natural phenomenon that is comparable with the sudden and apparently accidentally timed development of science, except perhaps the condensation of a super-saturated gas or the explosion of some unpredictable explosives. Will the fate of science show some similarity to one of these phenomena?

Actually, if one views detachedly the rapid growth of science, and of the power of man, one cannot help fearing the second alternative. Surely man has not been able to adjust his spiritual outlook to the responsibility which his increased power imposed on him and one has to fear a catastrophe as a consequence of this maladjustment. This has come to be so well recognized recently, particularly as a result of the development of atomic weapons and the subsequent failure of man to cope, or even to come to grips, with the problems created by these weapons, that it is almost a commonplace. Nevertheless, this possibility will be disregarded here, and the limits of the growth of science will be considered under the assumption that no cataclysmic effect will interrupt this growth. The following speculations therefore apply only if we should be able to avoid the cataclysm which threatens us, and science can develop in a relatively peaceful atmosphere. They will look for the inherent limitations of science, rather than the limitations imposed by external effects, whether or not these external effects are influenced by science.

What Can We Call "Our Science"

What might be considered as the natural limit of our science will become perhaps best apparent if we try to define what "our science" is. It is our store of knowledge of natural phenomena. The question then is, what is "our" store? This question will be approached by giving both too broad and too narrow definitions and then attempting an acceptable compromise. A set of volumes, containing information and theories, certainly does not become our store of knowledge by our mere possession of it: the renaissance, or rather the preceding dark ages, teach us that physical possession is not enough. Is it then necessary that anybody know all the contents of those volumes before they can be called "our science"? This may be a defensible point of view but if it were accepted science would have reached its limits already, might have reached them quite some time ago. Is it then enough that there be, in our society, for every volume a person who is fully familiar with it? No, because there may be contradictions between the statements of the various volumes which would remain hidden if everyone knew only part of them. Science is an edifice, not a pile of bricks, valuable as such a pile may be.

I would say that a store of knowledge can reasonably be called "our science" if there are people who are competent to learn and use any part of it, who would like to know each part of it even if they realize that they cannot, and if one has good assurance that the parts are not contradictory but form a whole. The section on elasticity must use the same picture of the structure of iron on which the section on magnetism is based.

Limits of "Our Science"

If the above is accepted as a fair description of what may be called "our science" then its limitations are in the human intellect, in its capacity for interest and learning, in its memory and facilities for communication. All these are surely related to the finite span of the human life. In fact, if we accept the above, science is already changing not only by acquiring new territories, but partly also by shifting from older to new fields. We forget things and focus our attention on more recent developments. Right now, the older parts of science cease to be parts of our science not so much because we have no assurance that they fit into the new picture—I believe they do—but rather because nobody has a strong desire to know them, at least nobody who is interested in the new parts.

Surely, the possibilities of this type of growth are very far from being exhausted. Today, we are neglecting the theory of solids, in which a student has to study perhaps six hundred papers before he reaches the frontiers and can do research on his own; we concentrate instead on quantum electrodynamics, in which he has to study six papers. Tomorrow, we may give up a whole science, such as chemistry, and concentrate on something that is less explored. These changes in interest are, furthermore, surely not arbitrary but in most cases well justified, inasmuch as the new subject is deeper than the abandoned one, starts from more fundamental realizations, and embraces the old one. The properties of solids follow from the principles of quantum electrodynamics and this discipline is, in addition, able to deal with many phenomena besides those important for solids.

One should realize, nevertheless, that the embracing of the old subject by the new discipline is somewhat illusory. Thus the theory of solids is relinquished by the student of quantum electrodynamics in a very real sense because the human intellect is not powerful enough to derive the important properties of solids from quantum theory, unless it has made a particular, both experimental and theoretical, study to develop the idealizations and approximations which are useful for the description of solids. Only an unusual intellect could guess on the basis of the principles of ordinary quantum theory that there are solids and that they consist of regular lattice-like arrangements of the atoms. No human intellect could overlook, as a matter of course, the significance and role of the defects of these lattices. The equations of quantum theory may form the words of a magic oracle which describes the phenomena of crystal physics in a wonderfully condensed fashion. However, no human intellect can understand this oracle without using a commentary to its words, the length of this commentary being in the same proportion to the condensate of the oracle as is the whole Bible to Leviticus 19: 18. There is clearly a limit beyond which condensation, elevating though it may be as a purpose *per se,* is not useful for storing information. Present day condensation in physics has certainly reached this limit.

Shift of the Second Type

The question now comes up whether science will at least be able to continue the type of shifting growth indefinitely in which the new discipline is deeper than the older one and embraces it at least virtually. The answer is, in my opinion, no, because the shifts in the above sense always involve digging one layer deeper into the "secrets of nature," and involve a longer series of concepts based on the previous ones, which are thereby recognized as "mere approximations." Thus, in the example above, first ordinary mechanics had to be replaced by quantum mechanics, thus recognizing the approximate nature and limitation of ordinary mechanics to macroscopic phenomena. Then, ordinary mechanics had to be recognized to be inadequate from another point of view and replaced by field theories. Finally, the approximate nature and limitation to small velocities of all of the above concepts had to be uncovered. Thus, relativistic quantum theory is at least four layers deep; it operates with three successive types of concepts all of which are recognized to be inadequate and are replaced by a more profound one in the fourth step. This is, of course, the charm and beauty of the relativistic quantum theory and of all fundamental research in physics. However, it also shows the limits of this type of development. The recognizing of an inadequacy in the concepts of the tenth layer and the replacing of it with the more refined concepts of the eleventh layer will be much less of an event than the discovery of the theory of relativity was. It will, furthermore, require a much more elaborate and a much longer study to arrive at an understanding of the roots of the evil than was the study needed to appreciate the discrepancies which were eliminated by the theory of relativity. It is not difficult to imagine a stage in which the new student will no longer be interested, perhaps will not be able any more, to dig through the already accumulated layers in order to do research at the frontier. The number of physics graduate students will then drop and the shift of science to new territories will be more drastic than the shifts we are accustomed to: the new discipline in fashion will not embrace physics any more in the same way, as, for instance, quantum theory embraces classical physics. I will call this type of shift the shift of the second type.

The above picture assumes that, in order to understand a growing body of phenomena, it will be necessary to introduce deeper and deeper concepts into physics and that this development will not end by the discovery of the final and perfect concepts. I believe that this is true: we have no right to expect that our intellect can formulate perfect concepts for the full understanding of inanimate nature's phenomena. However, the second type of shift will occur also if we do, because science does not seem to be viable if no research is being done on its outskirts and the interest will soon flag in a completed subject. It is possible also that neither of the two alternatives will come to pass, that it will never be decided whether the concepts of the tenth layer are adequate "in principle" for the understanding of the inanimate world. Absence of interest and the weakness of the human intellect may easily combine to postpone indefinitely the determination of the full adequacy of the nth layer of concepts. In that case physics will be left by the wayside, in a somewhat similar fashion to the way in which the phenomena connected with superconductivity are apparently being left by the wayside, most physicists not feeling an acute sense of unhappiness about it.

The second type of shift will not be all resignation. In fact, many feel nowadays that the life sciences and the science of the minds of both animals and men have already been neglected too long. Our picture of the world would surely be more rounded if we knew more about the minds of men and animals, their customs and habits. The second type of shift may mean, however, the acknowledgment that we are unable to arrive at the full understanding of even the inanimate world, just as, a few centuries ago, man came to the conclusion that he has no very good chance to foresee what will happen to his soul after the death of his body. We all continue to feel a frustration because of our inability to foresee our soul's ultimate fate. Although we do not speak about it, we all know that the objectives of our science are, from a general human point of view, much more modest than the objectives of, say, the Greek science were; that our science is more successful in giving us power than in giving us knowledge of truly human interest. The development of the natural sciences was, however, not less vigorous because of the ensuing sense of frustration. Similarly, the vigor of work in the fields to which shifts of the second type will lead will not be smaller because we shall have abandoned the full realization of our dreams concerning an earlier field.

However, the second type of shift will mean some new resignation and also mark a turning point in the existence of science, taking science in the sense of our definition. When shifts of the second type will have occurred in relevant numbers, science will lose some of the attraction on the young mind which it now holds. It will be something altogether different, a bit less fascinating. The wonderful elation which we scientists now are experiencing, and which comes from the new feeling of the power of our intellect, will be somewhat dampened by the recognition of the limits of that power. We will have to acquiesce in the fact that our intellect's toil cannot give us a satisfactory picture of the world such as the Greeks dreamed to attain, in an effortless way, by easy speculation.

Stabilizing Forces

Many of us will be inclined to make light of the preceding argument and say that science has a natural vitality by which it will overcome the limits which we, small minds of today, imagine to perceive in its path. There surely is much truth in this statement and we shall shortly turn to elements of elasticity in the whole picture which support it. However, I believe that the darker picture is the fundamentally correct one and that our instinctive desire not to believe it is the desire and ability of the human mind not to think of repugnant events in the future if their threat has no accurately foreseeable date. However, great changes, and often very unwanted changes, do take place and the elasticity of nature only delays them: buffaloes did die out as sources of food; the role of the individual warrior has vanished; the detailed explanation of the holy writings, once the only subject worthy of human studies, has ceased to be an element of our culture; Malthus' dire predictions are sure to come true at least in some respects. All the forecasts predicting these events were once resented by large groups just as we resent and resist the statement of the insufficiency of science.

Can we see even today signals of the crisis in science? Perhaps. The difficulty in penetrating to the frontiers of physics has been mentioned before. It is already so serious for the average human mind that only a negligible fraction of our contemporaries really feels the force of the arguments of quantum and relativity theories. Chemistry has grown so big that very few people can keep an even loose acquaintance with all its ramifications. Shifts of the first type are going on in these sciences constantly, some of them being the butts of constant jokes.

The clearest sign of the growing realization that the capacity of our intellect limits the volume of science is the number of queries which we hear every day, whether this or that piece of research "is worth doing." In almost all such cases, the problem posed is interesting, the proposed method of attack shows elements of ingenuity, and the answer, whatever it may turn out to be, can be expected to be worth remembering. However, the questioner realizes how great is the number of problems of similar importance, how limited the time and memory of those to whom the results will be of interest. He wonders whether his proposed work would not remain submerged in the mass of literature, with nobody taking time and energy fully to understand and appreciate it. Hence the query. Similar doubts on the "worth" of some proposed research must have arisen at all times. It seems to me doubtful, however, that they ever were as deep as they are now, and concerned as intrinsically interesting problems. I believe I have observed an increase in the frequency of these queries and doubts even during my own short scientific life.

Recently, M. Fierz, in a very thoughtful article, has pointed to what may well become in time a shift of the second type. He pointed out that both physics and psychology claim to be all-embracing disciplines: the first because it endeavors to describe all nature; the second because it deals with all mental phenomena, and nature exists for us only because we have cognizance of it. Fierz points out that the pictures of the world which these two disciplines project into us are not necessarily contradictory. However, it surely is difficult if not impossible to recognize the two pictures as only different aspects of the same thing. Furthermore, it is hardly an exaggeration to say that no psychologist understands the philosophy of modern physics. Conversely, only the exceptional physicist understands the language of the psychologist. Of course, psychology's philosophy is as yet too vague to draw definite conclusions. However, it is not impossible that we, or our students, are going to witness a real split of science right here.

It would be foolish to draw far reaching conclusions from the emergence of two sciences, both of which may claim to be all embracing and between the concepts and statements of which one cannot, at present, see any real similarity. Both may yet be united into a deeper common discipline without overtaxing our mind's capacity for abstraction. Altogether, there are many favorable stabilizing effects which can delay the balkanization of science for very long periods. Some of these are methodological: as we understand discoveries more fully, we will be able to explain them better. It is certainly no accident that we have scores of excellent books on thermodynamics but had surely until recently nothing comparable in quantum theory. Relativity theory was understood, so it was claimed, twenty-five years ago only by two—today we teach its principles to undergraduates. Examples of improving teaching techniques by both minor simplifications and by spectacular "condensations" and generalizations are in fact too obvious to bear enumeration.

Another important stabilizing effect will be the reduction of the size of disciplines by elimination of parts of it. An example which must have struck everyone of my age is that the theory of elliptic functions—a theory as spectacular in its methods and successes as any part of modern mathematics—is right now falling into oblivion. This is a shift of the first kind to which even the queen of sciences is not immune. As such it keeps mathematics more learnable.

Finally, it is not impossible that we'll breed during the coming centuries a human whose power of recollection, whose facility of abstraction, is greater than ours. Or at least that we make a greater and more aptly guided effort to select among the young those best suited for furthering science.

There is, on the other hand, a circumstance which will undoubtedly have an opposite effect. Thirst for knowledge, curiosity concerning the extent of one's mental faculties, and a healthy sense of rivalry, are strong stimulants of the young scientist and will continue to spur him along also in the future. They are, however, not his only motives: the desire to improve the lot of mankind, to extend its power, is also a traditional trait of scientists. These latter incentives are, however, waning, at least as far as the natural sciences are concerned, with the advent of man's full mastery of the elements, with the increasing realization that the economic welfare of man is a question of organization rather than a problem of production. The effect of the loss of this incentive will certainly be present; its magnitude is unpredictable.

Cooperative Research

If science is expected to grow so great, both in the comprehensiveness of its subject and also in depth, that the human mind will not be able to embrace it, that the life span of man will not be long enough to penetrate to its fringes in time to enlarge it, could several people not form a team and accomplish jointly what no single person can accomplish? Instead of returning with Shaw to Methuselah, can we find a new way to enlarge the capacity of human intellect by the juxtaposition of several individual intellects rather than by extending a single one? This is a possibility which is so little explored that all that one may say about it must remain highly speculative—much more speculative, I believe, than the rest of this article. The possibilities of cooperative research have to be explored, however, to a much greater extent than they have been so far because they form the only visible hope for a new lease on life for science when it will have grown too large for a single individual.

Most of us scientists are too individualistic to take cooperative research too seriously. As the founder of relativity theory once remarked, he cannot imagine how relativity theory could have been conceived by a group. Indeed, if we think of the present day research groups, working under a group leader who received his assignment through a section chief, the idea becomes amusingly absurd. Clearly, no fundamental change in our way of thinking can come about that way and no such fundamental change is intended by the groups referred to.

The case against group research can be stated more rationally on the basis of Poincaré's keen analysis of the nature of mathematical discovery. It is, I believe, our intuitive awareness of the facts which he and Hadamard have expressed so aptly which makes us smile at the idea of group research. Poincaré and Hadamard have recognized that, unlike most thinking which goes on in the upper consciousness, the really relevant mathematical thinking is not done in words. In fact, it happens somewhere so deep in the subconscious that the thinker is usually not even aware of what is going on inside of him.

It is my opinion that the role of subconscious thinking is equally important in other sciences, that it is decisive even in the solution of apparently trivial technical details. An experimentalist friend once told me (this was some twenty years ago) that if he could not find the leak in his vacuum system he usually felt like going for a walk, and very often, when he returned from the walk, he knew exactly where the leak was. The problem of group research is, therefore, to give free rein to the inventiveness of the subconscious of the individual but, at the same time, have available for him the whole store of knowledge of the group.

It is certainly impossible to tell now whether and how this can be accomplished. It will surely need a much more intimate symbiosis between collaborators than has been established to date. Part, but only part, of this more intimate symbiosis will be a higher faculty for the communication of ideas and information than we have developed so far. If group research is to be fully effective, it will also need a much deeper understanding of the functioning of the human mind than we now have. However, neither of these is impossible; in fact, we may be closer to both than we suspect.

Meanwhile, we should keep two facts in mind. The first is that the difficulty in the future development of science, which we have envisaged before, is based in the first place on the limited capacity of the human mind, not on its limited depth. Even if the depth, which is more intimately based on subconscious thinking, could not be increased, the first obstacle, the limitation of the capacity, might well be cut back by teamwork. Second, we should not forget that while it is true that relativity theory could not have been conceived by teamwork, the structure of the George Washington Bridge, and probably even that of the Hanford nuclear reactors, could not have been thought out by a single individual. The problem of group research is to avoid suppressing the subconscious thinking of the individual but to make available for him the information and to some degree even the unfinished ideas of his collaborators. Success of this may mean that the limitations of "our science" which were described above are limitations only for individualist science.

It is depressing for every scientist and for every person to have to conclude that his principal motive, or that of his epoch, is not here to stay. However, humanity's goals and ideals have shifted already several times during our known history. In addition, it must fill us with pride to believe that we are living in the heroic age of science, in the epoch in which the individual's abstract knowledge of nature, and, we may hope, also of himself, is increasing more rapidly and perhaps to a higher level than it ever has before or will afterwards. It is uncomfortable to believe that our ideals may pass as the Round Table's illusions disappeared. Still, we live in the heroic age of these ideals.

The Unreasonable Effectiveness of Mathematics in the Natural Sciences

E. P. Wigner

Symmetries and Reflections.
Indiana University Press, Bloomington, Indiana, 1967, pp. 222–237

"and it is probable that there is some secret here which remains to be discovered." —C. S. Peirce

There is a story about two friends, who were classmates in high school, talking about their jobs. One of them became a statistician and was working on population trends. He showed a reprint to his former classmate. The reprint started, as usual, with the Gaussian distribution and the statistician explained to his former classmate the meaning of the symbols for the actual population, for the average population, and so on. His classmate was a bit incredulous and was not quite sure whether the statistician was pulling his leg. "How can you know that?" was his query. "And what is this symbol here?" "Oh," said the statistician, "this is π." "What is that?" "The ratio of the circumference of the circle to its diameter." "Well, now you are pushing your joke too far," said the classmate, "surely the population has nothing to do with the circumference of the circle."

Naturally, we are inclined to smile about the simplicity of the classmate's approach. Nevertheless, when I heard this story, I had to admit to an eerie feeling because, surely, the reaction of the classmate betrayed only plain common sense. I was even more confused when, not

Richard Courant Lecture in Mathematical Sciences delivered at New York University, May 11, 1959. Reprinted by permission from *Communications in Pure and Applied Mathematics*, Vol. 13, No. 1 (February, 1960). Copyright by John Wiley & Sons, Inc.

many days later, someone came to me and expressed his bewilderment[1] with the fact that we make a rather narrow selection when choosing the data on which we test our theories. "How do we know that, if we made a theory which focuses its attention on phenomena we disregard and disregards some of the phenomena now commanding our attention, that we could not build another theory which has little in common with the present one but which, nevertheless, explains just as many phenomena as the present theory?" It has to be admitted that we have no definite evidence that there is no such theory.

The preceding two stories illustrate the two main points which are the subjects of the present discourse. The first point is that mathematical concepts turn up in entirely unexpected connections. Moreover, they often permit an unexpectedly close and accurate description of the phenomena in these connections. Secondly, just because of this circumstance, and because we do not understand the reasons of their usefulness, we cannot know whether a theory formulated in terms of mathematical concepts is uniquely appropriate. We are in a position similar to that of a man who was provided with a bunch of keys and who, having to open several doors in succession, always hit on the right key on the first or second trial. He became skeptical concerning the uniqueness of the coordination between keys and doors.

Most of what will be said on these questions will not be new; it has probably occurred to most scientists in one form or another. My principal aim is to illuminate it from several sides. The first point is that the enormous usefulness of mathematics in the natural sciences is something bordering on the mysterious and that there is no rational explanation for it. Second, it is just this uncanny usefulness of mathematical concepts that raises the question of the uniqueness of our physical theories. In order to establish the first point, that mathematics plays an unreasonably important role in physics, it will be useful to say a few words on the question, "What is mathematics?", then, "What is physics?", then, how mathematics enters physical theories, and last, why the success of mathematics in its role in physics appears so baffling. Much less will be said on the second point: the uniqueness of the theories of physics. A proper answer to this question would require elaborate experimental and theoretical work which has not been undertaken to date.

---

[1] The remark to be quoted was made by F. Werner when he was a student in Princeton.

What Is Mathematics?

Somebody once said that philosophy is the misuse of a terminology which was invented just for this purpose.[2] In the same vein, I would say that mathematics is the science of skillful operations with concepts and rules invented just for this purpose. The principal emphasis is on the invention of concepts. Mathematics would soon run out of interesting theorems if these had to be formulated in terms of the concepts which already appear in the axioms. Furthermore, whereas it is unquestionably true that the concepts of elementary mathematics and particularly elementary geometry were formulated to describe entities which are directly suggested by the actual world, the same does not seem to be true of the more advanced concepts, in particular the concepts which play such an important role in physics. Thus, the rules for operations with pairs of numbers are obviously designed to give the same results as the operations with fractions which we first learned without reference to "pairs of numbers." The rules for the operations with sequences, that is, with irrational numbers, still belong to the category of rules which were determined so as to reproduce rules for the operations with quantities which were already known to us. Most more advanced mathematical concepts, such as complex numbers, algebras, linear operators, Borel sets—and this list could be continued almost indefinitely—were so devised that they are apt subjects on which the mathematician can demonstrate his ingenuity and sense of formal beauty. In fact, the definition of these concepts, with a realization that interesting and ingenious considerations could be applied to them, is the first demonstration of the ingeniousness of the mathematician who defines them. The depth of thought which goes into the formulation of the mathematical concepts is later justified by the skill with which these concepts are used. The great mathematician fully, almost ruthlessly, exploits the domain of permissible reasoning and skirts the impermissible. That his recklessness does not lead him into a morass of contradictions is a miracle in itself: certainly it is hard to believe that our reasoning power was brought, by Darwin's process of natural selection, to the perfection which it seems to possess. However, this is not our present subject. The principal point which will have to be recalled later is that the mathematician could

---

[2] This statement is quoted here from W. Dubislav's *Die Philosophie der Mathematik in der Gegenwart* (Berlin: Junker and Dünnhaupt Verlag, 1932), p. 1.

formulate only a handful of interesting theorems without defining concepts beyond those contained in the axioms and that the concepts outside those contained in the axioms are defined with a view of permitting ingenious logical operations which appeal to our aesthetic sense both as operations and also in their results of great generality and simplicity.[3]

The complex numbers provide a particularly striking example for the foregoing. Certainly, nothing in our experience suggests the introduction of these quantities. Indeed, if a mathematician is asked to justify his interest in complex numbers, he will point, with some indignation, to the many beautiful theorems in the theory of equations, of power series, and of analytic functions in general, which owe their origin to the introduction of complex numbers. The mathematician is not willing to give up his interest in these most beautiful accomplishments of his genius.[4]

What Is Physics?

The physicist is interested in discovering the laws of inanimate nature. In order to understand this statement, it is necessary to analyze the concept, "law of nature."

The world around us is of baffling complexity and the most obvious fact about it is that we cannot predict the future. Although the joke attributes only to the optimist the view that the future is uncertain, the optimist is right in this case: the future is unpredictable. It is, as Schrödinger has remarked, a miracle that in spite of the baffling complexity of the world, certain regularities in the events could be discovered (1).[*] One such regularity, discovered by Galileo, is that two rocks, dropped at the same time from the same height, reach the ground at the same time. The laws of nature are concerned with such regularities. Galileo's regularity is a prototype of a large class of regularities. It is a surprising regularity for three reasons.

The first reason that it is surprising is that it is true not only in Pisa, and in Galileo's time, it is true everywhere on the Earth, was always

[3] M. Polanyi, in his *Personal Knowledge* (Chicago: University of Chicago Press, 1958), says: "All these difficulties are but consequences of our refusal to see that mathematics cannot be defined without acknowledging its most obvious feature: namely, that it is interesting" (page 188).

[4] The reader may be interested, in this connection, in Hilbert's rather testy remarks about intuitionism which "seeks to break up and to disfigure mathematics," *Abh. Math. Sem.*, Univ. Hamburg, 157 (1922), or *Gesammelte Werke* (Berlin: Springer, 1935), p. 188.

[*] The numbers in parentheses refer to the References at the end of the article.

true, and will always be true. This property of the regularity is a recognized invariance property and, as I had occasion to point out some time ago (2), without invariance principles similar to those implied in the preceding generalization of Galileo's observation, physics would not be possible. The second surprising feature is that the regularity which we are discussing is independent of so many conditions which could have an effect on it. It is valid no matter whether it rains or not, whether the experiment is carried out in a room or from the Leaning Tower, no matter whether the person who drops the rocks is a man or a woman. It is valid even if the two rocks are dropped, simultaneously and from the same height, by two different people. There are, obviously, innumerable other conditions which are all immaterial from the point of view of the validity of Galileo's regularity. The irrelevancy of so many circumstances which *could* play a role in the phenomenon observed has also been called an invariance (2). However, this invariance is of a different character from the preceding one since it cannot be formulated as a general principle. The exploration of the conditions which do, and which do not, influence a phenomenon is part of the early experimental exploration of a field. It is the skill and ingenuity of the experimenter which show him phenomena which depend on a relatively narrow set of relatively easily realizable and reproducible conditions.[5] In the present case, Galileo's restriction of his observations to relatively heavy bodies was the most important step in this regard. Again, it is true that if there were no phenomena which are independent of all but a manageably small set of conditions, physics would be impossible.

The preceding two points, though highly significant from the point of view of the philosopher, are not the ones which surprised Galileo most, nor do they contain a specific law of nature. The law of nature is contained in the statement that the length of time which it takes for a heavy object to fall from a given height is independent of the size, material, and shape of the body which drops. In the framework of Newton's second "law," this amounts to the statement that the gravitational force which acts on the falling body is proportional to its mass but independent of the size, material, and shape of the body which falls.

[5] See, in this connection, the graphic essay of M. Deutsch, *Daedalus*, 87, 86 (1958). A. Shimony has called my attention to a similar passage in C. S. Peirce's *Essays in the Philosophy of Science* (New York: The Liberal Arts Press, 1957), p. 237.

Unreasonable Effectiveness of Mathematics in Natural Sciences 227

The preceding discussion is intended to remind us, first, that it is not at all natural that "laws of nature" exist, much less that man is able to discover them.[6] The present writer had occasion, some time ago, to call attention to the succession of layers of "laws of nature," each layer containing more general and more encompassing laws than the previous one and its discovery constituting a deeper penetration into the structure of the universe than the layers recognized before (3). However, the point which is most significant in the present context is that all these laws of nature contain, in even their remotest consequences, only a small part of our knowledge of the inanimate world. All the laws of nature are conditional statements which permit a prediction of some future events on the basis of the knowledge of the present, except that some aspects of the present state of the world, in practice the overwhelming majority of the determinants of the present state of the world, are irrelevant from the point of view of the prediction. The irrelevancy is meant in the sense of the second point in the discussion of Galileo's theorem.[7]

As regards the present state of the world, such as the existence of the earth on which we live and on which Galileo's experiments were performed, the existence of the sun and of all our surroundings, the laws of nature are entirely silent. It is in consonance with this, first, that the laws of nature can be used to predict future events only under exceptional circumstances—when all the relevant determinants of the present state of the world are known. It is also in consonance with this that the construction of machines, the functioning of which he can foresee, constitutes the most spectacular accomplishment of the physicist. In these machines, the physicist creates a situation in which all the relevant coordinates are known so that the behavior of the machine can be predicted. Radars and nuclear reactors are examples of such machines.

The principal purpose of the preceding discussion is to point out that the laws of nature are all conditional statements and they relate only to a very small part of our knowledge of the world. Thus, classical mechanics, which is the best known prototype of a physical theory, gives the second derivatives of the positional coordinates of all bodies, on

---

[6] E. Schrödinger, in his *What Is Life* (Cambridge: Cambridge University Press, 1945), p. 31, says that this second miracle may well be beyond human understanding.

[7] The writer feels sure that it is unnecessary to mention that Galileo's theorem, as given in the text, does not exhaust the content of Galileo's observations in connection with the laws of freely falling bodies.

the basis of the knowledge of the positions, etc., of these bodies. It gives no information on the existence, the present positions, or velocities of these bodies. It should be mentioned, for the sake of accuracy, that we discovered about thirty years ago that even the conditional statements cannot be entirely precise: that the conditional statements are probability laws which enable us only to place intelligent bets on future properties of the inanimate world, based on the knowledge of the present state. They do not allow us to make categorical statements, not even categorical statements conditional on the present state of the world. The probabilistic nature of the "laws of nature" manifests itself in the case of machines also, and can be verified, at least in the case of nuclear reactors, if one runs them at very low power. However, the additional limitation of the scope of the laws of nature[8] which follows from their probabilistic nature will play no role in the rest of the discussion.

The Role of Mathematics in Physical Theories

Having refreshed our minds as to the essence of mathematics and physics, we should be in a better position to review the role of mathematics in physical theories.

Naturally, we do use mathematics in everyday physics to evaluate the results of the laws of nature, to apply the conditional statements to the particular conditions which happen to prevail or happen to interest us. In order that this be possible, the laws of nature must already be formulated in mathematical language. However, the role of evaluating the consequences of already established theories is not the most important role of mathematics in physics. Mathematics, or, rather, applied mathematics, is not so much the master of the situation in this function: it is merely serving as a tool.

Mathematics does play, however, also a more sovereign role in physics. This was already implied in the statement, made when discussing the role of applied mathematics, that the laws of nature must have been formulated in the language of mathematics to be an object for the use of applied mathematics. The statement that the laws of nature are written in the language of mathematics was properly made three hundred years ago[9]; it is now more true than ever before. In order to show the importance which mathematical concepts possess in the formulation

[8] See, for instance, E. Schrödinger, reference (1).
[9] It is attributed to Galileo.

Unreasonable Effectiveness of Mathematics in Natural Sciences 229

of the laws of physics, let us recall, as an example, the axioms of quantum mechanics as formulated, explicitly, by the great mathematician, von Neumann, or, implicitly, by the great physicist, Dirac (4, 5). There are two basic concepts in quantum mechanics: states and observables. The states are vectors in Hilbert space, the observables self-adjoint operators on these vectors. The possible values of the observations are the characteristic values of the operators—but we had better stop here lest we engage in a listing of the mathematical concepts developed in the theory of linear operators.

It is true, of course, that physics chooses certain mathematical concepts for the formulation of the laws of nature, and surely only a fraction of all mathematical concepts is used in physics. It is true also that the concepts which were chosen were not selected arbitrarily from a listing of mathematical terms but were developed, in many if not most cases, independently by the physicist and recognized then as having been conceived before by the mathematician. It is not true, however, as is so often stated, that this had to happen because mathematics uses the simplest possible concepts and these were bound to occur in any formalism. As we saw before, the concepts of mathematics are not chosen for their conceptual simplicity—even sequences of pairs of numbers are far from being the simplest concepts—but for their amenability to clever manipulations and to striking, brilliant arguments. Let us not forget that the Hilbert space of quantum mechanics is the complex Hilbert space, with a Hermitean scalar product. Surely to the unpreoccupied mind, complex numbers are far from natural or simple and they cannot be suggested by physical observations. Furthermore, the use of complex numbers is in this case not a calculational trick of applied mathematics but comes close to being a necessity in the formulation of the laws of quantum mechanics. Finally, it now begins to appear that not only complex numbers but so-called analytic functions are destined to play a decisive role in the formulation of quantum theory. I am referring to the rapidly developing theory of dispersion relations.

It is difficult to avoid the impression that a miracle confronts us here, quite comparable in its striking nature to the miracle that the human mind can string a thousand arguments together without getting itself into contradictions, or to the two miracles of the existence of laws of nature and of the human mind's capacity to divine them. The observation which comes closest to an explanation for the mathematical concepts' cropping up in physics which I know is Einstein's statement that

the only physical theories which we are willing to accept are the beautiful ones. It stands to argue that the concepts of mathematics, which
invite the exercise of so much wit, have the quality of beauty. However,
Einstein's observation can at best explain properties of theories which
we are willing to believe and has no reference to the intrinsic accuracy
of the theory. We shall, therefore, turn to this latter question.

Is the Success of Physical Theories Truly Surprising?

A possible explanation of the physicist's use of mathematics to formulate his laws of nature is that he is a somewhat irresponsible person. As a
result, when he finds a connection between two quantities which resembles a connection well-known from mathematics, he will jump at
the conclusion that the connection *is* that discussed in mathematics
simply because he does not know of any other similar connection. It is
not the intention of the present discussion to refute the charge that the
physicist is a somewhat irresponsible person. Perhaps he is. However,
it is important to point out that the mathematical formulation of the
physicist's often crude experience leads in an uncanny number of cases
to an amazingly accurate description of a large class of phenomena. This
shows that the mathematical language has more to commend it than
being the only language which we can speak; it shows that it is, in a very
real sense, the correct language. Let us consider a few examples.

The first example is the oft-quoted one of planetary motion. The laws
of falling bodies became rather well established as a result of experiments carried out principally in Italy. These experiments could not be
very accurate in the sense in which we understand accuracy today partly
because of the effect of air resistance and partly because of the impossibility, at that time, to measure short time intervals. Nevertheless, it
is not surprising that, as a result of their studies, the Italian natural
scientists acquired a familiarity with the ways in which objects travel
through the atmosphere. It was Newton who then brought the law of
freely falling objects into relation with the motion of the moon, noted
that the parabola of the thrown rock's path on the earth and the circle
of the moon's path in the sky are particular cases of the same mathematical object of an ellipse, and postulated the universal law of gravitation on the basis of a single, and at that time very approximate, numerical coincidence. Philosophically, the law of gravitation as formulated
by Newton was repugnant to his time and to himself. Empirically, it

Unreasonable Effectiveness of Mathematics in Natural Sciences 231

was based on very scanty observations. The mathematical language in which it was formulated contained the concept of a second derivative and those of us who have tried to draw an osculating circle to a curve know that the second derivative is not a very immediate concept. The law of gravity which Newton reluctantly established and which he could verify with an accuracy of about 4% has proved to be accurate to less than a ten thousandth of a per cent and became so closely associated with the idea of absolute accuracy that only recently did physicists become again bold enough to inquire into the limitations of its accuracy.[10] Certainly, the example of Newton's law, quoted over and over again, must be mentioned first as a monumental example of a law, formulated in terms which appear simple to the mathematician, which has proved accurate beyond all reasonable expectations. Let us just recapitulate our thesis on this example: first, the law, particularly since a second derivative appears in it, is simple only to the mathematician, not to common sense or to non-mathematically-minded freshmen; second, it is a conditional law of very limited scope. It explains nothing about the earth which attracts Galileo's rocks, or about the circular form of the moon's orbit, or about the planets of the sun. The explanation of these initial conditions is left to the geologist and the astronomer, and they have a hard time with them.

The second example is that of ordinary, elementary quantum mechanics. This originated when Max Born noticed that some rules of computation, given by Heisenberg, were formally identical with the rules of computation with matrices, established a long time before by mathematicians. Born, Jordan, and Heisenberg then proposed to replace by matrices the position and momentum variables of the equations of classical mechanics (6). They applied the rules of matrix mechanics to a few highly idealized problems and the results were quite satisfactory. However, there was, at that time, no rational evidence that their matrix mechanics would prove correct under more realistic conditions. Indeed, they say "if the mechanics as here proposed should already be correct in its essential traits." As a matter of fact, the first application of their mechanics to a realistic problem, that of the hydrogen atom, was given several months later, by Pauli. This application gave results in agreement with experience. This was satisfactory but still understandable because Heisenberg's rules of calculation were abstracted from prob-

[10] See, for instance, R. H. Dicke, *Am. Sci.*, 25 (1959).

lems which included the old theory of the hydrogen atom. The miracle occurred only when matrix mechanics, or a mathematically equivalent theory, was applied to problems for which Heisenberg's calculating rules were meaningless. Heisenberg's rules presupposed that the classical equations of motion had solutions with certain periodicity properties; and the equations of motion of the two electrons of the helium atom, or of the even greater number of electrons of heavier atoms, simply do not have these properties, so that Heisenberg's rules cannot be applied to these cases. Nevertheless, the calculation of the lowest energy level of helium, as carried out a few months ago by Kinoshita at Cornell and by Bazley at the Bureau of Standards, agrees with the experimental data within the accuracy of the observations, which is one part in ten million. Surely in this case we "got something out" of the equations that we did not put in.

The same is true of the qualitative characteristics of the "complex spectra," that is, the spectra of heavier atoms. I wish to recall a conversation with Jordan, who told me, when the qualitative features of the spectra were derived, that a disagreement of the rules derived from quantum mechanical theory and the rules established by empirical research would have provided the last opportunity to make a change in the framework of matrix mechanics. In other words, Jordan felt that we would have been, at least temporarily, helpless had an unexpected disagreement occurred in the theory of the helium atom. This was, at that time, developed by Kellner and by Hilleraas. The mathematical formalism was too clear and unchangeable so that, had the miracle of helium which was mentioned before not occurred, a true crisis would have arisen. Surely, physics would have overcome that crisis in one way or another. It is true, on the other hand, that physics as we know it today would not be possible without a constant recurrence of miracles similar to the one of the helium atom, which is perhaps the most striking miracle that has occurred in the course of the development of elementary quantum mechanics, but by far not the only one. In fact,.the number of analogous miracles is limited, in our view, only by our willingness to go after more similar ones. Quantum mechanics had, nevertheless, many almost equally striking successes which gave us the firm conviction that it is, what we call, correct.

The last example is that of quantum electrodynamics, or the theory of the Lamb shift. Whereas Newton's theory of gravitation still had obvious connections with experience, experience entered the formulation of matrix mechanics only in the refined or sublimated form of Heisenberg's prescriptions. The quantum theory of the Lamb shift, as conceived by Bethe and established by Schwinger, is a purely mathematical theory and the only direct contribution of experiment was to show the existence of a measurable effect. The agreement with calculation is better than one part in a thousand.

The preceding three examples, which could be multiplied almost indefinitely, should illustrate the appropriateness and accuracy of the mathematical formulation of the laws of nature in terms of concepts chosen for their manipulability, the "laws of nature" being of almost fantastic accuracy but of strictly limited scope. I propose to refer to the observation which these examples illustrate as the empirical law of epistemology. Together with the laws of invariance of physical theories, it is an indispensable foundation of these theories. Without the laws of invariance the physical theories could have been given no foundation of fact; if the empirical law of epistemology were not correct, we would lack the encouragement and reassurance which are emotional necessities, without which the "laws of nature" could not have been successfully explored. Dr. R. G. Sachs, with whom I discussed the empirical law of epistemology, called it an article of faith of the theoretical physicist, and it is surely that. However, what he called our article of faith can be well supported by actual examples—many examples in addition to the three which have been mentioned.

The Uniqueness of the Theories of Physics

The empirical nature of the preceding observation seems to me to be self-evident. It surely is not a "necessity of thought" and it should not be necessary, in order to prove this, to point to the fact that it applies only to a very small part of our knowledge of the inanimate world. It is absurd to believe that the existence of mathematically simple expressions for the second derivative of the position is self-evident, when no similar expressions for the position itself or for the velocity exist. It is therefore surprising how readily the wonderful gift contained in the empirical law of epistemology was taken for granted. The ability of the human mind to form a string of 1000 conclusions and still remain "right," which was mentioned before, is a similar gift.

Every empirical law has the disquieting quality that one does not know its limitations. We have seen that there are regularities in the events in the world around us which can be formulated in terms of mathematical concepts with an uncanny accuracy. There are, on the other hand, aspects of the world concerning which we do not believe in the existence of any accurate regularities. We call these initial conditions. The question which presents itself is whether the different regularities, that is, the various laws of nature which will be discovered, will fuse into a single consistent unit, or at least asymptotically approach such a fusion. Alternatively, it is possible that there always will be some laws of nature which have nothing in common with each other. At present, this is true, for instance, of the laws of heredity and of physics. It is even possible that some of the laws of nature will be in conflict with each other in their implications, but each convincing enough in its own domain so that we may not be willing to abandon any of them. We may resign ourselves to such a state of affairs or our interest in clearing up the conflict between the various theories may fade out. We may lose interest in the "ultimate truth," that is, in a picture which is a consistent fusion into a single unit of the little pictures, formed on the various aspects of nature.

It may be useful to illustrate the alternatives by an example. We now have, in physics, two theories of great power and interest: the theory of quantum phenomena and the theory of relativity. These two theories have their roots in mutually exclusive groups of phenomena. Relativity theory applies to macroscopic bodies, such as stars. The event of coincidence, that is, in ultimate analysis of collision, is the primitive event in the theory of relativity and defines a point in space-time, or at least would define a point if the colliding particles were infinitely small. Quantum theory has its roots in the microscopic world and, from its point of view, the event of coincidence, or of collision, even if it takes place between particles of no spatial extent, is not primitive and not at all sharply isolated in space-time. The two theories operate with different mathematical concepts—the four dimensional Riemann space and the infinite dimensional Hilbert space, respectively. So far, the two theories could not be united, that is, no mathematical formulation exists to which both of these theories are approximations. All physicists believe that a union of the two theories is inherently possible and that we shall find it. Nevertheless, it is possible also to imagine that no union of the two theories can be found. This example illustrates the two possibilities, of union and of conflict, mentioned before, both of which are conceivable.

In order to obtain an indication as to which alternative to expect ultimately, we can pretend to be a little more ignorant than we are and place ourselves at a lower level of knowledge than we actually possess. If we can find a fusion of our theories on this lower level of intelligence, we can confidently expect that we will find a fusion of our theories also at our real level of intelligence. On the other hand, if we would arrive at mutually contradictory theories at a somewhat lower level of knowledge, the possibility of the permanence of conflicting theories cannot be excluded for ourselves either. The level of knowledge and ingenuity is a continuous variable and it is unlikely that a relatively small variation of this continuous variable changes the attainable picture of the world from inconsistent to consistent.[11]

Considered from this point of view, the fact that some of the theories which we know to be false give such amazingly accurate results is an adverse factor. Had we somewhat less knowledge, the group of phenomena which these "false" theories explain would appear to us to be large enough to "prove" these theories. However, these theories are considered to be "false" by us just for the reason that they are, in ultimate analysis, incompatible with more encompassing pictures and, if sufficiently many such false theories are discovered, they are bound to prove also to be in conflict with each other. Similarly, it is possible that the theories, which we consider to be "proved" by a number of numerical agreements which appears to be large enough for us, are false because they are in conflict with a possible more encompassing theory which is beyond our means of discovery. If this were true, we would have to expect conflicts between our theories as soon as their number grows beyond a certain point and as soon as they cover a sufficiently large number of groups of phenomena. In contrast to the article of faith of the theoretical physicist mentioned before, this is the nightmare of the theorist.

---

[11] This passage was written after a great deal of hesitation. The writer is convinced that it is useful, in epistemological discussions, to abandon the idealization that the level of human intelligence has a singular position on an absolute scale. In some cases it may even be useful to consider the attainment which is possible at the level of the intelligence of some other species. However, the writer also realizes that his thinking along the lines indicated in the text was too brief and not subject to sufficient critical appraisal to be reliable.

Let us consider a few examples of "false" theories which give, in view of their falseness, alarmingly accurate descriptions of groups of phenomena. With some goodwill, one can dismiss some of the evidence which these examples provide. The success of Bohr's early and pioneering ideas on the atom was always a rather narrow one and the same applies to Ptolemy's epicycles. Our present vantage point gives an accurate description of all phenomena which these more primitive theories can describe. The same is not true any longer of the so-called free-electron theory, which gives a marvellously accurate picture of many, if not most, properties of metals, semiconductors, and insulators. In particular, it explains the fact, never properly understood on the basis of the "real theory," that insulators show a specific resistance to electricity which may be 10^{26} times greater than that of metals. In fact, there is no experimental evidence to show that the resistance is not infinite under the conditions under which the free-electron theory would lead us to expect an infinite resistance. Nevertheless, we are convinced that the free-electron theory is a crude approximation which should be replaced, in the description of all phenomena concerning solids, by a more accurate picture.

If viewed from our real vantage point, the situation presented by the free-electron theory is irritating but is not likely to forebode any inconsistencies which are unsurmountable for us. The free-electron theory raises doubts as to how much we should trust numerical agreement between theory and experiment as evidence for the correctness of the theory. We are used to such doubts.

A much more difficult and confusing situation would arise if we could, some day, establish a theory of the phenomena of consciousness, or of biology, which would be as coherent and convincing as our present theories of the inanimate world. Mendel's laws of inheritance and the subsequent work on genes may well form the beginning of such a theory as far as biology is concerned. Furthermore, it is quite possible that an abstract argument can be found which shows that there is a conflict between such a theory and the accepted principles of physics. The argument could be of such abstract nature that it might not be possible to resolve the conflict, in favor of one or of the other theory, by an experiment. Such a situation would put a heavy strain on our faith in our theories and on our belief in the reality of the concepts which we form. It would give us a deep sense of frustration in our search for what I called "the ultimate truth." The reason that such a situation is conceiv-

Unreasonable Effectiveness of Mathematics in Natural Sciences   237

able is that, fundamentally, we do not know why our theories work so well. Hence, their accuracy may not prove their truth and consistency. Indeed, it is this writer's belief that something rather akin to the situation which was described above exists if the present laws of heredity and of physics are confronted.

Let me end on a more cheerful note. The miracle of the appropriateness of the language of mathematics for the formulation of the laws of physics is a wonderful gift which we neither understand nor deserve. We should be grateful for it and hope that it will remain valid in future research and that it will extend, for better or for worse, to our pleasure, even though perhaps also to our bafflement, to wide branches of learning.

The writer wishes to record here his indebtedness to Dr. M. Polanyi, who, many years ago, deeply influenced his thinking on problems of epistemology, and to V. Bargmann, whose friendly criticism was material in achieving whatever clarity was achieved. He is also greatly indebted to A. Shimony for reviewing the present article and calling his attention to C. S. Peirce's papers.

References

(1) Schrödinger, E., *Über Indeterminismus in der Physik* (Leipzig: J. A. Barth, 1932); also Dubislav, W., *Naturphilosophie* (Berlin: Junker und Dünnhaupt, 1933), Chap. 4.

(2) Wigner, E. P., "Invariance in Physical Theory," *Proc. Am. Phil. Soc.*, 93, 521-526 (1949), reprinted in this volume.

(3) Wigner, E. P., "The Limits of Science," *Proc. Am. Phil. Soc.*, 94, 422 (1950); also Margenau, H., *The Nature of Physical Reality* (New York: McGraw-Hill, 1950), Ch. 8, reprinted in this volume.

(4) Dirac, P. A. M., *Quantum Mechanics*, 3rd ed. (Oxford: Clarendon Press, 1947).

(5) von Neumann, J., *Mathematische Grundlagen der Quantenmechanik* (Berlin: Springer, 1932). English translation (Princeton, N.J.: Princeton Univ. Press, 1955).

(6) Born, M., and Jordan, P., "On Quantum Mechanics," Z. *Physik*, 34, 858-888 (1925). Born, M., Heisenberg, W., and Jordan, P., "On Quantum Mechanics," Part II, Z. *Physik*, 35, 557-615 (1926). (The quoted sentence occurs in the latter article, page 558.)

The Growth of Science –
Its Promise and Its Dangers

E. P. Wigner

Symmetries and Reflections.
Indiana University Press, Bloomington, Indiana, 1967, pp. 267–280

The Success of Science

The statistics of the growth of science in the U. S. during the postwar years read like a success story. The total expenditures for research and development were 3.5 billion dollars in 1947; they were over 20 billion in 1963. It is true that the gross national product also greatly increased during that period: it grew from about 200 billion dollars to almost 600. Nevertheless, even the percentage of the gross national product devoted to research and development more than doubled by 1963 its value of 1¾ per cent in 1947. This means that more than three people in a hundred now work directly or indirectly for increasing our store of knowledge, or for developing new methods of production of commodities now available, or on the design and production of new commodities. The increase of the annual federal expenditure on science and development was even more spectacular: it grew from one billion in 1947 to its present value of 15 billion dollars. None of these figures includes the compensation of the science teachers in our high schools and colleges—those on whom we depend to produce the scientists of tomorrow.[1]

J. F. Carlson lecture, presented April 13, 1964, at Iowa State College.
[1] The best source for statistical information is, of course, government publications, such as the Statistical Yearbooks of the U.S. and the publications of the National Science Foundation on Federal Funds for Science. A quick survey can be obtained from Spencer Klaw's article in *Fortune*, 158 (Sept., 1964).

Some of the increase in the attention which science now receives was caused, unquestionably, by the increasing need for the protection of our nation, by the requirements of the so-called cold war. Yet this point can be easily exaggerated. Thus, research on biology and medicine, having little to do with defense, was responsible for about 20 per cent of the federal expenditures for research and development, and the government also supports other nondefense sciences generously. Privately financed research—which was on the average over the last fifteen years about half of the total effort—is not directly connected with defense needs. There is every indication that science would grow almost equally rapidly if the defense needs were less urgent.

Furthermore, this growth took place not only in the expenditure of money and in the number of people participating in the scientific and technological endeavor, but also in the interest and attention of the average citizen. Twenty or so years ago, scientific discoveries were briefly mentioned on the back pages of newspapers and the reporting on them often lacked competence. Today, prominent newspapers have a competent science writer on their staff, and it is not unusual to find the report on a discovery started on the front page and a careful description of it continued through several columns further back. There is also an increased emphasis on science in our schools and colleges—and this in the face of an acute shortage of science teachers. As a result of the increasing interest of our fellow citizens in science, scientific theories and the thinking of scientists greatly influence the outlook on life of most of us.[2] The confidence in divine justice on Earth, in a happy afterlife in Heaven, had all been shattered by the turn of the century. Even the recognition of the human will as a primitive reality was shaken under the influence of the success of deterministic physics, only to be restored when—and perhaps because—deterministic physics proved to be inadequate. Genetics taught us that we are all captives of our inheritance, and biophysics that the human spirit, once considered to be supremely independent, can be deeply affected by drugs. The effect of science on our most fundamental convictions, on our image of the human spirit, may well be even more significant than its effect on our everyday life. As I said, the story of science reads like a success story.

Evidently, the fraction of gross national product devoted to science

[2] Cf. in this connection P. W. Bridgeman's essay in *The Nature of Physical Knowledge* [Bloomington: Indiana University Press (1960)].

and development cannot double many more times. However, there is every indication that it will continue to grow for some time and our way of life will be increasingly affected directly by scientific developments and indirectly by the scientist's mode of thinking. It seems to me, therefore, that the scientist has an obligation to try to understand the effects of science on our way of life and to try to act in such a way that the influence of science be for the better.[3] Because it could be for the worse. Most, if not all, that I have to say on this subject may not be very original or striking. However, there has been until recently very little discussion on the subject and I would feel more than satisfied if this writing contributed a small amount of stimulation to an imaginative discussion of the effect which the growth of science will have on science itself and on society.

Past Contributions of Science to Human Welfare

The effect of science and civilization on human happiness, although certainly present, cannot be measured. Its direction and magnitude have been, nevertheless, the subject of much discussion and debate. Freud, in particular, wrote with feeling and deep insight about *Civilization and Its Discontents*.[4] It is undeniable that our increased knowledge enables us to provide food and shelter for several times more people than could live on the Earth only a few centuries ago, and that these have a much longer average life span than did the man of earlier days. People who complain about inequities and suffering caused by our progress remind me a little of the hard-working teacher, whose class showed splendid progress but who complained that half of the class was still below average. It seems evident that even the least developed nation suffers less from want today than any nation did before. It is true that some of the romance of old days has been destroyed by modern methods of production, but much more of the misery has been abolished. People who knew primitive tribes in their primitive days of pre-civilization, such as Peter Freuchen, who knew Eskimos and Indians intimately, are eloquent in their praise of the blessings which contact with more advanced civilizations brought to these people.[5]

[3] The Interim Committee on the Social Aspects of Science of the American Association for the Advancement of Science states this case very compellingly.

[4] S. Freud, *Civilization and Its Discontents,* translated by Joan Reviere [Garden City, N.Y.: Doubleday (1958)].

[5] P. Freuchen, *Book of the Eskimos* [Cleveland: World Publishing Co. (1961)], p. 417.

The preceding discussion refers, of course, only to those effects of our culture which have brought freedom from want and alleviation of suffering. Furthermore, it refers perhaps more to the effects of civilization and primitive science than to science as we now like to understand it. It is, indeed, difficult to go beyond this because, as has been mentioned before, happiness cannot be measured and we know the conditions therefore only from experience and introspection. At the present stage of our understanding of human emotions, we cannot say with certainty whether an educated man or a more simple child of nature would be happier "on the average," given the freedom from want now prevailing in our nation. There is at least one exception to the preceding statement: the scientists themselves. The smaller number of work hours required by the community to supply its daily needs in reasonable abundance has meant the release of a large number of the community for the arts and sciences. It has been said that the only occupations which bring true joy and satisfaction are those of poets, artists, and scientists, and, of these, the scientists are apparently the happiest. Scientists derive an immense amount of pleasure from the study and understanding of science, and even more from discovering or rediscovering relations between phenomena, and from the discussion and communication of these to students and to each other.[6]

If the joy of learning, understanding, and discovering could be extended over large groups of people, the positive contribution of science to human satisfaction could vie with the negative one, which is the alleviation of suffering and want. Since the fraction of people in scientific endeavor constantly increases, can we not hope for such an extension of the positive contribution?

What we have to be concerned about is whether the growth of science will not change its character, whether it will still be the same science, bringing the same satisfactions to its disciples, if a much larger fraction of mankind would try to partake in its cultivation. Such a situation would cause, first, great changes in the mode of operation and the character of the scientific world. Second, it would bring a much more rapid spread of the scientific effort into new areas than would be the case otherwise. The rest of the thoughts which will be presented here will be concerned with problems raised by these changes.

[6] M. Polanyi, *Personal Knowledge: Toward a Post-Critical Philosophy* [Chicago: University of Chicago Press (1958)], speaks of an "intellectual passion."

It is natural to ask whether it is desirable for us scientists to discuss these questions. I believe so. First, if we do not bring them up, others will, and their discussion, which may not be always kind, should not find us unprepared. More importantly, as was said above, we have a moral duty to try to visualize the effects of our endeavor. There is at least one example in man's history when a most noble conception became, for a while, a parody of that conception.° We should do our best that history should not repeat itself in this regard.

The Subject of Science—Now and Tomorrow

When a physicist speaks about science, he thinks about the so-called natural sciences, by which he means the sciences dealing with inanimate nature. He believes, and I share this belief, that only concerning the behavior of these have we succeeded in developing a coherent set of regularities which are, in some well-defined sense, complete.

No similarly well-rounded body of knowledge is available concerning the functioning of the mind, even though ministers, psychologists, philosophers, and many others know more about it than we physicists realize. This knowledge, such as it is, is separated from our knowledge of the inanimate world and the points of contact are few.[7] However, the desire to build a bridge between the two areas—nay, the need to eliminate the chasm between them—is natural not only for the generalist. Its need has become apparent to the physicist as a result of the understanding of the structure of quantum theory,[8] and I am sure similar tendencies exist also on the other side. The simple fact is that it becomes increasingly evident that the primitive idea of separating body and soul is not a valid one, that "inanimate" is a limiting case which could be understood more deeply if it were recognized as such.[9] Simi-

---

° Some of my friends, when reading this passage, thought that I had communism in mind. Actually, and very unfortunately, Christianity is an equally good example.

[7] This point has been popularized, perhaps even exaggerated, in the well-known writings of C. P. Snow. See *The Two Cultures* and *The Scientific Revolution* [Cambridge: Cambridge University Press (1959)]. Cf., on the other side, the Preface of R. E. Peierls' *The Laws of Nature*, which emphasizes the more encyclopedic interests of the scientist than *The Two Cultures* would lead one to believe.

[8] See, for instance, F. London and E. Bauer, *La Théorie de l'observation en mécanique quantique* [Paris: Hermann and Co. (1939)], or this writer's article in *Am. J. Phys.*, 31, 6 (1963).

[9] See the present writer's "Remarks on the Mind-Body Question" in *The Scientist Speculates*, I. J. Good, editor [New York: Basic Books (1962)], reprinted in this volume.

larly, the dependence of the character and soul on the chemical constitution of the body is becoming increasingly, and even frighteningly,
evident.

The extension of our understanding from the inanimate world to the
whole of nature is indeed a noble task. It may even, eventually, lead to
the determination whether and under what circumstances a greater
satisfaction from life could be derived on the average. For reasons on
which I wish to enlarge a little, it will be greatly stimulated by the increase in the number of those who devote their attention to science. If
this attention were confined to those parts of science which are now under
intense cultivation, there would be an overcrowding of these fields and
the same new realizations would be made independently by many.
There are already signs for this in the physical sciences.[10] Such new
realizations would not give the individuals the kind of satisfaction which
comes from having had an influence on the development of science, and
they would not clearly contribute to the growth of our knowledge. As
such situations develop, and, as I said, they may be developing right
now, people will drift into new fields of endeavor—and what field would
be more attractive than where the most obvious gap in our understanding is? The drift toward new areas of science will not be easy because
it requires less effort to add to a building, the foundations of which are
already firmly laid, than to start new foundations on ground which appears to be soft. As a result, if the number of scientists did not increase,
the drift toward new areas would be slow. However, the interest of
many other active minds in a field will be unquestionably a strong stimulation to seek satisfaction in another branch of science, and the proliferation of scientists will surely stimulate the cultivation of new areas,
the shift of the second kind, as I called it when thinking about the
subject some years ago.[11]

Evidently, the development which I just described will be a whole-
some and desirable one. Does it hide any dangers? Not if it takes place
slowly enough so that we can remain adjusted to it. However, our
penetration into new fields of knowledge will unquestionably give us
new powers, powers which affect the mind more directly than the

[10] One of the most important discoveries in physics, that of parity-violation, was
made practically simultaneously by three groups; the "correct" interaction for beta
decay was also proposed thereupon, independently, by three groups.

[11] "The Limits of Science," *Proc. Am. Phil. Soc.*, 94, 4221 (1950), reprinted in
this volume.

physical conditions which we now can alter. Although the power of nuclear weapons has been much exaggerated, it is true that they enable us to alter the inanimate nature around us to an unprecedented extent. However, the governments have acquired a monopoly of them, and only those whom we have entrusted with such a responsibility can cause their employment on a scale that would be globally significant. It is also helpful that there are many people who understand the nature of nuclear weapons—the only global weapons for the present. On the other hand, if there were many means which could alter the conditions of life on our planet, and if there were no one person who would understand all of them and their interrelations, the balance would be much more precarious. Already, we have more trouble in controlling the spread of dangerous and of habit-forming drugs than of nuclear weapons, and they have caused more unhappiness than the latter. Their more dangerous nature is due to two circumstances: their manufacture needs much less of a concerted effort and their use does not advertise itself with a big bang.

What one may fear even more is something more subtle. If the purpose of life is individual happiness, why not acquire it by any means available, by detachment from reality, if this is the easiest way? Surely, this must be possible. Few can avoid thinking about this possibility when learning about the effects of modern drugs and the frequent consequences of modern psychiatric treatment. The deeply disturbing fact is that no logical argument can be brought against detachment from reality if it abolishes our pains and sorrows, as no logical argument can be brought against drug-addiction. Fortunately, most of us have a deep revulsion against both. Is this based only on tradition, or will mankind be saved by the fact that what we truly strive for is influence on the real world, exercised through ourselves and our children, rather than abstract happiness? Will the world be populated by the progeny of those in whom the desire for influence is particularly strong? These are frightening questions, largely tabu in our circles. Unless we can come to an equilibrium with them before our power to change the world of the mind becomes overwhelming, our increased knowledge and understanding of wider areas may bring more suffering than joy, it may bring a deterioration of our genetic heritage rather than the advent of a nobler man.

There is another point that should be mentioned. The progress of science will give increasing power to the individual scientist, and to small groups of scientists. This has not yet come to the fore in the course of the progress of the physical sciences, such as the development of nuclear weapons. No such weapon can be produced by the individual scientist; a much larger group effort, involving not only scientists but also industrial enterprises, is needed. However, the situation may be, and probably will be, different with respect to the power which the study of the life sciences and psychology will yield. At present, we are used to trusting the individual scientist and cannot imagine his harboring diabolical schemes for increasing his power. We believe that his desire for influence is not excessive. The same may not be true of groups of scientists, and may not be true of the management of such groups. In these, the desire for influence on the real world, which we recognized as an important factor for the preservation of humanity, may be excessive. They may want to use the power yielded by their research or that of their group to extend their influence in ways as ruthless as those of a dictator. Will it be necessary to establish strict controls in scientific institutions similar to those which we now have in the mints and weapons factories? Again, one is repelled by the mere idea. I feel reassured that the directors of the scientific institutions and the management of the laboratories with which I am familiar show no particular desire to accumulate power and influence. They are all meditative, kind, and almost self-effacing people. It is true, however, that in order to manage an enterprise which depends on the cooperation of many individuals, the manager must be able, and occasionally willing, to wield power. This may form a bad habit and corrupt some of the scientific leaders, just as political leaders are often corrupted by similar circumstances and opportunities. Nevertheless, when I think of those who are now the directors of our national laboratories, it is impossible for me to take quite seriously the danger that their colleagues and successors may be beset by lust for power.

The role of large laboratories, and their management, naturally bring us back to the second question which I wanted to discuss: the changed mode of the cultivation of science as a consequence of its great expansion. However, before turning to this question, let me try to summarize what I tried to say about the effects of the increasing area of the subject matter of science. The increase in question will be surely stimulated by the increasing number of scientists because the pleasure which each scientist can derive from his work can be maintained at a high level only if the scientists scatter their interests over a wide field and do not tread, figuratively, on each other's toes. A rapid expansion of the area of scientific endeavor does give cause to some apprehension. Success in any field will bring not only increased insights, it will bring also new capabilities and powers. The great number and variety of capabilities and powers scattered widely all over the world threatens to have elements of instability. There is particular reason to fear the knowledge which the understanding of the mind will yield because this touches on the reasons and purpose of our life and existence. Mankind, on the whole, prospered under the traditional, not searching, attitude toward these questions. Will the more inquiring attitude, which is sure to come, give equal vitality?

The Mode of Science: Two Extreme Pictures

Evidently, the whole character and mode of operation of science will change if it becomes a significant endeavor of the community and if it claims a large fraction of the total effort of mankind.[12] Gone will be the days when the leaders of science could pursue their work playfully and with levity. The responsibility of scientific leaders will be quite similar to the present responsibility of leading members of the Administration or of the directors of large companies. An error in judgment may mean the frustration of thousands of hours work of others—it may mean, for the leader, embarrassment before a large and possibly unsympathetic audience. A leading position in science will entail responsibilities from which present-day scientists would and do shy away; its acquisition may be accompanied with as much intrigue and politicking as a political election or the election of the president of a large company. It is doubtful that scientists who will occupy such positions can derive the joy from their work which present-day scientists cherish. It is possible that many of these positions will not be occupied by scientists, but by a new breed of administrator. We have conjured up here the picture of big scientific institutions, of "big science," as Alvin Weinberg has called it.[13] It exists already today, though not in the extreme form here evisaged.

---

[12] This point has been well articulated, and perhaps a bit exaggerated, in a perceptive article by N. W. Storer [*Science*, 142, 464 (1963)].
[13] A. M. Weinberg, *Science*, 134, 161 (1961).

At the opposite end of the spectrum, science can be pursued by a large fraction of all people, not as a national effort, but playfully, each for his own amusement. This picture cannot, of course, be realized as long as national survival depends on a vigorous and well directed scientific effort, and indeed we are not now progressing in the direction of making science a source of pleasure and a means for recreation. However, if the leaders of all nations could somehow reconcile their desire for influence and domination with the true welfare of their subjects, science could become its own purpose by bringing satisfaction to many disciples. It would then play a role in human affairs somewhat similar to that of sports but much more powerful and intense. Arts and poetry, and perhaps some other avocations, can perhaps also play a similar role, and it would be desirable to devote more thought to them as sources of general human satisfaction. However, we are concerned with the role of science at present.

The first picture, that of big science, is the more grim one. It is possible under all circumstances. The second picture, that of little science, is possible only in a peaceful world. Assuming a peaceful world, therefore, two extreme possibilities of the role of science can be imagined. The first of these is science as a concerted effort of mankind, to augment the store of knowledge and to build, so to say, a superpyramid not of stone but of our understanding of nature. The second picture is individual science, pursued by those who want to pursue it, everyone for his own satisfaction and pleasure. At present, we are moving in the direction of the realization of the first picture, with the difference that it is not a pyramid but a shield that we are building. Otherwise, the picture has much reality; it was not necessary to invent it. The second picture is not new either: it could have been taken from the last act of George Bernard Shaw's *Back to Methuselah*.[14]

Both pictures are extremes and both have attractive as well as repulsive features. As to the first, it would provide humanity with a purpose and objective which seems worthy. It would call for collaboration toward a common goal, rather than competition—collaboration toward a goal which, considering the limitations of human intellect, could well utilize the best efforts of many people.

The trouble with this concept is that, if the goal of maximum scien-

[14] G. B. Shaw, *Back to Methuselah* [Oxford: Oxford University Press (1947) (World Classics)].

tific production is taken seriously—and only in this case could it command the loyalty of its members—its pursuit would require a strict organization of the scientific endeavor. This would be necessary because, even today, no single mind can understand all parts of our knowledge in depth. This will be even less possible as the content of scientific information continues to grow. Each scientist can now speak to a thousand others, but no scientist can listen to and understand more than a few others. Hence, in order to maintain the consistency of the whole scientific edifice, a constant review would be necessary. Since no single individual could undertake such a review, an organization would have to be created for the purpose.

As a member of Alvin Weinberg's Committee on Science and Information, I designed such an organization in outline.[15] It consisted of several layers of scientists, each layer being more highly specialized than the layer above it. Three or four teams of each layer would communicate their findings to a team of the layer above theirs, and this team would harmonize the findings of the reporting teams and evaluate them. The teams of the bottom layer would be the true scientists who "get their hands dirty," the teams of the top layer would consist of philosophers.

This concrete design is, of course, only one possibility, and quite possibly not the most effective one for the organization of science. In spite of this, it would be effective—if its members were in their hearts united toward the common goal. This requirement is, then, its *conditio sine qua non*, and it shares this, probably, with other designs which envisage science as a concerted, purposefully organized effort. Unfortunately, it is hard to see how such a science could remain the enjoyable study which it is now. One may well fear that it would become a heavy-handed enterprise and the participants therein might not be much better off than the builders of the Egyptian pyramids. They would no more retain the impression that they build their own science than the communist worker retains the impression that he builds his own factory. One does not have the satisfaction which creative work, as we know it today, provides, if one's activities are too closely directed by others. For one, the desire for influence remains frustrated. Even if there were enthusiasm in the first, and perhaps second, generation, this would wane in the third and later ones unless the human mind and human emotions

[15] *Science, Government, and Information* [Washington: Government Printing Office (1962)].

were deeply changed for the purpose. Even if this could be done, it appears doubtful whether really penetrating ideas would be conceived in a system the spirit of which is so contrary to the spirit of spontaneity of our best scientific atmosphere.

Let us now look at the opposite extreme: science for individual satisfaction. Shaw's picture, in which everyone goes off to the woods at the age of four to think in solitude about number theory, is unnecessarily grim or, put differently, presupposes an unnecessarily large change of human emotions and conditions of satisfaction. In order to avoid the regimented extreme of a hierarchical organization of science, it is not necessary to have cognition as one's only desire, or to try to satisfy this desire in complete solitude. Science could become the avocation of many. Even a somewhat dilettantish pursuit could give a similar but deeper and more elevating satisfaction than sports give nowadays.

The trouble which I see with this picture—and I admit that as a scientist I may be biased—is that, the more it deviates from Shaw's grim picture, the more does it treat science as a pastime and not as something that gives coherence to human society and fires it to a purpose. It does seem to me that society needs such a purpose and I can see nothing on the horizon that could constitute an equally strong purpose. The common purpose should, at the same time, leave enough freedom to the individual so that his desire for influence and uniqueness—a desire of which both the absence and the excess are so pernicious—should find reasonable satisfaction.

Both Alternatives: The Golden Middle Way?

The extremes which were discussed so far may be so unsatisfactory exactly because they are extremes. It seems to me that there is a need for both big science and for little science and the choice between the two, and their natural competition, would lend charm, significance, and vitality to both. Furthermore, both may find specific areas in which they may be more effective than the other. As to big science, let us not forget that, at least for the time being, there are specific capabilities which we would like to have—it is not only deeper insight that we wish to acquire. Very generally speaking, the capabilities which we wish to develop would render the Earth habitable by a larger number of people. For this, sources of more food and fresh water, and greater skill in combating disease, are necessary. Capabilities to acquire these can be effectively developed only by large enterprises and indeed, our large laboratories, our big science, were created to provide us with technological capabilities, for better living, and for better defense. On the other side, little science, the individual scientist, is also needed. Deep insights, radical departures from the consensus, rarely originate in closely knit groups, working in well organized laboratories. Such new insights—only Planck's conception of the elementary quantum needs to be mentioned—have had a decisive influence on science in the past and a similar influence can be expected in the future. It is hard to imagine how they can be developed other than in comparative solitude. As research will increasingly turn away from technology to search for insights, the role of little science should increase.

What is more difficult to visualize than either big or little science is the relation between them. I thought, some time ago, that they should be separated by working in different fields—big science where elaborate and costly equipment is necessary, little science where fundamental paradoxes have appeared. However, the two types of science need each other, both intellectually and emotionally. If separated, they would both show the weaknesses of the extremes which were discussed before.

At present, the trend is very strongly toward big science, and those who are convinced of the fertility of little science are often considered to be unable to understand the voice of the future and of progress. This, it seems to me, is unjustified. It is true that many of the young men are attracted by the big machines of big science and that it is difficult to resist the easy success which these machines promise. There are others, however, both among our young colleagues and our students, who prefer to work for themselves and to search devotedly and patiently for the deeper truth. It was said, with some malice, that some of the great physicists of the former group were appointed as great physicists more nearly by the government which gave them big accelerators than appointed by Providence by giving them a devoted interest in knowledge, an appreciation of the true problems and mysteries, and the humility necessary to search for the solution of these patiently.

Nevertheless, the present trend is toward big science. Furthermore, big science creates its organization spontaneously, whereas the individual scientist of little science is inclined to shun organization. I do wish, however, that little science be recognized for what it is: a necessary element for the vitality of science, a moral force which can act as the conscience of all science, and as public opinion when groups of big scientists disagree. The fact that it does not seek the easy glory but is willing to search under the less glamorous conditions of an earlier period should win for it respect and authority. Perhaps the middle road, that of supporting big science vigorously but extending encouragement and high esteem to little science also, even if it may not be a golden road, is at least the best.

Our theme has been the promise and the dangers inherent in the growth of science. We have seen that there is much of both. The promise of future science is to furnish a unifying goal to mankind rather than merely the means to an easy life, to provide some of what the human soul needs in addition to bread alone. If it can fulfill this function, it will play one of the great roles in the drama of mankind.

The danger with which the growth and increased significance of science may confront society stems from the increase in power and capabilities which results from increased knowledge and which necessitates new restraints. The growth of science will cause a dilemma also for science itself. The *Weltbild* it will generate when it is extended to more and more areas may outgrow the capacity of any single mind. This may result in the fragmentation of science or in a superstructure of organization. The organization of science could destroy the detachment and sublime satisfaction which is the reward of the scientist of our period and which keeps him from coveting power and undue influence. It will be easier to avoid the dangers posed by the growth of science if we do not try to forget them; its promise is more likely to be fulfilled if we keep it in mind.

Physics and the Explanation of Life*

E. P. Wigner

R. J. Seeger and R. S. Cohen (eds.) Philosophical Foundations of Science (Proc. of an AAAS 1969 Program). Boston Studies in the Philosophy of Science XI. D. Reidel, Dordrecht 1974, pp. 119–132

1. Preamble

I am a physicist, but the problem on which I wish to present some thoughts is not a problem of physics. It is, at present, a problem of philosophy, and I may well be told 'ne sutor ultra crepidam' – the shoemaker should stick to his last. I do have, however, several excuses for venturing into this difficult field. The first is that, if no solid knowledge is available in a field, it is good if representatives of neighbor sciences put forward the views which appear most natural from their own vantage point. The second reason for my speaking here today is that since I started to think and also to write on the subject, I have received many letters and verbal comments from colleagues, agreeing, on the whole, with my point of view. This means, I hope, that there is some interest in the subject among physicists and some consensus on it. It also means that much of what I will have to say will not be original but must have been conceived, at least in part, before me. My third excuse for putting forward views which do not have the solid foundation which one is used to expect from a physicist is that many others before me have done likewise and my fourth excuse is simply that the subject is of overwhelming interest and I like to speculate about it.

2. A bit of history

It would be difficult to review in this session even that small part of the philosophers' thinking on the problem of body and mind, physics and consciousness, with which I am familiar. Let me confine my attention to the ideas of three schools, all of which had a profound effect on our thinking.

Descartes seems to have been the first in modern times to have devoted a great deal of thought to our question. Descartes is, of course, well-known as the originator of the rectangular coordinate system, and for his pronouncement, 'cogito ergo sum'. This saying indicates that he recognized the thought, an evidence of the consciousness, as the primary concept. Descartes was also the first to recognize the nerves as transmitters of sensations and the brain as the depository of our emotions and our memory. He said that the brain is the body's link to the soul.[1]

Let me comment only on three characteristics of Descartes' thoughts. The first is the truly mechanistic picture which he used throughout. He considered the nerve impulses as a flow of a liquid through the nerves, which he imagined to be tubes. He imagined that the memory consists of an expansion of pores in the brain. I mention this because it shows how easily even a truly great thinker succumbs to the ideology of the contemporary state of science and makes too detailed images in accordance with that state of science. Of course, in Descartes' times nobody could dream of the travelling electric impulses which do constitute nerve action. The second point to which I wish to call attention is that Descartes considered mind and body to be two separate entities, the body acting on the mind. He did not think of body and mind as fused into an entity. Lastly, he maintained that only man has a soul, animals are mere machines or automata, devoid not only of thoughts and emotions, but also of sensations. One of his successors, Malebranche,[2] said:

Thus dogs, cats, and the other animals, have no intelligence, no soul in the sense in which this concept is usually understood. They eat without pleasure, cry without pain, grow without knowing this. They have no desires and no knowledge.

This sound fantastic to us, pupils of Darwin's recognition that man is an animal species. It sounds more fantastic than it should: some insects lack sensations to a surprising extent. Wasps, the thorax of which was cut off suddenly, did not appear to notice this but continued to eat – the food dropping out of the channel leading to the thorax. I'll return later to this illustration of the enormous differences between the inner lives of different animals.

The next philosopher whose ideas I wish to mention is Thomas Huxley.[3] He recognized the near absurdity of Descartes' view as far as the sharp and absolute difference between the states of consciousness of animals and man is concerned. He did accept, however, Descartes' view that animals are pure automata, and extended this to man. According to Huxley, and many other philosophers who followed him, man's and animals' volitions, their intentions, are consequences rather than causes of the physical circumstances. At first hearing, this appears absurd but, as will appear later on, in the deterministic framework of these philosophers it is more nearly meaningless than absurd. Causation is not a well-defined concept in a deterministic picture of the world – it may not have an unambiguous meaning in any known picture. This statement will be made more explicit and concrete later on; it will also be expressed in the physicist's language. However, you probably have heard the story of two Eskimos watching a water-skier. "Why boat go so fast?" asked the first. "Chased by fool on end of string," was the reply.

The last body of thought that I wish to refer to is that of the Gestaltslehre of Wertheimer, Köhler, and others. They point out that a steam engine, for instance, would be very inadequately described as a steel cylinder, covered at one end, having a closely fitting but movable disc on the inside and a rod, attached to this disc, protruding on the other side. Rather, in order to describe a steam engine, its purpose and the cooperation of its parts should be given. Similarly, an explanation of an animal in terms of the physical functioning of its parts will be inadequate; it is a whole, much more than the sum of its parts.[4]

The point of the Gestalt-theoreticians is undoubtedly correct and it is a valuable observation. However, it seems to me to be more a pedagogical than an ontological observation. Surely, we do not obtain a vivid picture of man by just describing his bones and muscles and how they are attached to each other. However, it is possible to describe the functioning of man's organs without answering the question of the relation between his body and soul, his emotions and his physical constitution, his volition and his movements. Hence, it appears to me that the statements of the Gestaltlehre, though both true and relevant, really avoid the principal issue which confronts us.[5]

Let me now attack our problem from the point of view of the physicist, the physicist familiar with the fundamental changes which quantum mechanics initiated in the physicist's picture of the world. I shall begin with the part of our title with which I should be familiar – with physics.

3. WHAT IS PHYSICS?
WILL IT FORM A UNION WITH THE LIFE SCIENCES?

One often hears the statement that the purpose of physics is the explanation of the behavior of inanimate objects. To most of us physicists this does not appear to be a very incisive statement. What our science is after is, rather, an exploration of the regularities which obtain between the phenomena, and an incorporation of these regularities – the laws of nature – into increasingly general principles (the theories of physics), thus establishing more and more encompassing points of view. I like to quote David Bohm[6] in this connection, who said that "science may be regarded as a means of establishing new kinds of contacts with the world, in new domains, on new levels." No ultimate explanation is possible and our science is rather a constant striving for more encompassing points of view than the provider of an explanation for one or another phenomenon. Furthermore, as Einstein often emphasized, the more encompassing point of view for which we strive must have a conceptual simplicity – otherwise it will not be credible. And, as Polanyi emphasized, it must be interesting.

If all this is accepted, it follows that the phenomena of life and mind will form a unit with our regularity-seeking physical sciences if regularities in the behavior of the thought processes can be discovered, and an encompassing point of view developed which embraces both the phenomena of the mind and those of matter. Surely, psychology has pointed to many regularities in our thought processes and has made many, many interesting observations. These are, however, at present entirely divorced from the regularities in the behavior of matter which are the subjects of present day physics.

My discussion of life and consciousness will be based on the assumption that these phenomena will become, along with ordinary physical phenomena, the subjects of a regularity-seeking science. It will be based on the assumption that a picture will be discovered which will provide us with a view encompassing both mental and physical phenomena and describes regularities in both domains from a unified point of view. Clearly, these are assumptions for which a proof is lacking at present.

Are there tendencies in the sciences of life to expand in the direction of physics and are there, conversely, tendencies in physics to consider the phenomena of life and consciousness? If the disciplines in question are considered broadly enough, both tendencies are present. It is true that physics, in the true sense of the word, is foreign to basic psychology. However, the life sciences, particularly those dealing with the lowest organisms, are endeavoring to acquire a base in chemistry and physics. They also hope to extend their interest, eventually, to organisms of which consciousness is an essential characteristic. Conversely, the basic principles of physics, embodied in quantum mechanical theory, are dealing with connections between observations, that is contents of consciousness. This is a difficult statement to accept at first hearing, and I must hope that most are familiar with it. In essence, it recalls that quantum mechanics is not a deterministic theory. The formulation of its laws in terms of our successive perceptions, between which it gives probability connections, is a necessity. Classical physics, of course, also can be formulated in terms of, this time deterministic, connections between perceptions and the true positivist may prefer such a formulation. However, it can also be formulated in terms of absolute reality; the *necessity* of the formulation in terms of perceptions, and hence the reference to consciousness, is characteristic only of quantum mechanics. In fact, the principal objection which your present speaker is inclined to raise against the epistemology of quantum mechanics is that it uses a picture of consciousness which is unrealistically schematized and barren. Nevertheless, there is a tendency in both physics, which we consider as the most basic science dealing with inanimate objects, and in the life sciences, to expand toward each other. Furthermore, the tendency is strongest in the modern parts of the two disciplines: in quantum mechanics on the one, in microbiology on the other, hand. Both feel that they cannot get along by relying solely on their own concepts.

Nevertheless, that the tendencies to which I just alluded will ultimately lead to a merger of the disciplines can be only a hope at present. Neither of them proposes in its present form a more encompassing point of view; both wish to use the concepts of the other only as the basic concepts in terms of which their own regularities can be formulated. That some such basic concepts are unavoidable is, I believe, clear: there must be some things which are the subjects of regularities. It is not clear, however, that these subjects must be either entirely in the realm of orthodox physical theory, or entirely in psychological subjects. Nevertheless, it is encouraging that there is a tendency on both sides of the chasm to take cognizance of the other side – even if both sides are forced to do so, or perhaps *because* both sides are forced to do so.

Let me now come to my last subject: the physicist's view of the

124   EUGENE P. WIGNER

relation between body and mind. I will try to give a rational discussion
of the two alternative roles which present-day physics can play in a future
regularity-seeking science the realm of which extends to the phenomena
of mind as well as to those of present-day physics.

4. A PHYSICIST'S VIEW ON THE MIND-BODY PROBLEM –
THE FIRST ALTERNATIVE

As particularly the historical discussion indicates, there is a possibility
that the laws of physics, formulated originally only for inanimate matter,
are valid also for the physical substance of living beings. To put it in a
somewhat vulgar fashion, even most physicists, if unexpectedly presented
with the question of the validity of the laws of physics for organic matter,
would affirm that validity. On the other side of the chasm, many, if not
most, microbiologists would concur in this view. The view does not lead
automatically to Huxley's view that we are automata because the
present laws of physics are not deterministic but have a probabilistic
character. Furthermore, if we are honest about it, we cannot now formulate laws of physics valid for inanimate objects under all conditions.
Hence, the statement which we are considering should be formulated
somewhat more cautiously: that laws of nature for the formulation of
which observations on inanimate matter suffice, are valid also for living
beings. In other words, physical laws, obtained by studying the traditional
subjects of physics, and perhaps not very different from those that physicists are trying to formulate now, will form the basis from which the
behavior of living matter can be derived – derived perhaps with a great
deal of effort and computing, but still correctly derived.

The assumption just formulated is surely logically possible. It is very
close to Huxley's views which were mentioned before. Would it mean
that, eventually, the whole science of the mind will become applied
physics? In my opinion, this would not be the case even if the assumption
which we are discussing is correct. What we are interested in is not only,
and not principally, the motion of the molecules in a brain but, to use
Descartes' terminology, the sensations which are experienced by the soul
which is linked to that brain, whether it is pain or pleasure, stimulation
or anxiety, whether it thinks of love or prime numbers. In order to
obtain an answer to these questions, the physical characterization of the

state of the brain would have to be translated into psychological-emotional terms.

It may be useful to give an example from purely physical theory for the need for such a translation. The example which I most like to present derives from the classical theory of the electromagnetic field in vacuum, that is, the simplest form of Maxwell's equations. These give the time derivative of the electric field in terms of the magnetic field, and the time derivative of the magnetic field in terms of the electric field. Both fields are free of sources. Although the actual form of the equations is not very relevant for our discussion, it may render this more concrete if I write down the equations in question for the electric and magnetic field, E and H:

$$\frac{\partial H}{\partial t} = -c \operatorname{curl} E \qquad \frac{\partial E}{\partial t} = c \operatorname{curl} H.$$

I shall refer to these equations, briefly, as Maxwell's equations; actually, they are Maxwell's equations for empty space; c is the velocity of light. If E and H are given at one instant of time, these equations permit their calculation for all later times, and for all earlier times. They will serve as model equations for the discussion which follows – they give both sides of the picture, the electric and the magnetic side, and do not prefer one over the other.

It is possible, however, to formulate an equation for the magnetic field alone. This is again, and should remain, free of sources and its time-dependence is regulated by the equation

$$\frac{\partial^2 H}{\partial t^2} = c^2 \left(\frac{\partial^2 H}{\partial x^2} + \frac{\partial^2 H}{\partial y^2} + \frac{\partial^2 H}{\partial z^2} \right).$$

I shall refer to this equation, briefly and somewhat incorrectly, as Laplace's equation. One can observe now that, if H and $\partial H / \partial t$ are given at one instant of time, this equation permits the calculation of H for all later times – and for all earlier times. Is now this equation, which is just as valid as Maxwell's original equations, a full substitute for the latter? The answer is no. If we want to obtain the force on a small charge at rest, the original form of the equations furnishes this directly: it is the electric field at the place where the charge is, multiplied by the magnitude of the charge. In order to obtain the force from the second, that is

Laplace's equation, referring only to the magnetic field H, one has to calculate first the electric field in terms of H. This can be done, though the formula is quite involved. The formula gives the translation of the magnetic field into the electric one and this is, in the case considered, more relevant than the magnetic field itself. We have, therefore, an example before us in which a theory – Laplace's equation for H alone – is completely valid but is not very useful without the translation which should go with it.

The example also shows that the translation into the more relevant quantity can be quite complicated – more complicated than the underlying theory, that is Laplace's equation. The translation equation is also more complicated than the set of equations, in this case Maxwell's equations, which uses both concepts: the one which turns out to be the more relevant one, that is E, along with the other, H, which does suffice for the formulation of the time-dependence. It is unnecessary to remark that, in the preceding illustration of a future theory of life, H plays the role of the purely physical variables, E plays the role of the psychological variables. In this illustration, the use of both types of variables in the basic equations is much preferable to the use of only one of them – the problem of translation does not arise in that case.

The example just given illustrates also the observation on the meaningless nature of the concept of causation in a deterministic theory – such as Maxwell's theory of the electromagnetic field in vacuum. Looking only at Laplace's equation, and the translation thereof, one will conclude that the magnetic field is the prime quantity, its development is determined by its magnitude in the past. The electric field will appear as a derived quantity, caused and generated by the magnetic field. Maxwell's original form of the equations shows, on the other hand, the possibility (and in the opinion of the physicists, the desirability) to consider the two to have equal rank and primitivity. One can also go to the other extreme and derive an equation similar to the last one, but involving only the electric field E and then claim the E is the primitive quantity, the magnetic field H the described one, the product of E.

I believe I have discussed the assumption that the laws of physics, in the sense described, are valid also for living matter. We also saw that this assumption need not imply, as is often postulated, that the mind and the consciousness are only unimportant derived concepts which need not enter the theory at all. It may be even possible to give them the privileged status. Let me now discuss the assumption opposite to the 'first alternative' considered so far: that the laws of physics will have to be modified drastically if they are to account for the phenomena of life. Actually, I believe that this second assumption is the correct one.

5. THE SECOND ALTERNATIVE: LIFE MODIFIES THE LAWS VALID FOR INANIMATE NATURE

I wish to begin this discussion by recalling how wonderfully actual situations in the world have helped us to discover laws of nature. The story may well begin with Newton and his law of gravitation. It is hard to imagine how he could have discovered this, had he not had the solar system before himself in which only gravitational forces play a significant role. The discovery, also due to Newton, that these forces also determine the motion of the Moon around the Earth, and the motion of freely falling bodies too heavy to be much affected by air resistance, was a wonderful example for science's power to create a unified point of view for phenomena which had, originally, widely differing characters. Newton, of course, recognized that there must be other forces in addition to the gravitational ones – forces which, however, remain of negligible importance as far as the motion of the planets is concerned.

Newton's discovery was followed by the discovery of most laws of macroscopic physics. Maxwell's laws of electromagnetism – the ones which we just considered in the special case of absence of matter – are perhaps the most remarkable among these. Again, these laws – those of macroscopic physics – could not have been discovered were not all the common objects which surround us of macroscopic nature, containing many millions of atoms, so that quantum effects, for instance, play no role in their gross behavior. Again, the unifying power of science manifested itself in a spectacular way: it turned out that Maxwell's equations also describe light and, as we now know, all electromagnetic radiation from radio waves to X-rays.

The next step of comparable, perhaps even greater, importance was the development of microscopic physics, starting with the theory of heat and soon leading to quantum theory. Most of this development took place in the first half of our century but, in some regards, the development is still

128   EUGENE P. WIGNER

incomplete. If we assume that it can and will be completed – most of us believe this – the question which we should face is whether our present microscopic theories also presuppose some special situation, the absence of certain forces or circumstances. The point of view which we are discussing maintains that this is the case. Just as gravitational theory can describe only the situation in which no other but gravitational forces play a role, and macroscopic physics describes only situations in which all bodies present consist of many millions of atoms, present microscopic theory describes only situations in which life and consciousness play no active role.

Similarly, just as macroscopic physics contains gravitational theory as a special case, applicable whenever only gravitational forces play a significant role, and just as microscopic physics contains macroscopic physics as a special case, valid for bodies which contain millions of atoms, in the same way the theory which is here anticipated should contain present microscopic physics as a special case, valid for inanimate objects. Thus, each successive theory is expected to be a generalization of the preceding one, to recognize the regularities which its antecedent postulated, but to recognize them as valid only under special conditions. This should apply also to the theory foreseen here, in the form of the 'second alternative'.

Naturally, the preceding story does not *prove* that the present, microscopic, physics will also have to be generalized, that the laws of nature as we now know them, or try to establish them, are only limiting cases, just as the planetary system, macroscopic physics, were limiting cases. In other words, it does not prove that our second alternative, rather than the first one, is correct. Can arguments be adduced to show the need for such modification? I know of two such arguments.

The first is that if one entity is influenced by another entity, in all known cases the latter one is also influenced by the former. The most striking and originally least expected example for this is the influence of light on matter, most obviously in the form of light pressure. That matter influences light is an obvious fact – if it were not so, we could not see the moon. We see it because it scatters the light emitted by the sun. The influence of light on matter is, however, a more subtle effect and is virtually unobservable under the conditions which surround us. Light pressure is, however, by now a well-demonstrated phenomenon and it plays a decisive role in the interior of stars. More generally, we do not know any case in which the influence is entirely one-sided. Since matter clearly influences the content of our consciousness, it is natural to assume that the opposite influence also exists, thus demanding a modification of the presently accepted laws of nature which disregard this influence.

The second argument which I like to put forward is that all extensions of physics to new sets of phenomena were accompanied by drastic changes in the theory. In fact, most were accompanied by drastic changes of the entities for which the laws of physics were supposed to establish regularities. These were the positions of bodies in Newton's theory and the developments which soon followed his theory. They were the intensities of fields as functions of position and time in Maxwell's theory. These were replaced then by the outcomes of observations (the perceptions referred to before) in modern microscopic physics, that is, quantum mechanics. In the development which we are trying to envisage, leading to the incorporation of life, consciousness, and mind into physical theory, the change of the basic entities indeed appears unavoidable: the observation, being the entity which plays the primitive role in the theory, cannot be further analyzed within that theory. Similarly, Newtonian theory did not further analyze the meaning of the position of an object, field-theory did not analyze further the concept of the field. If the concept of observation is to be further analyzed, it cannot play the primitive role it now plays in the theory and this will have to establish regularities between entities different from the outcomes of observations. An alteration of the basic concepts of the theory is necessary.[7]

These are the two arguments in favor of what I called the second alternative, that the laws of physics which result from the study of inanimate objects only are not adequate for formulating the laws for situations in which life and consciousness are relevant parts of the picture.

6. Conclusion and summary

I realize that the hope expressed in the last two sections, that man shall acquire deeper insights into mental processes, into the character of our consciousness, is only a hope. The intellectual capabilities of man may have their limits just as the capabilities of other animals have. The hope does imply, though, that the mental and emotional processes of man and animals will be the subjects of scrutiny just as processes in inanimate matter are subjects of scrutiny now. The knowledge of mind and consciousness may be less sharp and detailed than is the knowledge given by present day physics on the behavior of inanimate objects. The expectation is, nevertheless, that we can view mind and consciousness – at least those of other living beings – from the outside so that their perceptions will not be the primitive concepts in terms of which all laws and correlations are formulated. As to the loss in the sharpness and detail of the laws, this is probably unavoidable. It has taken place throughout the history of physics. Newton could determine all the initial conditions of the system of his interest and could foresee its behavior into the indefinite future. Maxwell's and his contemporaries' theories can be verified only by creating conditions artificially under which a verification is possible. Even then, it is possible only for limited periods of time. The laws of quantum mechanics, finally, neither make definite predictions under all conditions, nor have its equations of motion been verified in any detail similar to those of macroscopic theories. A further retrenchment of our demands for detail of verification is probably in the offing whenever we extend our interest to a wider variety of phenomena.

You will want to ask me, I believe, at least two questions. First, whether other physicists would agree with me, and second what good all this does, considering that I do not even specify the basic entities the behavior of which is subject to the new regularities to be established. My answer to the first question is that most physicists do not concern themselves too much with the questions I discussed. Their reason may well be given by the answer I'll give to the second question. However, Bohr, in his inimitable, profound, and somewhat ambiguous way, concurred in the view which I am embracing.[8] Also, I was just a few days ago reminded by Dr. Hartshorne, of the University of Texas, that Heisenberg spoke, in his *Philosophical Problems of Nuclear Science*, of the limited applicability of our present physics, of the necessity of broadening its laws if they are to apply to life.[9] Pascual Jordan, another founder of quantum mechanics, made a similar statement.

As to the usefulness of the considerations, I must admit that I do not see much of it. This may well be the reason for the lack of a more general interest on the part of physicists in the questions discussed. What I spoke about is philosophy and it would be presumptuous on my part to voice an opinion whether it shares the usefulness of the newborn child on which Abraham Lincoln commented. Even if not useful, I would like to summarize it when concluding my address.

I believe that the present laws of physics are at least imcomplete without a translation into terms of mental phenomena. More likely, they are inaccurate, the inaccuracy increasing with the increase of the role which life plays in the phenomena considered. The example of the wasp which does not seem to have sensations may indicate that even animals of considerable complexity are not far from being automata, largely subject to the present ideas of physics. On the other hand, the fact that the laws of physics are formulated in terms of observations is strong evidence that these laws become invalid for the description of observations whenever consciousness plays a decisive role. This also constitutes the difference between the view here represented and the views of traditional philosophers. They considered body and soul as two different and separate entities, though interacting with each other. The view given here considers inanimate matter as a limiting case in which the phenomena of life and consciousness play as little a role as the non-gravitational forces play in planetary motion, as fluctuations play in macroscopic physics. It is argued that, as we consider situations in which consciousness is more and more relevant, the necessity for modifications of the regularities obtained for inanimate objects will be more and more apparent.

ACKNOWLEDGMENT

I am much indebted to Dr. A. Shimony for his critical review of this article.

Princeton University

NOTES

* An article fully based on the address here presented appeared in *Foundations of Physics* 1 (1970).
[1] R. Descartes, *Oeuvres* (ed. by C. Adam and P. Tannery), Librairie Philosophique. Paris, 1967, Vol. XI, p. 119ff.
[2] Quoted by Thomas Huxley, ref. 3, p. 218.
[3] Thomas H. Huxley, *Selected Works*. Vol. 1: *Method and Results*, Appleton and Co.. New York, 1902, p. 199ff. I am greatly indebted to Dr. W. Schroebel for calling my attention

132   EUGENE P. WIGNER

to this essay. Ideas similar to those of Huxley were held by many others. P. B. Medawar, in *The Art of the Soluble* (Methuen, London, 1967) mentions, with approval, D'Arcy Thompson's very similar convictions.

[4] See, for instance. B. Petermann's *Gestaltslehre* (J. A. Barth, Leipzig, 1929), or W. Köhler's *The Task of Gestalt Psychology*, (Princeton University Press, 1969).

[5] An interesting account of the views of many philosophers, physicists, and biologists is presented in Chapter VII of S. L. Jaki's *The Relevance of Physics* (University of Chicago Press, Chicago, 1966).

[6] D. Bohm, *Special Theory of Relativity*, W. A. Benjamin, New York, 1965, p. 230.

[7] This is a point which was also brought out by G. G. Harris.

[8] N. Bohr, *Atomic Theory and the Description of Nature*. Cambridge University Press, 1934.

[9] W. Heisenberg, *The Philosophical Problems of Nuclear Science*, Faber and Faber, London, 1952. See also his *Physics and Philosophy*, Harper, New York, 1958, p. 155.

On Some of Physics' Problems

Eugene P. Wigner

Main Currents in Modern Thought *28*, 75–78 (1972)
(Reset by Springer-Verlag for this volume)

> The external problems of science
> are primarily societal;
> the internal problems
> are generated by the nature
> of the scientific enterprise.

The problems of physics which I would like to discuss have not created a crisis situation. It may be well to have a look at them, nevertheless, and to think about solutions for them. Problems are in the foreground of interest these days, and it may be refreshing to contemplate some which can be considered with a degree of detachment.

Similar to most other institutions, physics has both external and internal problems. The external are the more serious in this case and they are less pleasant to contemplate. Some of them are, indeed, on the verge of being threatening. Let me, therefore, begin with them. My discussion will not be scientifically ordered: I will talk about those questions which bother me particularly, which are in my mind most of the time. This applies to the discussion of both external and internal problems.

The First External Problem: The Problem of Communication

The problem of communication is not new. I myself first spoke about it almost twenty years ago, and it was not new then. Dr. Alvin Weinberg, Director of Oak Ridge National Laboratory, dealt with it most incisively in his *Reflections on Big Science*, and much of what I will say will be a personalized version of what I have learned from him.

In short, science begins to form such a large body that it threatens to fall to pieces. It has been, for some time, difficult to explain the intricacies of quantum theory to a chemist. The language which we have developed and the many concepts which we have created are difficult to assimilate for someone whose principal interest is in another field, with questions and with a special language of its own. Nevertheless, chemists, and in particular theoretical chemists, are trying to think in terms of quantum mechanical theory. Sometimes, it seems to us, they rediscover pictures which have long been familiar to us, and the novelty of their discovery is the projection of this picture in a language which we find cumbersome and unnatural. Sometimes they project pictures of aspects of nature about which they know less than we do, and we find these pictures abstruse and unnatural. I am sure that the theoretical chemists could render an equally critical judgment of our own work – both on metallic and on nuclear structure, the two subjects which were principally in my mind just now. Yet, if we want to maintain the unity of science, if we want it to provide a living and elevating picture of the world, physicists and chemists must learn to understand one another, and must understand each other not only in a half-hearted way. The same applies even to the relation between biologists and physicists, and to unite their thinking now appears an even more arduous, though *perhaps* not entirely hopeless task.

Alvin Weinberg has not only discussed the problem of the lack of communication among the sciences, he has also proposed some remedies. The first step in this direction, it seems to me, is to become fully aware of the problem and of its implications. Only if we understand the seriousness of the problem shall we be able to strive toward a solution wholeheartedly. Only then will we be willing, and perhaps even eager, to follow Alvin Weinberg's injunctions: to write our articles for a less specialized readership, to devote time and energy to the composition of reviews and to reading more of the reviews covering the results of sister sciences.

I shall add only a few words here to the foregoing. The schism about which I spoke, the communications gap to which I referred, separates not only the different natural sciences such as biology, chemistry and physics; it yawns also between different parts of physics. Only a few dispersion theorists understand those working in axiomatic field theory; virtually no one whose principal interest is in cosmology and the general theory of relativity knows what the nuclear physicist means by an analogue or even a two-particle-one-hole state. The problem of communication is not only an external problem of physics; it threatens to develop into an internal one. In fact, so many phenomena are being discovered and investigated, so many new ideas are advanced, that no one can fully keep track of them.

The Second External Problem: The Question of Purpose

What is the purpose of the study of physics or of the natural sciences? Every scientist must have asked himself this question, and most of us have answered it quietly, to the satisfaction of our souls. Our interest in science, and our devotion to the effort to make it deeper and more embracing, has continued as a result. Today, the same question of the purpose of engaging in research, the "relevance", as it is called, of the scientific effort, is asked in strident voices by hundreds.

It is not easy to articulate convictions of purpose, or of values, but I shall try. The first purpose is obvious to all of us: it gives us great pleasure to learn, to expand our knowledge and to add to the knowledge of the human race. I need hardly expand on this to the present audience. Some years ago, I expressed the hope that this pleasure will be shared by an increasing part of all humanity.

The second purpose is closely related to the first: it is a noble enterprise to see how far the human mind can go in "understanding" nature, that is, in uncovering hidden relations between the events that we experience. Stephen Weinberg, not a relative of Alvin's, articulated the underlying desire and motive very eloquently when pleading for additional support for high energy physics.

The third motive which fired scientists in the past was more altruistic than the first two. By expanding our knowledge and applying it to technical problems, we make it possible for more men to lead a carefree life, to devote more time to leisure and enjoyment. It is this purpose – the only truly selfless one – which is being questioned by the experiences of recent years. This is a bitter pill for all of us to swallow: for hundreds of years our spiritual ancestors and we ourselves have endeavored to make life easier for everybody. We are now wondering whether this was a valid objective, whether man can live if he does not have to struggle for tomorrow's bread. If his livelihood is assured, does man not begin to strive in too large numbers after other objectives, such as power; and will the striving of too many, too early in life, not destroy our dearest ideals? When I was your age, to earn a decent living was a valid purpose for me. Today, this is taken for granted and you are deprived of a purpose, the pursuit of which gave a great deal of satisfaction to me.

These are somber thoughts, deeply disturbing, and I do not want to pursue them now. Let me go over to the more technical internal problems of our discipline where I can speak as a physicist.

The two problems I have mentioned – the problem of communication and that of purpose – are not the only external problems of physics. We have probably many others which I have failed to recognize; we also have the problem of bigness, of impersonality, of much of modern experimental research. However, these are perhaps – and I do mean perhaps – less pressing than those which I chose to emphasize particularly.

Cosmology: The Problem of the Beginning

The simplest picture of cosmology is that of the big bang: initially, according to this theory, all matter was concentrated in a very small volume, with an enormously high energy density. Matter then spread with tremendous velocity in every direction, undergoing various transformations as the energy density decreased. What we now see is a rather mild later stage of this expansion: the volume of the universe is large, the energy density is small, but the expansion continues. There is much observational evidence to support this conception of the universe and of its origin. The most recent one that I know of is the cosmic black body radiation, corresponding to a few degrees absolute in temperature – the temperature which the big bang theory predicted. Its existence was first confirmed by colleagues of mine in Princeton, by R. H. Dicke's group.

The question then arises: what was the state of the universe before the big bang? This is neither a new, nor a popular, question. I understand that

St. Augustine was once asked: What did the Lord do before he created the world? St. Augustine's answer was, I am told, He created Hell for people who ask such questions. Nevertheless, as far as the big bang theory, in its primitive form, is concerned, the questions can be well asked.

According to the primitive form of the big bang theory, there was, originally, a singularity in the metric space and this dissolved at time zero. Few people are inclined to accept this fully. It is more reasonable to assume, instead, that the "original" density was very high but not infinite. One is led to believe, then, that it was the result of a contraction, that is, a collapse, of matter as a result of the gravitational attraction. Similarly, the final state is then not one of infinitely low density: the universe, after expanding for several billion years, will contract again, eventually to a very small volume and correspondingly high density. This will produce, in many billion years, a new big bang, and a renewed expansion, and so on. One is led in this way to the picture of a pulsating universe, alternating between states of very high and very low densities, corresponding to the very small and very large volumes which matter occupies.

The picture of a pulsating, that is, essentially periodic, universe seems a very satisfactory one. In particular, if we make a mistake today, our successors, in some billion years, will have a chance to do better, the absolute uniqueness of all events, the absence of any fresh start, is a profoundly perturbing idea.

The trouble is that the physical chemists, early in this century, raised valid objections against the concept of a periodic universe. They pointed out that, as a result of the continuing increase of the entropy, the world will have to come to a steady state, with uniform temperature everywhere, everything in perfect equilibrium. The solutions of the gravitational equations do permit a periodic universe, with infinitely many successive pulsations. However, the periodic solution of the gravitational equations does not correspond to reality because it does not take the inhomogeneities of the actual universe into account. Similarly, one can conceive of a gas in which all atoms fly with equal velocities radially outward from the center of a spherical container, strike this container all at the same time and are reflected back, and so on. However, actual gases do not have this structure, and the radial oscillations of actual gases in a spherical container eventually are damped out and the gas assumes the state of an equilibrium. Similarly, the final state of equilibrium, the state that the physical chemists called the Wärmetod, is an unavoidable consequence of our present concept. This, then, leads us to the picture of oscillations of the universe with diminishing amplitudes – not as satisfactory as the picture of a truly periodic universe, but a conceivable one.

The trouble is again that the initial state cannot be conceived very well. The entropy always increased – what was it originally? As a result of the quantum conditions, we know that the entropy has an absolute minimum, zero. Was is ever zero? If so, why and when did it start to increase? or, was it never actually zero, but infinitely close to this value in the infinitely distant past? We do not know. Both alternatives are perturbing and we simply do not have a cosmological picture "in the large" which does not contradict some simple and general law of physics. John A. Wheeler, also a colleague of mine in Princeton, has devoted a great deal of thought to questions similar to the one I raised, but not fully satisfactory answer has emerged so far.

Epistemology: The Role of the Observer

The epistemology forced upon us by quantum mechanics has proved, for some of the most eminent physicists, very difficult to accept. The problem can be formulated in many different ways. Perhaps the most simple formulation is as follows: The equations of motion of quantum mechanics, such as the time-dependent Schrödinger equation, are causal in the sense that knowing the state of an isolated system now (i.e., knowing its state vector) permits one to determine its state (its state vector) for all future time. Nevertheless, there is a fundamental stochastic element in the theory: the result of an observation (also called measurement) on the system can yield several results and only the probabilities of the various results can be predicted.

This situation, then, leaves several alternatives. One can assume, as did von Neumann, that the observation is a process which cannot be described by the equations of motion of quantum mechanics, that it involves an interaction with an individual, and such interaction is outside the domain of validity of the quantum equations. This is an entirely possible point of view but involves a duality – interactions which are and which are not subject to the quantum equations – which many dislike. It is possible to assume that the state vector of the observer is complicated, that it is never known, and that the result of the observation depends on the particular state, among the many undistinguishable states, in which the observer was before carrying out his measurement. This second alternative is mentioned only as a reasonable suggestion; it has been shown, both by d'Espagnat and me, that it cannot account for the probability laws postulated by the quantum theory of observation. Finally, it is possible to avoid von Neumann's dualism by eliminating the concept of the state vector from the theory, or equivalently by considering it to be only a mathematical tool, and formulating the laws to begin with as probability connections between subsequent observations. Thus, in the case of the simplest quantum mechanical system, that of a spin $1/2$, one can say that, if an observation yielded the result that the spin assumed a definite direction, rather than the opposite one, the probability that an observation whether it assumes another direction (rather than the direction opposite to it) is $\cos^2 1/2\,\vartheta$ where ϑ is the angle between the original and the second direction. Naturally, the probabilities for the different possible behaviors of more complicated systems, when subjected to more complicated observations, are not so easy to give in terms of equally explicit formulae. However, even for complicated systems it is possible to eliminate the concept of the state vector as a basic concept from the theory and formulate the laws of the theory directly in terms of the probabilities of the outcomes of observations – after all, these are the quantities in which one is ultimately interested.

Whether one adopts von Neumann's terminology or the one just suggested, the observer occupies a particular position in the theory. It is *his* observations that are the subject of the theory; the predicates of the theory are the probabilities with which he observes one or another outcome of the observation.

This separation of the world into two parts, the observer and the outside world, has many unpleasant consequences. These formed, unquestionably, the reasons for the reluctance of Einstein, of Schrödinger, to accept the quantum epistemology at face value. I myself, though convinced that no other epistemology is compatible with the principles of quantum mechanics, have often pointed to the difficulties and concluded that new principles will be needed when the phenomena of life and of consciousness will become subjects of the theory. Among the many difficulties, I now mention only the one which relates to the preceding discussion: since the theory is intended to foresee only results of observations, it is doubtful whether it considers cosmology to present meaningful questions. Yet we all feel that it does.

The particular problem which I want to point out now is a different one, however. It has been raised by the German physicist Zeh. He points out that, if the observer is macroscopic – and in the ultimate analysis he always is – he cannot be separated from the rest of the world. The energy levels of a macroscopic object are so close together that the electric charge of a single electron, one mile away, can cause transitions between these or, more generally, can change their state. Hence, in the quantum sense, an isolated macroscopic body does not exist in practice, and the dividing line between observer and the rest of the world is unrealistic.

Evidently, this is a new chink in the armor of quantum mechanics' epistemology. It is a chink, however, which is subject to investigation while one can still stand on the foundations of present theory. It should be solvable, though it is at present an unsolved problem.

Conclusion

I have spoken about two external and two internal problems of physics. The former give us pain, the latter pleasure. We hope the former will not proliferate; we could add many others to the latter. Will this ever end? Will we ever run out of problems? Or, what I fear much more, will we ever get tired of them? I hope and pray that we will not.

Eugene P. Wigner is Professor of Physics, Princeton University. The preceding paper is based upon an address given at Yale University on the occasion of Henry Margenau's retirement as Higgins Professors of Physics and Natural Philosophy; in its present form it was presented at the Department of Physics Honors Day at the University of Tennessee in 1969, and was published in the December 1969 issue of "The Tennessee Alumnus". It is scheduled for inclusion in "Physics, Philosophy and the Integration of Scientific Knowledge", a Festschrift for Henry Margenau, to be edited by Ervin Laszlo and Emily B. Sellon, presently in preparation.

Physics and Its Relation to Human Knowledge*

Eugene P. Wigner

Hellenike Anthropistike Hetaireia. Athens 1977, pp. 283–294
(Reset by Springer-Verlag for this volume)

The Success of the Natural Sciences

Physics and the natural sciences, including also chemistry and biology, have been amazingly successful in the past. Their success extends over two very distinct areas: on the one hand they have fostered, and in part made possible, an enormous technological development, on the other they have led to a surprising extension of our knowledge and influenced our whole outlook on life.

It is not necessary to elaborate the effect of the development of the natural sciences on technology and industry. We do not often think of it, but when we do, we cannot help being astounded. To mention only a few, we can hardly imagine living without automobiles or airplanes, without electric light, telephones, radio and even television. We take the existence of chromium tanned leather, the upper part of our shoes, of plastics, for granted. The acquisition of the daily bread, a very serious problem even as late as the early part of this century, is now a matter of routine – less than 5 percent of the working force in the United States produces much more food than the whole country consumes. The mean life time of man has increased phenomenally – partly because of the general greater affluence, but partly also as a result of the development of the medical science, based at least in part on the progress in biology. I hope it is not necessary to further enlarge on the beneficial effects of science and technology on our material well-being – it is enormous – and this time I won't discuss the negative effects of our increasing affluence; they are much smaller than the positive effects.

Let me mention, instead, again quite briefly, the spiritual effects of science, the way it has influenced the attitude of all of us toward life. This is equally important and equally strong. We no longer believe in magic or witches, we look for the causes of effects. If we hear thunder, we do not attribute it to the wrath of Zeus but to an electric discharge which suddenly heats up air and creates a sound wave. If we get sick, we do not attribute it to the effect of a hex. Our whole thinking has been profoundly influenced by science.

All that was said so far was intended as a reminder of the truly enormous effect that science, and also the resulting technology, has had on our material

---

* Ἡ Ἀνακοίνωσις ἀνεγνώσθη ὑπὸ τοῦ καθηγητοῦ PETER HODGSON, διότι ὁ εἰσηγητὴς κωλυόμενος δὲν ἠδυνήθη νὰ παραστῇ. Ὡς ἐκ τούτου, ἀπόντος τοῦ εἰσηγητοῦ, δὲν ἐπηκολούθησε συζήτησις.

and spiritual well-being. The question that raises itself then concerns the coherence and inner consistency of science – whether it is truly as strong as its many effects indicate. Being a physicist, I will choose my examples from the realm of that discipline. However, before engaging in a more systematic discussion, I wish to present an example for the apparent accuracy of the physical theories and the cause for that apparent accuracy.

The Nature of and Causes
for the Accuracies of Physical Theories

Surely, many of the conclusions of physical theories have a simply fantastic accuracy. Being here in Greece, we are particularly conscious of the oldest laws inherited in mathematical form, of Archimedes laws of the center of mass, of levers and, in particular, of the weight loss of objects immersed in a liquid. We are convinced, though chiefly on the basis of theoretical evidence, that this is valid to a much better accuracy than one part in a million – an accuracy surely not even dreamed of by Archimedes. Archimedes' law was, perhaps, the first quantitative experimental regularity. It had, at his time, no theoretical foundation and its accuracy, as obtained on the basis of our present theoretical ideas, is surely surprising. The first truly theoretical consideration which explained detailed observations was Newton's theory of the planetary orbitals. We know the accuracy of that: the axes of the orbit of the planet with the largest deviation from Newton's theory move ....? seconds per orbit, giving a deviation from Newton's theory, established 300 years ago, of one part in 30.000. Since I am not talking to physicists only, it may be well to mention that we now believe we have the explanation for this deviation – it is given by the general theory of relativity.

There is one further point well worth mentioning, still very much of the favorable side of present physics. This concerns the formulation of the laws of physics, the mathematical and simple language it uses. This became, indeed, a criterion for our acceptance of these laws – as Einstein said, we can believe laws of physics only if they have mathematical simplicity and, as he expressed it, beauty. Newton's law of gravitation tells us that the force between two celestial bodies is equal to the gravitational constant multiplied by the product of their masses, and divided by the square of the distance between them. Surely, the law could be much more complicated – the dependence on the distance, in particular, could follow any complicated function. It does not. This is, so-to-say, a gift from heaven for which we have no rational explanation. Surely, the laws of nature do not tell us everything: they give us only correlations between events. If we know the position of a planet at two different instants of time we can calculate it at a third instant of time – the laws of nature give correlations between positions at any three instants of time, but not the positions themselves. They do not give, either, the mass of the planets, their chemical composition, or anything similar. They tell us only that the mass does not change in time, and give rules for the changes in the chemical composition. They give these rules, however, with such accuracies, and in terms of expressions so simple in mathematical language, as could not have been expected a priori.

In spite of their accuracies, the laws of nature undergo constant modifications – the nature of these modifications will be, actually, one of the principal subjects of the present discussion. To quote one example: as was mentioned before, the actual orbit of one of the planets deviates from that given by Newton's theory by about one part in 30.000 and the actual orbit is more accurately described by a more advanced theory, Einstein's general theory of relativity. The basic concepts of this theory are entirely different from those of Newton's so that the very close, almost perfect, agreement between the two theories' conclusions appears very surprising. The situation is that the older theory, that of Newton, furnishes a mathematical approximation to that of Einstein, an approximation by means of which the consequences of the latter, a much more sophisticated and much more encompassing theory, can be obtained for certain situations with a very good approximation. For anyone familiar with both theories, this is truly surprising. The basic concept about which Newton's theory makes statements concerns the positions of objects at a given time. The correlations provided by the theory are correlations between the positions at different times. These determine the orbits of planets, for instance. The basic concept of the general theory of relativity is the coincidence of the positions of two objects (usually a light quantum is one of them), i.e. the occurrence of collisions. The possibility to obtain the description of the orbits of objects in terms of the collisions of these with light quanta is not unimaginable, that the resulting orbits can be characterized by equations as simple as those of Newton, is truly surprising. It applies if one can assume that some of the quantities, in the case of the planetary orbit, the masses of the "objects", are very large, others, such as the velocities of these, very small (compared with the velocity of light). Neither of these quantities is given by the laws of physics – it just happens that the situation is that way. The mathematician says that we are facing a limiting situation in which some quantities occurring can be assumed to be infinitely small, some others infinitely large. It is natural that such assumptions simplify the equations – though not natural that the simplification can reduce them to equations dealing with very different basic quantities. What is most surprising, however, is that nature has furnished us with situations representing such limiting cases. Had this not been true, Kepler would not have been able to derive his laws from Tycho Brahe's observations and Newton could not have given his interpretation to Kepler's laws. Newton's theory, the foundation block of our physical theories, would not have been created had nature not provided us with a limiting situation – actually limiting in more ways than specified above. The existence of such limiting situations is another gift which we should both appreciate and recognise. It is almost on a par with the existence of accurate laws of nature – some of the limiting cases of which constitute our present science.

A Very Cursory Story
of the Development of Physical Theories

Newton's theory of planetary motion was the first truly successful and encompassing theory in physics. It had already a characteristic which is still a sure sign of success: it established a relation between phenomena between which no relation was apparent before: the motion of the moon around the Earth and the free fall of objects on the Earth. The Earth's gravitational attraction which is responsible for the free fall of the objects around us which are not supported is responsible also for keeping the moon on a circular orbit rather than letting it move away from us on a straight line.

It was mentioned before that Newton's theory of the planetary orbits turned out to be an approximate theory because it assumes low velocities and large masses for the planets. It represents a limiting situation in other ways also: the theory assumes that only gravitational forces act between the sun and the planets, no other forces play a role. This was, of course, clear already to Newton. Surely, there are other forces, for instance those which cause a deformed spring to reassume its original shape, those of friction which cause a sliding object to come to a halt, and many others. Nevertheless, the fact of having a simple mathematical formalism which gave not only an excellent description of planetary motion but also related that to a phenomenon apparently completely unrelated thereto, made a profound impression on all interested in science in those days. It was clearly only a beginning – there was no doubt that the theory applies to a tiny fraction of all phenomena; it was a magnificent beginning nevertheless.

It may be well to repeat at this point that the "events" between which Newton's mechanics gave correlations were positions and velocities of objects. Hence, the state of a system was characterized by six times as many numbers as were objects present, at least as long as these objects could be considered point-like, as could the planets in the very extended solar system. However, mechanics was soon further generalised, partly to account for the fact that other than gravitational forces exist – Newton was already fully aware of this, even though he did not specify these other forces. The other aspect of the generalisation took care of the nonpointlike nature of most bodies, in particular also the existence of liquids. Bernoulli's Hydrodynamica already described the state of his systems by functions of the three space-variables, three such functions being the three components of the velocity of the liquid, each depending on the three coordinates determining the position at which the velocity components have the values given by the three aforementioned functions. Hence, in this theory, still a derivative of Newton's mechanics, the "events" were described not by a few numbers but by a few functions of the position, a much more complicated description. Actually, this was true in a sense already of Huygens' "Traité de la Lumière", given only three years after the appearance of Newton's "Philosophiae naturalis principia mathematica" but this did not deal with mechanical objects. Anyway, the description of the states of systems, and hence the equations giving the change of these states as functions of time, became much more complicated soon after Newton's pathbreaking contribution.

The development of hydrodynamics, and many reformulations of Newton's basic principles, demonstrated clearly that the same physical principles can be given in various mathematical forms, many of them of great generality and beauty. They did not have the significance of Newton's gravitational theory because they gave no definite expression for the forces acting between objects and hence did not lead to new and precise experimentally verifiable conclusions. The hydrodynamics of incompressible liquids could have been an exception since there are liquids (such as water) which are virtually incompressible. Once can derive the magnitude and the direction of the forces acting between different parts of the liquid from the fact of incompressibility, together with the viscosity. It was realised, however, that the lack of compressibility is an approximation and, anyway, no precise experimental confirmation of the equations of hydrodynamics was produced. What was accomplished in the period between Newton's and the next pathbreaking discovery in theoretical physics was, however, an enormous extension of the phenomena which came to be considered parts of physics. These included the phenomena of light, of heat, of electricity, of magnetism not to mention again the mechanics of liquids and of deformable bodies.

The next truly pathbreaking theory after Newton's thus came almost 200 years after Newton's Principia. It was Maxwell's electrodynamics, again a theory which postulated precise connections between its "events". These "events" were the magnitudes of the electric and magnetic field strengths, that is, six functions of space. If they were given at one instant of time, Maxwell's equations determined their magnitudes at all other times if no other but electromagnetic forces wer acting.

In restricting the validity of the theory to the presence of only certain types of forces, Maxwell's theory resembles Newton's. Also, similar to Newton's theory, it united several types of phenomena, in particular electric and magnetic phenomena, the connection between which was not known before in the generality established by Maxwell's equations. It was fundamentally different from the theory of gravitation by using, for the description of the state of the system not a set of numbers but a set of functions of position. Theories which describe the states by such functions have consequently come to be called "field theories". Being a field theory separated Maxwell's theory from the theory of gravitation and it was apparent that, if there is unity of nature, if there is a general theory describing all natural phenomena, this must encompass both gravitational and electromagnetic effects so that both theories, that of Newton and that of Maxwell, must be special cases thereof. We must admit that this problem remains largely open even at present we know of four or five types of forces (gravitational, weak, electromagnetic, strong) and the connections between them are not fully clarified.

Before proceeding to the next basic change, it may be worth mentioning one particular success of Maxwell's theory: Coulomb's law of the force between two pointlike charges is a consequence thereof. Coulomb's law has a great similarity to that of gravitation: the force between two point-like charges is equal to the product of the electric charges, divided by the square of the distance between them. It is almost the same as the gravitational law: the difference is only that the electric charges have to be substituted for the masses of the gravitationally interacting bodies. The Coulomb law appears in Maxwell's theory as a limiting case valid if the charges are stationary in time and are at a distance large compared with the spatial extension of the charges. Nevertheless, the similarity between the two interactions suggests that Newton's law of gravitation can also be deduced from a field theory and indeed, it is generally accepted now that all interactions, that is all forces, should be deduced from field theories. The explanation of the gravitational force given by the general theory of relativity is built up in this spirit.

The next breakthrough, a double one, succeeded Maxwell's much sooner than Maxwell's did Newton's. The two new theories, that of relativity and that of quanta, followed Maxwell's in hardly more than half a century. Einstein's special theory of relativity was, in fact, first proposed in 1905, only 32 years after Maxwell's. It is, as you know, a new picture of the connection between space and time, particularly important if some of the velocities are close to that of light. It is, as we all know, a bold and beautiful modification of the everyday concepts of space and time as completely separate. It does not have, however, the characteristics of the two earlier breakthroughs mentioned before: it does not extend physical theory to additional phenomena and does not propose fundamentally new "events" between which physics is to postulate correlations: it does not propose a new description of the "state" of physical systems. The general theory of relativity also has deep philosophical implications and is, as was mentioned before repeatedly, essentially a new theory of gravitation. It does propose, as was also mentioned before, a new class of events between which to look for correlations: coincidences of two objects in space and time, one could say collisions. However, since all these points were referred to before, this theory will not be further discussed now either.

Neither will I try to describe the development of quantum theory but will try to describe only its present status, the enormous contribution it has made. This is, essentially, its extension of the physical theory to the atomistic level.

Before the advent of quantum mechanics, the main line of physics was wholly on the macroscopic scale. The first physics book I read said that atoms and molecules may exist but this is uninteresting from the physicist's point of view. This meant that explanation of the magnitude of the material constants, such as density, viscosity, elasticity, was outside the realm of physics – these constants entered the theory from the outside, pretty much as the masses of the planets, their original distances from the sun, did. To be sure, there were attempts, though unpopular, to deduce the values of the constants from atomic theory but even these attempts took the properties of the atoms, their energy, size, and the forces between them, for granted, i.e. furnished from the outside. All this changed basically with the advent of quantum mechanics: as far as everyday physics is concerned, only the masses and electric charges of the electrons and the atomic nuclei are taken in from outside the basic equation, all the rest of the properties, dozens and dozens of them, then follow from the equation. Hence, chemistry became, at least in principle, united with physics; in a sense, it became part of it. Later on, attempts were made, quite but not completely successful ones, to deduce even the properties of nuclei, including their energy contents, from the basic equations.

The possibilities of the derivation of material constants, i.e. of the properties of materials, marks an even further reaching progress toward a common basis for all science than did the successes of Newton or of Maxwell's theories. Has the nature of the events between which quantum mechanics establishes correlations also changed from the nature of the events underlying the earlier theories? It changed in a most fundamental and surprising way, not easily accepted by scientists in other branches of science, not accepted even by some of the members of the theoretical physics community. Quantum mechanics' basic events are the outcomes of observations, it gives probabilities for these outcomes or, rather, correlations between the possible outcomes of subsequent observations on a system. The oldest but still most easily described example for this is the Stern-Gerlach experiment; if one obtains, by a measurement, the component of the angular momentum of a silver atom in one direction, its component in another direction can assume one of two values, and quantum mechanics gives the probabilities for the observation of this component to furnish one or the other of these values. The same applies for a third observation, in a third direction, and so on. It applies naturally also to other particles not only to silver atoms. Similarly, if we have ascertained the states of two atoms about to collide, the subsequent observation of the direction of motion of the atoms after the collision is subject to probabilistic laws and quantum mechanics gives only the probabilities for the different directions, not a definite direction. But if this direction is ascertained by an observation, the probabilities of the outcomes of a second collision can be calculated, and so on. Naturally, if the subject of observation is not a microscopic but a macroscopic object, the range of the possible outcomes of the observation are becoming very narrow. The energies of single molecules evaporating from a liquid can be, percentage wise, very different from each other. If a great number of molecules evaporates, the percentage wise variation of the total energy becomes very small: it will be very close to the number of molecules multiplied by the average energy of the single molecule. Hence the values which quantum mechanics furnishes for truly macroscopic material constants, such as density, energy, etc. are sharply defined. However, the fact that observations on the microscopic, that is atomic, level do not have a uniquely defined outcome, remains important not only for the philosopher, for the scientist also.

I will conclude here the very cursory discussion of the past development of physical theories – it turned out even more cursory than I expected it – and proceed to the discussion of the problems which now loom largest, including the problem of the essence of an observation, of the basic event in terms of which quantum mechanics' laws are formulated.

The Consistency and Completeness of Sciences

As far as the future is concerned, these are difficult questions, rarely discussed, and often avoided in view of the enormous influence which science has on our material well-being as well as on our spiritual life. In my opinion, however, the present situation is well worth serious attention, serious attention with regard to every scientific discipline. However, I am able to discuss only the natural sciences' situation, and that probably inadequately.

As far as the consistency of present day physics is concerned, I have serious reservations. Though quantum mechanics has been successfully applied to the determination of many macroscopic constants, its ultimate validity for macroscopic systems is not clear. Quantum mechanics describes the behaviour of isolated systems, as do all physical theories and as did Newton's theory of the solar system. However, quantum mechanics makes statements also about the outcomes of such intricate observations the outcomes of which, in the case of macroscopic objects, can be influenced by tiny objects at very great distances. This means that macroscopic bodies can not constitute isolated systems as far as intricate quantum mechanical observations are concerned. The full body of quantum mechanics, as applied to macroscopic systems can not be completely verified – a conclusion one arrives at very reluctantly.

The general theory of relativity appears to represent the opposite extreme. In order to define or observe an orbit, in order to establish a metric in the absolute sense, an infinite number of collisions is necessary. If light were not quantised, infinitely weak light signals could be used for determining the position of objects and the metric. Since light is quantised and the energies and momenta of the quanta increases with the accuracy of measurement they permit, the position and the metric can be obtained only on a macroscopic scale. For the planets, this condition is vastly overfulfilled – no one thinks that the light needed to observe them, very much less than is actually available, influences their motion. This would not be so for microscopic, that is atomic, object, so that we must conclude that the general theory of relativity is, in contrast to quantum mechanics, a macroscopic theory.

I realise that I was discussing an inner problem of physics, of lesser interest to my audience, but I could not omit this discussion in good conscience.

Let us now come to the question of the completeness of physical theory. The basic event of quantum mechanics is the observation and this is not described by the equations thereof. It involves a living being and, if quantum mechanics is valid at least for inanimate objects, it is not completed until its result enters a consciousness. And this is outside the realm of quantum mechanics, just as the means of observation of planets, that is light, was outside the realm of gravitational theory. It seems to me, therefore, that present physics, vastly more encompassing as it is than earlier physical theories, is still incomplete: it still describes only a limiting situation, a situation in which life plays no role. This conclusion is surely difficult to swallow for those who were told, by some scientist-philosophers familiar only with macroscopic pre-quantum physics, that the physical laws rule the universe, that the behavior of men and animals can be fully described if a knowledge of inanimate objects is available. The conclusion I mentioned is contrary to dialectial materialism and this is enough reason for some to dislike it.

Actually, the importance of life should always have been apparent. Even if the laws which can be derived from the study of the limiting case of no life were valid for the matter of living beings, a translation of the description of the state of this matter into the sentiments of the living being would have to be furnished. After all, one is most interested in the experiences of a person: whether he experiences pain or pleasure, whether he is thinking of love or prime numbers, is of more interest than the position of the molecules in his body. Even if the latter could be predicted from some physical observation, we would not know, but would want to know, what he feels. If the "events" of present quantum mechanics remain the description of true reality, man's impressions, the content of his consciousness, are the basic quantities. If physics reverts to more materialistic events as basic quantities between which it tries to find correlations, the connection between these and the content of consciousness becomes the question of most profound importance.

Is it likely that the laws of nature, valid in the limiting case of no life, are valid also when life is present? Past experience seems to speak against this assumption. When physics was extended beyond mechanics to describe also the effects and propagation of electromagnetic phenomena, of light, the basic concepts underwent a drastic change, field theory was introduced. An even more drastic change took place when the attention turned from macroscopic to microscopic physics, with the advent of quantum mechanics. Except for the case of the incorporation of thermal phenomena, every extension of the realm of physics was accompanied by a fundamental change of the basic concepts. The earlier theory remained a useful and interesting limiting case, but it became a limiting case, not valid under more general conditions. It seems reasonable to assume that this will repeat itself when the description of life will be part of the same science which describes inanimate objects, when physics and psychology will be united. When? And are there other phenomena, at present unknown to us, which influence life? These are questions which we cannot answer and which, perhaps, we should not even ask.

The last section of this discussion, dealing with the consistency and completeness of sciences, is not very complimentary to the present stage of our knowledge. I wish to recall, therefore, what I said first, when trying to describe the past beneficial influence of science and to suggest that one consider the possible future beneficial effects thereof, by maintaining our interest in learning, by stimulating our interest in thinking. These factors may, and I am convinced will, assume increasing importance.

There is one more, very general, point that should not remain unmentioned, particularly not at the present conference. The most precise and most compact description of the statements which quantum mechanics furnishes is: probabilistic connections between the outcomes of subsequent observations on a system. This was mentioned before. The question then arises: who is the observer and how does he observe? There is no encompassing answer to the second question and the only consistent answer that can be given to the first one is that I am the observer, quantum mechanics tells me with what probability I will obtain one or another impression from my observation. In other words if we try to describe nature fully by quantum mechanics, we are led to a solipsistic philosophy. This was recognised long ago, though little advertised and, to be frank, led most of us to the undesired conclusion that quantum mechanics does not describe nature fully.

It is, of course, quite true that our impressions, the outcomes of our observations in the language of quantum mechanics, constitute for all of us the ultimate reality. "Life is a dream", says the poet. Yet he adds: "Find beauty in your dreams", acknowledging that another person's dreams also have reality and the wish to influence them. And, indeed, all of us believe in the "reality" (whatever this word may mean) of the world, in the existence of the impressions of others (in contrast to Malebranche, even of animals) and even in the real existence of material objects. Is this compatible with our belief in quantum mechanics? Not if we consider this to give a full description of "reality". And though I am hesitant to say this in the presence of colleagues so much more versed in philosophy, this also stimulates the hope in me that quantum mechanics will also turn out to be a limiting case, limiting in more than one regard, and that the philosophy which an even deeper theory of physics will support will give a more concrete meaning to the word "reality", will not embrace solipsism, much truth as this may contain, and will let us admit that the world really exists.

The Problems, Future and Limits of Science

E. P. Wigner

The Search for Absolute Values in a Changing World (Proc. 6th ICUS, 1977).
International Cultural Foundation Press, New York 1978, pp. 869–877

The indefinite extension of scientific knowledge faces two kinds of grave obstacles. The first one is of emotional nature. Society at large may lose interest in the progress of science and may cease to support it. Also an aversion may develop in the minds of young people toward the devotion of their lives to increasing the knowledge of man. The second obstacle which may develop is of cognitive nature. It may arise from the finiteness of the capacity of the human mind and human intelligence. The human intelligence may not be able to cope with the problems which the future development of science will present. The two obstacles will be discussed in some detail below.

It is hoped that both obstacles can be overcome in the sense that science will continue to benefit mankind, to give pleasure to its cultivators, and enrich man's picture of the world. Some changes in the role and method of the cultivation of science will probably occur to enable it to achieve these objectives; proposals in this direction will be made. Actually, some significant changes in science's role are probably unavoidable and will at least be alluded to.

The Miracle of Present-day Science

What is science? It is knowledge, much of it abstract and surely not obvious, which does not serve immediately any simple purpose, such as acquisition of food, protection of life, etc. As we well know, it contributed and continues to contribute effectively toward these and other natural objectives but it is acquired and produced not directly for such purposes. It is, as David Bohm put it, the establishment of "new contacts with the world, in new domains, on new levels."

As we know, people always were interested in the acquisition of knowledge, even abstract knowledge, and this is one of the characteristics

869

870   Eugene P. Wigner

of man. Polanyi said that "Man's capacity to think is his most outstanding attribute." And indeed we know of no animal that wishes to acquire abstract knowledge—they are unfamiliar even with the small multiplication table.

In view of man's innate ability to acquire knowledge, and of so many people's taking pleasure in such acquisition, it is surprising that what we call "our science" was not developed earlier. We know, of course, of scientific accomplishments created before what I call "our science" was developed. The Egyptians' description of the seasons is perhaps not pure science—it served them directly to make use of the Nile's changing flow. But the Greek's development of geometry is surely true science—and beautiful science. Yet what I call "our science" is only about 300 years old. Newton's *Principia* appeared in 1687, Boyle's *Skeptical Chemist* in 1661. Since then, science became a magnificent edifice, with new stories being constantly added to those already present. What initiated this is not clear. According to the experts, language is more than 100,000 years old and a development similar to the one we are witnessing today, and in fact participating in, could have been started tens of thousands of years ago. The simplest but not fully convincing explanation of the sudden but late appearance of today's science structure is that the development of science is autocatalytic. Once it becomes known that scientific observations can be made, many people are attracted to this endeavor. Also, though this is not its initial incentive, science does make life easier so that, as it develops, more people can find free time which they can devote to the endeavour to contribute to human knowledge. And, if more try, more will succeed. In other words, this explanation of the late but rapid development of science in the past 300 years attributes this to the resolution of an instability. In the words of René Thom, it is a "catastrophe."

If "our science" was created so suddenly and truly unexpectedly, is there not a danger that it behaves like an adventure and ceases to exist after the excursion has come to the end of its path so that a return is in the offing? As Goethe said "Alles was entsteht ist Wert dass es zu Grunde geht." If such a danger exists, should we not think whether it would be good to avoid it, and if so, how we could avoid it?

The existence of the danger was often pointed out by scientists. I like to mention the observation by George Marx, a Hungarian physicist. He pointed to the existence of billions of stars, with billions of planets. A certain fraction of such planets should be able to support life and a certain percentage of these should have produced intelligent beings—similar to *homo sapiens* here on our Earth. Some of the cultures of these should be further advanced than ours and should have noticed that life exists on our Earth. Why did they not try then to establish a contact with us, why did they not send some message or even a committee to us? If we proceed with the development of our technical abilities for another century or so in the way we did in the last few decades, we should be able to do that. The absence of any evidence of any communication from other stars' planetary systems is indeed very puzzling—it has been noted also by others (including Roy Ringo). Marx' explanation of the puzzle is that if science, culture, and technology progress to about the level which they have reached here on Earth, they become in some way self-destructive. Marx' observation is indeed evidence—not easily refutable evidence—for the non-existence of much higher technical, and perhaps also scientific, civilizations anywhere in outer space. It does not give an explanation for the universal lack of technological progress anywhere which would be much higher than ours.

Our first problem is, therefore, to find such an explanation. The only explanation that I could think of and that I was not reluctant to accept is that once science and technology have reached a stage similar to ours, attention increasingly turns toward the mental sciences—I had in fact proposed such a shift of the prime interest without reference to, or knowledge of, Marx' argument. If we accept this, we should give reasons for such a shift of interest and this will be the next subject. I will propose two pairs of such reasons, the first pair referring to the relation between the society at large and the scientific community; the second pair to the natural interests of the members of this community, to the change of the subjects which may increasingly attract the interests of the scientists, particularly those at the beginnings of their careers. Let me begin with the first of these pairs, society's changed expectations of benefits which the scientific community can furnish.

The Emotional Obstacles to the Indefinite Growth of Science—The First One.

Why might society at large cease to wish to support the present scientific and technological effort, the effort which has brought so many benefits? There may be two reasons for this. The first one is that the direction of science which is most prominent at present, leading to our wealth and affluence, "has done its job";* it has made it physically possible for everyone to have a carefree life. By this I mean that everyone can be well provided with food and shelter. This is indeed already the case: in our country about 4 per cent of the population provides all the food we need, even a good deal more, and if the population does not increase, as it hardly does at present, it will be very easy to provide adequate shelter for everyone. Actually, this fact influences not only society, it can have a serious effect also on the community of scientists. Even though science, as I defined it, does not directly foster the material welfare, the fact that its later effects are useful also in this regard always had a stimulating effect on the actual or prospective scientist. In fact, as I mentioned at last year's meeting, everyone of Wilhelm Ostwald's Grosse Männer (Great Men) contributed to the material welfare of man also directly. In Hungary, when I was young, people did not believe that science does make such a contribution and the few scientists they had were looked upon as queer people; they were almost outcasts. The fact that it is already physically possible for everyone to have an easy life is the first possible, in fact not unlikely, reason for an adverse change of attitude of society toward making sacrifices for, or to approve of, our present science.

The second reason may be a fear from the dangers which science may produce. The nuclear weapons present one such danger, not much in the foreground now, the recombinant DNA production another and this is much discussed. And surely, many other dangers can turn up.

How could we alleviate these two problems? The most natural way that I could think of, and a way suggested also by other considerations, is to turn the scientific effort into a new direction, more toward the mental and emotional phenomena. Surely, what present science has accomplished is magnificent, having virtually eliminated the material problems of man. We would be simply flabbergasted if we woke up one morning into the early age of this century, not finding electric lights, telephones, surely no radios, no airplanes, none of the present kitchen equipment—we would miss many, many other things. I'd better not even consider how we would feel if we had only the facilities which existed 1000 or 10,000 years ago. Yet has science contributed to what we call human wisdom? Has the happiness of man much advanced? Schrödinger, in his article "Mind and Matter" says that science does not even deal with the mind. Einstein also

*Soon after the meeting, I found this same idea expressed by G. Boniecki in the December 1977 issue of the Japanese journal of PHP (Peace, Happiness, Prosperity).

doubted that our technical developments enhanced man's happiness. Can we even define what happiness consists of? No, as Schrödinger said, it is entirely outside the range of present science. It should not be. Present-day physics, in particular the philosophical ideas forced on us by quantum mechanics, postulate the study of mental phenomena just as strongly as the fact that we obtain our information on the motion of objects, in particular of planets, by means of light rays, postulated the study of light phenomena.

Would such a diversion of the attention from the present subject of physics to mental phenomena also alleviate the second aversion of society toward science? At least temporarily, surely. The study of mental phenomena will not produce explosives as powerful as that of nuclear weapons, nor will it lead to the production of materials the poisonous character of which might create fears similar to those of recombinant DNAs. It must be admitted, though, that the study of mental phenomena, if it becomes as successful as physics' mastery of inanimate structures is by now, can lead to dangerous and highly undesirable possibilities. For example, it may enable dictators to influence the mental structure of their subjects in such a way that their society assumes the characters of an ant society. I do not want to deny this danger, not even close my eyes to it. It is however, a danger only, and in spite of it I fully expect that it can be avoided just as nuclear wars, I hope, can. Hence I still feel that a turning of science's interest a bit toward mental phenomena would be in the interests of both society and science.

The Second Emotional Obstacle to the Indefinite Growth of Science.

Let me now turn to the other non-cognitive menace to science, a decrease of the interest in its cultivation. It was, actually, a fear from such a decrease of the interesting and elevating nature of science which created my interest in our subject: The Limits of Science.

During my work on the nuclear chain reaction, I already became scared by the increasing specialization of our knowledge. That we have, even in our studies of inanimate nature, physics and chemistry as separate disciplines is an old fact and is generally accepted. But the increasing specialization in chemistry, and the same phenomenon in physics, are truly disturbing. I do not even want to talk about the specialization in chemistry but it has grown immensely even in physics. The *Physical Review* now has five editions and most of those who contribute to the progress of physics are not simply physicists but either soild state physicists, or particle physicists, or field theorists, or general relativists, they may be interested in statistical mechanics or in one of the several other branches. Surely, it is much less wonderful to be a solid state physicist, to contribute to our knowledge of that subject and to know it, than to be a "physicist" quite generally, and to be familiar with and understand our knowledge of inanimate nature. Surely, there are a hundred times more physicists than there were when I was young, in the 20s and 30s. But I am afraid they derive less pleasure from their work, and from reading the work of others, than we did at their age—and the ratio of their devotion to their job to their devotion to their subject is greater than it was for physicists fifty years ago. It was possible, at that time, to know physics, today it is difficult to know nuclear physics.

Another feature which also has a discouraging effect is that, as a consequence of the development of our knowledge it is increasingly difficult to reach the boundaries of science. In order to contribute to nuclear physics, to publish an article on the subject, one has to learn an awful lot, beginning with elementary classical physics and mathematics. And when one has an idea, it is very difficult to make sure that the idea is not in conflict with some generally accepted theory, and also that it has not been put forward before and discussed positively or negatively. Surely, this problem could be greatly alleviated if better review articles were available—and this would make the study of the subject also easier. It is good, therefore, that the appreciation of such articles, and the credit attached to them, has greatly increased in the last one or two decades. The problem would also be alleviated if the frontier of science were spread out a bit so that the number of scientists working in the same narrow area would be smaller. This is another reason for favoring an increased cultivation of the mental sciences. Also, in these, at least at present, the contributor would not have the impression that he is inserting the ninth cog of a wheel while not knowing how many cogs it will have altogether.

These are then the two reasons which I fear will render in the future the pleasure which the cultivators of science will find in its cultivation smaller than was the pleasure at earlier times. It is true, on the other hand, and should not be forgotten, that the number of people to whom science gives some pleasure, the number of scientists, has increased enormously. The American Physical Society has about 30,000 members now—it had about 3,000 in the thirties, it had 96 in 1900. Of course, even though the number of published articles has also increased tremendously during the same period, many of the members do not contribute very actively to the research effort.

Could this mode of the usefulness of science—to give a real though perhaps not overwhelming pleasure to an increasing number of people—be further advanced? When I encounter this question, I always think of the pleasure which I myself and my colleagues derived in high school from attending the meeting of a semi-scientific society. We gathered every two weeks for a session, one of us presented a paper, the rest of us listened and discussed it afterwards. Of course, there was no pretense on the part of the speaker to have invented a new idea—he did understand one with which his colleagues were not familiar and tried to explain it. When I visited Hungary about a year ago, I met one of my high school classmates and asked him about his recollections on this circle. He had as fond recollections as I have.

What I am proposing is, of course, not that our high schools establish circles similar to that which I just recalled, even though this may be very useful also. What I have in mind are similar circles for colleagues in the enterprises where they work, that is get-togethers after their work during which one of them presents some subject or ideas in which he is interested, and which is discussed afterwards. Surely, this would be an indirect contribution of science to the pleasure of man but a contribution nevertheless. The big scope of the present science guarantees that they would not run out of subjects—and they may make some real contributions. Just the same, even the fact that it is natural to make such a proposal indicates that the future contribution of science to mankind will significantly differ from the past one. Science may become less exciting but more widely spread.

The Cognitive Problems of Science.

I considered, so far, the social problems of science and of the scientific community and about possible solutions to their problems. Let me now admit that I am a scientist and interested also, in fact mainly, in scientific problems. Let me make, therefore, a few remarks on the problem of advances in the mental sciences, of finding eventually a common basis for the physical and mental sciences, just as a common basis for electric, magnetic and even mechanical phenomena was found. The common basis differed in all these cases from the bases of both types of phenomena which were united. It is natural to wonder, therefore, which parts of our present ideas of physics will have to be replaced.

As every physicist knows, but few mention, the laws of physics by themselves cannot predict events or phenomena. I know where I'll see the moon tonight but I know it in part because I looked at it yesterday. Its position yesterday, and its velocity if I am to be accurate, are initial conditions; the derivation of its position of today from these initial conditions is based on a law of nature, that is physics. Present-day physics postulates a sharp distinction between initial conditions—about which it makes no statements—and laws of nature which enable us to make predictions on the basis of these initial conditions. The laws of nature therefore in themselves give only *correlations between events,* for instance a correlation between a thrown rock's three positions at three different times. These correlations are almost unbelievably accurate and, in the words of Einstein, their mathematical formulation is of great beauty. The initial conditions, on the contrary, are essentially random.

The question which the hope for a true science of the mind raises is whether the distinction between initial conditions and laws of nature will apply also in their domain. It is hard to believe that it will—even harder to believe that it can be verified. One, though not very cogent, reason for the doubt in this is based on the fact that experiments in the mental domain cannot be really repeated—the initial conditions cannot be precisely reproduced. When an experiment is made on me, I probably remember it and will not be the same person when a repetition is attempted. This is the first cognitive problem of science.

The question which must be raised, therefore, concerns the kind of information which the sciences of the mind will furnish. Will this information always be approximate, perhaps only qualitative, or are we going to discover some new types of laws, some new type of information, just as unforeseen now as was the separation of initial conditions and laws of nature to the Greek philosophers. We must not forget that the new science, embracing the existence of life and consciousness, must contain, as a limiting case, the laws of present-day physics, just as quantum electrodynamics contains, as a limiting case, the laws of classical optics. But it goes much beyond it.

A second question of cognitive nature which demands attention is whether man's mind is powerful enough to establish a theory, embracing mind and consciousness, a theory as powerful, as relevant, and also as

attractive as is present-day physics which is restricted to situations in which
life and consciousness play no role? We do not know. If Darwin was right,
man's comprehension is surely limited as is that of other animals. Nevertheless, there is hope. It has often happened that scientists doubted the
ability of the human mind to produce a satisfactory theory for a set of
phenomena but such a theory was established soon after the doubts arose.
I remember very well that at the Berlin physics colloquia, at which Einstein,
Planck, von Laue, and many others participated, I gained the impression
that they doubted the existence of a satisfactory atomic theory which takes
the quantum phenomena into account. I just read an article by Dirac who
remembers similar doubts of himself. Yet quantum mechanics was discovered less than two years after my visiting those colloquia. We can
well hope that a theory embracing not only inanimate nature but also life
will come into being, that it will, in the words of Bohm, establish "new
contacts with the world, in new domains, on new levels," and that the new
domains will embrace also phenomena of the mind. We cannot be sure but
we can hope.

Would that complete science? Surely not, new phenomena are being
constantly discovered and our theories are bound to remain incomplete.
If this stimulates man to think and ponder, and let his imagination go, it is
a good thing—it occupies minds and gives pleasure, even if only restricted
pleasure, to many.

Let me apologize in conclusion for the pessimism which is contained
in my contribution. It is more important to foresee problems than pleasures—we can do more by trying to overcome future problems than by
increasing future pleasures. I was trying to do the former.

The Extension of the Area of Science

Eugene P. Wigner

Jahn, R.G. (ed.) The Role of Consciousness in the Physical World.
AAAS Symposium No. 57. Westview Press, Boulder 1981, pp. 7–16
(Reset by Springer-Verlag for this volume)

Physicists very rarely boast about the immense extension of the area of physics which took place in the very short period of about 300 years since the birth of "our" discipline. But we emphasize equally rarely that our discipline is as yet very far from complete. Not only are the two branches discovered in our century not fully consistent, there are also phenomena which can not even be described by means of the present concepts of physics. These two observations will be the subjects of this discussion and I will begin with a short review of the past extensions of the area of physics, with a short description of the extension of the circumstances under which the physical theory did and does give information about the succession of, and correlation between, the events which we observe.

The Past Extension of the Area of Physics

What I call "our science" started about 300 years ago when Newton discovered the relation between the law of freely falling bodies here on our Earth and the motion of the moon around the Earth. He was concerned with situations in which electric or magnetic forces play no role, only gravitational forces do, and discovered a simple rule to describe the effects of these forces. This surely was a great accomplishment and is generally appreciated as such. Yet I consider another contribution of his, the clear separation of the information which physics does *not* provide from the information which it *does* furnish, even more important. The former we call initial conditions, the latter the laws of nature. That the moon is now in that direction, that there is a moon, that I let loose this pencil now from this point, are initial conditions entirely outside the scope of physics. But if you give me the positions of the moon at two different instants of time, physics, the law of nature, permit me to calculate its position at all the other times. The same applies to a falling body. If I have two pieces of information about its position such as initial position and, let us say, its initial velocity, its later positions and velocities are given by the laws of nature which permit us to obtain the aforementioned information about the moon.

This is the first point I wanted to make. The point is that physics does not even try to give us complete information about the events around us – it gives us information about the *correlations* between these events. Thus, it gives a correlation between the positions of the moon at any three instants of time, but, though we may know it, its position at one instant of time is not the consequence of the laws of nature. I shall return to this point later.

The second point I wanted to make is implicit already in what was said before. While the preceding point seems to apply permanently, the observation which follows is characteristic of the status of science at definite periods. As was mentioned before, Newton's theory is valid only if gravitational forces alone play a role, if the electric, magnetic, and many other types of forces do not affect the system appreciably. This is true for the planetary system of the sun, for the moon, for satellites and comets but is, for instance, not true for the insides of stars. Light pressure, for instance, produces an important additional force there. And Newton surely did not believe that the laws describing the motions of the planets and of freely falling objects here on Earth can adequately describe all motions.

The next big step in the physics was initiated by Faraday and Maxwell, describing the time variation of the electromagnetic field. Maxwell's theory also gave a common basis for two types of phenomena which appear distinct: the electric and magnetic effects commonly observed, and the phenomenon of light. But as we know today, the theories of his time also represented a limiting case, that of macroscopic bodies, i.e., they were restricted to situations in which all objects contain billions of atoms. The transition to the theory which describes also the behavior of systems containing only very few atoms, or of single atoms, took place in the present century with the advent of quantum theory. The question then arises, whether this theory, that is the present-day physics, still represents a limiting case, whether it applies under all conditions or only when some effect is so small that it can be neglected or disregarded. I will try to give evidence that this is the case; at least life and consciousness are not described by any present theory of physics, in particular not by quantum mechanics.

I wish to add three observations to the preceding discussion. First, that it is wonderful that nature provided us with situations in which grossly simplified laws of nature have an almost perfect validity. The system of the sun's planets, their satellites, etc., are in such a situation – a theory which neglects all non-gravitational forces describes their motion with an almost incredible accuracy. The gravitational macroscopic bodies similarly obey laws, those of classical physics, with a similar accuracy. The inanimate objects seem to obey the laws of quantum theory also very closely. The fact that nature provided us with such situations is a wonderful gift – if these situations did not exist, it might have surpassed human intelligence to discover laws of nature because the general consequences of these are very complicated in all situations except in "limiting cases", when greatly simplified laws have high validity. The point will be made below that present-day physics also deals with such a "limiting case", with situations in which life and consciousness do not influence the events. And perhaps other effects are also disregarded.

My second observation points to the enormous increase of the area in which the known laws of nature have validity. Newton's equations described the motion of planets, satellites, of falling bodies here on Earth – nothing else.

Maxwell's equations described all electromagnetic phenomena and, together with the equations of Newton and his successors, all macroscopic objects' common behavior. But the physical properties of these objects, the hardness, specific weight, electrical conductivity, etc., etc., of these materials was not given by these theories, these properties played a role similar to the initial conditions – they had to be determined experimentally. Quantum theory, in the course of its development changed this – not only does it give equations valid also, and particularly so, for microscopic systems, atoms, molecules, etc., – it also permits the derivation of the *properties* of macroscopic bodies, at least as long as no living system is involved. Will this enormous and, on a historical scale, immensely rapid expansion of the area of validity of physical laws continue? Can we hope that it will eventually encompass the phenomena of life? We don't know but we hope.

This was my second remark. Lastly I wish to call attention to the way the second revolution in physics, that initiated by Faraday and Maxwell, was related to Newton's physics. As was mentioned before, this latter one was based on the knowledge of the motion of the planets and satellites – the knowledge being obtained by observing the sunlight scattered by these planets and reaching us. Yet the description of light was not part of the theory – this, the communicator of the information, was largely outside physics' range almost until the Maxwellian revolution. It is remarkable that the present situation is similar. Quantum mechanics' function is, from a basic point of view, to give the probabilities of the outcomes of observations. Yet the nature of the observation, the description of the observer, is just as foreign to present-day quantum mechanics as was the nature of light to Newton's theory. Are we going to have a revolution, similar to that initiated by Faraday and Maxwell, to incorporate the observer into physics, as Maxwell's theory incorporated light into that discipline? We can hope for it and surely such an extension of the area of "natural science" will modify the fundamental structure of this science at least as much as Maxwellian theory changed the basic concepts of the earlier physics. This is the third remark.

Internal Problems of Our Physics

Is present-day physics, dealing solely with inanimate nature, entirely free from inner contradictions? This also is a question not often discussed. Yet, it may be of interest to call attention to the fact that our century's two greatest additions to earlier physics, general relativity and quantum theory, are not in harmony. Interestingly enough, the special theory of relativity can be harmonized with either.

The general theory of relativity is a macroscopic theory. Its greatest success is in the area of astronomy and cosmology. Its basic quantity is the metric tensor, that is, the relativistic space–time distance between events. Yet, on the microscopic scale the events are not so closely spaced that continuous functions of the space–time coordinates could be established, representing the components of the metric tensor. In fact, what are the events? They are either a crossing of the world-lines of two objects, that is collisions of them, or splittings of a world-line into two, that is the disintegration of a particle (or, conversely, the "absorption" of a light quantum or a particle by another). On a macroscopic scale, these events do occur at definite points at definite times, but not so on the microscopic scale. In fact, quantum mechanical collision theory does not use the concepts of the time or place of the collision, it is interested only in the *outcome* of the collision. That space–time distance measurements are subject to quantum uncertainties has been demonstrated long ago, first I believe by Salecker and myself in 1958. The point here, however, is different: it is that a metric tensor cannot be defined at all on the microscopic scale. There is no way to define it, even less to measure it.

How about the converse, the validity of quantum mechanics for macroscopic systems? The first point that should be considered in this connection is that physics deals with isolated systems, that is systems so far removed from other bodies that these do not exert noticeable influences on the system considered. This, in fact, is a necessary condition for the validity of what is called "causality" that is, for the postulate that the initial conditions of the system determine its later behavior. Clearly, if the system is under the influence of other systems, the behavior of the latter will affect it, and its future is no longer determined solely by its own state.

Yet, there are strong indications that macroscopic systems, even of inanimate nature, can not be truly isolated in our world. The German physicist D. Zeh has called attention to the fact that the quantum theoretical energy levels of a reasonably macroscopic body are so close to each other that even a single atom or electron, though spatially well separated from it, will influence it in the quantum theoretical sense. Thus the impossibility of true isolation begins much before the interference of life and consciousness with such isolation. Actually your present speaker is proposing an equation describing the probabilistic behavior of a not isolated system. Naturally, the equation proposed is not deterministic but gives probabilities for several possible modes of development. Equally naturally, its validity and usefulness are very much open to question.

What was just said remains true even if one admits that Zeh's calculation overestimates the effect of a distant electron or atom on a macroscopic body – his macroscopic body of a cubic centimeter of gas may have to be replaced by a much larger body. The lack of the possibility of isolating a truly macroscopic body from the environment remains, and limits the validity of our microscopic theory, of quantum mechanics. Of course, we are, as a rule, not truly interested in the quantum states of macroscopic bodies and are satisfied by the fact that quantum mechanics permits the derivation of the classical equations describing the macroscopic behavior of macroscopic bodies. Yet the fundamental difficulty remains.

I should mention, perhaps, two other basic problems of our so very successful quantum mechanics. The first of these is generally known: as Fleming and Hegerfeldt have demonstrated, it is questionable whether the position of a particle can be precisely measured. This also renders it questionable whether the strength of any field, electromagnetic in particular, can be measured pointwise. Yet this is implicitly assumed in quantum field theories. The other problem concerns the gravitational effects of the measuring apparata on the measurement processes. This is practically never considered, yet it should be. That these apparata rarely if ever are "isolated" from the rest of the world was mentioned before and so was a proposal to adjust to this fact.

The preceding discussion of the present section appears to be very critical of the present status of physics. It may be good, therefore, to recall the almost unbelievable accuracy which the present laws of physics exhibit in the areas of their validity, that is under conditions which are close to the "limiting situations" in which they are supposed to hold. The largest known deviation from Newton's original theory of planetary motion, that is from Kepler's laws, amounts to only 1 part in about 29,000 – the elliptical orbit of the planet Mercury rotates by this amount during one trip around the ellipse. As we known from listening to the radio or watching the television, the accuracy of Maxwell's equations is astounding. The greatest known deviation comes from the effect of gravitational forces on the light emitted by stars. The sun also deflects the light rays – by an angle which is about a half millionth part of a right angle. Furthermore, all these deviations from the classical theories are postulated, "explained" as it is often said, by the general theory of relativity. Under suitable conditions, even the present theories are valid and point to new areas to which they should, and perhaps might be, extended. Our first section sketched the past development in such directions.

The Most Important Phenomenon
Outside the Present Area of Physics

Evidently, the most important phenomena not dealt with by our physical theories are those of life and consciousness. I will not demonstrate in all technical detail that present-day physics does not describe the phenomenon of consciousness. Even if the physical theories could completely describe the motions of the atoms in our bodies, they would not give a picture of the content of our consciousnesses, they would not tell us whether we experience pain or pleasure, whether we are thinking of prime numbers or of our granddaughters. This fact is, in my opinion, the most obvious but also the most convincing evidence that life and consciousness are outside the area of present day physics. There is a more technical and, for some people, even more convincing proof which I will not give here (and again) in detail. This points out that it follows from the present theory of quantum mechanics that, after an observation which could have two different outcomes, in one of which the observer does see a light flash, in the other of which he does not, after such an observation his mind would be in a superposition of the two corresponding states. But, of course, no mind is in such a superposition – one either sees, or does not see a light flash. It follows that the equations of quantum mechanics do not describe the process of observation, hence do not describe the mind of the observer. This was, actually, recognized already by von Neumann in the late twenties who therefore postulated a "collapse of the wave function" to describe the process of observation. Actually, the non-deterministic theory, mentioned in the preceding section, could take care of this problem – but not of the problem first mentioned: the absence of any description of the context of the mind. I will therefore say a few words about the question of which of the fundamental principles of our present-day physics would have to be modified if this were to be extended to encompass consciousness.

The first, the necessity of the abandonment of which is shown already in the preceding section, is the postulate of determinism. This applies to isolated systems and it has been found that truly macroscopic systems can not be isolated in our world. The abandonment of this principle is even more clear for living beings with consciousness – a being with a highly developed mind can not be fully isolated from its environment. If we put him into interstellar space, this fact will influence his thoughts. And of course, even if we could isolate him, it would be beyond man's capacity to determine, or even to register, the totality of his "initial conditions", that is the total state of his body – not to mention his mind.

The preceding discussion also indicates that we may have to abandon, in our quest for laws of nature embodying life, what I called Newton's greatest discovery: the separation of initial conditions and laws of nature. This postulate can be meaningful only if the same initial conditions can be reproduced. And as far as living organisms of any complexity are concerned, the same initial state can hardly be realized several times. There are no two identical people and if we repeat the same experiment on the same individual the initial conditions are no longer the same – the individual will remember at the second experiment the event of the first one – his mental outlook will have changed thereby. This means that the relevant statements of the theory encompassing life will be terribly different from those of the present natural sciences. There are, of course, signs for some degree of causality also in the description of living beings – the last Frontiers of Science statement (by Marvin Harris) reads:

Taboos on the consumption of the flesh of certain edible animals are often used to uphold the view that cultural evolution is not a causally determined process. Examination of such taboos in an evolutionary–ecological context suggests, however, that *deterministic* processes shape many of these apparently random processes and aversions.

This shows that a certain amount of determinism may be possible on the macroscopic level, just as such determinism exists also in the realm of inanimate nature. But, of course, the two "macroscopic" levels are very different. In the case just considered it implies the behavior of many living beings; in the case of present-day physics, it implies the consideration of such properties of a single macroscopic object which are macroscopically definable and observable.

Nevertheless, it is surprising that the analysis of the postulates of present-day physics indicates future changes of these which are even more clearly necessary if we want to extend them to the area of life and consciousness. I may mention, in this connection, that the impossibility of the precise repetition of experiments even on the microscopic scale is indicated by a new theory of Dirac. The reason for this is, according to him, that the fundamental constants of nature, such as the ratio of the gravitational and electric forces between two particles, change in time. The change is very small – about 10^{-10} per year – but even a small change shows that, even as far as inanimate nature is concerned, we are far from a complete understanding and a logically clear formulation of the laws. There are many indications for this but it is not necessary to discuss them on this occasion.

It is natural to ask, at this point, whether the extension of science into the area of life will lead to the realization of such, *a priori* unbelievable, conclusions as is, for instance, the rotation of our Earth, or the possibility of the instantaneous transmission of information to the other side of the Globe. The next two speakers will describe such "unbelievable" phenomena. I am afraid that their arguments did not fully convince me because of some vagueness of the description of their experiments, possibly motivated by the brevity of the time available for their presentation. But I believe that, even though we should not close our ears to what they are saying, we should examine their experiments with a critical mind.

Let me finish by saying more concisely what I believe the relation of our present science, and in particular of physics, is to an encompassing theory which describes also life. I do not believe there are two entities: body and soul. I believe that life and consciousness are phenomena which have a varying effect on the events around us – just as light pressure does. Under many circumstances, those with which present-day physics is concerned, the phenomenon of life has an entirely negligible influence. There is then a continuous transition to phenomena, such as our own activities, in which this phenomenon has a decisive influence. Probably, the behavior of viruses and bacteria could be described with a high accuracy with present theories. Those of insects could be described with a moderate approximation, those of mammals and men are decisively influenced by their minds. For these, present physical theory would give, I believe, a false picture even as far as their physical behavior is concerned.

Will man ever have as good a picture of life and consciousness as we now have of the situations in which the role of these is negligible? Of course, I do not know but I am a bit pessimistic. We are, after all, animals and there is probably a continuous transition from the capabilities of other mammals to our capabilities. And the other animals were not able even to develop classical physics – our abilities probably also have boundaries. And this is good – what would man do if his scientific effort were completed? He wants to strive for something and an increase of his understanding is a worthy goal and I hope he will be successful in continuing to increase his knowledge further and further but never making it complete.

One may also ask whether there are other phenomena, even less obvious than consciousness and life? I do not know but I do know that it would be wrong to deny its possibility. I will not.

The Glorious Days of Physics

Eugene P. Wigner

Zichichi, Antonio (ed.) The Unity of Fundamental Interactions.
Plenum Press, New York 1983, pp. 765–774
(Reset by Springer-Verlag for this volume)

Introduction

The title "Glorious Days of Physics" implies that our science's present days are less glorious. If we accept this, we are induced to think of the changes both in the development of physics, and in the attitude of physicists toward science, which have taken place in the last 40 to 50 years and these will be the subject of my talk.

I'll discuss the subject under three headings. The first of these will refer to the more modest nature of our earlier discipline, that it did not imagine that it can explain all phenomena – in fact, the explanations of an increasing number of phenomena often came as a pleasant surprise – thus contributing to the glorious nature of those days. My second topic will be the change in the subjects of physics research – partly the result of our increasing confidence in the generality of our theories' validity. Lastly, I will speak about the changed relation of physics and of physicists to society and also the somewhat changed relation of ourselves to science.

I. The Increase in the Area of Physics –
Our Decreased Modesty

It is surely unnecessary to recall to this audience that, ever since Newton, physics makes a sharp distinction between initial conditions and laws of nature, and that theoretical physics deals almost exclusively with the latter, that is, not with isolated events but with correlations between events, called laws of nature. However, the area for which such laws of nature were found has increased enormously in the course of time – Newton's theory enables us to calculate the future positions and velocities are given, but even before the "glorious days" our science was extended to a wealth of other phenomena, electromagnetic, thermodynamic and some other. And that was wonderful.

Yet, while I was studying chemical engineering and visited the Berlin University's physics colloquium every week, I gained the impression that many of the participants had doubts whether the human mind is strong enough to extend physics to the microscopic domain, to the behavior of atoms, molecules, but particularly also to the constituents of these, nuclei and electrons. Perhaps

I should mention that the first physics book I read said "atoms and molecules may exist but this is irrelevant from the point of view of physics". And this was true – physics dealt at that time only with macroscopic phenomena. But by the time of my studies of chemical engineering the interest in microscopic phenomena was very alive – partly as a result of N. Bohr's work – even though it was doubted by many, if not most, that we can cope with them.

The first big change in this pessimistic attitude came with Heisenberg's paper, modest though the aim of that was. He adopted a positivistic philosophy and felt that the theory should not deal with the positions and motions of the electrons in the atom – these are not observable – but only with the energy levels, and transition probabilities between these. And he proposed a method for calculating these which, interestingly enough, turned out to be correct. His ideas were almost immediately adopted by M. Born and P. Jordan and the joint work of these authors fundamentally changed the pessimistic attitude toward the future of microscopic physics.

An equally large change came with Schrödinger's paper which proposed not only a method to calculate the energies of the energy levels – he proposed an equation which turned out to be identical to that of Born, Heisenberg and Jordan – but also described the time development of all states – those corresponding to the energy levels were stationary. His theory was, essentially, an extension of Einstein's idea of the "guiding field" but guided not the individual particles but the whole configuration ("Schrödinger's second equation"). This resulted in the fact that his equation is consistent with the conservation laws of energy and momentum – which laws cannot be preserved if each particle is guided by a separate field. It also resulted in the very important fact that his equation permitted the description of collision phenomena the importance of which became increasingly realized. It may be worth mentioning at this point that Einstein never published his idea of the guiding field (Führungsfeld) because he realized that it is in conflict with the conservation laws of energy and momentum.

The glory of the aforementioned articles introduced the delightful atmosphere of the following years. It was a pleasure to apply their theory to a vast number of phenomena and produce an understanding of these. This will be discussed in the next section. We did realize the remaining inconsistencies of the theory but were glad and surprised that we had any theory which described a great variety of phenomena, phenomena with which physicists were familiar for a long time but which could not be incorporated into physics, for which their theory provided no description.

The first and most obvious weakness of the theory just acquired by physics was the non-relativistic nature of Schrödinger's equation, its enormous distance from the theory of general relativity. The Klein-Gordon equation, and even more the Dirac equation of the electron, were admired. But they applied to single-particle systems and did not encompass the greatest miracle which was inherent in Schrödinger's equation: this described also many-particle systems. It was able, therefore, to describe also collision phenomena – the most important phenomena according to a later article of Heisenberg.

Another limitation of the theory which we accepted, and which persists also to the present, was also recognized: the impossibility to describe the process of observation, the "measurement", by means of the equation of our theory, by means of the equations of quantum mechanics which were just praised. This was recognized already by von Neumann, the founder of the mathematical formulation of the theory of measurement. He realized that the process of observation cannot be described by the Schrödinger equation – it is in fact conflicting with it. This eventually led to a set of new questions and ideas but, at that time, we accepted the existence of this difficulty.

In spite of recognizing the limitations of "our theory" just mentioned – some of which are still with us – we were proud of the development we witnessed, most of us considered it a miracle. It was a development which occurred even though it was not necessary for the assurance of the survival of the human race, as Darwin postulated to be the reason of developments.

To repeat it: we were much more modest, we accepted and worked with a theory which we knew to be incomplete, the limitations of which we recognized. We were both proud and surprised when we could extend it to a new set of phenomena. Now, instead, the work of theoretical physicists is directed principally toward establishing a perfect theory, in particular a relativistic one, and both the older generation, in particular Dirac, and the younger one have significantly contributed to this effort.

II. The Change in the Subject and Spirit of Physics Research

As was mentioned before, one of the changes that have taken place since the Glorious Days is our increased confidence in the validity of quantum mechanics and, even more importantly, the increased confidence that we'll make it "perfect". The second and equally significant change concerns the phenomena to which we wish to apply it.

That quantum mechanics will reproduce the atomic spectra was largely taken for granted, particularly when Hylleraas' calculation of the ground stage energy of helium was found to reproduce the experimental value of this quantity with an amazing accuracy. Hylleraas' calculation referred to a two-electron system – a system for which no equation could be formulated before the advent of quantum mechanics.

We rather expected that the theory will describe the molecular spectra also. But, perhaps strangely, when the article of Heitler and London on the binding energy of the hydrogen atoms into a hydrogen molecule became known, this was a surprise to most of us.* This started the extension of physics to the area of chemistry – a wonderful extension which resulted in our belief that we possess the basic information from which all chemistry could be deduced. This does, of course, not mean that chemistry became uninteresting – just as the

---

* Dr. Teller tells me that he was not surprised.

moon's position and state of illumination remain interesting even though we could calculate it. But chemistry is also immensely useful – and it continues to develop on its own, not on the basis of quantum mechanical calculations. Just the same, it is truly wonderful that two parts of science, physics and chemistry, acquired a common basis.

Other phenomena which were described by means of equations of quantum mechanics include the electrical, and also the thermal, conductivity of metals, even superconductivity, the Van der Waals forces between molecules, the cohesion of solids, their strength. In fact, all the properties of normal matter were expected to be derivable from the basic equations.

All this indicates that the interest was centered on the well known phenomena and we were pleased at the ease these could be described. I will admit that I do not consider all these descriptions fully satisfactory – in particular I consider the phenomena of the permanence of the currents in superconductors a remaining mystery. But the success was great and pleased us greatly. In contrast, the present principal interest of physics concerns the prediction of new phenomena, the formulation of new theories, and produces new and interesting but often hardly acceptable ideas. This is largely due to the fact that it is believed that the common phenomena are "understood", and that the uncommon phenomena, about most of which we know very little, if anything, require new basic approaches.

The difficulty to describe the measurement process and hence "life" and "consciousness" was realized but not much emphasized. This remains a real difficulty also for the present, affecting the foundation of quantum mechanics. In classical physics we describe the state of the system by giving the positions and velocities of the objects – these are natural quantities, acceptable also from a positivistic point of view, at least for macroscopic objects. The wave function or state vector has no similar observable meaning and an interpretation of the quantum mechanical state, by a description of the observation, is a necessity – an unsatisfactory one.

I cannot resist at this point to mention the impossibility to describe the cosmological events by quantum theory. The collision of two stars leads to a branching of their wave function, i.e. to different probabilities of different directions; what has happened to these branches? Only one branch remained and it is not reasonable to say that the others were eliminated by "observations". I'll try to give an explanation of this kind of phenomena in a couple of weeks at a meeting in Bad Windsheim, but am not sure it will be correct.

Perhaps I should reiterate that the greater modesty of the "Glorious Days" was much fostered by the earlier fear, and the realization of the possibility, that man may not be able to describe atomic phenomena in an interesting way, that he is not gifted enough to do that. And the pleasure and glory of the time resulted from the recognition that we are able to do so. The concentration of the theorists on showing that the theory does describe the bulk of the well known phenomena of inanimate nature was largely the result of induced modesty. Of course, it can be claimed that the present physics is even more glorious –

we have overcome our fear, perhaps to a too large extent, and are willing to speculate wildly. Let me now turn to my last subject:

III. The Increased Number of Physicists and Their Changed Attitude Toward Science

I often tell a story about my childhood when my father asked me what I'd like to become as a grown-up. I said that I'd like to become a scientist and, if possible, a physicist. But when he asked me about the number of physics jobs in Hungary, I had to admit that there were only four. (We decided that I'd study chemical engineering.) At present there are about 400 physics jobs in Hungary. And even in the United States, the number of members of the Physical Society quadrupled since I joined it. In the "glorious days" we knew a much larger fraction of our colleagues than we can hope to know now. And the same holds for articles – we can read only a small fraction of those we should read and, in fact, the Physical Review of the earlier days has split into Physical Review Letters, and Physical Review A, B, C, and D journals.

All this influenced the attitude of the physicist greatly. To be a scientist was, in the glorious days, "crazy". It was difficult to find a job, and the salary was low. There was, of course, a wonderful compensation: the pleasure we derived from our work from both research and teaching. Today we have good salaries and we also have public influence – we are no longer considered to be "crazy". As a result, we also have obligations, in particular to inform the public on technical problems, and we try to satisfy these.

Another reason which justifies our calling the old times "glorious" is that we did know physics. We knew a much larger fraction of the accumulated information than we know now. Our present knowledge of physics as a whole is very similar to the chemists' knowledge of their science in our glorious days. I have listened to many lectures at this meeting, on a great variety of subjects, including the origin of our universe, queer particles unknown so far, including magnetic monopoles, new symmetry postulates and others. And even though I am convinced they were well presented, quite a few of them I did not really understand. And I do not believe there was only a small minority to which this applies.

It must be admitted that this constitutes a danger to science which was not really present in the glorious days. It is much more attractive to be a physicist than to be a solid state physicist or a low energy nuclear physicist or anything equally specialized. The need for specialization decreases the attractiveness of our discipline – as it decreases the attractiveness of all science including chemistry. Yet it is very important for the future of man that attractive interests be available to him which can divert him from the search for power even if his physical needs are satisfied for practically everybody, particularly in the developed nations. We must do all we can to keep science interesting and truly attractive – if all people could see only power as a goal our societies would be demoralized. And art has a problem similar to that of science – we have enough wonderful pictures and beautiful music to give pleasure to everyone – there is no need for new ones to ornament all houses and satisfy all tastes. It remains wonderful to hear about new ones, and of producers of new ones, yet not as wonderful as it was in Michelangelo's times.

What I just touched upon is a serious problem for our societies and a serious problem for physics. Yet, I believe that at least the present days of physics are also glorious. After all, it is good that we can have so many more colleagues than before and perhaps also good that so many of us acquired social influence. And that the interest now extends to almost unreasonably large energy ranges, to the origin of our world and to the many other questions which were discussed at our meeting. To some degree this kind of diversification does counteract the problems I mentioned. It was considered somewhat unreasonable when I proposed the consideration of the $SU(4)$ symmetry in nuclear physics – now we heard about the consequences and breakup of an $SU(7)$ symmetry! And it is very good also that we have conferences like the one last year!

Let me, finally, support the more optimistic side of the preceding discussion by telling you that I believe, and I am convinced, that our basic concepts need, and will receive, as basic modifications as were those furnished by Planck, Einstein, Heisenberg, Dirac (and perhaps some others) in the glorious days. It is not likely that they will be attributed to a single physicist, but they will be basic, interesting and will contribute to the beauty of our discipline. To give an indication, I mention that, rather recently, H.D. Zeh has demonstrated that the principle of determinism cannot be maintained for the microscopic (that is quantum mechanical) description of macroscopic bodies. This means that Newton's greatest accomplishment, the theorem that initial conditions and the laws of nature are separate concepts, will have to be modified. (I'll propose such a modification at the conference in Bad Windsheim). Another example is the observation of Fleming and of Hegerfeldt (actually based on an article of Ted Newton and myself) that the idea of space-time points should be modified – there is no state vector which corresponds to the situation that a particle occupies at a definite time a definite point in space. Clearly, the corresponding set of state vectors would be invariant with respect to Lorentz transformations about that point of space-time. There are some proposals in the literature for the modification of the concept of space-time points but it is not clear whether they are final. The solutions of the aforementioned two problems, and probably of others, are probably in the near future but I believe that they will be the result of the ideas and collaboration of several people, not of single individuals. They will become of fundamental importance for physics – physics' glorious days are not over!

And, as I said before, we should maintain glorious days for science. We should love science and not power and extend its area not only to high energy phenomena but also to life and psychology, of ourselves and of animals, also to the social problems of man and his happiness!

Some Problems of Our Natural Sciences

E. P. Wigner

International Journal of Theoretical Physics 25, 467–476 (1986)

Received July 15, 1985

Two categories of limitations of present-day physical science are discussed. The first relates to the inadequacy of physics to deal with the phenomena of life and consciousness. The second relates to weaknesses within physical theory itself, with regard to the definition of space-time points in general relativity and quantum mechanics as well as with regard to problems of quantum mechanical measurement theory.

1. A LIMITATION OF THE AREA OF PRESENT-DAY PHYSICS

The reductionistic materialism reintroduced into Western thought during the Enlightenment was taken up by several scientists and philosophers in the middle of the 19th century, including Comte and Marx. They all had, of course, strong political interests and convictions. According to their theories, man is simply a machine, and his life and behavior are governed by the laws of physics, which are deterministic. Their ideas, which also had political implications, were largely abandoned in our century. Early in the century it was recognized that the area of physics is very restricted—the first physics book I read said "atoms and molecules may exist but this is irrelevant from the point of view of physics." Chemistry was, at that time, entirely separate from physics. As a result of the achievements of quantum mechanics, this has changed completely and it is fair to say that, fundamentally, chemistry is now part of physics, but of course, only fundamentally. However, the phenomena of life, the existence of emotions, pain, pleasure, desires, are still entirely outside the area of physics. We do not know whether the description of these phenomena of life will ever become truly united with physics. We can hope that this will happen even if we admit that this is not now the case, and that the basic idea of materialism is not now valid, and was much less valid when that idea was proposed. The hope that it becomes valid is very much supported by the past success

467

of the fundamental unification of chemistry and physics. It is weakened by the recognition of Darwin that man is fundamentally an animal; we all believe that an animal's knowledge is very limited and will remain so. And we all realize that the phenomena of our life, of our emotions, desires, and cognitions, are outside the area of our physics. Their existence and their interaction with the phenomena of physics constitute a limitation of the area and also of the validity of the present laws of physics. Let me discuss now earlier limitations, particularly of quantum mechanics, and then the inner weaknesses of our theory.

2. EARLIER BOUNDARIES OF QUANTUM MECHANICS

One of the most important, perhaps the most important, observation which led to the founding of quantum mechanics is that of Heisenberg, who proposed that quantum theory not consider the motion of the electrons in atoms, but be restricted to the consideration of energy levels and of transition probabilities. This proposal of a very positivistic nature led to the early development of quantum mechanics by M. Born and P. Jordan—it surely had very narrow boundaries, since not even collisions or the motions of particles were recognized by it. It was confined to the description of the interaction of atoms with electromagnetic radiation.

The next marvelous progress was embodied in Schrödinger's "second equation," which implicitly introduced the wave function in configuration space (a function depending on all the position coordinates of all the particles under consideration) as the description of the state of the system. The "second equation" (the content of the "first equation" turned out to be identical with that of the Born–Heisenberg–Jordan theory) gave the time dependence of the wave function in configuration space. I will call it the Schrödinger equation.

Our present quantum mechanics differs in its philosophical aspects little from that of Schrödinger. The spin's description was introduced (mostly by Pauli) soon after Schrödinger's equation became known. Somewhat later, an attempt was made to make the quantum equations relativistically invariant. The function depending on the variables of configuration space was replaced by a much more complicated function in order to assure real relativistic invariance. This is called "field theory." Unless some of the velocities are close to light velocity, however, the Schrödinger equation (with the introduction of the spin variables) is still valid and is used extensively. Its principal limitation, that of excluding a description of the phenomenon of life, has not been overcome by the field theories.

It is important to notice in this connection that the verifiability of Schrödinger's theory is very different from that of classical theory. In classical theory, it is in principle possible to determine the initial state of the system considered, and also that of a later stage. This renders the verification of the equation giving the later stages of the system on the basis of the knowledge of its initial state clearly possible. In quantum mechanics this is entirely impossible—there is no way to ascertain what the wave function at any given time was. In addition, it is recognized that any "measurement" carried out on the system will change its state. This is, of course, true in general, but is disregarded in classical theory—disregarded because the systems considered are macroscopic—and if the measuring apparatus, carefully designed, is very small, it will have little, and perhaps negligible, effect on the system.

In addition, the description of classical systems, such as a set of planets, is simpler than that of quantum mechanical ones; it can be given by a finite set of numbers, the positions and velocities of the objects. This is not true for the quantum mechanical description—it involves a function, the wave function. The classical description of a deformable object, such as a vibrating elastic body, is also given by functions, and the theory is difficult to verify, but in such cases very simple equations of motion are usually valid and assumed to be correct. In quantum mechanics the description of the interaction of the system considered with the measuring apparatus is not so simple. To describe it, let us start with the situation in which the system is in a state in which the outcome of the measurement is definite. Let us denote its wave function by $\psi_k(X)$, the X standing for all its variables. Let us further denote the wave function of the measuring apparatus by $m(\mu)$, μ being the variables of the apparatus. Then the interaction of the two will change the initial state of the two, $\psi_k(X)m(\mu)$, into

$$\psi_k(X)m(\mu) \to \psi_k(X)m_k(\mu) \tag{1}$$

where $m_k(\mu)$ is a state of the apparatus which shows the proper value k of the measured quantity. The postulation of (1) is both natural and necessary.

But if we consider a general state of the system, which is a linear combination $\sum a_k\psi_k$ of the states ψ_k, because of the linear nature of the time development equation the state of the two will become, as a result of the measurement interaction,

$$\sum a_k\psi_k(X)m(\mu) \to \sum a_k\psi_k(X)m_k(\mu) \tag{2}$$

This means that the outcome of the apparatus–system interaction does not lead with a definite probability $|a_k|^2$ to the state m_k of the apparatus, but to a rather complicated wave function of apparatus and system. It can be—and is—claimed, of course, that if we look at the apparatus, we see its pointer with a probability $|a_k|^2$ pointing to the value k of the observable, but if we claim this, we implicitly admit that our interaction with the apparatus is not described by quantum mechanics. If it were, our final state after interacting with the apparatus would also involve a linear combination of states each describing us as having acquired a particular impression of the state of the apparatus, i.e., the state vector of the system, apparatus plus observer, would be something like

$$\sum a_k \psi_k(X) m_k(\mu) \sigma_k \qquad (2a)$$

where σ_k is a state of the observer. But this is not so. After the observation, the observer is in one of the states σ_k, not in a linear combination. If we admit that the observer actually sees the apparatus' pointer in one of the directions k, we admit that this state is not described by quantum mechanics which leads to (2a).

It is good to admit at this point that, originally, both the apparatus and also the observer may not be in a definite state, but have various probabilities in different states. However, each of the original states of apparatus plus observer transforms, as a result of the interaction with the object, into a superposition of the type (2a) and this is in contradiction with the fact that after the interaction with the object their state is only one of the $m_k(\mu)\sigma_k$, not a superposition of them.

The preceding is a generally accepted argument against the possibility of describing life by the laws of present-day quantum mechanics. Surely, our mind is never in a superposition state; we either see a flash or do not see it—our mind is never a superposition of the two states. Our impressions and our emotions are not described by quantum mechanics—in fact the latter even less than the former.

Inspired by an observation of Bell (1965), I proposed an equation for macroscopic bodies which admits that they cannot be isolated from the environment, and hence that a probabilistic equation will better describe their behavior than a deterministic one. Such an equation can also account for the fact that our mind is not in a superposition of different impressions— that its state breaks up into several states, each describing only one impression. Of course, the proposed equation does not alter the fact that our mental states, our impressions and emotions, are not described by quantum mechanics; hence even if the proposed changes of the present equations should prove useful—which is not at all sure—the area of our present physics would not at all extend to the phenomena of life and emotions. There would still be no wave function for pain, pleasure, knowledge, or desire, and this remains a significant limitation of its area of applicability.

As to the possibility of having a macroscopic isolated system, I considered a 1 cm^3 of tungsten and put it into intergalactic space. But even tungsten, which does not evaporate at the temperature of intergalactic space

(3 K), is influenced by the cosmic radiation of the apparently empty space. The number of light quanta per cm³ is

$$\int \frac{8\pi h \nu^3}{c^3} (e^{-h\nu/kT} \cdots) \frac{1}{h\nu} \, d\nu \approx 16\pi \left(\frac{kT}{hc}\right)^3 \tag{3}$$

which is a few hundred at $T = 3$ K. This means that per second some 10^{13} light quanta strike the tungsten cube. Most of these are ineffective, but, according to my estimate, its quantum state changes after about 1 msec. This change is induced by the cosmic radiation and it would be impossible to introduce the state of this into the initial conditions, since it is influenced itself by many other parts of the world. Hence, the time development of the tungsten cube, and similarly that of all macroscopic objects particularly if they are not in intergalactic space, cannot be fully described by deterministic equations and I have proposed a probabilistic one (Wigner, 1970) which admits the fact that "the future is uncertain," and that the time development of macroscopic objects also depends on the state of the distant environment. This equation is, of course, not useful, since its character and constants depend also on the character and density of the environment, but it does show that our present physics is far from final, that it will undergo fundamental modifications, perhaps also by philosophers. Perhaps I should reiterate that the equation I proposed, and the problem I considered, has nothing to do with the truly basic problem of life and consciousness; it shows only that the basic deterministic idea of our present-day physics has to be modified if applied to macroscopic systems. Such a modification, even if it could be effectively carried out, may not yet touch the problem of life—the truly fundamental problem.

The modification of the quantum mechanical equations just described has no practical effect as long as we stay in the present area of physics. In practice, it deals almost exclusively with time intervals shorter than 10^{-3} sec and with no macroscopic objects in the microscopic sense, i.e., not with the total wave function of objects, which would demand 10^{23} variables—not even with objects that demand a good deal fewer variables.

Having discussed the limitations of the area of present-day physics at some length, I now turn to its inner problems—to problems within the area over which it does claim validity.

3. GENERAL RELATIVITY VS. QUANTUM MECHANICS

3.1. The Definition of Space–Time Points

Quantum mechanics told us that we should describe situations or events only in terms of quantities that can be observed. This is also the reason that the wave function's variables are only position coordinates—the momentum coordinates cannot be measured together with these. The question then arises whether the basic quantities of the general relativity theory, the metrics g_{ik} can be measured. These are functions of space-time coordinates and give, in the form of the expression $\sum g_{ik}\, dx_i\, dx_k$, the space-time distance between the point defined by the variables of the g_{ik} and the point with coordinates increased by dx_i. Can the g_{ik} be measured? How can space-time points be defined? This is a difficult question and, as we will see, it also plays an important role outside the general theory of relativity. But in general relativity, it is a basic question.

In classical theories, space-time points are best defined as crossing points of the paths of two objects—naturally infinitely small ones. And in general relativity, it is implicitly assumed that there are infinitely many such very light objects, so that the intersections of their world-lines define a sufficiently dense set of space-time points. This is, evidently, a very wild assumption and one must admit that the general relativity theory is not really positivistic.

The situation is worse in quantum mechanics. The objects have no paths and the coincidence of two is not defined—there is no "point of collision." The collision matrix, which can be determined by many repeated experiments, does not define the point of collision. It is implicitly assumed, both in general relativity and in quantum mechanics, that there are macroscopic measuring systems which enable the determination of the coordinates of space-time points, but the influence of these systems on the systems under observation can be neglected. Altogether, as will be discussed further, the real existence of space-time points and the possibility of determining their coordinates is an assumption both in general relativity theory and in quantum mechanics—particularly in the field theories of the latter—but is very questionable in both. I believe that even the probability of the system's particles to have given positions at definite times is not determinable—the magnitude of the field strengths at a space like surface even less. The point will be further supported below. Its realization will, I believe, fundamentally change our quantum mechanics and probably all fundamental concepts of our physics.

3.2. Another Problem of the Specification of Space–Time Points

Can a wave function be defined which specifies the position of a particle at a definite point of space at a definite time? As was mentioned before, in classical physics space-time points can be specified as crossing points of two world-lines—perhaps also by the pointer of a clock proceeding on one world-line—but this surely would not be the specification of a microscopic point. The situation is much worse in quantum theory. We will consider the assertion that a particle of spin 0 is, at time 0, at the origin of the coordinate system. What is its state vector then? This and related questions have been thoroughly investigated by M. H. L. Pryce as well as others, including E. Schrödinger. The following discussion is based on an article by T. D. Newton and myself (Newton and Wigner, 1949). This article shows, implicitly, that no state vector gives a truly localized state in true agreement with the special theory of relativity. For the sake of simplicity, I will consider a particle of spin 0.

It is simplest to describe the states of such a particle by a wave function with coordinates in momentum space—it is then easy to make sure that the wave function has no negative energy components. It depends on the three spatial momentum coordinates—the components of p. The energy p_0 is determined by these and the mass m of the particle

$$p_0 = c(p^2 + m^2 c^2)^{1/2} \qquad (4)$$

where the positive sign of the square root is to be taken. The scalar product of two wave functions ψ and φ is then given by

$$(\psi, \varphi) = \int \psi(p)^* \varphi(p) \, dp_x \, dp_y \, dp_z / p_0 \qquad (4a)$$

This is invariant under the transformations of the group of the special relativity theory. The wave function of the state at the origin of the coordinate system is clearly invariant under rotations, hence the corresponding ψ depends only on the length of the momentum p. It should be orthogonal to the wave functions of states displaced within the light cone. This includes, of course, purely spacelike displacements. Displacement by the vector x is represented by multiplication of the wave function by $e^{ix \cdot p / \hbar}$ in which $x \cdot p$ is the scalar product of x and the spatial part of the momentum vector. This leads to

$$\psi(p) = \sqrt{p_0} \qquad (5)$$

which is indeed the wave function proposed by Newton and myself and discussed also by others. The position-dependent form of (5), that is, its proper Fourier transform, is

$$\psi(r) = (h/r)^{5/4} H_{5/4}(imcr/h) \qquad (5a)$$

which is a rather complicated expression. But the $\psi(p)$ should be orthogonal not only to the wave functions that arise from it by purely spatial displacement, i.e., to

$$\psi_x(p) = \sqrt{p_0} \, e^{ip \cdot x / h} \qquad (6)$$

[which is correct according to (4a)], but also to those that result from it by an additional time displacement by t,

$$\psi_{x,t}(p) = \sqrt{p_0}\, e^{i(p \cdot x - p_0 t)/h} \tag{6a}$$

as long as the space-time vector is within the light cone, i.e., as long as $ct < x$. This is not the case and this shows that at least particles of spin 0 cannot be truly localized. And the situation is pretty much the same for higher spins. This is the other reason I believe our present idealized space-time concept will undergo modification.

It may be good to admit, though, that the preceding argument applies only to elementary systems. If we have a system consisting of many particles—perhaps a macroscopic system—some of its parts may be at rest having the state vector of (5); but it is possible to assume that others may be "at rest" if viewed from a moving coordinate system. The total state vector then contains a product of (5) and the state vectors of these other particles, which are really in motion. The resulting product state vector will then be orthogonal to the state vectors of a similar character over practically the whole light cone. Clearly, if we want the composite system to be truly localized with a good approximation, its state vector must be a product of many factors of the character (5), but moving—it must be the state vector of a macroscopic body. And the position of a macroscopic body is not so easily defined.

The preceding remark applies also to other fundamental difficulties of quantum mechanics. If it is applied to macroscopic systems, it becomes equivalent with the classical theory and its problems diminish in significance, but only diminish.

4. TWO OTHER PROBLEMS OF QUANTUM MECHANICAL MEASUREMENT THEORY

The preceding argument shows that the determination of the presence of an object at a space-time point is possible, at best, if the object consists of many particles, that is, if it is essentially macroscopic. This already suggests that all measuring instruments must be macroscopic and since it is virtually impossible to determine the total state vector of a macroscopic object, it raises the suspicion that there may be a fundamental distinction between microscopic and macroscopic systems, between the objects within quantum mechanics' validity and the measuring objects that verify the statements of that theory. This was indicated already in Section 1.

There is other evidence for this limitation and the desirability of introducing changes in the laws of quantum mechanics if applied to macroscopic objects, such as measuring instruments. It was demonstrated some time ago (Wigner, 1952), by considering the interaction between object and measuring instrument, that only those quantities can be precisely measured that commute with all additive conserved quantities, such as the three components of the momentum and of the angular momentum. This includes, naturally, the difficulty of the measurement of the position and even the components of the angular momentum. If one is satisfied with a measurement of limited accuracy, that can be achieved, but requires a measuring apparatus in a state of superposition of many values of momentum and angular momentum—it has to be a macroscopic system. Since the measurement of position momentum and angular momentum coordinates is a natural objective of measurements, and since all these demand a macroscopic measuring system, this again indicates that there is not only a separation of the description of living systems from the present area of physics but also a separation of the physics for macroscopic from that of microscopic systems—natural but disappointing.

The last remark in this connection is another criticism of our "measurement theory." This refers to the wave function at a definite time, which is virtually impossible, since signals from the distant parts of the object under measurement cannot be obtained instantaneously. The measurement takes time and it would be desirable to take this into account when describing it. This is not easy and may create a new problem—the time extension of the measurement process will depend on the space extension of the wave function of the object of measurement.

The discussion of the present section is rather pessimistic as to the philosophical value of our present-day physics. One may be inclined to agree with Heisenberg, who said, originally, that quantum mechanics describes only the energy values of the possible states of atoms (and molecules) and transition probabilities between these. Later he said that the principal subject—I would say, principal other subject—is the collision matrix, which is, as Goldrich and I have shown (Goldrich and Wigner, 1972), observable to a very large extent. And these make, one can claim, the most important contributions of quantum mechanics.

5. FINAL, AND MORE OPTIMISTIC, REMARKS

The titles of the preceding sections indicate that these are principally concerned with the limitations of the validity of our present science and with its weaknesses. It may be good therefore to recall how much the development of our science has done for mankind, how much it has changed our life, and how much it has made it more interesting.

The first observation in this connection refers to a change that has taken place in the present century—in the last third of the development of our science and of physics, as I define it. The change I am referring to is in the life expectancy of a newborn baby. As it is usually, though not quite perfectly, calculated, this was 47 years in 1900; it is 74 years now. The change has been even greater in less advanced countries.

This change is largely due to the greater availability of physicians, but, I believe, even more to the development of the medical sciences, which were, and are, greatly supported by physics. X rays are one example.

Another great change is that individuals can choose their occupation much more freely. A few hundred years ago this was prescribed by the situation into which people were born. And the production of food required a much larger fraction of the working force than it does now. This has had, I must admit, one quite unfavorable consequence: a fraction of young people are not attracted to any truly useful occupation but are striving mainly for power and influence. But the favorable effect is much greater: many more of us can be devoted to learning and to the development of knowledge. This country has about 2500 universities and more than half of them were founded in this century. The 2500 universities have, on the average, about 100 teachers each—there are about a quarter million teachers at universities. Most of them are devoted not only to teaching, but also to expanding our knowledge, that is, to research. And from this point of view it is very good that our science is, as we saw, very, very incomplete; there are many problems awaiting solution and this will probably also be so in the distant future. It may be true that most of the solutions of problems that will be presented by the large majority of scientists will be of a very special nature and restricted to narrow areas, but even such contributions will give pleasure and satisfaction to the contributor. To illustrate the enormous expansion of interest in science, in particular, I may mention that the American Physical Society had 96 members in 1900—it has more than 33,000 members now. We hope that all find pleasure in their work.

REFERENCES

Bell, J. S. (1965). *Physics.*
Goldrich, F. E., and Wigner, E. P. (1972). In *Magic withough Magic*, J. Klauder, ed., W. H. Freeman, New York.
Newton, T. D., and Wigner, E. P. (1949). *Reviews of Modern Physics.*
Wigner, E. P. (1952). *Zeitschrift für Physik.*
Wigner, E. P. (1970). *American Journal of Physics.*

Bibliography

1949

Invariance in Physical Theory. Proc. Amer. Phil. Soc. *93*, 521–526 (1949); Symmetries and Reflections, Indiana University Press, Bloomington 1967, pp. 3–13

1950

The Limits of Science. Proc. Amer. Phil. Soc. *94*, 422–427 (1950); Symmetries and Reflections, Indiana University Press, Bloomington 1967, pp. 211–221

1952

Die Messung quantenmechanischer Operatoren. Z. Physik *133*, 101–108 (1952)

On the Law of Conservation of Heavy Particles. Proc. Natl. Acad. Sci. USA *38*, 449–451 (1952)

1955

On the Development of the Compound Nucleus Model. Am. J. Phys. *23*, 371–380 (1955); Symmetries and Reflections, Indiana University Press, Bloomington 1967, pp. 93–109

1957

Relativistic Invariance and Quantum Phenomena. Rev. Mod. Phys. *29*, 255–268 (1957); Symmetries and Reflections, Indiana University Press, Bloomington 1967, pp. 51–81

1960

The Unreasonable Effectiveness of Mathematics in the Natural Sciences. Comm. Pure and Applied Math. *13*, 1–14 (1960); Symmetries and Reflections, Indiana University Press, Bloomington 1967, pp. 222–237

1961

Remarks on the Mind-Body Question. The Scientist Speculates (I.J. Goog, ed.), William Heinemann, Ltd., London 1961; Basic Books, Inc., New York 1962, pp. 284–302; Symmetries and Reflections, Indiana University Press, Bloomington 1967, pp. 171–184; Quantum Theory and Measurement (J.A. Wheeler and W.H. Zurek, eds.), Princeton University Press, Princeton 1983, pp. 168–181

1962

Theorie der quantenmechanischen Messung. Physikertagung, Wien 1961. Physik Verlag, Mosbach/Baden 1962, pp. 1–8

Discussion: Comments on Prof. Putnam's Comments. Phil. Sci. *29*, 292–293 (1962)

1963

The Problem of Measurement. Am. J. Phys. *31*, 6–15 (1963); Symmetries and Reflections, Indiana University Press, Bloomington 1967, pp. 153–170; Theory of Measurement in Quantum Mechanics (Yanase, Namicki and Machida, eds.), Phys. Soc. Japan, 1978, pp. 123–132; Quantum Theory and Measurement (J.A. Wheeler and W.H. Zurek, eds.), Princeton University Press, Princeton 1983, pp. 324–341

1964

Two Kinds of Reality. The Monist *48*, 248–264 (1964); Symmetries and Reflections, Indiana University Press, Bloomington 1967, pp. 185–199

Symmetry and Conservation Laws. Proc. Natl. Acad. Sci. USA *51*, 956–965 (1964); Symmetries and Reflections, Indiana University Press, Bloomington 1967, pp. 14–27

The Role of Invariance Principles in Natural Philosophy. Dispersion Relations and their Connection with Causality, Intern'l School of Physics "Enrico Fermi" 1963 (E.P. Wigner, ed.), Academic Press, New York 1964, pp. ix–xvi; Symmetries and Reflections, Indiana University Press, Bloomington 1967, pp. 28–37

Events, Laws of Nature, and Invariance Principles. Science *145*, 995–998 (1964); Les Prix Nobel en 1963, Nobel Foundation, Stockholm 1964; Symmetries and Reflections, Indiana University Press, Bloomington 1967, pp. 38–50

1965

Violations of Symmetry in Physics. Scientific American *213*, 28–36 (1965)

1967

(With J.M. Jauch and M.M. Yanase) Some Comments Concerning Measurement in Quantum Mechanics. Il Nuovo Cimento *48B*, 144–151 (1967)

The Growth of Science – Its Promise and Its Dangers. Symmetries and Reflections, Indiana University Press, Bloomington 1967, pp. 267–280

1968

Symmetry Principles in Old and New Physics. Bull. Amer. Math. Soc. *74*, 793–815 (1968)

1969

Epistemology of Quantum Mechanics. Contemporary Physics. Vol. II, Atomic Energy Agency, Vienna 1969, pp. 431–437

Summary of the Conference. Proc. Intern'l Conference on Properties of Nuclear States, Les Presses de l'Université de Montréal, Montréal 1969, pp. 633–647

1970

Physics and the Explanation of Life. Found. Phys. *1*, 33–45 (1970); Philosophical Foundations of Science, Proc. of an AAAS 1969 Program, (R.J. Seeger and R.S. Cohen, eds.), Boston Studies in the Philosophy of Science XI, D. Reidel, Dordrecht 1974, pp. 119–132

(With A. Frenkel and M. Yanase) On the Change of the Skew Information in the Process of Quantum Mechanical Measurements. Mimeographed notes, ca. 1970

1971

The Subject of Our Discussions. Foundations of Quantum Mechanics, Intern'l School of Physics "Enrico Fermi" 1970 (B. d'Espagnat, ed.). Academic Press, New York 1971, pp. 1–19

The Philosophical Problem. Foundations of Quantum Mechanics, Intern'l School of Physics "Enrico Fermi" 1970 (B. d'Espagnat, ed.). Academic Press, New York 1971, pp. 122–124

Questions of Physical Theory. Foundations of Quantum Mechanics, Intern'l School of Physics "Enrico Fermi" 1970 (B. d'Espagnat, ed.). Academic Press, New York 1971, pp. 124–125

Summary of the Conference. Polarization Phenomena in Nuclear Reactions (H.H. Barschall and W. Haeberli, eds.). University of Wisconsin Press, Madison 1971, pp. 389–395

1972

The Place of Consciousness in Modern Physics. Conciousness and Reality (C.A. Muses and A.M. Young, eds.). Outerbridge and Lazard, New York 1972, chap. 9, pp. 132–141; Avon Books, 1974

Introductory Talk. Statistical Properties of Nuclei (J.B. Garg, ed.). Plenum Press, New York 1972, pp. 7–23

On Some of Physics' Problems. Main Currents in Modern Thought *28*, 75–78 (1972); Vistas in Physical Reality (E. Laszlo and E.B. Sellon, eds.), Plenum Press, New York 1976, pp. 3–9

1973

Epistemological Perspective on Quantum Theory. Contemporary Research in the Foundations and Philosophy of Quantum Theory (C.A. Hooker, ed.). D. Reidel, Dordrecht 1973, pp. 369–385

(With S. Freedman) On Bub's Misunderstanding of Bell's Locality Argument. Found. Phys. *3*, 457–458 (1973)

Symmetry in Nature. Proc. R.A. Welsh Foundation on Chemical Research XVI, Theoretical Chemistry, Houston 1972 (W.O. Milligan, ed.), 1973, pp. 231–260

Relativistic Equations in Quantum Mechanics. The Physicist's Conception of Nature (J. Mehra, ed.). D. Reidel, Dordrecht 1973, pp. 320–330

1974

Concluding Remarks. Symmetry Properties of Nuclei (Proc. of the 1970 Solvay Conference). Gordon and Breach, New York 1974, pp. 351–362

1977

Physics and Its Relation to Human Knowledge. Hellenike Anthropistike Hetaireia, Athens 1977, pp. 283–294

1978

New Dimensions of Consciousness. Mimeographed notes, ca. 1978

The Problems, Future and Limits of Science. The Search for Absolute Values in a Changing World (Proc. 6th ICUS, 1977), International Cultural Foundation Press, New York 1978, pp. 869–877

1979

The Existence of Conciousness. The Reevaluation of Existing Values and the Search for Absolute Values (Proc. 7th ICUS, 1978), International Cultural Foundation Press, New York 1979, pp. 135–143; Modernization (R.L. Rubinstein, ed.), Paragon House, New York 1982, pp. 279–285

1980

Events, Laws of Nature, and Invariance Principles. Mimeographed notes, ca. 1980

1981

The Extension of the Area of Science. The Role of Consciousness in the Physical World, AAAS Symposium No. 57 (R.G. Jahn, ed.). Westview Press, Boulder 1981, pp. 7–16

1982

Realität und Quantenmechanik. Address in Lindau and Tutzing 1982, pp. 7–17

1983

Interpretation of Quantum Mechanics. Quantum Theory and Measurement (J.A. Wheeler and W.H. Zurek, eds.). Princeton University Press, Princeton 1983, pp. 260–314

The Limitations of Determinism. Absolute Values and the Creation of the New World (Proc. 11th ICUS, 1982), International Cultural Foundation Press, New York 1983, pp. 1365–1370

Review of the Quantum-Mechanical Measurement Problem. Quantum Optics, Experimental Gravity and Measurement Theory (P. Meystre and M.G. Scully, eds.), Plenum Press, New York 1983, pp. 43–63; Science, Computers and the Information Onslaught (D.M. Kerr et al., eds.), Academic Press, New York 1984, pp. 63–82

The Glorious Days of Physics. The Unity of Fundamental Interactions, 19th Course of the Intern'l School of Subnuclear Physics, 1981 (A. Zichichi, ed.), Plenum Press, New York 1983, pp. 765–774; Quantum Optics, Experimental Gravity and Measurement Theory (P. Meystre and M.G. Scully, eds.), Plenum Press, New York 1983, pp. 1–7

1986

The Non-Relativistic Nature of the Present Quantum-Mechanical Measurement Theory. Ann. NY Acad. Sci. *480*, 1–5 (1986)

Some Problems of Our Natural Sciences. Intern'l Jour. Theor. Phys. *25*, 467–476 (1986)